AIR POLLUTION CONTROL EQUIPMENT CALCULATIONS

AIR POLLUTION CONTROL EQUIPMENT CALCULATIONS

Louis Theodore

An Introduction by
Humberto Bravo Alvarez

A JOHN WILEY & SONS, INC., PUBLICATION

Published by John Wiley & Sons, Inc., Hoboken, New Jersey
Published simultaneously in Canada

Library of Congress Cataloging-in-Publication Data:

Theodore, Louis.
Air pollution control equipment/Louis Theodore.
p. cm.
ISBN 978-0-470-20967-7 (cloth)
1. Air—Purification—Equipment and supplies. I. Title.
TD889.T49 2008
628.5′3—dc22 2007032133

Printed in the United States of America

10 9 8 7 6 5 4 3 2 1

CONTENTS

NOTE

Additional problems for Chapters 3–12 are available for all readers at www.wiley.com. The problems may be used for homework purposes. Solutions to these problems plus six exams (three for each year or semester) are available to those who adopt the text for instructional purposes. Visit www.wiley.com and follow links for this title for details.

PREFACE

I fear the Greeks, even when bearing gifts.
—Virgil (70–19 B.C.), *Aeneid*, Book II

In the last four decades, the technical community has expanded its responsibilities to society to include the environment, with particular emphasis on air pollution from industrial sources. Increasing numbers of engineers, technicians, and maintenance personnel are being confronted with problems in this most important area. The environmental engineer and scientist of today and tomorrow must develop a proficiency and an improved understanding of air pollution control equipment in order to cope with these challenges.

This book serves two purposes. It may be used as a textbook for engineering students in an air pollution course. It may also be used as a reference book for practicing engineers, scientists, and technicians involved with air pollution control equipment. For this audience, it is assumed that the reader has already taken basic courses in physics and chemistry, and should have a minimum background in mathematics through calculus. The author's aim is to offer the reader the fundamentals of air pollution control equipment with appropriate practical applications and to provide an introduction to design principles. The reader is encouraged through references to continue his or her own development beyond the scope of the presented material.

As is usually the case in preparing any text, the question of what to include and what to omit has been particularly difficult. However, the problems and solutions in this book attempt to address calculations common to both the science and engineering professions. The book provides the reader with nearly 500 solved problems in the air pollution control equipment field. Of the 12 chapters, 4 are concerned with gaseous control equipment and 6 with airborne particulate pollutants. The interrelationship between both classes of pollutants is emphasized in many of the chapters, Each chapter contains a number of problems, with each set containing anywhere from 30 to 50 problems and solutions.

As indicated above, the book is essentially divided into two major parts: air pollution control equipment for gaseous pollutants (Chapters 3–6), and control equipment for particulate pollutants (Chapters 7–12). Following two introductory chapters, the next four chapters examine control equipment for gaseous pollutants, including incineration, absorption, and adsorption. The last six chapters are devoted to gravity settlers, cyclones, electrostatic precipitators, scrubbers, and baghouses. Each chapter contains a short introduction to the control device, which is followed by problems dealing with performance equations, operation and maintenance, and recent developments. The Appendix contains writeups on hybrid systems, the SI system (including conversion constants), and a cost equipment model.

This project was a unique undertaking. Rather than prepare a textbook in the usual format—essay material, illustrative examples, nomenclature, bibliography, problems, and so on—the author considered writing a calculations book that could be used as a self-teaching aid. One of the key features of this book is that the solutions to the problems are presented in a near stand-alone manner. Throughout the book, the problems are laid out in such a way as to develop the reader's understanding of the control device in question; each problem contains a title, problem statement and data, and the solution, with the more difficult problems located at or near the end of each chapter set. (Additional problems and solutions are available at a Website for all readers, but particularly for classroom/training purposes.) Thus, this book offers material not only to individuals with limited technical background but also to those with extensive industrial experience. As such, this book can be used as a text in either a general environmental and engineering science course and (perhaps primarily) as a training tool for industry.

Knowledge of the information developed and presented in the various chapters is essential not only to the design and selection of industrial control equipment for atmospheric pollutants but also to their proper operation and maintenance. It will enable the reader to obtain a better understanding of both the equipment itself and those factors affecting equipment performance.

Hopefully, the text is simple, clear, to the point, and imparts a basic understanding of the theory and mechanics of the calculations and applications. It is also hoped that a meticulously accurate, articulate, and practical writing style has helped master the difficult task of explaining what was once a very complicated subject matter in a way that is easily understood. The author feels that this delineates this text from others in this field.

The author cannot claim sole authorship to all the problems and material in this book. The present book has evolved from a host of sources, including notes, homework problems, and exam problems prepared by J. Jeris for graduate environmental engineering courses; notes, homework problems, and exam problems prepared by L. Theodore for several chemical and environmental engineering graduate and undergraduate courses; problems and solutions drawn (with permission) from numerous Theodore Tutorials; and, problems and solutions developed by faculty participants during National Science Foundation (NSF) Undergraduate Faculty Enhancement Program (UFEP) workshops.

During the preparation of this book, the author was ably assisted in many ways by a number of graduate students in Manhattan College's Chemical Engineering Master's Program. These students, particularly Agogho Pedro and Alex Santos, contributed much time and energy researching and classroom testing various problems in the book.

My sincere thanks go to Anna Daversa, Andrea Paciga, and Kevin Singer for their invaluable help and assistance in proofing the manuscript.

Louis Theodore

April 2008

INTRODUCTION

By Humberto Bravo Alvarez

Two fundamental reasons for the cleaning of gases in industry, particularly waste gases, are *profit* and *protection*. For example, profits may result from the utilization of blast furnace gases for heating and power generation, but impurities may have to be removed from the gases before they can be burned satisfactorily. Some impurities can be economically converted into sulfur, or solvent recovery systems can be installed to recover valuable hydrocarbon emissions. Protection of the health and welfare of the public in general, of the individual working in industry, and of property is another reason for cleaning gases.

The enactment of air pollution control regulations (see Chapter 2) reflects the concern of government for the protection of its people. For example, waste gases containing toxic constituents such as arsenic or lead fumes constitute a serious danger to the health of both plant operators and the surrounding population. Other waste gases, although not normally endangering health in the concentrations encountered, may kill plants, damage paintwork and buildings, or discolor wallpaper and curtains, thus making an industrial location a less pleasant area in which to live.

The extent to which industry cleans polluted gas streams depends largely on the limits imposed by four main considerations:

1. Concentration levels harmful to humans, physical structures, and plant and animal life

---

Air Pollution Control Equipment Calculations. By Louis Theodore

2. Legal limitations imposed by the country, state, county, or city for the protection of the public health and welfare
3. Reduction of air pollution to establish civic goodwill
4. The reduction and/or elimination of potential liability concerns

These considerations are not necessarily independent. For example, the legal limits on emissions are also closely related to the degree of cost needed to prevent concentrations that can damage the ecosystem.

Earth is a huge sphere covered with water, rock, and soil, and is surrounded by a mixture of gases. These gases are generally referred to as *air*. Earth's gravity holds this blanket of air—the atmosphere—in place. Without gravity, these gases would drift into space. Pristine or "clean" air, which is found in few (if any) places on Earth, is approximately composed of nitrogen (78.1%), oxygen (20.9%), argon (0.9%), and other components (0.1%). Other components include carbon dioxide [330 parts per million by volume (ppmv)], neon (18 ppmv), helium (5 ppmv), methane (1.5 ppmv), and very small amounts (less than 1.0 ppmv) of other gases. Air can also include water droplets, ice crystals, and dust, but they are not considered part of the composition of the air. Also, the nitrogen, oxygen, etc., content of air almost always refers to the composition of dry air at ground level.

The aforementioned air pollutants may be divided into two broad categories, *natural* and *human-made* (synthetic). Natural sources of air pollutants include the following:

1. Windblown dust
2. Volcanic ash and gases
3. Ozone from lightning and the ozone layer
4. Esters and terpenes from vegetation
5. Smoke, gases, and fly ash from forest fires
6. Pollens and other aeroallergens
7. Gases and odors from natural decompositions
8. Natural radioactivity

Such sources constitute background pollution and that portion of the pollution problem over which control activities can have little, if any, effect. Human-made sources cover a wide spectrum of chemical and physical activities, and are the major contributors to urban air pollution. Air pollutants in the United States pour out from over 100 million vehicles, from the refuse of 300 million people, from the generation of billions of kilowatts of electricity, and from the production of innumerable products demanded by everyday living.

Air pollutants may also be classified by *origin* and *state of matter*. Under the classification by origin, the following subdivisions pertain: *primary*—emitted to the atmosphere from a process; and *secondary*—formed in the atmosphere as a result of a chemical reaction. Under the state of matter, there exist the classifications *particulate* and *gaseous*. Although gases need no introduction, particulates have been defined as solid or liquid matter whose effective diameter is larger than a molecule but smaller than approximately 1000 μm (micrometers). Particulates dispersed in a gaseous medium may be collectively termed an *aerosol*. The terms *smoke*, *fog*, *haze*, and *dust* are commonly used to describe particular types of aerosols, depending on the size, shape, and characteristic behavior of the dispersed particles. Aerosols are rather difficult to classify on a scientific basis in terms of their fundamental properties such as their settling rate under the influence of external forces, optical activity, ability to absorb electric charge, particle size and structure, surface-to-volume ratio, reaction activity, physiological action, etc. In general, the combination of particle size and settling rate has been the most characteristic properties employed. For example, particles larger than 100 μm may be excluded from the category of dispersions because they settle too rapidly. On the other hand, particles on the order of 1 μm or less settle so slowly that, for all practical purposes, they are regarded as permanent suspensions.

When a liquid or solid substance is emitted to the air as particulate matter, its properties and effects may be changed. As a substance is broken up into smaller and smaller particles, more of its surface area is exposed to the air. Under these circumstances, the substance—whatever its chemical composition—tends to physically or chemically combine with other particulates or gases in the atmosphere. The resulting combinations are frequently unpredictable. Very small aerosol particles ranging from 1.0 to 150 nm (nanometers) can act as condensation nuclei to facilitate the condensation of water vapor, thus promoting the formation of fog and ground mist. Particles less than 2 or 3 μm in size—about half (by weight) of the particles suspended in urban air—can penetrate into mucous membranes and attract and convey harmful chemicals such as sulfur dioxide. By virtue of the increased surface area of the small aerosol particles, and as a result of the adsorption of gas molecules or other such activities that are able to facilitate chemical reactions, aerosols tend to exhibit greatly enhanced surface activity.

Many substances that oxidize slowly in a given state can oxidize extremely rapidly or possibly even explode when dispersed as fine particles in air. Dust explosions, for example, are often caused by the unstable burning or oxidation of combustible particles, brought about by their relatively large specific surfaces. Adsorption and catalytic phenomena can also be extremely important in analyzing and understanding particulate pollution problems. For example, the conversion of sulfur dioxide to corrosive sulfuric acid assisted by the catalytic action of iron oxide particles, demonstrates the catalytic nature of certain types of particles in the atmosphere.

The technology of control (as it applies to this book) consists of all the sciences and techniques that can be brought to bear on the problem via air pollution control equipment. These include the analysis and research that enter into determinations of technological and economic feasibility, planning, and standard-setting, as well as the application of specific hardware, fuels, and materials of construction. Technology also includes the process of evaluating and upgrading the effectiveness of air pollution control practices.

At the heart of the control strategy process is the selection of the best air pollution control measures from among those available. To eliminate or reduce emissions from a polluting operation, four major courses of action are open:

1. Eliminate the operation
2. Regulate the location of the operation

3. Modify the operation
4. Reduce or eliminate discharges from the operation by applying control devices and systems

The ability to achieve an acceptable atmosphere in a community often requires a combination of these measures aimed at all or a major fraction of the contaminant sources within any control jurisdiction.

Control technology is self-defeating if it creates undesirable side effects in meeting (limited) air pollution control objectives. Air pollution control should be considered in terms of both the total technological system and ecological consequences. The former considers the technology that can be brought to bear on not only individual pieces of equipment but also the entire technological system. Consideration of ecological side effects must also take into account, e.g., the problem of disposal of possibly unmanageable accumulations of contaminants by other means. These may be concentrated in the collection process, such as groundwater pollution resulting from landfill practices or pollution of streams from the discharges of air pollution control systems.

Gaseous and particulate pollutants discharged into the atmosphere can be controlled. The five generic devices available for particulate control include gravity settlers, cyclones (centrifugal separators), electrostatic precipitators, wet scrubbers, and baghouses (fabric filtration). The four generic devices for gases include absorbers, adsorbers, and enumerators. These control devices are discussed in individual chapters later in the text.

There are a number of factors to be considered prior to selecting a particular piece of air pollution control hardware. In general, they can be grouped in three categories: environmental, engineering, and economic. These three categories are discussed below.

1. Environmental
   a. Equipment location
   b. Available space
   c. Ambient conditions
   d. Availability of adequate utilities (power, compressed air, water, etc.) and ancillary systems facilities (waste treatment and disposal, etc.)
   e. Maximum allowable emission (air pollution codes)
   f. Aesthetic considerations (visible steam or water vapor plume, etc.)
   g. Contribution of air pollution control system to wastewater and land pollution
   h. Contribution of air pollution control system to plant noise level
2. Engineering
   a. Contaminant characteristics [physical and chemical properties, concentration, particulate shape and size distribution (in the case of particulates), chemical reactivity, corrosivity, abrasiveness, toxicity, etc.]

   b. Gas stream characteristics (volumetric flow rate, temperature, pressure, humidity, composition, viscosity, density, reactivity, combustibility, corrosivity, toxicity, etc.)
   c. Design and performance characteristics of the particular control system [size and/or weight fractional efficiency curves (in the case of particulates), mass transfer and/or contaminant destruction capability (in the case of gases or vapors), pressure drop, reliability and dependability, turndown capability, power requirements, utility requirements, temperature limitations, maintenance requirements, flexibility of complying with more stringent air pollution codes, etc.]
3. Economic
   a. Capital cost (equipment, installation, engineering, etc.)
   b. Operating cost (utilities, maintenance, etc.)
   c. Expected equipment lifetime and salvage value

Prior to the purchase of control equipment, experience has shown that the following points should be emphasized:

1. Refrain from purchasing any control equipment without reviewing *certified independent test data* on its performance under a similar application. Request the manufacturer to provide performance information and design specifications.
2. In the event that sufficient performance data are unavailable, request that the equipment supplier provide a small pilot model for evaluation under existing conditions.
3. Request participation of the appropriate regulatory authorities in the decision-making process.
4. Prepare a good set of specifications. Include a *strong performance guarantee* from the manufacturer to ensure that the control equipment will meet all applicable local, state, and federal codes/regulations at specific process conditions.
5. Closely review the process and economic fundamentals. Assess the possibility for emission trade-offs (offsets) and/or applying the "bubble concept" (see Chapter 2). The bubble concept permits a plant to find the most efficient way to control its emissions as a whole. Reductions at a source where emissions can be lessened for the least cost can offset emissions of the same pollutant from another source in the plant.
6. Make a careful material balance study before authorizing an emission test or purchasing control equipment.
7. Refrain from purchasing any equipment until *firm* installation cost estimates have been added to the equipment cost. *Escalating installation costs are the rule rather than the exception.*
8. Give operation and maintenance costs high priority on the list of equipment selection factors.

9. Refrain from purchasing any equipment until a solid commitment from the vendor(s) is obtained. Make every effort to ensure that the new system will utilize fuel, controllers, filters, motors, etc., that are compatible with those already available at the plant.
10. The specification should include written assurance of *prompt* technical assistance from the equipment supplier. This, together with a completely understandable operating manual (with parts list, full schematics, consistent units, and notations, etc.), is essential and is too often forgotten in the rush to get the equipment operating.
11. Schedules, particularly on projects being completed under a court order or consent judgment, can be critical. In such cases, delivery guarantees should be obtained from the manufacturers and penalties identified.
12. The air pollution equipment should be of fail-safe design with built-in indicators to show when performance is deteriorating.
13. Withhold 10–15% of the purchase price until compliance is clearly demonstrated.

The usual design, procurement, construction, and/or startup problems can be further compounded by any one or a combination of the following:

1. Unfamiliarity of process engineers with air pollution engineering
2. New and changing air pollution codes/regulations
3. New suppliers, frequently with unproven equipment
4. Lack of industry standards in some key areas
5. Interpretations of control by agency field personnel
6. Compliance schedules that are too tight
7. Vague specifications
8. Weak guarantees for the new control equipment
9. Unreliable delivery schedules
10. Process reliability problems

Proper selection of a particular system for a specific application can be extremely difficult and complicated. In view of the multitude of complex and often ambiguous pollution control regulations, it is in the best interest of the prospective user (as noted above) to work closely with regulatory officials as early as possible in the process.

The final choice in equipment selection is usually dictated by that piece of equipment capable of achieving compliance with regulatory codes at the lowest uniform annual cost (amortized capital investment plus operation and maintenance costs). More recently, there have been attempts to include liability problems, neighbor/consumer goodwill, employee concerns, etc., in the economic analysis, but these effects—although important—are extremely difficult to quantify.

In order to compare specific control equipment alternatives, knowledge of the particular application and site is also essential. A preliminary screening, however, may be performed by reviewing the advantages and disadvantages of each type of air pollution control equipment. For example, if water or a waste treatment system is not available at the site, this may preclude the use of a wet scrubber system and instead focus particulate removal on dry systems such as cyclones, baghouses, and/or electrostatic precipitators. If auxiliary fuel is unavailable on a continuous basis, it may not be possible to combust organic pollutant vapors in an incineration system. If the particulate-size distribution in the gas stream is relatively fine, gravity settlers and cyclone collectors most probably would not be considered. If the pollutant vapors can be reused in the process, control efforts may be directed to adsorption systems. There are many other situations where knowledge of the capabilities of the various control options, combined with common sense, will simplify the selection process.

1

AIR POLLUTION HISTORY

BANG! The Big Bang. In 1948 physicist George Gamow proposed the Big Bang theory on the origin of the universe. He believed that the universe was created in a gigantic explosion as all mass and energy were created in an instant of time. On the basis of this thesis, estimates on the age of the universe at the present time range between 7 and 20 billion years with 12 billion years often mentioned as the age of planet Earth.

Gamow further believed that the various elements present today were produced within the first few minutes after the Big Bang when near-infinitely high temperatures fused subatomic particles into the chemical elements that now constitute the universe. More recent studies suggest that hydrogen and helium would have been the primary products of the Big Bang, with heavier elements being produced later within the stars. The extremely high density within the primeval atom caused the universe to expand rapidly. As it expanded, the hydrogen and helium cooled and condensed into stars and galaxies. This explains the expansion of the universe and the physical basis of Earth.

As noted in Dr. Bravo's Introduction, one might assume that the air surrounding Earth has always been composed primarily of nitrogen and oxygen, but that is not the case. Since Earth's atmosphere was first formed, its composition undoubtedly has undergone great changes. The "normal" composition of air today is not likely the same as it was when the first primitive living cells inhabited this planet. Some scientists believe that

Earth's earliest atmosphere probably contained almost no free oxygen. The oxygen in today's atmosphere is probably the result of several million of years of photosynthesis.

Over the history of Earth, plants and animals have adapted—albeit very slowly—to changes in the environment. When environmental changes occur more rapidly than a species' ability to adapt, however, the species oftentimes either does not thrive or does not survive. Human contributions to environmental changes in recent history, e.g., global warming, have come relatively quickly compared to the natural rate of change, and Earth's and its inhabitants' natural adaptation capabilities might not be adequate to meet this challenge.

Air pollution has been around for a long time. Natural phenomena such as volcanoes, windstorms, forest fires, and decaying organic matter contribute substantial amounts of air pollutants. Plants and trees also emit organic vapors and particles. For the most part, Earth, which has a well-balanced natural "cleansing" system, is able to keep up with natural pollution.

Air pollution has bedeviled humanity since the first person discovered fire. However, humans did not significantly affect the environment until relatively recent times. This is due to two reasons: (1) the human population has been large for only a small part of recorded history, and (2) the bulk of human-made produced air pollution is intimately related to industrialization. In fact, humans did not begin to alter the environment until they began to live in communities.

From the fourteenth century until recently, the primary air pollutants have been released in industrialized areas. Unfortunately, the control of pollutants rarely takes place prior to public outcry, even though the technology for controlling pollutants may be available. Early recognition of pollutants as health hazards have not resulted in pollution reduction; traditionally, only when personal survival is at stake has effective action been taken.

During the reign of the English King Edward I (1271–1307), there was a protest by the nobility against the use of "sea" coal. In the succeeding reign of Edward II (1307–1327), a man was put to torture for filling the air with a "pestilential dust" resulting from the use of coal. Under Richard III (1377–1399), and later under Henry V (1413–1422), England took steps to regulate and restrict the use of coal. Both taxation and regulation of the movement of coal in London were employed. Other legislations, parliamentary studies, and literary comments appeared sporadically during the next 250 years. In 1661, a pamphlet was published by the Royal Command of Charles II entitled "Fumifugium; or the Inconveniences of Air and Smoke in London Dissipated; Together with Some Remedies Humbly Proposed." The paper was written by John Evelyn, one of the founding fathers of the Royal Society. Later, in 1819, a Select Committee of the British Parliament was formed to study smoke abatement. As is the case of most civic actions, by the time the committee submitted its report, the problem had subsided and no action was taken.

Air pollution was a fact of life during the first half of the twentieth century. Comments such as "good, clear soot," "it's our lifeblood," "the smell of money," "an index to local activity and enterprise," and "God bless it" were used to describe air pollution. However, society began to realize that air pollution was a "deadly" problem. The term "smog" originated in Great Britain, where it was used to describe the over 1000 smoke–fog deaths that occurred in Glasgow, Scotland in 1909. The smoke problem in London reached its peak in December 1952; during this "air pollution episode" approximately 4000 people died, primarily of respiratory problems. In 1948, 20 people died and several hundred became ill in the industrial town of Donora, Pennsylvania. New York City, Birmingham, the entire state of Tennessee, Columbia River, St. Louis, Cincinnati, and Pittsburgh have had similar problems. Additional details of these often-referenced episodes are briefly summarized below.

1. On Friday December 5, 1952, static weather conditions turned the air of London, England into a deadly menace. A prolonged temperature inversion held in the city's air close to the ground and an anticyclonic high pressure system prevented the formation of winds that would have dispersed the pollutants that were accumulating heavily at ground level. For 5 days the greater London area was blanketed in airborne pollution. Few realized it at the time, but there were 4000 more deaths than normal for a 5-day period, hospital admissions were 48% higher, and sickness claims to the national health insurance system were 108% above the average, and 84% of those who died had preexisting heart or lung diseases. Hospital admissions for respiratory illness increased 3-fold, and deaths due to chronic respiratory disease increased 10-fold.
2. The same static atmospheric conditions in London caused a similar incident in Donora, Pennsylvania in 1948. A town of only 14,000, it had 15–20 more deaths than normal during the episode. More than 6000 of its residents were adversely affected, 10% of them seriously. Among those with preexisting illnesses, 88% of the asthmatics, 77% of those with heart diseases, and 79% of those with chronic bronchitis and emphysema, were adversely affected. Allowing for the difference in population, Donora paid a much higher price for air pollution than did London.
3. New York City has experienced similar periods of atmospheric stagnation on numerous occasions since the mid-1940s. During one such episode in 1953, the city reported more than 200 deaths above normal.
4. Birmingham, Alabama is another high-exposure area whose residents have frequently exhibited a greater than average incidence of respiratory irritation symptoms such as coughing, burning throats or lungs, and shortness of breath. EPA monitoring studies indicated that nonsmokers in these two cities developed respiratory symptoms 2 or 3 times more frequently than did nonsmokers in cleaner communities.
5. In the early 1900s, gases from short stacks at two copper smelters near the Georgia border of Tennessee caused widespread damage to vegetation in the surrounding countryside. When taller stacks were built, damage extended 30 miles into the forests of Georgia. An interstate suit resulted, which was finally carried to the United States Supreme Court. The problem was eventually solved by means of a byproduct sulfur dioxide recovery plant.
6. Two decades later, a similar case involved the lead and zinc smelter of the Consolidated Mining and Smelting Company of Canada at Trail, BC (British

Columbia). The smelter was located on the west bank of the Columbia River, 11 miles north of the international boundary between Canada and the United States. When extensive damage to vegetation occurred on the U.S. side of the border, a damage suit was filed and finally settled by an international tribunal. In this case, after damages were assessed, the problem was solved partly by sulfur recovery and partly by operating the smelter according to a plan based on meteorological considerations.

Unfortunately, the climatic conditions and human activities that combine to form critical buildups of pollutants are by no means uncommon in the United States. They occur periodically in various parts of the country and will continue to threaten public health as long as air pollutants are emitted into the atmosphere in amounts sufficient to accumulate to dangerous levels.

Approximately 200 million tons of waste gases are released into the air annually. Regarding sources, slightly over half of the pollution comes from the internal-combustion engines of cars and other motor vehicles. Roughly 25% comes from fuel burned at stationary sources such as power-generating plants, and another 15% is emitted from industrial processes.

The average person breathes 35 lb of the air containing these discharges each day—6 times as much as the food and drink normally consumed in the same period of time. While low levels of air pollution can be detrimental or even deadly to the health of some people, extremely high levels can be detrimental to large numbers of people. Dangerously high concentrations of air pollutants can occur during *air pollution episodes* described above and *air pollution accidents* such as those that occurred in Flixborough (England), Seveso (Italy), Three Mile Island, Chernobyl, Bhopal, etc. (Details on these accidents are available in the text/reference book by A. M. Flynn and L. Theodore, *Health, Safety and Accident Management in the Chemical Process Industries*, CRC Press/Taylor & Francis, Boca Raton, FL, 2002.) These episodes and accidents continue to occur in various parts of the world, and are well documented.

Perhaps the federal government of the United States could have done more earlier to protect the land and resources as well as public health. But for most of the nineteenth century, the government was still a weak presence in most areas of the country. There was, moreover, no body of laws with which the government could assert its authority. By the end of that century there was a growing body of information about the harm being done and some new ideas on how to set things straight. Yet, there was no acceptable ethic that would impel people to treat the land, air, and water with wisdom and care.

As the nineteenth century was drawing to a close, three very special individuals made their entrance on the national stage. Gifford Pinchot, John Muir, and Theodore Roosevelt were to write the first pages of modern environmental history in the U.S., which in turn led to the birth of the modern environmental movement early in the twentieth century. The federal government ultimately entered into the environmental and conservation business in a significant and somewhat dramatic fashion when Teddy Roosevelt's second cousin Franklin entered the White House in 1933. It was his political ideology, as much as his love of nature, that led Roosevelt to include major conservation projects in his New Deal reforms. The Civilian Conservation

Corps, the Soil Conservation Service, and the Tennessee Valley Authority were among the many New Deal programs created to serve both the environment and the people.

At this point in time, muscle, animal, and steam power had been replaced by electricity, internal-combustion engines, and nuclear reactors. During this period, industry was consuming natural resources at an incredible rate. All of these events began to escalate at a dangerous rate after World War II. In 1962, a marine biologist named Rachel Carson, author of *Silent Spring* (Houghton-Mifflin, 1962), a best-selling book about ocean life, opened the eyes of the world to the dangers of ignoring the environment. It was perhaps at this point that America began calling in earnest for environmental reform and constraints on environmental degradation. Finally, in the 1970s, Congress began turning out environmental laws that addressed these issues. It all began in 1970 with the birth of the Environmental Protection Agency.

[For additional literature regarding early history and the environmental movement, the interested reader is referred to the book by Philip Shabecoff, titled *A Fierce Green Fire* (Farrar-Strauss-Giroux, 1993). This outstanding book is a "must" for anyone whose work is related to or is interested in the environment.]

2

AIR POLLUTION REGULATORY FRAMEWORK

2.1 INTRODUCTION

It is now 1970, a cornerstone year for modern environmental policy. The National Environmental Policy Act (NEPA), enacted on January 1, 1970, was considered a "political anomaly" by some. NEPA was not based on specific legislation; instead, it referred in a general manner to environmental and quality of life concerns. The Council for Environmental Quality (CEQ), created by NEPA, was one of the councils mandated to implement legislation. April 22, 1970 brought Earth Day, where thousands of demonstrators gathered all around the nation. NEPA and Earth Day were the beginning of a long, seemingly never-ending debate over environmental issues.

The Nixon Administration at that time became preoccupied with not only trying to pass more extensive environmental legislation but also implementing the laws. Nixon's White House Commission on Executive Reorganization proposed in the Reorganizational Plan 3 of 1970 that a single, independent agency be established, separate from the CEQ. The plan was sent to Congress by President Nixon on July 9, 1970, and this new US Environmental Protection Agency (EPA) began operation on December 2, 1970. The EPA was officially born.

In many ways, the EPA is the most far-reaching regulatory agency in the federal government because its authority is so broad. The EPA is charged by the Congress of

Air Pollution Control Equipment Calculations. By Louis Theodore

the United States of America to protect the nation's land, air, and water systems. Under a mandate of national environmental laws, the EPA strives to formulate and implement actions that lead to a compatible balance between human activities and the ability of natural systems to support and nurture life.

The EPA works with the states and local governments to develop and implement comprehensive environmental programs. Amendments to federal legislations such as the Clean Air Act, the Safe Drinking Water Act, the Resource Conservation and Recovery Act, and the Comprehensive Environmental Response, Compensation and Liability Act, all mandate more involvement by state and local governments in the details of implementation.

This chapter presents the regulatory framework governing air management. It provides an overview of environmental laws and regulations used to protect human health and the environment from the potential hazards of air pollutants.

2.2 THE REGULATORY SYSTEM

Since the early 1970s, environmental regulations have become a system in which laws, regulations, and guidelines have become interrelated. Requirements and procedures developed under previously existing laws may be referenced to in more recent laws and regulations. The history and development of this regulatory system has led to laws that focus principally on only one environmental medium, i.e., air, water, or land. Some environmental managers feel that more needs to be done to manage all of the media simultaneously since they are interrelated. Hopefully, the environmental regulatory system will evolve into a truly integrated, multimedia management framework in the future.

Federal laws are the product of Congress. Regulations written to implement the law are promulgated by the Executive Branch of government, but until judicial decisions are made regarding the interpretations of the regulations, there may be uncertainty about what regulations mean in real situations. Until recently, environmental protection groups were more frequently the plaintiffs in cases brought to court seeking interpretation of the law. Today, industry has become more active in this role. Forum shopping, the process of finding a court that is more likely to be sympathetic to the plaintiffs' point of view, continues to be an important tool in this area of environmental regulation. Many environmental cases have been heard by the Circuit Court of the District of Columbia.

Enforcement approaches for environmental regulations are environmental management–oriented in that they seek to remedy environmental harm, not simply a specific infraction of a given regulation. All laws in a legal system may be used in enforcement to prevent damage or threats of damage to the environment or human health and safety. Tax laws (e.g., tax incentives) and business regulatory laws (e.g., product claims, liability disclosures) are examples of laws not directly focused on environmental protection, but that may also be used to encourage compliance and discourage noncompliance with environmental regulations.

Common law also plays an important role in environmental management. Common law is the set of rules and principles relating to the government and security of persons and property. Common law authority is derived from the usages and customs that are recognized and enforced by the courts. In general, no infraction of the law is necessary when establishing a common law court action. A common law "civil wrong" (e.g., environmental pollution) that is brought to court is called a *tort*. Environmental *torts* may arise because of nuisance, trespass, or negligence.

Laws tend to be general and contain uncertainties relative to the implementation of principles and concepts they contain. Regulations derived from laws may be more specific, but are also frequently too broad to allow clear translation into environmental technology practice. Permits may be used in the environmental regulation industry to bridge this gap and provide specific, technical requirements imposed on a facility by the regulatory agencies for the discharge of pollutants or on other activities carried out by the facility that may impact the environment.

Most major federal environmental laws (perhaps unfortunately) provide for citizen lawsuits. This empowers individuals to seek compliance or monetary penalties when these laws are violated and regulatory agencies do not take enforcement action against the violator.

2.3 LAWS AND REGULATIONS: THE DIFFERENCES

The following (W. Matystik: private communications, 1995) are some of the major differences between a federal *law* and a federal *regulation*, as briefly discussed in the previous section.

1. A law (or Act) is passed by both houses of Congress and signed by the President. A regulation is issued by a government agency such as the EPA or the Occupational Safety and Health Administration (OSHA).
2. Congress can pass a law on any subject it chooses. It is limited only by the restrictions in the Constitution. A law can be challenged in court if it is unwise, unreasonable, or even silly. If, for example, a law was passed that placed a tax on burping (belching), it could not be challenged in court just because it was unenforceable. A regulation can be issued by an agency only if the agency is authorized to do so by the law passed by Congress. When Congress passes a law, it usually assigns an administrative agency to implement that law. A law regarding radio stations, for example, may be assigned to the Federal Communications Commission (FCC). Sometimes a new agency is created to implement a law. This was the case with the Consumer Product Safety Commission (CPSC). OSHA is authorized by the Occupational Safety and Health Act to issue regulations that protect workers from exposure to the hazardous chemicals they use in manufacturing processes. If those hazardous chemicals are emitted by the plant and affect the surrounding community but do not expose the workers in the plant, OSHA is not authorized to issue an order to stop the practice. (*Note*: The EPA is authorized to regulate such practices.)
3. Laws can include a Congressional mandate directing EPA to develop a comprehensive set of regulations. Regulations, or rulemakings, are issued by an agency, such as EPA, that translates the general mandate of a statute into a set of requirements for the Agency and the regulated community.

4. Regulations are developed by EPA in an open and public manner according to an established process. When a regulation is formally proposed, it is published in an official government document called the *Federal Register* to notify the public of EPA's intent to create new regulations or modify existing ones. EPA provides the public, which includes the potentially regulated community, with an opportunity to submit comments. Following an established comment period, EPA may revise the proposed rule on the basis of both an internal review process and public comments.
5. The final regulation is published, once promulgated, in the *Federal Register*. Included with the regulation is a discussion of the Agency's rationale for the regulatory approach, known as *preamble language*. Final regulations are compiled annually and incorporated in the Code of Federal Regulations (CFR) according to a highly structured format based on the topic(s) of the regulation. This latter process is called *codification*, and each CFR title corresponds to a different regulatory authority. For example, EPA's regulations are in Title 40 of the CFR. The codified RCRA regulations can be found in Title 40 of the CFR, Parts 240–282. These regulations are often cited as 40 CFR, with the part listed afterward (e.g., 40 CFR Part 264), or the part and section (e.g., 40 CFR §264.10).
6. A regulation may be challenged in court because the issuing agency exceeded the mandate given it by Congress. If the law requires the agency to consider costs versus benefits of the regulation, the regulation could be challenged in court if the cost/benefit analysis were not correctly or adequately done. If OSHA issues a regulation limiting a worker's exposure to a hazardous chemical to 1 part per million (ppm), OSHA could be called on to prove in court that such a low limit was needed to prevent a worker from being harmed. Failure to prove this would mean that OSHA exceeded its mandate under the law, as OSHA is charged to develop standards only as stringent as those required to protect worker health and provide worker safety.
7. Laws are usually brief and general. Regulations are usually lengthy and detailed. The Hazardous Materials Transportation Act, for example, is only approximately 20 pages long. It speaks in general terms about the need to protect the public from the dangers associated with transporting hazardous chemicals and identifies the Department of Transportation (DOT) as the agency responsible for issuing regulations implementing the law. The regulations issued by the DOT are several thousand pages long and are very detailed down to the exact size, shape, design, and color of the warning placards that must be used on trucks carrying any of the thousands of regulated chemicals.
8. Generally, laws are passed infrequently. Often years pass between amendments to an existing law. A completely new law on a given subject already addressed by an existing law is unusual. Laws are published as a "Public Law— - —" and are eventually codified into the United States Code.
9. Regulations are issued and amended frequently. Proposed and final new regulations and amendments to existing regulations are published daily in the *Federal Register*. Final regulations have the force of law when published.

Annually, see (5) above, the regulations are codified in the Code of Federal Regulations (CFR). The CFR is divided into 50 volumes called *Titles*. Each Title is devoted to a subject or agency. For example, labor regulations are in Title 29, while environmental regulations, as noted above, are in Title 40.

2.4 THE CLEAN AIR ACT

The Clean Air Act defines the national policy for air pollution abatement and control in the United States. It establishes goals for protecting health and natural resources, and delineates what is expected of federal, state, and local governments to achieve those goals. The Clean Air Act, which was initially enacted as the Air Pollution Control Act of 1955, has undergone several revisions over the years to meet the ever-challenging needs and conditions of the nation's air quality. On November 15, 1990, President George H. W. Bush signed the most recent amendments to the Clean Air Act, referred to as the *1990 Clean Air Act Amendments*. Embodied in these amendments were several progressive and creative new themes deemed appropriate for effectively achieving the air quality goals and for reforming the air quality control regulatory process. Specifically the amendments:

1. Encouraged the use of market-based principles and other innovative approaches similar to performance-based standards plus emission banking and trading.
2. Promoted the use of clean low-sulfur coal and natural gas, as well as innovative technologies to clean high-sulfur coal through the acid rain program.
3. Reduced energy waste and creates enough of a market for clean fuels derived from grain and natural gas to cut/reduce dependence on oil imports by one million barrels/day.
4. Promoted energy conservation through an acid rain program that gave utilities flexibility to obtain needed emission reductions through programs that encouraged customers to conserve energy.

These Amendments provided the framework for air quality regulations in the United States, which remain in effect today.

The earlier Amendments of 1970 differentiated areas of the country with relatively good air quality (areas meeting established standards) and those with relatively poor air quality, and created different rules to regulate air pollution in these different areas. The law also established schedules under which areas with poor air quality would come into compliance with the established standards.

By the mid-1970s, it was generally recognized that many areas of the country would not be able to meet the established schedules for improving air quality. Congress passed the Clean Air Act Amendments of 1977 to address this fact. These laws established new schedules and introduced more stringent means to meet the schedules. Even though the Amendments of 1977 contained stringent pollution measures, many areas of the county continued to experience difficulty in meeting established standards. Despite this fact, development of new air quality legislation on the federal level was stalled until

November 15, 1990, when congress finally passed the aforementioned Clean Air Act Amendments of 1990.

Several of the key provisions of the 1990 Act are reviewed below (see L. Stander and L. Theodore, *Environmental Regulatory Calculations Handbook*, John Wiley & Sons, Inc., Hoboken, NJ, 2008, for additional details).

Provisions for Attainment and Maintenance of National Ambient Air Quality Standards

Although the Clean Air Act brought about significant improvements in the nation's air quality, the urban air pollution problems of ozone (smog) and particulate matter continue to persist in certain areas. In 1955, approximately 70 million US residents were living in counties with ozone levels exceeding the EPA's current ozone standard. The present National Ambient Air Quality Standards (NAAQS) are provided in Table 2.1.

The Clean Air Act, as amended in 1990, created a more balanced strategy for the nation to address the problem of urban smog. Overall, the amendments revealed Congress' high expectations of the states and the federal government. While it gave states more time to meet the air quality standard (up to 20 years for ozone in Los Angeles), it also required states to make constant progress in reducing emissions. It required the federal government to reduce emission from cars, trucks, and buses; from consumer products such as hairspray and window-washing compounds; and, from ships and barges during loading and unloading of petroleum products. The federal government also developed the technical guidance that states need to control stationary sources.

TABLE 2.1 National Ambient Air Quality Standards

Criteria Pollutant	Averaging Period	Primary NAAQS[a]	Secondary NAAQS[a]
PM10 (particulate matter $<10\ \mu m$)	Annual	$50\ \mu g/m^3$	$50\ \mu g/m^3$
	24-hr	$150\ \mu g/m^3$	$150\ \mu g/m^3$
PM2.5 (particulate matter $<2.5\ \mu m$)	Annual	$15\ \mu g/m^3$	$15\ \mu g/m^3$
	24-hr	$65\ \mu g/m^3$	$65\ \mu g/m^3$
SO_2 (sulfur dioxide)	Annual	0.030 ppm ($80\ \mu g/m^3$)	—
	24-hr	0.14 ppm ($365\ \mu g/m^3$)	—
	3-hr	—	0.050 ppm ($1300\ \mu g/m^3$)
NO_2 (nitrogen dioxide)	Annual	0.053 ppm ($100 \mu g/m^3$)	0.053 ppm ($100\ \mu g/m^3$)
Ozone	1-hr	0.12 ppm ($235\ \mu g/m^3$)	0.12 ppm ($235\ \mu g/m^3$)
	8-hr	0.08 ppm ($157\ \mu g/m^3$)	0.08 ppm 0.12 ppm
CO (carbon monoxide)	8-hr	9 ppm ($10\ \mu g/m^3$)	—
	1-hr	35 ppm ($40\ \mu g/m^3$)	—
Lead	Quarterly	$1.5\ \mu g/m^3$	$1.5\ \mu g/m^3$

[a]NAAQS concentrations are expressed by EPA in different units of measurement—micrograms per cubic meter ($\mu g/m^3$), milligrams per cubic meter (mg/m^3), or parts per million by volume (ppmv)—depending on the pollutant and the standard. Values in the parentheses in this table are approximate equivalent concentrations.

The Clean Air Act addresses the urban air pollution problems of ozone (smog), carbon monoxide (CO), and particulate matter (PM). Specifically, it clarifies how areas are designated and redesignated "attainment." It also allows the EPA to define the boundaries of "nonattainment" areas—geographic areas whose air quality does not meet federal air quality standards designed to protect public health. The law also establishes provisions defining when and how the federal government can impose sanctions on areas of the country that have not met certain conditions.

The Clean Air Act established nonattainment area classifications ranked according to the severity of the area's air pollution problem for the pollutant ozone. These classifications are marginal, moderate, serious, severe, and extreme. The EPA assigns each nonattainment areas one of these categories, thus triggering varying requirements that area must comply with in order to meet the ozone standard.

As mentioned, nonattainment areas have to implement different control measures, depending on their classification. Marginal areas, for example, are the closest to meeting the standard. They are required to conduct an inventory of their ozone-causing emissions and institute a permit program. Nonattainment areas with more serious air quality problems must implement various control measures. The worse the air quality, the more controls these areas will have to implement.

The Clean Air Act also established similar programs for areas that do not meet the federal health standards for carbon monoxide and particulate matter. Areas exceeding the standards for theses pollutants are divided into "moderate" and "serious" classifications. Depending on the degree to which they exceed the carbon monoxide standard, areas are then required to implement programs introducing oxygenated fuels and/or enhanced emission inspection programs, among other measures. Depending on their classification, areas exceeding the particulate matter standard have to implement either reasonably available control measures (RACMs) or best available control measures (BACMs), among other requirements.

To summarize, the NAAQS represents a maximum concentration or "threshold level" of a pollutant in the air above which humans or the environment may experience some adverse effects. The actual threshold levels are based on years of epidemiological, health, and environmental effects research conducted by the EPA. There are two types of NAAQS: primary standards, which are set at levels that are designed to protect the public health, and secondary standards, which are designed to protect the public welfare (such as vegetation, livestock, building materials, and other elements in the environment). The NAAQS also differentiate between effects from short-term exposure and longer-term exposure to air pollutants; there are short-term NAAQS, based on 1-hr, 3-hr, 8-hr averages, or 24-hr concentrations, and long-term NAAQS, based on quarterly or annual concentrations.

If monitoring indicates that the concentration of a pollutant exceeds the NAAQS in any area of the country, that area (as noted earlier) is labeled a *nonattainment area* for that pollutant, meaning that the area is not meeting the ambient standard. Conversely, any area in which the concentration of a criteria pollutant is below the NAAQS is labeled an *attainment area*, indicating that the NAAQS is being met. The attainment/nonattainment designation is made on a pollutant-by-pollutant basis. Therefore, the air quality in an area of the country may be designated attainment for some pollutants and nonattainment for other pollutants at the same time. For example, many cities are designated nonattainment for ozone, but are in attainment for the other criteria pollutants.

The NAAQS provide target levels of concentrations of pollutants in the atmosphere, but do not set pollutant emission limitations for individual pollution sources. There are four key federal regulations that govern emissions from individual sources, each of which is described below.

1. Prevention of significant deterioration (PSD) permitting program
2. Nonattainment new source review (NA-NSR) permitting program
3. New source performance standards (NSPS)
4. Maximum achievable control technology (MACT) for hazardous air pollutants (HAPs)

Prevention of Significant Deterioration (PSD). These major federal rules that govern air quality in attainment areas are designed to ensure that air quality in "clean" areas (i.e., attainment areas) will not degrade, but will remain clean, even as new sources of pollution are constructed. The PSD program applies to new major sources and major modifications to existing major sources.

Nonattainment Area New Source Review (NA-NSR). This set of rules and regulations applies to new or modified emissions sources in nonattainment areas, the areas of the country where NAAQS are not being met. Restrictions on emissions and control technology requirements under NA-NSR provisions are more stringent than under PSD, because the goal of the NA-NSR rules is to improve the air quality until the NAAQS are met.

The New Source Performance Standards (NSPS). These provisions were established under the amendments of 1970, relatively early in the history of air quality regulation, in recognition of the fact that newly constructed sources should be able to operate more "cleanly" than existing, older sources. The NSPS establish the minimum level of control of certain pollutants that specific categories of industrial sources constructed since 1971 must achieve. The emissions limits under NSPS are based on the best technological system of continuous emission reduction available, taking into account annual costs and other factors of applying the technology.

Hazardous Air Pollutants (HAPs). The HAP program regulates, in two phases, routine emissions of 189 specific toxic compounds. The first phase takes a "technology-based" approach to regulating pollutants, rather than the risk-based approach used in the NESHAP (National Emission Standards for Hazardous Air Pollutants) program. This requires that "major" HAP sources install maximum achievable control technology (MACT). MACT standards are defined by EPA and cover selected categories of industrial sources. To date, EPA has promulgated MACT standards for nearly 10 different source categories. The second phase of the HAP program will require certain facilities, to be identified by EPA, to determine the *residual risk* to the public health and the environment of the small amounts of HAPs that may still be released into the atmosphere after MACT is applied (see "Air Toxics" subsection for more details).

Provisions Relating to Mobile Sources

While motor vehicles built today emit fewer pollutants (60–80% less, depending on the pollutant) than those built in the 1960s, cars and trucks still account for almost half the emissions of the ozone precursors that include volatile organic carbons (VOC) and nitrogen oxides (NO_x), and up to 90% of the CO emissions in urban areas. The principal reason for this problem is the rapid growth in the number of vehicles on the roadways and the total miles driven. This growth has offset a large portion of the emission reductions gained from motor vehicle controls.

In view of the continuing growth in automobile emissions in urban areas, combined with the serious air pollution problems in many urban areas, Congress made significant changes to the motor vehicle provisions in the Clean Air Act and established tighter pollution standards for emissions from automobiles and trucks. These standards were set so as to reduce tailpipe emissions of hydrocarbons, carbon monoxide, and nitrogen oxides on a phased-in basis beginning in the model year 1994. Automobile manufacturers were also required to reduce vehicle emissions resulting from the evaporation of gasoline during refueling.

Fuel quality was also controlled. Scheduled reductions in gasoline volatility and sulfur content of diesel fuel, for example, were required. Programs requiring cleaner (called "reformulated") gasoline were initiated in 1995 for the nine cities with the worst ozone problems. Higher levels (2.7%) of alcohol-based oxygenated fuels were to be produced and sold in those areas that exceed the federal standard for carbon monoxide during the winter months.

The 1990 amendments to the Clean Air Act also established a clean fuel car pilot program in California, requiring the phase-in of tighter emission limits for 150,000 vehicles in model year 1996 and 300,000 by the model year 1999. These standards were to be met with any combination of vehicle technology and cleaner fuels. The standards became even stricter in 2001. Other states were able to "opt in" to this program, through incentives, not sales or production mandates.

Air Toxics

Toxic air pollutants are those pollutants that are hazardous to human health or the environment. These pollutants are typically carcinogens, mutagens, and reproductive toxins.

The toxic air pollution problem is widespread. Information generated in 1987 from the Superfund "right to know" rule (SARA Section 313), indicated that more than 2.7 billion pounds (lb) of toxic air pollutants were emitted annually in the United States. The EPA studies indicated that exposure to such quantities of toxic air pollutants may result in 1000–3000 cancer deaths each year.

Section 112 of the Clean Air Act includes a list of 189 substances that are identified as hazardous air pollutants. As noted earlier, a list of categories of sources that emit these pollutants was prepared. [The list, of course, included the categories (1) major sources, or sources emitting 10 tons per year of any single hazardous air pollutants; and, (2) area sources (smaller sources, such as dry cleaners and autobody refinishing).] In turn, EPA promulgated emission standards, referred to as the aforementioned *maximum achievable control technology* (MACT) standards, for each listed source category. These standards were based on the best demonstrated control technology or practices utilized by sources that make up each source category. Within 8 years of promulgation of a MACT standard, EPA must evaluate the level of risk that remains (residual risk), due to exposure to emissions from a source category, and determine whether the residual risk is acceptable. If the residual risks are determined to be unacceptable, additional standards are required.

Acid Deposition Control

Acid rain occurs when sulfur dioxide and nitrogen oxide emissions are transformed in the atmosphere and return to Earth in rain, fog, or snow. Approximately 20 million tons of sulfur dioxide are emitted annually in the United States, mostly from the burning of fossil fuels by electric utilities. Acid rain damages lakes, harms forests and buildings, contributes to reduced visibility, and is suspected of damaging health.

It was hoped that the Clean Air Act would bring about a permanent 10 million ton reduction in sulfur dioxide (SO_2) emissions from 1980 levels. To achieve this, the EPA allocated allowances in two phases, permitting utilities to emit one ton of sulfur dioxide. The first phase, which became effective January 1, 1995, required 110 power plants to reduce thier emissions to a level equivalent to the product of an emissions rate of 2.5 lb of SO_2/MM Btu (British thermal unit) $\times$ an average of their 1985–1987 fuel use. Emissions data indicate that 1995 SO_2 emissions at these units nationwide were reduced by almost 40% below the required level.

The second phase, which became effective January 1, 2000, required approximately 2000 utilities to reduce their emissions to a level equivalent to the product of an emissions rate of 1.2 lb of SO_2/MM Btu $\times$ the average of their 1985–1987 fuel use. In both phases, affected sources were required to install systems that continuously monitor emission in order to track progress and assure compliance.

The Clean Air Act allowed utilities to trade allowances within their systems and/or buy or sell allowances to and from other affected sources. Each source must have had sufficient allowances to cover its annual emissions. If not, the source was subject to a \$2000/ton excess emissions fee and a requirement to offset the excess emissions in the following year.

The Clean Air Act also included specific requirements for reducing emissions of nitrogen oxides.

Operating Permits

The Act requires the implementation of an operating permit program modeled after the National Pollution Discharge Elimination System (NPDES) of the Clean Water

Act. The purpose of the operating permits program is to ensure compliance with all applicable requirements of the Clean Air Act. Air pollution sources, subject to the program, must obtain an operating permit; states must develop and implement an operating permit program consistent with the Act's requirements; and, EPA must issue permit program regulations, review each state's proposed program, and oversee the state's effort to implement any approved program. The EPA must also develop and implement a federal permit program when a state fails to adopt and implement its own program.

In many ways, this program is the most important procedural reform contained in the 1990 Amendments to the Clean Air Act. It enhanced air quality control in a variety of ways and updated the Clean Air Act, making it more consistent with other environmental statutes. The Clean Water Act, the Resource Conservation and Recovery Act and the Federal Insecticide, Fungicide, and Rodenticide Act all require permits.

Stratospheric Ozone Protection

The Clean Air Act requires the phase-out of substances that deplete the ozone layer. The law required a complete phase-out of chlorofluorocarbons (CFCs) and halons, with stringent interim reductions on a schedule similar to that specified in the Montreal Protocol–CFCs, halons, and carbon tetrachloride by 2000 and methyl chloroform by 2002. Class II chemicals (HCFCs) will be phased out by 2030.

The law required nonessential products releasing Class I chemicals to the banned. This ban went into effect for aerosols and noninsulating foams using Class II chemicals in 1994. Exemptions are included for flammability and safety.

2.5 PROVISIONS RELATING TO ENFORCEMENT

The Clean Air Act contains provisions for a broad array of authorities to make the law readily enforceable. EPA has authority to

1. Issue administrative penalty orders up to \$200,000 and field citations up to \$5000
2. Obtain civil judicial penalties
3. Secure criminal penalties for knowing violations and for knowing and negligent endangerment
4. Require sources to certify compliance
5. Issue administrative subpoenas for compliance data
6. Issue compliance orders with compliance schedules of up to one year

Citizen suit provisions are also included to allow citizens to seek penalties against violators, with penalties going to a United States Treasury fund for use by the EPA for compliance and enforcement activities.

The following EPA actions represent recent regulations promulgated to implement the requirements of the Clean Air Act:

1. Clean Air Interstate Rule published on May 12, 2005 (70 FR 25161) amends requirements for State Implementation Plans and for provisions for Acid Rain Program.
2. Mercury Rules published on May 18, 2005 (20 FR 28605) amends New Source Performance Standards for electric utility steam generating units and some provisions of the Acid Rain Program.
3. Non-road Diesel Rule published on May 11, 2004 (69 FR 38957) amends provisions for mobile sources and for highway vehicles and engines.
4. Ozone Rules identified those areas that are designated as not attaining the ambient air quality standards for ozone.
5. Fine Particle Rules identified those areas that are designated as not attaining the ambient air quality standards for particulate matter.

2.6 CLOSING COMMENTS AND RECENT DEVELOPMENTS

Today, more than 35 years after the 1970 Clean Air Act was adopted, and 18 years after its last major revision, many areas of the country continue to experience difficulty in meeting established ambient air standards, and the EPA is embarking on new programs to ensure that the air is clean.

The Agency is currently working on several programs to manage a range of developing air quality issues. Two of the more prominent of these issues are greenhouse gases (such as carbon dioxide) that affect global climate, and fine particulate matter, referred to also as PM2.5 (or particulate matter less than 2.5 μm in diameter) that can produce regional haze and reduce visibility in otherwise pristine regulatory environments. Action on nanoparticles (less than 0.1 μm) has been discussed but appears (at the time of the preparation of this text) to be in limbo. These and other developing air issues (e.g., vapor intrusion)—and new issues not yet on the horizon—represent challenges for the EPA that may prompt Congress to review and amend the Clean Air Act again in the coming years.

Finally, the reader should note that EPA's early "Command-and-Control" regulatory standards of the 1970s and 1980s have been replaced by less costly and more flexible standards as well as risk-based standards. In addition, "surrogate" regulators, such as lawyers, bankers, and accountants, have (perhaps fortunately) also contributed to insure corporate environmental responsibility.

3

FUNDAMENTALS: GASES

3.1 INTRODUCTION

This chapter provides a review of some basic concepts from physics, chemistry, and engineering in preparation for material that is covered in later chapters. These basic concepts include units and dimensions, some physical and chemical properties of substances, the ideal gas law, the Reynolds number, and phase equilibria. Because many of these topics are unrelated to each other, this chapter admittedly lacks the cohesiveness that chapters covering a single topic might have. This is usually the case when basic material from such widely differing areas of knowledge as physics, chemistry, and engineering is surveyed. Although these topics are widely divergent and covered with varying degrees of thoroughness, all of them will find later use in this book.

3.2 MEASUREMENT FUNDAMENTALS

Units and Dimensions

The units used in this text are consistent with those adopted by the engineering session in the United States. For engineering work, SI (*Système International*) and *English* units

Air Pollution Control Equipment Calculations. By Louis Theodore

are most often employed; in the U.S., the *English engineering* units are generally used, although efforts are still in place or underway to obtain universal adoption of SI units for all engineering and science applications. The SI units have the advantage of being based on the decimal system, which allows for more convenient conversion of units within the system. However, the English engineering units will primarily be used here. Conversion factors between SI and English units and additional details on the SI system are provided in the Appendix.

Conversion of Units

Converting a measurement from one unit to another can conveniently be accomplished by using *unit conversion factors*; these factors are obtained from a simple equation that relates the two units numerically. For example, for

$$12\,\text{inches (in)} = 1\,\text{foot (ft)} \tag{3.1}$$

the following conversion factor can be obtained:

$$12\,\text{in}/1\,\text{ft} = 1 \tag{3.2}$$

Since this factor is equal to unity, multiplying some quantity (e.g., 18 ft) by this factor cannot alter its value. Hence

$$18\,\text{ft}\,(12\,\text{in}/1\,\text{ft}) = 216\,\text{in} \tag{3.3}$$

Note that in Equation (3.3), the original units of *feet* on the left-hand side cancel out leaving only the desired units of *inches.*

Physical equations must be dimensionally consistent. For the equality to hold, each term in the equation must have the same dimensions. This condition can be and should be checked when solving engineering problems. Throughout this book, great care is exercised in maintaining the dimensional formulas of all terms and the dimensional homogeneity of each equation. Equations will generally be developed in terms of specific units rather than general dimensions (e.g., feet, rather than length). This approach should help the reader to more easily attach physical significance to the equations presented.

Significant Figures and Scientific Notation

Significant figures provide an indication of the precision with which a quantity is measured or known. The last digit represents, in a qualitative sense, some degree of doubt. For example, a measurement of 8.32 inches implies that the actual quantity is somewhere between 8.315 and 8.325 inches. This applies to calculated and measured quantities; quantities that are known exactly (e.g., pure integers) have an infinite number of significant figures.

The significant digits of a number are the digits from the first nonzero digit on the left to either (1) the last digit (whether it is nonzero or zero) on the right if there is a decimal point, or (2) the last nonzero digit of the number if there is no decimal point. For example:

370	has 2 significant figures
370.	has 3 significant figures
370.0	has 4 significant figures
28,070	has 4 significant figures
0.037	has 2 significant figures
0.0370	has 3 significant figures
0.02807	has 4 significant figures

In general, whenever quantities are combined by multiplication and/or division, the number of significant figures in the result should equal the lowest number of significant figures of any of the quantities. In long calculations, the final result should be rounded off to the correct number of significant figures. When quantities are combined by addition and/or subtraction, the final result cannot be more precise than any of the quantities added or subtracted. Therefore, position (relative to the decimal point) of the last significant digit in the number that has the lowest degree of precision is the position of the last permissible significant digit in the result. For example, the sum of 3702, 370, 0.037, 4, and 37. should be reported as 4110 (without a decimal). The least precise of the five numbers is 370, which has its last significant digit in the *tens* position. The answer should also have its last significant digit in the *tens* position. The author has *attempted* to abide by these rules.

In the process of performing engineering calculations, very large and very small numbers are often encountered. A convenient way to represent these numbers is to use *scientific notation*. Generally, a number represented in scientific notation is the product of a number (<10 but $>$ or $=1$) and (multiplied by) 10 raised to an integer power. For example:

$$28{,}070{,}000{,}000 = 2.807 \times 10^{10}$$

$$0.000\ 002\ 807 = 2.807 \times 10^{-6}$$

A nice feature of using scientific notation is that only the significant figures need appear in the number.

3.3 CHEMICAL AND PHYSICAL PROPERTIES

Temperature

Whether in the gaseous, liquid, or solid state, all molecules possess some degree of kinetic energy, i.e., they are in constant motion—vibrating, rotating, or translating.

The kinetic energies of individual molecules cannot be measured, but the combined effect of these energies in a very large number of molecules can. This measurable quantity is known as *temperature*; it is a macroscopic concept only and as such does not exist on the molecular level.

Temperature can be measured in many ways; the most common method makes use of the expansion of mercury (usually encased inside a glass capillary tube) with increasing temperature. However, thermocouples or thermistors can also be employed.

The two most commonly used temperature scales are the Celsius (or Centigrade) and Fahrenheit scales. The Celsius is based on the boiling and freezing points of water at 1 atmosphere pressure; to the former, a value of 100°C is assigned, and to the latter, a value of 0°C. On the older Fahrenheit scale, these temperatures correspond to 212 and 32°F, respectively. Equations (3.4) and (3.5) provide the conversion from one scale to the other.

$$^{\circ}\text{F} = 1.8(^{\circ}\text{C}) + 32 \tag{3.4}$$

$$^{\circ}\text{C} = (^{\circ}\text{F} - 32)/1.8 \tag{3.5}$$

where $^{\circ}\text{F}$ = a temperature on the Fahrenheit scale
$^{\circ}\text{C}$ = a temperature on the Celsius scale

Experiments with gases at low to moderate pressures (up to a few atmospheres) have shown that, if the pressure is kept constant, the volume of a gas and its temperature are linearly related (Charles' law) and that a decrease of 0.3663% or $\left(\frac{1}{273}\right)$ of the initial volume is experienced for every temperature drop of 1°C. These experiments were not extended to very low temperatures, but if the linear relationship were extrapolated, the volume of the gas would *theoretically* be zero at a temperature of approximately −273°C or −460°F. This temperature has become known as *absolute zero* and is the basis for the definition of two *absolute* temperature scales. (An *absolute* scale is one which does not allow negative quantities.) These absolute temperature scales are the Kelvin (K) and Rankine (°R) scales; the former is defined by shifting the Celsius scale by 273°C so that 0 K is equal to −273°C; Equation (3.6) shows this relationship:

$$\text{K} = {^{\circ}\text{C}} + 273 \tag{3.6}$$

The Rankine scale is defined by shifting the Fahrenheit scale 460°, so that

$$^{\circ}\text{R} = {^{\circ}\text{F}} + 460 \tag{3.7}$$

Pressure

In the gaseous state, the molecules possess a high degree of translational kinetic energy, which means that they are able to move quite freely throughout the body of the gas. If the gas is in a container of some type, the molecules are constantly bombarding the walls of the container. The macroscopic effect of this bombardment by a tremendous number of molecules—enough to make the effect measurable—is called *pressure*. The natural units of pressure are force per unit area. In the example of the gas in a container, the *unit area* is a portion of the inside solid surface of the container wall while the *force*, measured perpendicularly to the unit area, is the result of the molecules hitting the unit area and swing up momentum during the sudden change of direction.

There are a number of different methods used to express a pressure measurement. Some of them are natural units, i.e., based on a force per unit area, e.g., pound (force) per square inch (abbreviated lb_f/in^2 or psi) or dyne per square centimeter (dyn/cm^2). Others are based on a fluid height, such as inches of water (in H_2O) or millimeters of mercury (mm Hg); units such as these are convenient when the pressure is indicated by a difference between two levels of a liquid as in a *manometer or barometer*. *Barometric pressure* and *atmospheric pressure* are synonymous and measure the ambient air pressure. *Standard barometric pressure* is the average atmospheric pressure at sea level, 45° north latitude at 32°F. It is used to define another unit of pressure called the atmosphere (atm). Standard barometric pressure is 1 atm and is equivalent to 14.696 psi and 29.921 in Hg. As one might expect, barometric pressure varies with weather and altitude.

Measurements of pressure by most gauges indicate the difference in pressure either above or below that of the atmosphere surrounding the gauge. *Gauge pressure* is the pressure indicated by such a device. If the pressure in the system measured by the gauge is greater than the pressure prevailing in the atmosphere the gauge pressure is expressed positively; if lower than atmospheric pressure the gauge pressure is a negative quantity; the term *vacuum* designates a negative gauge pressure. Gauge pressures are often identified by the letter *g* after the pressure unit; for example, psig (pounds per square inch gauge) is a gauge pressure in psi units.

Since gauge pressure is the pressure relative to the prevailing atmospheric pressure, the sum of the two gives the *absolute pressure*, indicated by the letter *a* after the unit [e.g., psia (pounds per square inch absolute)]:

$$P = P_a + P_g \tag{3.8}$$

where P = absolute pressure (psia)
P_a = atmospheric pressure (psia)
P_g = gauge pressure (psig)

The absolute pressure scale is absolute in the same sense that the absolute temperature scale is absolute, i.e., a pressure of zero psia is the lowest possible pressure theoretically achievable—a perfect vacuum.

Moles and Molecular Weights

An atom consists of protons and neutrons in a nucleus surrounded by electrons. An electron has such a small mass relative to that of the proton and neutron that the weight of the atom (called the *atomic weight*) is approximately equal to the sum of the weights of the particles in its nucleus. Atomic weight may be expressed in *atomic mass units (amu) per atom* or in *grams per gram-atom*. One gram-atom contains 6.02×10^{23} atoms (Avagadro's number). The atomic weights of all the elements are available in the literature.

The *molecular weight* (MW) of a compound is the sum of the atomic weights of the atoms that make up the molecule. Units of atomic mass units per molecule (amu/molecule) or grams per gram-mole (g/gmol) are used for molecular weight. One gram-mole (gmol) contains an Avogadro number of molecules. For the English system, a pound-mole (lbmol) contains $454 \times 6.023 \times 10^{23}$ molecules.

Molal units are used extensively in air pollution control calculations as they greatly simplify material balances where chemical (including combustion) reactions are occurring. For mixtures of substances (gases, liquids, or solids), it is also convenient to express compositions in mole fractions or mole percentages instead of mass fractions. The mole fraction is the ratio of the number of moles of one component to the total number of moles in the mixture.

Mass and Volume

The *density* (ρ) of a substance is the ratio of its mass to its volume and may be expressed in units of pounds per cubic foot (lb/ft^3), kilograms per cubic meter (kg/m^3), and so on. For solids, density can be easily determined by placing a known mass of the substance in a liquid and determining the displaced volume. The density of a liquid can be measured by weighing a known volume of the liquid in a volumetric flask. For gases, the ideal gas law, discussed later in Section 3.4, can be used to calculate the density from the pressure, temperature, and molecular weight of the gas.

Densities of pure solids and liquids are relatively independent of temperature and pressure, and can be found in standard reference books. The *specific volume* of a substance is its volume per unit mass (ft^3/lb, m^3/kg, etc.) and is, therefore, the inverse of its density.

An important density term that finds applications with particles (see later chapters) is the bulk density. For particles, a distinction should always be made between *true* density and *bulk* density. The true density is the actual density of a discrete particle or solid, whereas the bulk density is the density with the void volume between the particles is included in the determination. In lieu of data, the bulk density may be assumed, for purposes of engineering calculations, to be approximately 60% of the true density (L. Theodore: personal notes, 1971).

The *specific gravity* (SG) is the ratio of the density of a substance to the density of a reference substance at a specific condition:

$$SG = \frac{\rho}{\rho_{ref}} \tag{3.9}$$

The reference most commonly used for *solids* and *liquids* is water at its maximum density, which occurs at 4°C; this reference density is $1.000\, g/cm^3$, $1000\, kg/m^3$, or $62.43\, lb/ft^3$. Note that, since the specific gravity is a ratio of two densities, it is dimensionless. Therefore, any set of units may be employed for the two densities as long as they are consistent. The specific gravity of *gases* is used only rarely; when it is, air at the same conditions of temperature and pressure as the gas is usually employed as the reference substance.

Another dimensionless quantity related to density is the API (American Petroleum Institute) gravity, which is often used to indicate densities of fuel oils and some liquid hazardous wastes. The relationship between the API scale and the specific gravity is

$$\text{Degree API} = \frac{141.5}{\text{SG}(60/60^\circ\text{F})} - 131.5 \quad (3.10)$$

where SG$(60/60^\circ$F$)$ = specific gravity of the liquid at 60°F using water at 60°F as the reference.

Viscosity

Viscosity is a property associated with a fluid's resistance to flow; more precisely, this property accounts for energy losses which result from shear stresses occurring between different portions of the fluid that are moving at different velocities. The *absolute viscosity* (μ) has units of mass per length · time; the fundamental unit is the *poise*, which is defined as 1 g/cm · s. This unit is inconveniently large for many practical purposes and viscosities are frequently given in *centipoises* (0.01 poise), which is abbreviated cP. The viscosity of pure water at 68.6°F is 1.00 cP. In English units, absolute viscosity is expressed either as pounds (mass) per foot-second (lb/ft · s) or pounds per foot-hour (lb/ft · hr). The absolute viscosity depends primarily on temperature and to a lesser degree on pressure.

The *kinematic viscosity* (ν) is the absolute viscosity divided by the density of the fluid and is useful in certain fluid flow problems; the units for this quantity are length squared per time, e.g., square foot per second (ft^2/s) or square meters per hour (m^2/hr). A kinematic viscosity of 1 cm^2/s is called a *stoke*. For pure water at 70°F, ν = 0.983 cS (centistokes). Because fluid viscosity changes rapidly with temperature, a numerical value of viscosity has no significance unless the temperature is specified.

Liquid viscosity is usually measured by the amount of time it takes for a given volume of liquid to flow through an orifice. The *Saybolt universal viscometer* is the most widely used device in the United States for the determination of the viscosity of fuel oils and liquid wastes.

The viscosities of air at atmospheric pressure and water as functions of temperature are presented in Tables 3.1 and 3.2 respectively. Viscosities of other substances are available in the literature.

Heat Capacity

The *heat capacity* of a substance is defined as the quantity of heat required to raise the temperature of that substance by 1°. The term *specific heat* is frequently used in place of *specific heat capacity*. This is not strictly correct, because specific heat has been traditionally defined as the ratio of the heat capacity of a substance to the heat capacity

TABLE 3.1 Viscosity of Air at 1 Atmosphere

T, °C	Viscosity, Micropoise (μP)[a]
0	170.8
18	182.7
40	190.4
54	195.8
74	210.2
229	263.8

[a] 1 P = 100 cP = 10^6 μP; 1 cP = 6.72×10^{-4} lb/ft · s.

TABLE 3.2 Viscosity of Water

T, °C	Viscosity, Centipoise (cP)
0	1.792
10	1.308
20	1.000
25	0.894
30	0.801
40	0.656
50	0.594
60	0.469
70	0.406
80	0.357
90	0.317
100	0.284

cP = 6.72×10^{-4} lb/ft · s.

of water. However, since the specific heat of water is approximately 1 cal/g · °C or 1 Btu/lb · °F, the term *specific heat* is often interchanged with heat capacity. For gases, the addition of heat to cause the 1° temperature rise may be accomplished either at constant pressure or at constant volume. Since the amounts of heat necessary are different for the two cases, subscripts are used to identify which heat capacity is being used—c_p for constant pressure and c_v for constant volume. For liquids and solids, this distinction does not have to be made since there is little difference between the two. Values of heat capacities are available in Chapter 4 and in the literature.

Heat capacities are often used on a *molar* basis instead of a *mass* basis, in which case the units become cal/gmol · °C or Btu/lbmol · °F. To distinguish between the two, uppercase letters (C_p, C_v) are used in this text to represent the molar-based heat capacities, and lowercase letters (c_p, c_v) are used for the mass-based heat capacities.

Heat capacities are functions of both the temperature and pressure, although the effect of pressure is generally small and is neglected in almost all engineering calculations. The effect of temperature on C_p has been described by

$$C_p = \alpha + \beta T + \gamma T^2 \tag{3.11}$$

$$C_p = a + bT + cT^{-2} \tag{3.12}$$

where α, β, γ, a, b, and c are constants specific to a particular chemical.

Reynolds Number

The *Reynolds number* (Re) is a dimensionless number that indicates whether a fluid flowing is in the laminar or turbulent regime. *Laminar* flow is characteristic of fluids flowing slow enough so that there are no eddies (whirlpools) or macroscopic mixing of different portions of the fluid. (*Note*: In any fluid, there is always *molecular* mixing due to the thermal activity of the molecules; this is distinct from *macroscopic* mixing due to the swirling motion of different portions of the fluid.) In laminar flow, a fluid can be imagined to flow like a deck of cards, with adjacent layers sliding past one another. *Turbulent* flow is characterized by eddies and macroscopic currents. In practice, moving gases are generally in the turbulent region. For flow in a *pipe*, a Reynolds number above 2100 is an indication of turbulent flow.

The Reynolds number is dependent on the fluid velocity, density, viscosity, and some characteristic *length* of the system or conduit; for pipes, the characteristic length is the inside diameter, and for particles it is the particle diameter:

$$\mathrm{Re} = \frac{Dv\rho}{\mu} = \frac{Dv}{\nu} \tag{3.13}$$

where Re = Reynolds number (dimensionless)
D = inside diameter of the pipe (ft)
v = fluid velocity (ft/s)
ρ = fluid density (lb/ft^3)
μ = fluid viscosity (lb/ft $\cdot$ s)
ν = fluid kinematic viscosity (ft^2/s)

A consistent set of units (other than the above) may be used with Equation (3.13).

pH

An important chemical property of an aqueous solution is its pH. The pH measures the acidity or basicity of the solution. In a neutral solution, such as pure water, the hydrogen (H^+) and hydroxyl (OH^-) ion concentrations are equal. At ordinary temperatures, this concentration is

$$C_{\mathrm{H^+}} = C_{\mathrm{OH^-}} = 10^{-7}\ \mathrm{g \cdot ion/L} \tag{3.14}$$

where $C_{\mathrm{H^+}}$ = hydrogen ion concentration
$C_{\mathrm{OH^-}}$ = hydroxyl ion concentration

The unit g · ion stands for gram·ion, which represents an Avogadro number of ions. In all aqueous solutions, whether neutral, basic, or acidic, a chemical equilibrium or balance is established between these two concentrations, so that

$$K_{eq} = (C_{H^+})(C_{OH^-}) = 10^{-14} \tag{3.15}$$

where K_{eq} = equilibrium constant.

The numerical value for K_{eq} given in Equation (3.15) holds for room temperature and only when the concentrations are expressed in gram-ion per liter (g · ion/L). In acid solutions, C_{H^+} is $> C_{OH^-}$; in basic solutions, C_{OH^-} predominates. The OH^- is a direct measure of the hydrogen ion concentration and is defined by

$$\text{pH} = -\log C_{H^+} \tag{3.16}$$

Thus, an acidic solution is characterized by a pH below 7 (the lower the pH, the higher the acidity)—a basic solution, by a pH above 7—and, a neutral solution by a pH of 7.

Boiling and Freezing Points

The boiling point is the temperature at which the vapor pressure of a liquid is equal to its surrounding pressure. Since the vapor pressure remains constant during boiling, so does the temperature. The higher the system pressure, the higher the temperature must be to induce boiling. The boiling point is specific to each individual substance and is a strong function of pressure. The freezing point is the temperature at which the liquid and solid state of a substance coexist in equilibrium at a given pressure. At this point the rate at which the substance leaves the solid state equals the rate at which it leaves the liquid state.

Vapor Pressure

This section concludes with a discussion of *vapor pressure* (denoted p'), which is an important property of liquids, and, to a much lesser extent, of solids. If a liquid is allowed to evaporate in a confined space, the pressure in the vapor space increases as the amount of vapor increases. If there is sufficient liquid present, a point is eventually reached at which the pressure in the vapor space is exactly equal to the pressure exerted by the liquid at its own surface. At this point, a dynamic equilibrium exists in which vaporization and condensation take place at equal rates and the pressure in the vapor space remains constant. The pressure exerted at equilibrium is called the vapor pressure of the liquid. The magnitude of this pressure for a given liquid depends on the temperature, but not on the amount of liquid present. Solids, like liquids, also exert a vapor pressure. Evaporation of solids (called *sublimation*) is noticeable only for those solids with appreciable vapor pressures.

3.4 IDEAL GAS LAW

Observations based on physical experimentation often can be synthesized into simple mathematical equations called *laws*. These laws are never perfect and hence are only an approximate representation of reality. The *ideal gas law* (IGL) was derived from experiments in which the effects of pressure and temperature on gaseous volumes were measured over moderate temperature and pressure ranges. This law works well in the pressure and temperature ranges that were used in taking the data; extrapolations outside of the ranges have been found to work well in some cases and poorly in others. As a general rule, this law works best when the molecules of the gas are far apart, i.e., when the pressure is low and the temperature is high. Under these conditions, the gas is said to behave *ideally*, and its behavior is a close approximation to the so-called *perfect* or *ideal gas*, a hypothetical entity that obeys the ideal gas law perfectly. For engineering calculations, the ideal gas law is almost always assumed to be valid since it generally works well (usually within a few percent of the correct result) up to the highest pressures and down to the lowest temperatures used in air pollution applications.

The two precursors of the ideal gas law were *Boyle's* and *Charles'* laws. Boyle found that the volume of a given mass of gas is inversely proportional to the absolute pressure if the temperature is kept constant:

$$P_1 V_1 = P_2 V_2 \tag{3.17}$$

where V_1 = volume of gas at absolute pressure P_1 and temperature T
V_2 = volume of gas at absolute pressure P_2 and temperature T

Charles found that the volume of a given mass of gas varies directly with the absolute temperature at constant pressure:

$$\frac{V_1}{T_1} = \frac{V_2}{T_2} \tag{3.18}$$

where V_1 = volume of gas at pressure P and absolute temperature T_1
V_2 = volume of gas at pressure P and absolute temperature T_2

Boyle's and Charles' laws may be combined into a single equation in which neither temperature nor pressure need be held constant:

$$\frac{P_1 V_1}{T_1} = \frac{P_2 V_2}{T_2} \tag{3.19}$$

For Equation (3.19) to hold, the mass of gas must be constant as the conditions change from (P_1, T_1) to (P_2, T_2). This equation indicates that for a given mass of a specific gas, PV/T has a constant value. Since, at the same temperature and pressure, volume and mass must be directly proportional, this statement may be extended to

$$\frac{PV}{mT} = C \tag{3.20}$$

where m = mass of a specific gas
C = constant that depends on the gas

Moreover, experiments with different gases showed that Equation (3.20) could be expressed in a far more generalized form. If the number of moles (n) is used in place of the mass (m), the constant is the same for all gases:

$$\frac{PV}{nT} = R \tag{3.21}$$

where $R =$ universal gas constant.

Equation (3.21) is called the *ideal gas law*. Numerically, the value of R depends on the units used for P, V, T and n (see Table 3.3). In this text, all gases are assumed to approximate ideal gas behavior. As is generally the case in engineering practice, the ideal gas law is assumed to be valid for all problems to follow later.

Other useful forms of the ideal gas law are shown in Equation (3.22) and (3.23). Equation (3.22) applies to gas flow rather than to gas confined in a container:

$$Pq = \dot{n}RT \tag{3.22}$$

where $q =$ gas volumetric flow rate (ft^3/hr)
$P =$ absolute pressure (psia)
$\dot{n} =$ molar flow rate (lbmol/hr)
$T =$ absolute temperature (°R)
$R = 10.73$ psia · ft^3/lbmol · °R

Equation (3.23) combines n and V from Equation (3.21) to express the law in terms of density:

$$P\,(\text{MW}) = \rho RT \tag{3.23}$$

where MW = molecular weight of gas (lb/lbmol)
$\rho =$ density of gas (lb/ft^3)

TABLE 3.3 Values of R in Various Units

R	Temperature Scale	Units of V	Units of n	Units of P
10.73	°R	ft^3	lbmol	psia
0.7302	°R	ft^3	lbmol	atm
21.85	°R	ft^3	lbmol	in Hg
555.0	°R	ft^3	lbmol	m Hg
297.0	°R	ft^3	lbmol	in H_2O
0.7398	°R	ft^3	lbmol	bar
1545.0	°R	ft^3	lbmol	psfa
62.361	K	L	gmol	mm Hg
0.08205	K	L	gmol	atm
0.08314	K	L	gmol	bar
8314	K	L	gmol	Pa
8.314	K	m^3	gmol	Pa
82.057	K	cm^3	gmol	atm

TABLE 3.4 Common Standard Conditions

System	Temperature	Pressure	Molar Volume
SI	273 K	101.3 kPa	22.4 m^3/kmol
Universal scientific	0°C	760 mm Hg	22.4 L/gmol
Natural gas industry	60°F	14.7 psia	379 ft^3/lbmol
American engineering	32°F	1 atm	359 ft^3/lbmol
Hazardous waste	60°F	1 atm	379 ft^3/lbmol
incinerator industry	70°F	1 atm	387 ft^3/lbmol

Volumetric flow rates are often given not at the actual conditions of pressure and temperature but at arbitrarily chosen standard conditions (*STP, standard temperature and pressure*). To distinguish between flow rates based on the two conditions, the letters *a* and *s* are often used as part of the unit. The units acfm and scfm stand for actual cubic feet per minute and standard cubic feet per minute, respectively. The ideal gas law can be used to convert from *standard* to *actual* conditions, but since there are many standard conditions in use, the *STP* being employed must be known or specified. Standard conditions most often used are shown in Table 3.4.

The reader is cautioned on the incorrect use of acfm and/or scfm. The use of standard conditions is a convenience; *when predicting the performance of or designing equipment, the actual conditions must be employed.* Designs based on standard conditions can lead to disastrous results, with the unit usually underdesigned. For example, for a flue gas stream at 2140°F, the ratio of acfm to scfm (standard temperature = 60°F) is 5.0. Equation (3.24), which is a form of Charles' law, can be used to correct flow rates from standard to actual conditions:

$$q_{\mathrm{a}} = q_{\mathrm{s}}\left(\frac{T_{\mathrm{a}}}{T_{\mathrm{s}}}\right) \tag{3.24}$$

where q_{a} = volumetric flow rate at actual conditions (ft^3/hr)
q_{s} = volumetric flow rate at standard conditions (ft^3/hr)
T_{a} = actual absolute temperature (°R)
T_{s} = standard absolute temperature (°R)

The reader is again reminded that absolute temperatures and pressures must be employed in all ideal gas law calculations.

Partial Pressure and Partial Volume

In engineering practice, mixtures of gases are more often encountered than single or pure gases. The ideal gas law is based on the *number* of molecules present in the gas volume; the *kind* of molecules is not a significant factor, only the number. This law applies equally well to mixtures and pure gases alike. Dalton and Amagat both applied the ideal gas law to mixtures of gases. Since pressure is caused by gas molecules colliding with the walls of the container, it seems reasonable that the total pressure of a gas mixture is made up of pressure contributions due to each of the component gases. These pressure contributions are called *partial pressures*. Dalton defined the partial pressure of a component as the pressure that would be exerted if the same mass of the component gas occupied the same total volume alone at the same temperature as the mixture. The sum of these partial pressures equals the total pressure:

$$P = p_a + p_b + p_c + \cdots + p_n = \sum_{i=1}^{n} p_i \tag{3.25}$$

where $P =$ total pressure
$n =$ number of components
$p_i =$ partial pressure of component i

Equation (3.25) is known as *Dalton's law*. Applying the ideal gas law to one component (a) only,

$$p_a V = n_a RT \tag{3.26}$$

where $n_a =$ number of moles of component a. Eliminating R, T, and V between Equations (3.21) and (3.26) yields

$$\frac{p_a}{P} = \frac{n_a}{n} = y_a$$

or

$$p_a = y_a P \tag{3.27}$$

where $y_a =$ mole fraction of component a.

Amagat's law is similar to Dalton's law. Instead of considering the total pressure to be made up of partial pressures where each component occupies the total container volume, Amagat considered the total volume to be made up of partial volumes in which each component is at the total pressure. The definition of the *partial volume* is therefore the volume occupied by a component gas alone at the same temperature and pressure as the mixture:

$$V = V_a + V_b + V_c + \cdots + V_n = \sum_{i=1}^{n} V_i \tag{3.28}$$

Applying Equation (3.21), as before yields

$$\frac{V_a}{V} = \frac{n_a}{n} = y_a \tag{3.29}$$

or

$$V_a = y_a V$$

where $V_a =$ partial volume of component a.

It is common in the air pollution control industry to describe low concentrations of pollutants in gaseous mixtures in parts per million by volume (ppmv). Since partial volumes are proportional to mole fractions, it is only necessary to multiply the mole fraction of the pollutant by 1 million to obtain the concentration in parts per million. [For liquids and solids, parts per million (ppm) is also used to express concentration, although it is usually on a *mass* (ppmv) basis rather than a *volume* basis. The terms ppmv and ppmw are sometimes used to distinguish between the volume and mass bases, respectively.]

3.5 PHASE EQUILIBRIUM

In air pollution control equipment applications (particularly in absorption), the most important equilibrium phase relationship is that between liquid and vapor. Henry's law theoretically describes liquid–gas behavior and under certain conditions is applicable in practice.

Raoults's law states that the partial pressure of a component over a solution is the product of the vapor pressure of that component and the mole fraction of that component. Raoult's law is expressed as

$$p_{\mathrm{A}} = p'_{\mathrm{A}} x_{\mathrm{A}} \tag{3.30}$$

where p_{A} = partial pressure of A, Pa
p'_{A} = vapor pressure of A, Pa
x_{A} = mole fraction of A in solution

Raoult's law applies over the entire concentration range from 0 to 1.0, but is applicable only if the solution behaves in an ideal manner. Unfortunately few solutions behave ideally. A special case of Raoult's law is Henry's law.

Henry's law is an empirical relationship used for representing data on many systems:

$$p_i = H_i x_i \tag{3.31}$$

where H = Henry's law constant for component i (in units of pressure).

If the gas behaves ideally, Equation (3.31) may be written as

$$y_i = m_i x_i \tag{3.32}$$

where m_i = constant (dimensionless).

This is a more convenient form of Henry's law to work with. The constant m_i (or H_i) has been determined experimentally for a large number of compounds and is usually valid at low concentrations.

Another phase equilibrium application involves the psychrometric or humidity chart. A humidity chart is used to determine the properties of moist air and to calculate moisture content in air. The ordinate of the chart is the absolute humidity $\mathcal{H}$, which is defined as the mass of water vapor per mass of bone-dry air. (Some charts base the ordinate on moles instead of mass.) On the basis of this definition, Equation (3.33) gives $\mathcal{H}$ in terms of moles and also in terms of partial pressure:

$$\mathcal{H} = \frac{18n_{H_2O}}{29(n - n_{H_2O})} = \frac{18p_{H_2O}}{29(P - p_{H_2O})} \tag{3.33}$$

where n_{H_2O} = number of moles of water vapor
n = total number of moles in gas
p_{H_2O} = partial pressure of water vapor
P = total system pressure

The *wet-bulb* temperature (T_{WB}) is another measure of humidity; it is the temperature at which a thermometer with a wet wick wrapped around the bulb stabilizes. As water evaporates from the wick to the ambient air, the bulb is cooled; the rate of cooling depends on how humid the air is. No evaporation occurs if the air is saturated with water; hence, T_{WB} and T_{DB} (dry bulb) are the same. Furthermore, the lower the humidity, the greater the difference between these two temperatures.

The *humid volume* is the volume of wet air per mass of dry air and is linearly related to the humidity. The product of the humid volume and the absolute humidity gives the volume of the moist air. The *humid enthalpy* (also called *humid heat*) is the enthalpy of the moist air on bone-dry air basis. The term *enthalpy* is a measure of the energy content of the mixture and is defined later in this chapter. Steam tables, charts, and diagrams may also be required in air pollution control equipment calculations. Abbreviated and more comprehensive information is available in the literature.

3.6 CONSERVATION LAWS

Conservation of Mass

The conservation law for mass can be applied to any process or system. The general form of this law is given in Equation (3.34):

$$\text{mass}_{\text{in}} - \text{mass}_{\text{out}} + \text{mass}_{\text{generated}} = \text{mass}_{\text{accumulated}} \tag{3.34}$$

Equation (3.34) may be applied to the total mass involved or to a particular species, on either a mole or mass basis. The conservation law for mass can be applied to steady-state or unsteady-state processes and to batch or continuous systems. To isolate a system for study, one separates it from the surroundings by a boundary or envelope. This boundary may be real (e.g., the walls of a vessel) or imaginary. Mass crossing the boundary and entering the system is part of the mass-in term in Equation (3.34), whereas that leaving the system is part of the mass-out term. Equation (3.34) may be written for any compound whose quantity is not changed by chemical reaction, or for any chemical element, regardless of whether it has participated in a chemical reaction. It may be written for one piece of equipment, around several pieces of equipment, or around an entire process. It may be used to calculate an unknown quantity directly, to check the validity of experimental data, or to express one or more of the independent relationships among the unknown quantities in a particular problem.

A steady-state process is one in which there is no change in conditions (temperature, pressure, etc.) or rates of flow with time at any given point in the system. The accumulation term in Equation (3.34) is then zero. If there is no chemical reaction, the generation term is also zero. All other processes are classified as unsteady-state. In a batch process, the container holds the product or products. In a continuous process, reactants are fed in an unending flow to a piece of equipment or to several pieces in series, and products are continuously removed from one or more points. A continuous process may or may not be steady-state. As indicated previously, Equation (3.34) may be applied to the total mass of each stream (referred to as an *overall* or *total material balance*) or to the individual components of the streams (referred to as *componential* or *component material balance*). Often, the primary task in preparing a material balance is to develop the quantitative relationships among the streams.

Conservation of Energy

A presentation of the conservation law for energy would be incomplete without a brief review of some introductory thermodynamic principles. *Thermodynamics* is defined as the science that deals with the relationships between the various forms of energy.

A system may possess energy as a result of its temperature, velocity, position, molecular structure, surface, and so on. The energies corresponding to these conditions are internal, kinetic, potential, chemical, surface, and so on, respectively. Engineering thermodynamics is founded on three basic laws. Energy, like mass, is conserved. Application of the conservation law for energy gives rise to the first law of thermodynamics. (The reader is directed to the literature for details on the second and third law.) This law, in steady-state form for batch processes, is presented here (potential, kinetic, and other energy effects are not considered):

$$\Delta U = Q + W \tag{3.35}$$

For flow processes

$$\Delta H = Q + W_s \tag{3.36}$$

where Q = heat energy transferred across the system boundary to the system
W = work energy transferred across the system boundary to the system
W_s = mechanical work energy transferred across the system boundary to the system
U = internal energy of the system
H = enthalpy of the system
ΔU = change in internal energy during the process
ΔH = change in enthalpy during the process

The internal energy and enthalpy in Equations (3.35) and (3.36), as well as in the other equations in this discussion, may be on a mass or a mole basis, or they may represent the total internal energy and enthalpy of the entire system. Most air pollution equipment, as well as industrial facilities, operate in a steady-state flow mode. If no significant mechanical or shaft work is added or withdrawn from the system, Equation (3.36) reduces to

$$Q = \Delta H \tag{3.37}$$

If a unit or system is operated adiabatically, where $Q = 0$, Equation (3.37) then becomes

$$\Delta H = 0 \tag{3.38}$$

Although the topics of material and energy balances are covered separately in this subsection, it should be emphasized that this segregation does not occur in reality; one must often work with both energy and material balances simultaneously.

PROBLEMS

3.1 Basic Units

The basic units employed in the SI system are (select one):

(a) Mass, length, and force
(b) Force, length, time, and mass
(c) Mass, length, and time
(d) None of the above

Solution: The reader is referred to the Appendix for this answer. The basic units in the SI system are the kilogram (mass), meter (length), second (time), Kelvin (temperature), ampere (electric current), candela (the unit of luminous intensity), and radian (angular measure). All have come to be used by the engineer. The Celsius scale of temperature (0°C = 273.15 K) is commonly used with the absolute Kelvin scale. The correct answer is therefore (d).

3.2 Temperature Conversions

Perform the following temperature conversions:

(a) Convert 55°F to (1) Rankine, (2) Celsius, and (3) Kelvin
(b) Convert 55°C to (1) Fahrenheit, (2) Rankine, and (3) Kelvin

Solution

(a) (1) $^\circ\text{R} = {}^\circ\text{F} + 460 = 55 + 460 = 515$

(2) $^\circ\text{C} = \frac{5}{9}({}^\circ\text{F} - 32) = \frac{5}{9}(55 - 32) = 12.8$

(3) $\text{K} = \frac{5}{9}({}^\circ\text{F} + 460) = \frac{5}{9}(55 + 460) = 286$

(b) (1) $^\circ\text{F} = 1.8({}^\circ\text{C}) + 32 = 1.8(55) + 32 = 131$

(2) $^\circ\text{R} = 1.8({}^\circ\text{C}) + 492 = 1.8(55) + 492 = 591$

(3) $\text{K} = {}^\circ\text{C} + 273 = 55 + 273 = 328$

3.3 Definition of Absolute Zero

At absolute zero temperature a gas has:

(a) A temperature of -273.15°C

(**b**) No kinetic energy
(**c**) No pressure
(**d**) All of the above

Solution: Interestingly, a gas at absolute zero temperature (0K) satisfies answers (a)–(c). The correct answer is therefore (d).

3.4 Definition of Absolute Pressure

Absolute pressure is:

(**a**) Measured above a perfect vacuum
(**b**) 29.92 in Hg
(**c**) Required for proper gauge pressure readings
(**d**) $P_{abs} = P_{atm} + P_{gauge}$

Solution: This is a tricky question. Answers (c) and (d) can be eliminated immediately. Normally, but *not* always, absolute pressure is approximately 29.92 in Hg. However, by definition, absolute pressure is that measured above a perfect vacuum. The correct answer is therefore (a).

3.5 Pressure Calculations

The height of a liquid column of mercury open to the atmosphere is 2.493 ft. Calculate the absolute pressure at the base of the column in lb_f/ft^2, psia, mm Hg, and in H_2O. Assume that the density of mercury is 848.7 lb/ft^3 and atmospheric pressure is 2116 lb_f/ft^2 absolute.

Solution: Expressed in various units, the standard atmosphere is equal to

1.0	Atmospheres (atm)
33.91	Feet of water (ft H_2O)
14.7	Pounds-force per square inch absolute (psia)
2116	Pounds-force per square foot absolute (psfa)
29.92	Inches of mercury (in Hg)
760.0	Millimeters of mercury (mm Hg)
1.013×10^5	Newtons per square meter (N/m^2)

The equation describing the pressure at the base of the column height due to the liquid is

$$P = \frac{\rho g h}{g_c} \tag{3.39}$$

where P = gauge pressure
ρ = liquid density
h = height of column
g = acceleration of gravity
g_c = conversion constant

Thus,

$$P = (848.7\ \text{lb/ft}^3)\left(1\,\frac{\text{lb}_f}{\text{lb}}\right)(2.493\ \text{ft})$$

$$= 2116\,\text{lb}_f/\text{ft}^2\ \text{gauge}$$

The pressure in lb_f/ft^2 absolute is

$$\begin{aligned} P_{\text{absolute}} &= P + P_{\text{atmosphere}} \\ &= 2116\ lb_f/ft^2 + 2116\ lb_f/ft^2 \\ &= 4232\ lb_f/ft^2 \text{ absolute (psfa)} \end{aligned}$$

The pressure in psia is

$$P(\text{psia}) = (4232 \text{ psfa})\left(\frac{1 \text{ ft}^2}{144 \text{ in}^2}\right) = 29.4 \text{ psia}$$

The corresponding gauge pressure in psi is

$$P(\text{psig}) = 29.4 - 14.7 = 14.7 \text{ psig}$$

3.6 Effect of Volume Change on Viscosity

As the volume of a gas increases at the same pressure and temperature, the viscosity of the gas generally

(**a**) Increases
(**b**) Decreases
(**c**) Remains the same
(**d**) Varies sinusoidally

Solution: Since the temperature and pressure are the same, the volume change of the gas has no effect on the viscosity. The correct answer is therefore (c).

3.7 Engineering Conversion Factors

Given the following data for liquid methanol, determine its density in lb/ft^3 and convert heat capacity, thermal conductivity, and viscosity from the International System of Units (SI) to English units:

Specific gravity = 0.92 (at 60°F)
Density of reference substance (water) = 62.4 lb/ft^3 (at 60°F)
Heat capacity = 0.61 cal/g·°C) (at 60°F)
Thermal conductivity = 0.0512 cal/m · s · °C)
Viscosity = 0.64 cP (at 60°F)

Solution: As noted earlier, the definition of specific gravity for liquids and solids is

$$\text{Specific gravity SG} = \frac{\text{density}}{\text{density of water at } 4^\circ\text{C}} \tag{3.9}$$

Note that the density of water at 4°C is 62.4 lb/ft^3 in English engineering units or 1.0 g/cm^3.

Calculate the density of methanol in English units by multiplying the specific gravity by the density of water:

$$\begin{aligned} \text{Density of methanol} &= \text{(specific gravity) (density of water)} \\ &= (\text{SG})\ (\rho_{\text{ref}}) \qquad (3.9) \\ &= (0.92)\ (62.4) = 57.4\ \text{lb/ft}^3 \end{aligned}$$

The procedure is reversed if one is interested in calculating specific gravity from density data. The notation for density is usually ρ.

Now convert the heat capacity from units of cal/(g·°C) to Btu/(lb·°F).

$$\left(\frac{0.61\ \text{cal}}{\text{g}\cdot{}^\circ\text{C}}\right)\left(\frac{454\ \text{g}}{\text{lb}}\right)\left(\frac{\text{Btu}}{252\ \text{cal}}\right)\left(\frac{{}^\circ\text{C}}{1.8^\circ\text{F}}\right) = 0.61\ \text{Btu/(lb}\cdot{}^\circ\text{F)}$$

Note that 1.0 Btu/(lb·°F) is equivalent to 1.0 cal/(g·°C). This also applies on a mole basis so that:

$$1\ \text{Btu/(lbmol}\cdot{}^\circ\text{F)} = 1\ \text{cal/(gmol}\cdot{}^\circ\text{C)}$$

The usual notation for heat capacity is C_p. As noted earlier, C_p represents the heat capacity on a mole basis, while c_p will generally indicate a mass basis.

Now convert the thermal conductivity of methanol from cal/(m·s·°C) to Btu/(ft·hr·°F):

$$\left(\frac{0.0512\ \text{cal}}{\text{m}\cdot\text{s}\cdot{}^\circ\text{C}}\right)\left(\frac{\text{Btu}}{252\ \text{cal}}\right)\left(\frac{0.3048\ \text{m}}{\text{ft}}\right)\left(\frac{3600\ \text{s}}{\text{hr}}\right)\left(\frac{{}^\circ\text{C}}{1.8^\circ\text{F}}\right)$$
$$= 0.124\ \text{Btu/(ft}\cdot\text{hr}\cdot{}^\circ\text{F)}$$

The usual engineering notation for thermal conductivity is k.

Finally, convert viscosity from centipoises to lb/(ft·s):

$$(0.64\ \text{cP})\left(\frac{6.72\times10^{-4}\ \text{lb}}{\text{ft}\cdot\text{s}\cdot\text{cP}}\right) = 4.3\times10^{-4}\ \text{lb/ft}\cdot\text{s}$$

The notation for viscosity is typically μ. The kinematic viscosity ν is defined by the ratio of viscosity to density, i.e., $\nu = \mu/\rho$ with units of length^2/time. Also note that this physical property is a strong function of the temperature but weak function of the pressure. Interestingly, the viscosity of a gas increases with increasing temperature, while the viscosity of a liquid decreases with an increase in temperature.

3.8 Kinematic Viscosity

What is the kinematic viscosity in ft^2/s of a gas of the specific gravity and absolute viscosity are 0.8 and 0.02 cP, respectively?

Solution: First convert cP to lb/ft·s:

$$\mu = \frac{0.02\ \text{cP}}{1} \times \frac{6.720 \times 10^{-4}\ \text{lb/ft} \cdot \text{s}}{1\ \text{cp}} = 1.344 \times 10^{-5}\ \text{lb/ft} \cdot \text{s}$$

By definition,

$$\rho = (\text{SG})\,(\rho_{\text{ref}}) = (0.8)\ (62.43\ \text{lb/ft}^3) = 49.94\ \text{lb/ft}^3 \tag{3.9}$$

The kinematic viscosity is then

$$\begin{aligned}\nu &= \mu/\rho \\ &= (1.344 \times 10^{-5}\ \text{lb/ft} \cdot \text{s})/(49.94\ \text{lb/ft}^3) \\ &= 2.691 \times 10^{-7}\ \text{ft}^2/\text{s}\end{aligned}$$

3.9 Atomic Number, Atomic Weight, and Mass Number

Discuss atomic number, atomic weight, and mass number.

Solution: What differentiates an atom of one element from an atom of another element? The answer is based on the number of protons in the nucleus of the atom in question. Further, all atoms of an element have the same number of protons in the nucleus with the specific number of protons different for different elements. As noted earlier, since an atom has no net electrical charge, the number of electrons in it must equal its number of protons. For example, all atoms of the element carbon have six protons and six electrons. Note that it is customary to speak of "atomic weights," although "atomic masses" would perhaps be more accurate. Mass is a measure of the quantity of matter in a body, whereas weight is the force exerted on the body under the influence of gravity. Furthermore, "atomic weight" is measured in amu. The standard now used for the calculation of atomic weights has completely replaced the two earlier standards. The new standard is particularly appropriate because carbon-12 is often used as a reference standard in computations of atomic masses. The mass number of an atom is defined as the integer closest to the atomic weight of the atom. To summarize, the average mass of each element (expressed in amu) is also known as the atomic weight. Although the term *average atomic mass* or atomic mass may be more appropriate, the term *atomic weight* has become common. The atomic weights of the elements are listed in the periodic table.

3.10 Chemical Conversion I

Answer the following:

(**a**) What is the molecular weight of nitrobenzene ($C_6H_5O_2N$)?
(**b**) How many moles are there in 50.0 g of nitrobenzene?

Solution

(**a**) Pertinent atomic weights are listed below:

Carbon = 12

Hydrogen = 1

Oxygen = 16

Nitrogen = 14

The molecular weight of nitrobenzene is then

$$\begin{aligned}\text{MW} &= (6)\,(12) + (5)\,(1) + (2)\,(16) + (1)\,(14)\\ &= 123\ \text{g/gmol}\end{aligned}$$

(**b**) To convert a mass to moles, divide by the molecular weight:

$$n = (50.0\ \text{g})\left(\frac{\text{gmol}}{123\ \text{g}}\right) = 0.407\ \text{gmol}$$

3.11 Chemical Conversion II

Refer to Problem 3.10. How many molecules are contained in 50.0 g of nitrobenzene?

Solution: There are 6.02×10^{23} (Avagadro's number) molecules/gmol. Therefore,

$$(0.406\ \text{gmol})\ (6.02 \times 10^{23}\ \text{molecules/gmol}) = 2.44 \times 10^{23}\ \text{molecules}$$

3.12 Concentration Conversion

Express the concentration 72 g of HCI in 128 cm^3 of water into terms of fraction and percent by weight, parts per million, and molarity.

Solution: The fraction by weight can be calculated as follows:

$$w = 72\ \text{g}/200\ \text{g} = 0.36$$

The percent by weight can be calculated from the fraction by weight:

$$w = (0.36)\,(100\%) = 36\%$$

The ppm (parts per million by mass) can be calculated as follows:

$$\text{ppmw} = (72\ \text{g}/128\ \text{cm}^3)\ (10^6) = 562{,}500$$

Note that 128 cm^3 water is assumed to weigh 128 g.

The molarity (M) is defined as follows:

$$M = \text{moles of solute/volume of solution (L)}$$

Using atomic weights, one obtains

$$\text{MW of HCl} = 1.0079 + 35.453 = 36.4609$$

$$M = \left[(72\,\text{g HCl})\left(\frac{1\,\text{mol HCl}}{36.4609\,\text{g HCl}}\right)\right] \Big/ \left(\frac{128\,\text{cm}^3}{1000\,\text{cm}^3/\text{L}}\right) = 15.43\,\text{mol/L}$$

3.13 Air Density

What is the density of air at 75°F and 14.7 psia? The molecular weight of air is 29.

Solution: This example is solved using the ideal gas law:

$$PV = nRT = \left(\frac{m}{MW}\right)RT$$

$$\rho = \frac{P(MW)}{RT} \tag{3.23}$$

$$= \frac{(14.7\ \text{psia})\,(29\,\text{lb/lbmol})}{(10.73\,\text{ft}^3 \cdot \text{psi/lbmol} \cdot {}^\circ\text{R})\ (75 + 460)^\circ\text{R}}$$

$$= 0.0743\,\text{lb/ft}^3$$

This is an important air density value to remember.

3.14 Ideal Gas Law Volume Calculation

Calculate the volume (in ft^3) of 1.0 lbmol of any ideal gas at 60°F and 14.7 psia.

Solution: Solve the ideal gas law for V and calculate the volume:

$$V = \frac{nRT}{P} \tag{3.21}$$

$$= \frac{(1)\,(10.73)\,(60 + 460)}{14.7}$$

$$= 379\ \text{ft}^3$$

This result is another important number to remember in air pollution calculations—1 lbmol of any (ideal) gas at 60°F and 1 atm occupies 379 ft^3.

3.15 Gas Density

Calculate the density of a gas (MW = 29) in g/cm^3 at 20°C and 1.2 atm using the ideal gas law.

Solution: Calculate the density of the gas again using the ideal gas law:

$$PV = nRT = \left(\frac{m}{\text{MW}}\right)RT \tag{3.23}$$

$$\frac{m}{V} = \rho = \frac{P(\text{MW})}{RT} = \frac{(1.2)(29)}{(82.06)(20+273)}$$

$$= 0.00145\ \text{g/cm}^3$$

The effects of pressure, temperature, and molecular weight on density can be obtained directly from the ideal gas law equation. Increasing the pressure and molecular weight increases the density; increasing the temperature decreases the density.

3.16 Incinerator Flow Rate

Data from an incinerator indicate a volumetric flow rate of 30,000 acfm (60°F, 1 atm). If the operating temperature and pressure of the unit are 1100°F and 1 atm, respectively, calculate the flow rate in standard cubic feet per minute.

Solution: Since the pressure remains constant, calculate the standard cubic feet per minute using Charles' law:

$$q_s = q_a\left(\frac{T_s}{T_a}\right) = 30{,}000\left(\frac{60+460}{1100+460}\right) \tag{3.24}$$

$$= 10{,}000\ \text{scfm}$$

The reader is again cautioned on the use of acfm and/or scfm. Predicting the performance of and designing equipment should always be based on *actual* conditions. Designs based on standard conditions can lead to disastrous results, with the unit underdesigned. The reader is also reminded that absolute temperatures and pressures must be employed in all ideal gas law calculations.

3.17 Vent Discharge Velocity

The exhaust gas flow rate from a facility is 1000 scfm. All of the gas is vented through a small stack that has an inlet area of 1.0 ft^2. The exhaust gas temperature is 300°F. What is the velocity of the gas through the stack inlet in feet per second? Assume standard conditions to be 70°F and 1.0 atm. Neglect the pressure drop across the stack.

Solution: Calculate the actual flow rate, in acfm, using Charles' law:

$$q_{\text{a}} = q_s\left(\frac{T_a}{T_s}\right) \tag{3.24}$$

$$= 1000\left(\frac{460+300}{460+70}\right)$$

$$= 1434\ \text{acfm}$$

Note that since the gas in vented through the stack to the atmosphere, the pressure is 1.0 atm. Calculate the velocity of the gas:

$$v = \frac{q_a}{S}$$
$$= \frac{1434}{1.0}$$
$$= 1434 \text{ ft/min}$$

3.18 Flue Gas Analysis

The mole percent (gas analysis) of a flue gas is given below:

$N_2 = 79\%$
$O_2 = 5\%$
$CO_2 = 10\%$
$CO = 6\%$

Calculate the average molecular weight of the mixture.

Solution: First write the molecular weight of each component:

$MW(N_2) = 28$
$MW(O_2) = 32$
$MW(CO_2) = 44$
$MW(CO) = 28$

Table 3.5 can be prepared by multiplying the molecular weight of each component by its mole percent.

TABLE 3.5 Mixture Molecular Weights

Compound	Molecular Weight	Mole Fraction	Weight, lb
N_2	28	0.79	22.1
O_2	32	0.05	1.6
CO_2	44	0.10	4.4
CO	28	0.06	1.7
Total		1.00	

Finally, calculate the average molecular weight of the gas mixture:

$$\text{Average molecular weight} = \overline{\text{MW}} = 22.1 + 1.6 + 4.4 + 1.7 = 29.8$$

The sum of the weights in pounds represents the average molecular weight because the calculation above is based on 1.0 mol of the gas mixture. The reader should also note that in a gas, mole percent equals volume percent and vice versa. Therefore, a volume percent can be used to determine weight fraction as illustrated in Table 3.5. The term y is used in engineering practice to represent mole (or volume) fraction of gases; the term x is often used for liquids and solids.

3.19 Partial Pressure I

The exhaust to the atmosphere from an incinerator has a SO_2 concentration of 0.12 mm Hg partial pressure. Calculate the parts per million of SO_2 in the exhaust.

Solution: First calculate the mole fraction y. By Dalton's law

$$y = \frac{p_{SO_2}}{P} \tag{3.27}$$

Since the exhaust is discharged to the atmosphere, the atmospheric pressure, 760 mm Hg, is the total pressure P. Thus,

$$y = (0.12\,\text{mm Hg})/(760\,\text{mm Hg}) = 1.58 \times 10^{-4}$$

$$\begin{aligned}\text{ppm} &= (y)(10^6) = (1.58 \times 10^{-4})(10^6)\\ &= 158\ \text{ppm}\end{aligned}$$

3.20 Partial Pressure II

A storage tank contains a gaseous mixture consisting of 30% CO_2, 5% CO, 5% H_2O, 50% N_2, and 10% O_2, by volume. What is the partial pressure of each component if the total pressure is 2 atm? What are their pure-component volumes if the total pressure is 2 atm? What are the pure-component volumes if the total volume is 10 ft^3? What are the concentrations in ppm (parts per million)?

Solution: Dalton's law states that the partial pressure p_a of an ideal gas is given by

$$p_a = y_a P \tag{3.27}$$

where y_a = mole fraction of component a
P = total pressure

Thus

$$\begin{aligned}p_{CO_2} &= 0.30(2) = 0.60\,\text{atm}\\ p_{CO} &= 0.05(2) = 0.10\,\text{atm}\\ p_{H_2O} &= 0.05(2) = 0.10\,\text{atm}\\ p_{N_2} &= 0.50(2) = 1.00\,\text{atm}\\ p_{O_2} &= 0.10(2) = \underline{0.20\,\text{atm}}\\ P &= 2.00\,\text{atm}\end{aligned}$$

Amagat's law states that the pure component volume V_a of an ideal gas is given by

$$V_a = y_a V \tag{3.29}$$

where V is the total volume.

Thus

$$\begin{aligned}
V_{CO_2} &= 0.30(10) = 3.00 \text{ ft}^3 \\
V_{CO} &= 0.05(10) = 0.50 \text{ ft}^3 \\
V_{H_2O} &= 0.05(10) = 0.50 \text{ ft}^3 \\
V_{N_2} &= 0.50(10) = 5.00 \text{ ft}^3 \\
V_{O_2} &= 0.10(10) = \underline{1.00 \text{ ft}^3} \\
V &= 10.00 \text{ ft}^3
\end{aligned}$$

By definition, the parts per million (ppm) is given by

$$\text{ppm}_a = y_a 10^6$$

Thus

$$\begin{aligned}
\text{ppm}_{CO_2} &= 0.30(10^6) = 3.00 \times 10^5 \text{ ppm} \\
\text{ppm}_{CO} &= 0.05(10^6) = 0.50 \times 10^5 \text{ ppm} \\
\text{ppm}_{H_2O} &= 0.05(10^6) = 0.50 \times 10^5 \text{ ppm} \\
\text{ppm}_{N_2} &= 0.50(10^6) = 5.00 \times 10^5 \text{ ppm} \\
\text{ppm}_{O_2} &= 0.10(10^6) = 1.00 \times 10^5 \text{ ppm}
\end{aligned}$$

Unless otherwise stated, ppm for gases is by mole or volume and is usually designated as ppmv. When applying the term to liquids or solids, the basis is almost always by mass; the notation may appear as ppmw or ppmm.

3.21 Vapor Pressure Calculation

Two popular equations that are used to estimate the vapor pressure of compounds are the Clapeyron and Antoine equations. The Clapeyron equation is given by

$$\ln p' = A - \frac{B}{T} \tag{3.40}$$

where p' and T are the vapor pressure and temperature, respectively, and A and B are the experimentally determined Clapeyron coefficients. The Antoine equation is given by

$$\ln p' = A - \frac{B}{T + C} \tag{3.41}$$

where A, B, and C are the experimentally determined Antoine coefficients.

Use the Clapeyron and Antoine equations to estimate the vapor pressure of acetone at 0°C. The Clapeyron coefficients have been experimentally determined to be

$$A = 15.03$$
$$B = 2817$$

for p' and T in mm Hg and K, respectively. The Antoine coefficients are

$$A = 16.65$$
$$B = 2940$$
$$C = -35.93$$

with p' and T in the same units.

Solution: First, calculate the vapor pressure p' of acetone at 0°C using the Clapeyron equation:

$$\begin{aligned} \ln p' &= 15.03 - [2817/(0 + 273)] \\ &= 4.7113 \\ p' &= 111.2\,\text{mm Hg} \end{aligned}$$

Calculate the vapor pressure of acetone at 0°C using the Antoine equation:

$$\begin{aligned} \ln p' &= 16.65 - [2940/(273 - 35.93)] \\ &= 4.2486 \\ p' &= 70.01\,\text{mm Hg} \end{aligned}$$

The Clapeyron equation generally overpredicts the vapor pressure at or near ambient conditions. The Antoine equation is widely used in industry and usually provides excellent results. Also note that, contrary to statements appearing in the *Federal Register* and some EPA publications, vapor pressure is not a function of pressure.

3.22 Reynolds Number

The Reynolds number (Re)

(**a**) Describes fluid flow and is equal to $\mu C_p/\rho DQ$

(**b**) Equals 6.02×10^{23}

(**c**) Describes how a fluid behaves while flowing and is defined as the inertial forces divided by the viscous forces ($Dv\rho/\mu$)

(**d**) Is generally used only for liquids

Solution: Answers (a), (b), and (d) are obviously incorrect. The definition of Re is given in Equation (3.13). The correct answer is therefore (c).

3.23 Reynolds Number Calculation

Calculate the Reynolds number for a gas flowing through a 5-inch-diameter pipe at 10 fps (feet per second) with a density of 0.050 lb/ft^3 and a viscosity of 0.065 cP? Is the flow turbulent or laminar?

Solution: By definition

$$\mathrm{Re} = \frac{Dv\rho}{\mu} \tag{3.13}$$

Substitution yields

$$\begin{aligned}\mathrm{Re} &= \left(\frac{5\,\mathrm{in}}{1}\right)\left(\frac{1\,\mathrm{ft}}{12\,\mathrm{in}}\right)\left(\frac{10\,\mathrm{ft}}{\mathrm{s}}\right)\left(\frac{0.050\,\mathrm{lb}}{\mathrm{ft}^3}\right)\Big/\left(\frac{1}{0.065\,\mathrm{cP}}\right)\left(\frac{1\,\mathrm{cP}}{6.720\times10^{-4}\,\mathrm{lb/ft}\cdot\mathrm{s}}\right)\\ &= (5/12\ \mathrm{ft})\,(10\,\mathrm{ft/s})\,(0.050\,\mathrm{lb/ft}^3)/(0.065\times6.72\times10^{-4}\,\mathrm{lb/ft}\cdot\mathrm{s})\\ &= 4770\end{aligned}$$

The Reynolds number is >2100; therefore, this gas flow is turbulent. Generally, moving gases are in the turbulent flow regime.

3.24 Process Calculation

An external gas stream is fed into an air pollution control device at a rate of 10,000 lb/hr in the presence of 20,000 lb/hr of air. Because of the energy requirements of the unit, 1250 lb/hr of a vapor conditioning agent is added to assist the treatment of the stream. Determine the rate of product gases exiting the unit in pounds per hour (lb/hr). Assume steady-state conditions.

Solution: Apply the conservation law for mass to the control device on a rate basis:

$$\begin{aligned}&\text{Rate of mass in} - \text{rate of mass out} + \text{rate of mass generated}\\ &\quad = \text{rate of mass accumulated}\end{aligned} \tag{3.34}$$

Note that mass is not generated and steady conditions (no accumulation) apply. Rewrite this equation subject to the conditions in the problem statement.

$$\text{Rate of mass in} = \text{rate of mass out}$$

or

$$\dot{m}_{\mathrm{in}} = \dot{m}_{\mathrm{out}}$$

Now refer to the problem statement for the three inlet flows:

$$\dot{m}_{\text{in}} = 10{,}000 + 20{,}000 + 1250 = 31{,}250\ \text{lb/hr}$$

Determine $\dot{m}_{\text{out}}$, the product gas flow rate. Since $\dot{m}_{\text{in}} = \dot{m}_{\text{out}}$, it follows that

$$\dot{m}_{\text{out}} = 31{,}250\ \text{lb/hr}$$

As noted earlier, the conservation law for mass may be written for any compound whose quantity is not changed by chemical reaction and for any chemical element whether or not it has participated in a chemical reaction. It may be written for one piece of equipment around several pieces of equipment, or around an entire process. It may be used to calculate an unknown quantity directly, to check the validity of experimental data, or to express one or more of the independent relationships among the unknown quantities in a particular problem situation.

3.25 Collection Efficiency

Given the following inlet loading and outlet loading of an air pollution control unit, determine the collection efficiency of the unit:

$$\text{Inlet loading} = 0.02\ \text{gr/ft}^3$$

$$\text{Outlet loading} = 0.001\ \text{gr/ft}^3$$

Solution: Collection efficiency is a measure of the degree of performance of a control device; it specifically refers to the degree of removal of a pollutant and may be calculated through the application of the conversation law for mass. *Loading*, generally refers to the concentration of pollutant. The equation describing collection efficiency (fractional) E in terms of inlet and outlet loading is

$$E = \frac{(\text{inlet loading}) - (\text{outlet loading})}{\text{inlet loading}} \tag{3.42}$$

Calculate the collection efficiency of the control unit in percent for the rates provided:

$$E = \frac{2 - 0.1}{2} 100 = 95\%$$

The term η is also used as a symbol for efficiency E. The reader should also note that the collected amount of pollutant by the control unit is the product of E and the inlet loading. The amount discharged to the atmosphere is given by the inlet loading minus the amount collected.

3.26 Penetration Definition

Define penetration.

Solution: By definition, the penetration P is given by

$$\begin{aligned} P &= 100 - E; \text{ percent basis} \\ P &= 1.0 - E; \text{ fractional basis} \end{aligned} \tag{3.43}$$

Note that there is a 10-fold increase in P as E goes from 99.9 to 99%. For a multiple series of n collectors, the overall penetration is simply given by

$$P = P_1 P_2, \ldots, P_{n-1} P_n \tag{3.44}$$

For particulate control in air pollution units, penetrations and/or efficiencies can be related to individual size ranges. The overall efficiency (or penetration) is then given by the contribution from each size range, i.e., the summation of the product of mass fraction and efficiency for each size range. This is treated in more detail in the chapters on particulates (Chapters 7–12).

3.27 Spray Tower Application

A proposed incineration facility design requires that a packed column and a spray tower be used in series for the removal of HCl from the flue gas. The spray tower is to operate at an efficiency of 65% and the packed column at an efficiency of 98%. Calculate the mass flow rate of HCl leaving the spray tower, the mass flow rate of HCl entering the packed tower, and the overall efficiency of the removal system if 76.0 lb of HCl enters the system every hour.

Solution: As defined in Problem 3.25

$$E = \frac{\dot{m}_{\text{in}} - \dot{m}_{\text{out}}}{\dot{m}_{\text{in}}} \tag{3.45}$$

Then,

$$\dot{m}_{\text{out}} = (1 - E)\ (\dot{m}_{\text{in}})$$

For the spray tower

$$\dot{m}_{\text{out}} = (1 - 0.65)\ (76.0) = 26.6 \text{ lb/hr HCl}$$

The mass flow rate of HCl leaving the spray tower is equal to the mass flow rate of HCl entering the packed column. For the packed column

$$\dot{m}_{\text{out}} = (1 - 0.98)\ (26.6) = 0.532 \text{ lb/hr HCl}$$

The overall efficiency can now be calculated:

$$E = \frac{\dot{m}_{\text{in}} - \dot{m}_{\text{out}}}{\dot{m}_{\text{in}}} = \frac{76.0 - 0.532}{76.0}$$

$$= 0.993$$

$$= 99.3\%$$

3.28 Compliance Determination

A proposed incinerator is designed to destroy a hazardous waste at 2100°F and 1 atm. Current regulations dictate that a minimum *destruction and removal efficiency* (DRE) of 99.99% must be achieved. The waste flow rate into the unit is 960 lb/hr while that flowing out of the unit is measured as 0.08 lb/hr. Is the unit in compliance?

Solution: Select as a basis the 1 hourly rate of operation. The mass equation employed for efficiency may also be used to calculate the minimum destruction and removal efficiency.

$$\text{DRE} = \frac{\dot{m}_{\text{in}} - \dot{m}_{\text{out}}}{\dot{m}_{\text{in}}}(100) = \frac{960 - 0.08}{960}(100)$$

$$= 99.992\% \tag{3.45}$$

Thus the unit is operating in compliance with present regulations. The answer is *yes*.

3.29 Velocity Determination

Given 20,000 ft^3/min of air at ambient conditions exiting a system through a pipe whose cross-sectional area is 4 ft^2, determine the mass flow rate in lb/min and the exit velocity in ft/s.

Solution: The continuity equation is given by

$$\dot{m} = \rho S v \tag{3.46}$$

where ρ = liquid density
S = cross-sectional area
v = velocity

Since 20,000 ft^3/min of air enters the system, then

$$q_1 = S_1 v_1$$

$$= 20{,}000\ \text{ft}^3/\text{min}$$

where subscript 1 refers to inlet conditions. Now, assuming that $\rho = 0.075\ \text{lb/ft}^3$, then

$$\dot{m} = \rho_1 S_1 v_1 = (0.075)\ (20{,}000) = 1500\ \text{lb/min}$$

3.30 Outlet Temperature

Heat at 18.7×10^6 Btu/hr is transferred from the flue gas of an incinerator. Calculate the outlet temperature of the gas stream using the following information:

Average heat capacity c_p of gas = 0.26 Btu/(lb · °F)
Gas mass flow rate $\dot{m}$ = 72,000 lb/hr
Gas inlet temperature T_1 = 1200°F

Solution: The first law of thermodynamics states that energy is conserved. For a flow system, neglecting kinetic and potential effects, the energy transferred Q to or from the flowing medium is given by the enthalpy change ΔH of the medium. The enthalpy of an ideal gas is solely a function of temperature; enthalpies of liquids and most real gases are almost always assumed to depend on temperature alone. Changes in enthalpy resulting from a temperature change for a single-phase material may be calculated from the equation

$$\Delta H = mc_p\,\Delta T \tag{3.47}$$

or

$$\Delta\dot{H} = \dot{m}c_p\,\Delta T$$

where ΔH = enthalpy change
m = mass of flowing medium
c_p = average heat capacity per unit mass of flowing medium across the temperature range of ΔT
$\Delta\dot{H}$ = enthalpy change per unit time
$\dot{m}$ = mass flow rate of flowing medium

(*Note*: The symbol Δ means "change in.")

Solve the conservation law for energy for the gas outlet temperature T_2:

$$\dot{Q} = \Delta\dot{H} = \dot{m}c_p\,\Delta T = \dot{m}c_p(T_2 - T_1)$$

where $\dot{Q}$ is the rate of energy transfer:

$$T_2 = \frac{\dot{Q}}{\dot{m}C_P} + T_1$$

The gas outlet temperature is therefore

$$T_2 = [-18.7 \times 10^6/\{(72{,}000)\ (0.26)\}] + 1200$$
$$= 200°\text{F}$$

This equation is based on adiabatic conditions, i.e., the entire heat load is transferred from the flowing gas. The unit is assumed to be perfectly insulated so that no heat is transferred to the surroundings. However, this is not the case in a real-world application. As with mass balances, an enthalpy balance may be performed within any properly defined boundary, whether real or imaginary. For example, an enthalpy balance can be applied across the entire unit or process. The enthalpy of the feed stream(s) is equated with the enthalpy of the product stream(s) plus the heat loss from the process. All the enthalpy terms must be based on the same reference temperature. Finally, the enthalpy has two key properties that should be kept in mind:

1. Enthalpy is a point function, i.e., the enthalpy change from one state (say, 200°F, 1 atm) to another state (say 400°F, 1 atm) is a function only of the two states and not the path of the process associated with the change.
2. Absolute values of enthalpy are not important. The enthalpy of water at 60°F, 1 atm, as recorded in some steam tables is 0 Btu/lbmol. This choice of zero is arbitrary. However, another table may indicate a different value. Both are correct! Note that changing the temperature of water from 60 to 100°F results in the same change in enthalpy using either table.

Enthalpy changes may be expressed with units (English) of Btu, Btu/lb, Btu/lbmol, Btu/scf, or Btu/time depending on the available data and calculation required.

3.31 Process Cooling Water Requirement

Determine the total flow rate of cooling water required for the services listed below assuming that a cooling tower system supplies the water at 90°F with a return temperature of 115°F. How much freshwater makeup is required if 5% of the return water is sent to "blowdown?" Note that the cooling water heat capacity is 1.00 Btu/(lb · °F), the heat of vaporization at cooling tower operating conditions is 1030 Btu/lb, and the density of water at cooling tower operating conditions is 62.0 lb/ft^3.

TABLE 3.6 Cooling Water—Heat Duty Data

Process Unit	Heat Duty, Btu/hr	Required Temperature, °F
1	12,000,000	250
2	6,000,000	200–276
3	23,500,000	130–176
4	17,000,000	300
5	31,500,000	150–225

Solution: The required cooling water flow rate, $\dot{m}_{CW}$ is given by the following equation:

$$\dot{m}_{CW} = \frac{Q_{HL}}{(\Delta T)(c_p)(\rho)} \tag{3.48}$$

where Q_{HL} = heat load, Btu/min
ΔT = change in temperature $= 115°F - 90°F = 25°F$

$$c_p = \text{heat capacity} = 1.00\ \text{Btu}/(\text{lb}\cdot{}^\circ\text{F})$$
$$\rho = \text{density of water} = (62.0\ \text{lb}/\text{ft}^3)\ (0.1337\ \text{ft}^3/\text{gal}) = 8.289\ \text{lb}/\text{gal}$$
$$\dot{m}_{\text{CW}} = \text{cooling water flow rate, mass}/\text{time}$$

The heat load is

$$Q_{\text{HL}} = (12 + 6 + 23.5 + 17 + 31.5)\ (10^6\ \text{Btu}/\text{hr})/\ (60\ \text{min}/\text{hr})$$
$$= 1{,}500{,}000\ \text{Btu}/\text{min}$$

Thus,

$$q_{\text{CW}} = \frac{1{,}500{,}000\ \text{Btu}/\text{min}}{(25^\circ\text{F})\ (1.00\ \text{Btu}/\text{lb}\cdot{}^\circ\text{F})\ (8.289\ \text{lb}/\text{gal})} = 7250\ \text{gpm}$$

where q_{CW} = water volumetric flow rate, gal/min.

The blowdown flow q_{BD} is given by

$$q_{\text{BD}} = (\text{BDR})\ (q_{\text{CW}})$$

where BDR is the blowdown rate = 5% = 0.05.

Thus

$$q_{\text{BD}} = (0.05)\ (7250\ \text{gpm}) = 362.5\ \text{gpm}$$

The amount of water vaporized by the cooling tower q_{v} is given by

$$q_{\text{v}} = (Q_{\text{HL}}/[(\rho)(\Delta H_{\text{v}})] \tag{3.49}$$

where H_{V} is the heat of vaporization = 1030 Btu/lb. Substitution yields

$$q_{\text{v}} = \frac{(1{,}500{,}000\ \text{Btu}/\text{min})}{(8.289\ \text{lb}/\text{gal})\ (1030\ \text{Btu}/\text{lb})} = 175.7\ \text{gpm}$$

3.32 Steam Requirement Options

Determine how many pounds per hour of steam are required for the following two cases: (1) if steam is provided at 500 psig and (2) if steam is provided at both 500 and 75 psig pressures. The plants heating requirements are listed in Table 3.7.

TABLE 3.7 Process Heat Duty Data

Process Unit	Unit Heat Duty (UHD), Btu/hr	Required Temperature, °F
1	10,000,000	250
2	8,000,000	450
3	12,000,000	400
4	20,000,000	300

The properties of saturated steam are listed in Table 3.8.

TABLE 3.8 Steam Data

Pressure Provided, psig	Saturation Temperature, °F	Enthalpy of Vaporization, ΔH_v, Btu/lb
75	320	894
500	470	751

Solution: The total required flow rate of 500 psig steam $\dot{m}_{BT}$ is given by

$$\dot{m}_{BT} = \dot{m}_{B1} + \dot{m}_{B2} + \dot{m}_{B3} + \dot{m}_{B4}$$

For this equation:

$\dot{m}_{B1}$ (mass flow rate of 500 psig steam through unit 1)
$= \text{UHD}_1/\Delta H_v = 13{,}320\,\text{lb/hr}$

$\dot{m}_{B2}$ (mass flow rate of 500 psig steam through unit 2)
$= \text{UHD}_2/\Delta H_v = 10{,}655\,\text{lb/hr}$

$\dot{m}_{B3}$ (mass flow rate of 500 psig steam through unit 3)
$= \text{UHD}_3/\Delta H_v = 15{,}980\,\text{lb/hr}$

$\dot{m}_{B4}$ (mass flow rate of 500 psig steam through unit 4)
$= \text{UHD}_4/\Delta H_v = 26{,}635\,\text{lb/hr}$

Thus,

$$\dot{m}_{BT} = 66{,}590\,\text{lb/hr}$$

The required combined total flow rate of 500 and 75 psig steam $\dot{m}_{CT}$ is given by

$$\dot{m}_{CT} = \dot{m}_{75.1} + \dot{m}_{B2} + \dot{m}_{B3} + \dot{m}_{75.4}$$

For this situation

$\dot{m}_{75.1}$ (mass flow rate of 75 psig steam through unit 1)
$= \text{UHD}/\Delta H_v = 11{,}185\ \text{lb/hr}$

$\dot{m}_{75.4}$ (mass flow rate of 75 psig steam through unit 4)
$= \text{UHD}/\Delta H_v = 22{,}371.4\,\text{lb/hr}$

Thus,

$$\dot{m}_{\mathrm{CT}} = 60{,}192\ \mathrm{lb/hr}$$

Note that since the saturation temperature of 75 psig steam is lower than those of two of the process units, the 500 psig steam must be used for process units 2 and 3. What conclusions can be drawn from this calculation?

3.33 Humidity Effect

A flue gas [molecular weight (MW) = 30, dry basis] is being discharged from a scrubber at 180°F (dry bulb) and 125°F (wet bulb). The gas flow rate on a dry basis is 10,000 lb/hr. The absolute humidity at the dry-blub temperature of 180°F and wet-bulb temperature of 125°F is 0.0805 lb H_2O/lb dry air.

(**a**) What is the mass flow rate of the wet gas?

(**b**) What is the actual volumetric flow rate of the wet gas?

Solution: Curves showing the relative humidity (ratio of the mass of the water vapor in the air to the maximum mass of the water vapor the air can hold at that temperature, i.e., if the air were saturated) of humid air appear on the psychrometric chart (see Figure 3.1). The curve for 100% relative humidity is also referred to as the *saturation curve*. The abscissa of the humidity chart is air temperature, also known as the *dry-bulb temperature* (T_{DB}). As discussed in Section 3.2, the wet-bulb temperature (T_{WB}) is another measure of humidity; it is the temperature at which a thermometer with a wet wick wrapped around the bulb stabilizes. As water evaporates from the wick to the ambient air, the bulb is cooled; the rate of cooling depends on how humid the air is. No evaporation occurs if the air is saturated with water; hence, T_{WB} and T_{DB} are then the same. The lower the humidity, the greater the difference between these two temperatures. On a psychrometric chart, constant wet-bulb temperature lines are

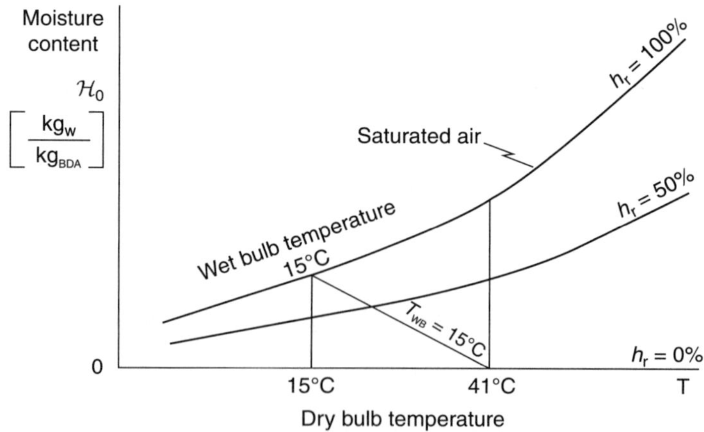

Figure 3.1 Diagram of a psychrometric chart.

straight with negative slopes. The value of T_{WB} corresponds to the value of the abscissa at the point of intersection of this line with the saturation curve.

Based on the problem statement, calculate the flow rate of water in the air. Note that both the given flow rate and humidity are on a dry basis.

$$\text{Water flow rate} = (0.0805)\ (10{,}000) = 805\ \text{lb/hr}$$

Calculate the total flow rate by adding the dry gas and water flow rates:

$$\text{Total flow rate} = 10{,}000 + 805 = 10{,}805\ \text{lb/hr}$$

The molar rate of water and dry gas are thus

$$\text{Moles gas} = 10{,}000/30 = 333.3\ \text{lbmol/hr}$$
$$\text{Moles water} = 805/18 = 44.7\ \text{lbmol/hr}$$

Calculate the mole fraction of water vapor using the above two results.

$$y_{\text{water}} = \frac{44.7}{(44.7 + 333.3)} = 0.12$$

The average molecular weight of the mixture becomes

$$\begin{aligned} \text{MW} &= (1.0 - 0.12)\ (30) + (0.12)\ (18) \\ &= 28.6\ \text{lb/lbmol} \end{aligned}$$

The molar flow rate of the wet gas may now be determined:

$$\dot{n} = 10805/28.6 = 378\ \text{lbmol/hr}$$

The ideal gas law may be applied to calculate the volumetric flow rate of the wet gas:

$$\begin{aligned} q &= \dot{n}RT/P \qquad (3.22) \\ &= (378)\ (0.73)\ (460 + 180)/1.0 \\ &= 1.77 \times 10^5\ \text{ft}^3/\text{hr} \end{aligned}$$

The following are some helpful points on the use of psychrometric charts:

1. In problems involving the use of the humidity chart, it is convenient to choose a mass of dry air as a basis since the chart uses this basis.
2. Heating or cooling at temperatures above the dew point (temperature at which the vapor begins to condense) corresponds to a horizontal movement on the

chart. As long as no condensation occurs, the absolute humidity stays constant.

3. If the air is cooled, the system follows the appropriate horizontal line to the left until it reaches the saturation curve and follows this curve thereafter.

3.34 Saturated Water Discharge

A flue gas is discharged at 120°F from an HCl absorber in a hazardous waste incinerator (HWI) facility in which carbon tetrachloride (CCl_4) is being incinerated. If 9000 lb/hr (MW = 30) of gas enters the absorber essentially dry (negligible water) at 560°F, calculate the moisture content, the mass flow rate, and the volumetric flow rate of the discharged gas. The discharge gas from the absorber may safely be assumed to be saturated with water vapor. The discharge humidity of the flue gas is approximately.

$$\mathcal{H}_{\text{out}} = 0.0814 \text{ lb } H_2O/\text{lb bone-dry air}$$

Solution: The $\mathcal{H}_{\text{out}}$ represents the moisture content of the gas at outlet conditions in lb H_2O/lb dry air. If the gas is assumed to have the properties of air, the discharge water vapor rate is

$$\begin{aligned} \dot{m}_{H_2O} &= (0.0814)\,(9000) \\ &= 733 \text{ lb/hr} \end{aligned}$$

The total flow rate leaving the absorber is

$$\begin{aligned} \dot{m}_{\text{total}} &= 733 + 9000 \\ &= 9733 \text{ lb/hr} \end{aligned}$$

The volumetric (or molar) flow rate can be calculated only if the molecular weight of the gas is known. The average molecular weight of the discharge flue gas must first be calculated from the mole fraction of the flue gas (fg) and water vapor (wv).

$$\begin{aligned} y_{\text{fg}} &= \frac{9000/30}{(9000/30) + (733/18)} = 0.88 \\ y_{\text{wv}} &= \frac{733/18}{(733/18) + (9000/30)} = 0.12 \\ \overline{\text{MW}} &= (0.88)\,(30) + (0.12)\,(18) = 28.6 \text{ lb} \end{aligned}$$

The ideal gas law is employed to calculate the actual volumetric flow rate, q_{a}.

$$q_{\mathrm{a}} = \frac{\dot{m}}{\mathrm{MW}}(RT/P)$$

$$q_{\mathrm{a}} = \frac{9733}{28.6}\frac{(0.73)\,(460 + 140)}{1.0}$$

$$= 1.49 \times 10^5\ \mathrm{ft^3/hr} \qquad (3.22)$$

NOTE: Additional problems are available for all readers at www.wiley.com. Follow links for this title.

4

INCINERATORS

4.1 INTRODUCTION

Combustion is often used to control the emissions of organic compounds from process industries. At a sufficiently high temperature and adequate residence time, any hydrocarbon can be oxidized to carbon dioxide and water by the combustion process.

Combustion systems are often relatively simple devices capable of achieving very high removal destruction efficiencies. They consist of burners, which ignite the fuel and organic, and a chamber, which provides appropriate residence (or detention) time for the oxidation process. Because of the high cost and decreasing supply of fuels, combustion systems may be designed to include some type of heat recovery. Combustion is also used for the more serious emission problems that require high destruction efficiencies, such as the emission of toxic or hazardous gases. There are, however, some problems that may occur when using combustion. Incomplete combustion of many organic compounds results in the formation of aldehydes and organic acids, which may create an additional pollution problem. Oxidizing organic compounds containing sulfur or halogens produce unwanted pollutants such as sulfur dioxide, hydrochloric acid, hydrofluoric acid, or phosgene. If present, these pollutants

Air Pollution Control Equipment Calculations. By Louis Theodore

would require a scrubber (see Chapter 11) to remove them prior to release into the atmosphere.

Several basic combustion systems are in use. Although these devices are physically similar, the conditions under which they operate may be different. Choosing the proper device depends on many factors, including the type of hazardous contaminants in the waste stream, concentration of combustibles in the steam, process flow rate, control requirements, and an economic evaluation.

Combustion is a chemical process occurring from the rapid combination of oxygen with various elements or chemical compounds resulting in the release of heat. The process of combustion is also referred to as *oxidation* or *incineration*. Most fuels used for combustion along with the waste are composed essentially of carbon and hydrogen, but can include other elements such as sulfur, nitrogen, and chlorine. Although combustion seems to be a very simple process that is well understood, in reality it is not. The exact manner in which a fuel or waste is oxidized occurs in a series of complex, free radical chain reactions. The precise set of reactions by which combustion occurs is termed the *mechanism of combustion*. By analyzing the mechanism of combustion, the rate at which the reaction proceeds and the variables affecting the rate can be estimated. For most combustion devices, the rate of reaction proceeds extremely fast compared to the mechanical operation of the device. Maintaining efficient and complete combustion is somewhat of an art rather than a science. Therefore, this section focuses on the factors that influence the completeness of combustion, rather than analyzing the mechanism involved.

To achieve complete combustion once the air (oxygen), waste, and fuel have been brought into contact, the following conditions must be provided: a temperature high enough to ignite the waste/fuel mixture, turbulent mixing of the air and waste/fuel; and, sufficient residence time for the reaction to occur. These three conditions are referred to as the "three Ts of combustion." Time, temperature, and turbulence govern the speed and completeness of reaction. They are not independent variables since changing one can affect the other two.

The rate at which a combustible compound is oxidized is greatly affected by temperature. The higher the temperature, the faster the oxidation reaction will proceed. The chemical reactions involved in the combustion of a fuel and oxygen can occur even at room temperature, but very slowly. For this reason, a pile of oily rags can be a fire hazard. Small amounts of heat are liberated by the slow oxidation of the oils. This, in turn, raises the temperature of the rags and increases the oxidation rate, liberating more heat. Eventually, a full-fledged fire can break out.

For combustion processes, ignition is accomplished by adding heat to speed up the oxidation process. Heat is needed to combust any mixture of air and fuel until the ignition temperature of the mixture is reached. By gradually heating a mixture of fuel and air, the rate of reaction and energy released will gradually increase until the reaction no longer depends on the outside heat source. In effect, more heat is being generated than is lost to the surroundings. The ignition temperature must be reached or exceeded to ensure complete combustion. To maintain combustion of a waste, the amount of energy released by the combusted waste must be sufficient to heat up the incoming waste (and air) up to its ignition temperature; otherwise, a fuel must be added. The ignition temperature of various fuels and compounds can be found in combustion handbooks. These temperatures are dependent on combustion conditions and therefore should be used only as a guide. Ignition depends on

1. Concentration of combustibles in the waste stream
2. Inlet temperature of the waste stream
3. Rate of heat loss from combustion chamber
4. Residence time and flow pattern of the waste stream
5. Combustion chamber geometry and materials of construction

Most incinerators operate at higher temperatures than the ignition temperature, which is a minimum temperature. Thermal destruction of most organic compounds occurs between 590 and 650°C (1100 and 1200°F). However, most hazardous waste incinerators are operated at 1800 to 2200°F to ensure near-complete conversion of the waste.

Time and temperature affect combustion in much the same manner as temperature and pressure affect the volume of a gas. When one variable is increased, the other may be decreased with the same end result. With a higher temperature, a shorter residence time can achieve the same degree of oxidation. The reverse is also true; a higher residence time allows the use of a lower temperature. In describing incinerator operation, these two terms are always mentioned together. This effect is described in graphical form in Figure 4.1. The choice between higher temperature or longer residence time is based on economic considerations. Increasing residence time involves using a larger combustion chamber resulting in a higher capital cost. Raising the operating temperature increases fuel usage which also adds to the operating costs, and fuel costs are the major operating expense for most incinerators. Within certain limits, lowering the temperature and adding volume to increase residence time can be a cost-effective alternative method of operation.

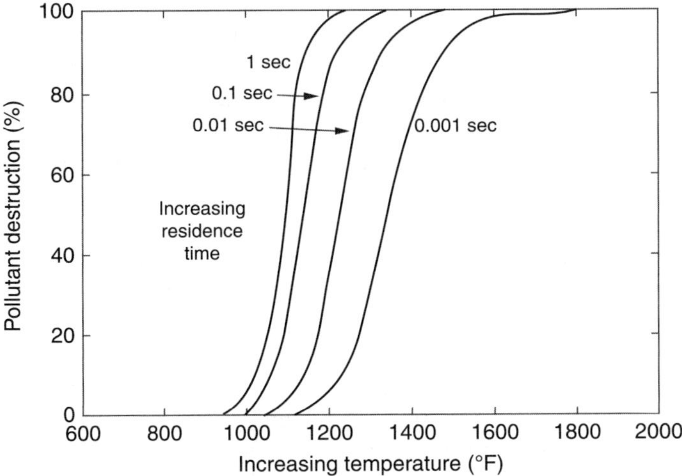

Figure 4.1 Coupled effects of temperature and time on rate of pollutant oxidation.

The residence time of gases in the combustion chamber may be calculated from

$$t = \frac{V}{q} \tag{4.1}$$

where t is the residence time, s; V is the chamber volume, ft^3; and q is the gas volumetric flow rate at combustion conditions, ft^3/s. i.e., q is the total flow of hot (flue) gases in the combustion chamber. Adjustments to the flow rate may include outside air added for combustion.

Proper mixing is important in combustion processes for two reasons: (1) for complete combustion to occur every part of the waste and fuel must come in contact with air (oxygen)—if not, unreacted waste and fuel will be exhausted from the stack; and (2) not all of the fuel or waste stream is able to be in direct contact with the burner flame. In most incinerators, a portion of the waste stream may bypass the flame and be mixed at some point downstream of the burner with the hot products of combustion. If the two streams are not completely mixed, a portion of the waste stream will not react at the required temperature and incomplete combustion will occur. A number of methods are available to improve mixing the air and waste (combustion) streams. Some of these include the use of refractory baffles, swirl fired burners, and baffle plates. The problem of obtaining complete mixing is not easily solved. Unless properly designed, many of these mixing devices may create "dead spots" and reduce operating performance. Merely inserting obstructions to increase turbulence is often not the answer. The process of mixing waste stream and the flame to obtain a uniform temperature for decomposition of wastes is often the most difficult part in the design of the incinerator.

Oxygen is necessary for combustion to occur. To achieve complete combustion of a compound, a sufficient supply of oxygen must be present to convert all of the carbon to CO_2. This quantity of oxygen is referred to as the *stoichiometric* or *theoretical amount*. The stoichiometric amount of oxygen is determined from a balanced chemical equation summarizing the oxidation reactions. If an insufficient amount of oxygen is supplied, the mixture is referred to as "rich"; there is not enough oxygen to combine with all the fuel and waste so that incomplete combustion occurs. If more than the stoichiometric amount of oxygen is supplied, the mixture is referred to as "lean." The added oxygen then plays no part in the oxidation reaction and passes through the incinerator. Oxygen for a combustion process is almost always supplied by using air. Since air is essentially 79% nitrogen and 21% oxygen by volume, a larger volume of air is required than if pure oxygen were used. Stoichiometric calculations are treated in the "Problems" section of this chapter.

In most applications, more than the stoichiometric amount of air is used to ensure complete combustion. This extra volume is referred to as *excess air*. If ideal mixing were achievable, no excess air would be necessary. However, most combustion devices are not capable of achieving ideal mixing of the fuel and air/waste streams. The amount of excess air is sometimes held to a minimum in order to reduce heat losses; excess air takes no part in the reaction but does absorb some of the heat produced. To raise the excess air to the combustion temperature, additional fuel must be used to make up for any loss of heat. Operating at a high volume flow rate of excess air can thus be very costly in terms of the added fuel required.

In addition to the theoretical air required, one also needs to consider the volume of combustion products produced from oxidizing a substance. This is an important term used to determine the size of the combustion chamber. For example, when 1 ft^3 of methane is combusted with the theoretical amount of air, 10.53 ft^3 of flue gas is produced. Natural gas is unique since its chemical composition can vary. When 1 ft^3 of natural gas is burned with a stoichiometric amount of air, it produces *approximately* 11.4 ft^3 (average value) of flue gas. This is an important number to remember in incinerator calculations.

Afterburning is a term used to describe the combustion process used to control gaseous emissions. This term is appropriate only for describing a thermal oxidizer used to control gases coming from a process where combustion was not complete. Incinerators are used to combust solid, liquid and gaseous materials. When used in this chapter, the term *incinerator* will refer to combusting waste streams.

Equipment used to control waste gases by combustion can be divided into three categories: direct combustion or flaring, thermal oxidation, and catalytic oxidation. A direct combustor or flare is a device in which air and all the combustible waste gases react at the burner (see Figure 4.2). Complete combustion must occur instantaneously since there is no residence chamber. Therefore, the flame temperature is the most important variable in flaring waste gases. By contrast, in thermal oxidation, the combustible

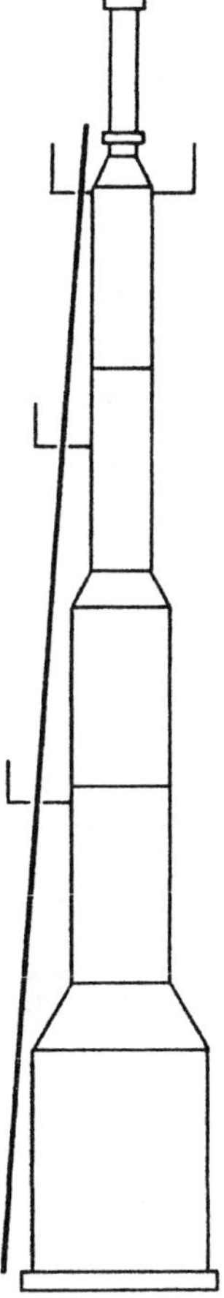

Figure 4.2 Elevated flare schematic.

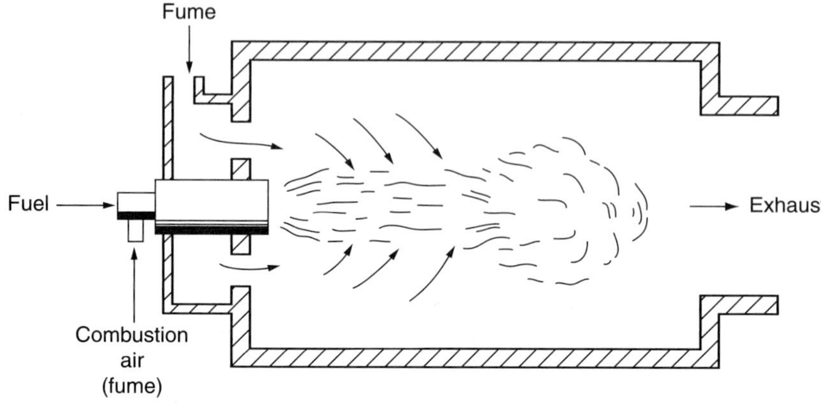

Figure 4.3 Schematic of a thermal incinerator.

waste gases pass over and/or around a burner flame into a residence chamber where oxidation of the waste gases is completed (see Figure 4.3). Catalytic oxidation is very similar to thermal oxidation. The main difference is that after passing through the flame area, the gases pass over a catalyst bed that promotes oxidation at a lower temperature than does thermal oxidation (see Figure 4.4). Details on these three control devices, emphasizing afterburners, are given below.

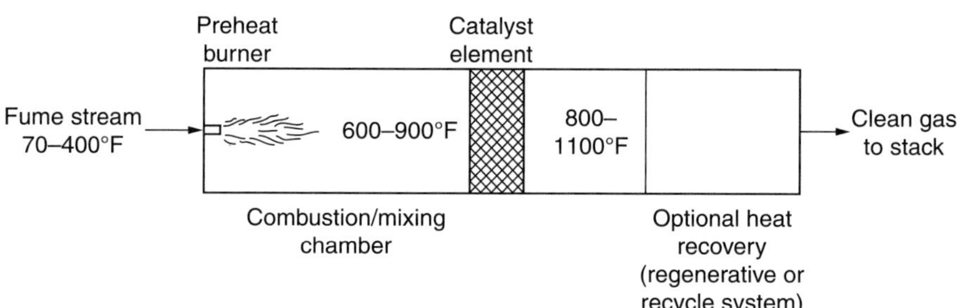

Figure 4.4 Schematic of a catalytic incinerator.

Afterburners can be used over a fairly wide, but low, range of organic vapor concentrations. The concentration of the organics in air must be substantially below the lower explosive (or flammable) limit (LEL). As a rule, a factor of four is employed for safety precautions. Reactions are conducted at elevated temperatures in order to ensure high chemical reaction rates for the organics. To achieve this temperature it is necessary to preheat the feed stream with auxiliary energy. Along with the contaminant-laden gas stream, air and fuel are continuously delivered to the reactor where the fuel is combusted with air in the firing unit (burner). The burner may utilize the air in the process waste stream as the combustion air for auxiliary fuel, or it may use a separate source of outside air for this purpose. The products of combustion and the unreacted feed stream are intensely mixed and enter the reaction zone of the unit. The pollutants in the process gas stream are then reacted at the elevated temperature. The unit requires operating temperatures in the 1200–2000°F range for combustion of most pollutants.

A residence time of 0.2–2.0 s is recommended in the literature, but this factor is dictated primarily by kinetic considerations. A length-to-diameter ratio of 2.0–3.0 is usually employed. The end products are continuously discharged at the outlet of the reactor. The average gas velocity can range from as low as 10 ft/s to as high as 50 ft/s. (The velocity increases from inlet to outlet owing to the increase in the number of moles of the reacting fluid and the increase in temperature due to reaction.) These high velocities are required to prevent settling of particulates (if present) and to minimize the dangers of flashback and fire hazards. The fuel is usually natural gas. The energy liberated by reaction may be directly recovered in the process, or indirectly recovered by suitable external heat exchange. This should be included in a design analysis since energy is the only commodity of value that is usually derived from the combustion process.

Because of the high operating temperatures, the unit must be constructed of metals capable of withstanding this condition. Combustion devices are usually constructed with an outer shell that is lined with refractory material. However, refractory material is heavy, with densities as low as 50 lb/ft^3 for lightweight insulating firebrick, and as high as 175 lb/ft^3 for castable refractories. Refractory wall thickness is in the 3–9 inch range. This weight adds considerably to the cost. Because of its light weight, firebrick wall construction is being used in some units. All-metal construction has found limited application. These combustion reactors also vary in shape from tanks to tubular pipes. The latter type is usually used since it has a surface-to-volume ratio which is advantageous for energy recovery and suitable for continuous operation.

The incineration process may be thought of as occurring in two separate stages:

1. Combustion of fuel
2. Combustion of pollutants

The combustion process in the first stage is an extremely rapid and highly irreversible chemical reaction. The oxygen supplied by the (primary) air may be in excess or obtained directly from the process gas stream (secondary air). The carbon content of the fuel burns almost quantitatively to carbon dioxide and the hydrogen to water. As noted above, combustion reactors should be operated with the least amount of excess air compatible with the need for complete combustion and/or fuel requirements. The mixing of the air and the fuel in the burner section of the reactor also determines the completeness of combustion.

One can show that any fuel completely burned without excess air produces a discharge gas containing the maximum CO_2 possible for that fuel. Thus, with no excess air and perfect mixing, combustion will be complete with the resulting gas containing maximum CO_2 and little to no O_2, CO, and H_2. Since the reaction is extremely rapid, the combustion of fuel occurs in a rather narrow zone in the reactor, perhaps within a foot of reactor length at entrance conditions. The average residence time thus approaches zero. The energy liberated on combustion of the fuel is then used to heat the combustion temperature. Under these conditions, the rate of the combustion reaction is dictated by the fuel rate; kinetics effects need not be considered. The calculations reduce to one of stoichiometry, and overall mass and energy balances. Some of the procedures set forth in Chapter 3 can be applied. However, the mixing of the fuel combustion gases and the process gas stream does not occur instantaneously. A finite residence time (or reactor length) is required to achieve "complete" system mixing that will yield a uniform temperature profile through the cross section of the reactor. This is a critical design and operational consideration, for failure to achieve the above can result in incomplete combustion, high(er) fuel requirements, or both. In fact, variations in performance can in many cases be either directly or indirectly attributed to this mixing process. Care must be exercised in this part of the design and operation since flame quenching can result if the mixing process is too severe. Once mixing is complete and the temperature uniform, the so-called residence requirement is applicable to the combustion of the pollutant in the process gas stream. Thus, the residence time specified in the codes and regulations applies to both the mixing process, where the process gas stream and fuel combustion products achieve a uniform temperature, and the combustion process for the pollutant(s), which is assumed to occur at this elevated temperature.

In the second stage of the process, the heated gases from the burner pass through the reactor where the pollutants (organics) are reacted (oxidized) to harmless end products. Although the pollutants also serve as a source of fuel, enthalpy of reaction (heat) effects are often small enough to be neglected so that the calculation of conversion reduces to a kinetic problem. However, the reaction mechanism is quite complex. It is difficult to obtain and interpret experimental data for combustion reactions. In fact, it is common practice to treat a mixture of hydrocarbons in terms of a single hydrocarbon component. Since many chemical (combustion) reacting systems of interest to the environmental engineer can be closely approximated by first-order or pseudo-first-order reactions (L. Theodore: personal notes, 1976), industry usually employs a simple first-order, irreversible reaction mechanism in the design calculation. (It is well known that even zero- or second-order reactions can often be satisfactorily represented by first-order reactions.) This is a reasonable assumption since the oxygen concentration can approach a few parts per million. The reaction velocity constant is an empirically determined parameter that is applicable over the desired range of operating conditions. Although this approach is displaced from earlier presentations, the procedure has merit and is thoroughly justifiable if the resulting equation works.

Catalytic reactors are an alternative to thermal reactors. If a solid catalyst is added to the reactor, the reaction is said to be heterogeneous. For simple reactions, the effect of the presence of a catalyst is to

1. Increase the rate of reaction
2. Permit the reaction to occur at a lower temperature
3. Permit the reaction to occur at a more favorable pressure
4. Reduce the reactor volume
5. Increase the conversion of a product(s) and/or yield of a reactant(s) relative to the other(s)

In a typical catalytic reactor, the gas stream is delivered to the reactor continuously by a fan at velocity in the 10–30 ft/s range but at a lower temperature—usually in the 650–800°F range—than a thermal unit. A length-to-diameter ratio less than 0.5 is usually employed. The gases, which may or may not be preheated, pass through the catalyst bed where the reaction occurs. The combustion products, which are again made up of water vapor, carbon dioxide, inerts, and unreacted vapors, are continuously discharged from the outlet at a higher temperature. Energy savings can again be achieved with heat recovery from the exit stream.

Metals in the platinum family are recognized for their ability to promote combustion at low temperatures. Other catalysts include various oxides of copper, chromium, vanadium, nickel, and cobalt. These catalysts are subject to poisoning, particularly from halogens, halogen and sulfur compounds, zinc, arsenic, lead, mercury, and particulates. High temperatures can also reduce catalyst activity. It is therefore important that catalyst surfaces be clean and active to ensure optimum performance. Catalysts can usually be regenerated with superheated steam.

Catalysts may be porous pellets, usually cylindrical or spherical in shape, ranging from $\frac{1}{16}$ to $\frac{1}{2}$ inch in diameter. Small sizes are recommended, but the pressure drop through the reactor increases. Other shapes include honeycombs, ribbons, wire mesh, etc. Since catalysis is a surface phenomena, a physical property of these particles is that the internal pore surface is near infinitely greater than the outside surface. The reader is referred to the literature for more information on catalyst preparation, properties, comparisons, costs, and impurities.

From a macroscopic perspective, the following sequence of steps is involved in the catalytic conversion of reactants to products:

1. Transfer of reactants to and products from the outer catalyst surface
2. Diffusion of reactants and products within the pore of the catalyst
3. Active adsorption of reactants and the desorption of the products on the active pore centers of the catalyst
4. Reaction(s) on active centers on the catalyst surface

At the same time, energy effects arising due to chemical reaction can result in the following:

1. Heat transfer to or from active centers to the catalyst particle surface
2. Heat transfer to and from reactants and products within the catalyst particle
3. Heat transfer to and from moving streams in the reactor
4. Heat transfer from one catalyst particle to another within the reactor
5. Heat transfer to or from the walls of the reactor

The basic problem in the design of a heterogeneous reactor, as in the case of a catalytic combustion unit, is to determine the quantity of catalyst and/or reactor size required for a given conversion and flow rate. In order to obtain this, information on the rate equation(s) and their parameter(s) must be made available. A rigorous approach to the evaluation of these reaction velocity constants has yet to be accomplished. At this time, industry still relies on the procedures set forth earlier.

Although not treated in any detail in this text, flares provide another means of control. If the concentration of the organics in air equals or exceeds the lower flammability limit, a flare unit may be employed. Elevated, ground-level, and open-pit flares may be used. However, ground-level flares have been outlawed in most states; open-pit burning is used only under special conditions. Elevated flares are usually employed to ensure sufficient dilution and subsequent dispersion from adjacent structures and potential receptors of the energy and end products of combustion. Flares find their primary application in the petroleum and petrochemical industries. The suggested design procedures are available in the literature.

In operation, the gas containing the organics is continuously fed to and discharged from a stack, with the combustion occurring near the top of the stack and characterized by a flame at the end of the stack. The discharge temperature is in the 1500–3000°F range. The combustion flame is located at a point where the velocity of flow equals the velocity of flame propagation, provided the mixture is at or above the ignition temperature. Good mixing and a H/C (hydrogen/carbon) ratio in excess of 0.3 in the process gas stream can help insure proper combustion. A blue flame, appearing colorless against a blue-sky background, indicates good operation; a yellow-orange flame with a trail of black smoke indicates poor operation. The addition of steam (jets) at the top of the stack can help remove operational problems resulting from incomplete combustion. Steam rates in the 0.1–0.5 lb steam per pound of process gas range are often employed. Operating stack velocities are in excess of the flame propagation rate, and usually in excess of 200 ft/s. The design of this unit includes flame arresters in the stack to help remove some of the flashback and explosion possibilities. The diameter of the stack is obtained from flow (velocity) considerations and the height from atmospheric dispersion calculations.

In general, flare performance depends on such factors as flare gas exit velocity, emission stream heating value, residence time in the combustion zone, waste gas/oxygen mixing, and flame temperature. Steam-assisted smokeless flares are the most frequently used. In these units, process off-gases enter the flare through the collection header. When water or organic droplets are present, passing the off-gases through a knockout drum may be necessary since these droplets can create problems; water droplets can extinguish the flame and organic droplets can result in burning particles. Once the off-gases enter the flare stack, the aforementioned flame flashback can occur if the emission stream flow rate is too low. Flashback may be prevented, however, by passing the gas through a gas barrier, a water seal, or a stack seal. Purge gas is another option. At the flare tip, the emission stream is ignited by pilot burners. If conditions in the flame zone are optimum (oxygen availability, adequate residence time, etc.), the volatile organic compounds (VOCs) in the emission stream may be completely burned with nearly 100% efficiency. In some cases, it may be necessary to add supplementary fuel (natural gas) to the emission stream to achieve destruction efficiencies of 98% and greater if the net heating value of the emission stream is less than 300 Btu/scf.

Another important topic is combustion limits. Not all mixtures of fuel and air are able to support combustion. The flammable or explosive limits for a mixture are the maximum and minimum concentrations of fuel in air that will support combustion. The *upper explosive limit* (UEL) is defined as the concentration of fuel which produces a nonburning mixture due to a lack of oxygen. The aforementioned *lower explosive limit* (LEL) is defined as the concentration of fuel which combustion will not be self-sustaining. For example, consider a mixture of gasoline vapors and air that is at atmospheric conditions. The LEL is 1.4% by volume of gasoline vapors and the UEL is 7.6%. Any concentration of gasoline in air within these limits will support combustion, i.e., once a flame has been ignited, it will continue to burn. Concentrations of gasoline in air below or above these limits will not burn and can quench the flame.

The lower explosive limit is the more important of the two terms in describing gas streams with combustible contaminants. Industrial processes that handle combustible vapors, such as paint or solvent vapors, are usually required by insurance companies to operate at less than 25% of the LEL of the vapor in the ducts to minimize fire hazards. By using gas analyzers and an alarm system, the concentration of vapor may be allowed to be as high as 50% of the LEL by an insurance company covering the plant. Also note that flammability limits are strong functions of temperature.

4.2 DESIGN AND PERFORMANCE EQUATIONS

In describing any combustion process, there are numerous terms used to define heat. These terms can be split into two categories: combustion and thermodynamic. Combustion terms are included to aid in standardizing fuel usage calculations; these terms are applied to heat that is produced by combustion methods. Thermodynamic terms apply to all systems. Since the combustion terms are specific examples of the thermodynamic terms, some overlap is involved in defining them.

The following are important terms describing heat thermodynamically:

1. *Heat content or enthalpy* (H)—the sum total of latent and sensible heat present in a substance (gas, liquid or solid) minus that contained at an arbitrary set of conditions chosen as the base or zero point. Values for various gases are listed in Table 4.1.
2. *Sensible heat* (H_s)—heat that when added or removed causes a change in temperature.
3. *Latent heat* (H_l)—heat given off by a vapor condensing to a liquid or gained by a liquid evaporating to a vapor, without a change in temperature. The latent heat of vaporization of water 212°F is 970.3 Btu/lb.

Some useful terms describing the heat produced by the combustion of a fuel are:

1. *Gross heating value* (HV_G)—the total heat obtained from the complete combustion of a fuel at 60°F. Constant pressure, normally 1 atm, is maintained throughout the entire combustion process. Gross heating values are also referred to as total or higher heating values (HHV).
2. *Net heat value* (HV_N)—the gross heating value minus the latent heat of vaporization of the water formed by the combustion of the hydrogen in the fuel. For a fuel containing no hydrogen, the net and gross heating values are the same. Net heating values are also referred to as the lower heating values (LHV).

TABLE 4.1 Heat Contents (Enthalpy) of Various Gases

	Heat Content H; Btu/lb at 1 atm								
Temperature, °F	O_2	N_2	Air	CO	CO_2	SO_2	H_2	CH_4	H_2O
60	0	0	0	0	0	0	0	0	0
100	8.8	9.9	9.6	10.0	8.0	5.9	137	21.0	—
200	30.9	34.8	33.6	34.9	29.3	21.4	484	76.1	—
300	53.3	59.9	57.7	59.9	52.0	37.5	832	136.4	1165
400	76.2	85.0	81.8	85.0	75.3	54.4	1182	202.1	1212
500	99.4	110.3	106.0	110.6	99.8	71.8	1532	272.6	1259
600	123.1	136.1	130.2	136.3	125.1	89.8	1882	347.8	1307
700	147.2	161.7	154.5	162.4	149.6	108.2	2233	427.4	1355
800	171.7	187.7	178.9	188.7	177.8	127.0	2584	511.2	1404
900	196.6	213.9	203.4	215.6	205.6	146.1	2935	599.2	1454
1000	221.7	240.7	235.0	242.7	233.6	165.5	3291	691.1	1505
1200	272.5	294.7	288.5	297.8	290.9	205.1	4007	886.2	1609
1400	324.3	350.8	343.0	354.3	349.7	245.4	4729	1094.1	1717
1600	377.3	407.3	398.0	407.5	416.3	286.4	5460	1313.0	1829
1800	430.7	465.0	455.0	465.3	470.9	327.8	6198	1542.6	—
2000	484.0	523.8	513.0	523.8	532.8	369.1	6952	—	—
2200	539.3	583.2	570.7	583.3	596.1	411.1	7717	—	—
2400	594.4	642.3	628.5	643.0	659.2	452.7	8490	—	—
2600	649.0	702.8	687.3	703.2	723.2	495.2	9272	—	—
2800	702.8	763.1	746.6	771.3	787.4	557.5	10060	—	—
3000	758.6	824.1	806.3	832.6	852.0	580.0	10870	—	—
3200	816.4	885.8	866.0	894.0	916.7	622.5	11680	—	—
3400	873.4	947.6	925.9	956.0	981.6	665.0	12510	—	—
3600	931.0	1010.3	986.1	1018.3	1047.3	707.5	13330	—	—

Source: *North American Combustion Handbook*, North American Manufacturing Co., Cleveland, OH, 1952.

3. *Available heat* (HA)—the gross quantity of heat released within a combustion chamber minus (1) the sensible heat carried away by the dry flue gases and (2) the latent heat and sensible heat carried away in water vapor contained in the flue gases. The available heat represents the net quantity of heat remaining for useful heating. Figure 4.5 shows the available heat from the complete combustion (no excess air) of various fuels at various flue gas temperatures.

Depending on the user, the above terms can also have more than one definition. For example, a laboratory chemist may describe latent heat as the energy used in the chemical combustion of a fuel to carbon dioxide and water; while a boiler operator may describe latent heat as the difference between the gross and net heating values.

As discussed in Chapter 3, another important term used in performing combustion calculations is the specific heat, or more appropriately the heat capacity c_p, of a substance. One set of units for heat capacity is Btu/lb · °F in English units. Heat capacity also depends on temperature (see Chapter 3).

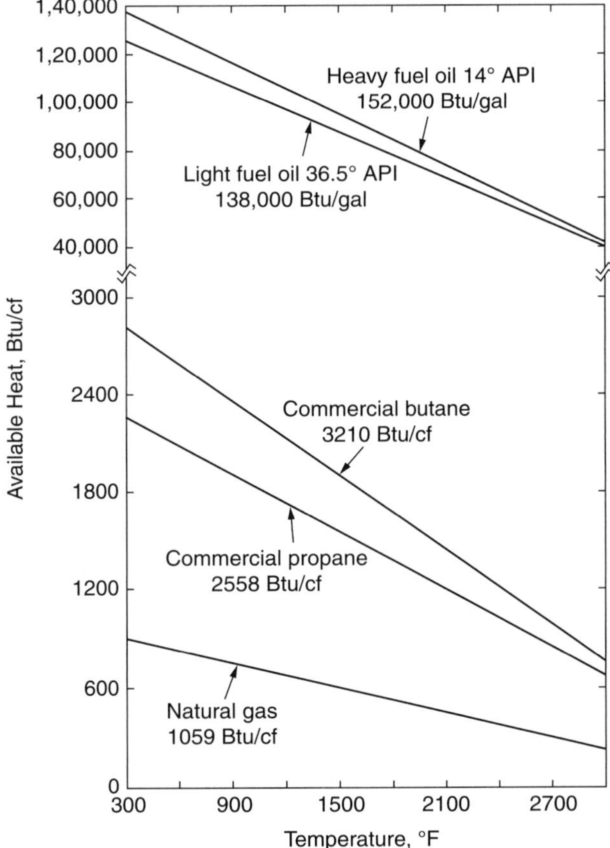

Figure 4.5 Available heat for some typical fuels (referred to 60°F). From *North American Combustion Handbook*, North American Manufacturing Co., Cleveland, OH, 1952.

An area of concern in incineration deals with calculations involving sensible heat and heat (energy) balances. Some approaches with key concepts and equations are presented below.

1. Compute the heat required to perform a heat balance around the combustion system. From the law of conservation of energy (for steady-state conditions):

$$\text{heat in} = \text{heat out} \tag{4.2}$$

As discussed earlier, heat is a relative term that is compared at a reference temperature. The heat content of a substance is arbitrarily taken as zero at a specified reference temperature. In the gas industry (natural gas), the reference temperature employed is normally 60°F.

2. The heat content can be computed from Equation (4.3) as well as using tables such as Table 4.1.

$$\Delta H = c_p(T - T_0) \tag{4.3}$$

where ΔH = enthalpy, Btu/lb
c_p = heat capacity, Btu/lb · °F (heat capacity over temperature range from T to T_0)
T = temperature of the substance, °F
T_0 = reference temperature, °F

3. Subtracting the heat going in from the heat leaving the system gives the heat that must be supplied by the fuel. This is referred to as a *change in enthalpy* or *heat content*. Using Equation (4.3), the enthalpy going in (T_1) is subtracted from the enthalpy leaving (T_2), giving.

$$\Delta H = \left[(c_{p2})(T_2 - T_0) - (c_{p1})(T_1 - T_0)\right] \tag{4.4}$$

where ΔH = change in enthalpy
c_{p2} = heat capacity over the temperature range T_2 to T_0
c_{p1} = heat capacity over the temperature range T_1 to T_0

4. To simplify this calculation, an average specific heat value ($\bar{c}_p$) can be used over the temperature range involved. This reduces Equation (4.4) to

$$\Delta H = \bar{c}_p(T_2 - T_1) \tag{4.5}$$

where $\bar{c}_p$ = heat capacity over the temperature range T_1 to T_2.

The heat capacity varies with temperature and composition of the gas stream. Therefore, Equation (4.5) is used to obtain an approximate value. For most incineration systems, the waste gases are considered to be essentially air. For air, the average heat capacity value $\bar{c}_p$ is 0.26 Btu/lb · °F for typical temperature ranges normally encountered in fume incinerators.

5. Equation (4.5) depicts the amount of heat required to raise a set quantity of gas from T_1 to T_2. Of more importance is the heat rate Q that is required. The heat rate is easily determined by multiplying either side of Equation (4.5) by the mass flow rate ($\dot{m}$) of the process gas stream. The heat rate required is given by

$$\dot{Q} = \dot{m}\Delta H = \dot{m}c_p(T_2 - T_1) \tag{4.6}$$

where $\dot{Q}$ = heat rate, Btu/hr
$\dot{m}$ = mass flow rate of gases, lb/hr

Equation (4.6) can therefore be used to compute the heat rate required to raise the gas temperature from T_1 to T_2. These equations are simple heat (enthalpy) balances. They do not account for any heat losses in the system. Heat losses from refractory or ducting are usually accounted for by assuming a fixed percent (or fraction) of the total theoretical heat is lost. For example, if an afterburner is required to supply heat at the rate of 1×10^6 Btu/hr and there is a 10% heat loss from the combustion chamber, the total heat rate would have to be 1.1×10^6 Btu/hr to account for the losses.

The following procedure for calculating fuel requirements and physical design of an incinerator are available in the literature and are presented below.

Note: This design procedure was originally developed by Dr. Louis Theodore in 1985 and later published by others. This was done without properly acknowledging the author, a violation of the copyright laws.

1. Employing Equations (4.2–4.6), calculate the heat load required to raise the process shown from its inlet temperature to the operating temperature of the combustion device:

$$Q = \Delta H \tag{4.7}$$

2. Correct the heat load term for any radiant losses (RL):

$$Q = (1 + \text{RL})(\Delta H) \tag{4.8}$$

where RL = fractional basis.

3. Assuming natural gas at a known heating value HV_G as fuel, calculate the available heat at the operating temperature:

$$\text{HA}_\text{T} = (\text{HV}_\text{G})\left(\frac{\text{HA}_\text{T}}{\text{HV}_\text{G}}\right)_\text{ref} \tag{4.9}$$

where the subscript "ref" refers to a reference fuel. The available heat for natural gas with a reference HV_G of 1059 Btu/scf, (assuming stoichiometric air) is given by (L. Theodore: personal notes, 1979)

$$(\text{HA}_\text{T})_\text{ref} = -0.237(T) + 981 \tag{4.10}$$

where T = incinerator operating temperature, °F.

4. Calculate the natural gas required (NG):

$$\mathrm{NG} = \frac{Q}{\mathrm{HA_T}}; \quad \text{consistent units} \tag{4.11}$$

5. Determine the volumetric flow rate of both the process gas stream q_p and the flue products of combustion of the natural gas q_c at the operating temperature

$$q_T = q_p + q_c \tag{4.12}$$

A good estimate for q_c is

$$q_c = (11.5)(\mathrm{NG}) \tag{4.13}$$

6. The cross-sectional area of the combustion device is given by

$$S = q_T/v \tag{4.14}$$

where v = average throughput velocity of the hot gases in the combustion device.

7. The diameter of the (tubular) combustion device is given by

$$D = (4S/3.14)^{0.5} \tag{4.15}$$

8. The length can be obtained from information on the residence time t_r or the length-to-diameter ratio L/D. Each is considered below:

a. Since $t_r = L/v$, then

$$L = (t_r)(v) \tag{4.16}$$

b. Typical L/D ratios are in the 2–3 range. For $L/D = 2.5$, then

$$L = (2.5)(D) \tag{4.17}$$

The reader can refer to the literature for additional details, including L. Theodore, "Ask the Experts: Designing Thermal Afterburners," *Chem. Eng. Progress*, pp. 18–19, April 2005.

4.3 OPERATION AND MAINTENANCE, AND IMPROVING PERFORMANCE

Normal operation of an incinerator should be quite simple. A controller should be incorporated into the design to maintain the outlet temperature at a fixed value by varying the auxiliary fuel input. Combustion air (assuming limited air in the fume stream) is usually controlled by the average amount of fumes to be incinerated. This adjustment is normally set manually if the fume flow rate and fume heat content are fairly constant. If the fume heat content or flow rate are likely to be highly variable, a more sophisticated control system may be appropriate—one that analyzes combustion efficiency at the outlet as well as fume flow and outlet temperature, and varies both auxiliary fuel input and combustion air accordingly. But, normal operation of even a complicated control system usually requires nothing more than an occasional monitoring of the instruments.

Most incinerators are custom-designed within certain basic parameters. Therefore, they are likely to be accompanied by a very complete instruction manual that should include the manufacturer's basic maintenance instructions from all the subsuppliers, i.e., the incinerator manufacturer will have purchased equipment, such as the refractory, valves, and controls, from other suppliers. The operating and maintenance instructions for this equipment will be quite extensive and complete because it has been written by the original manufacturer, who is concerned only with particular items. The instruction manual is therefore a very useful document from a maintenance viewpoint and should be followed explicitly. (The instruction manual may not be so useful from a system operations viewpoint because fume incinerator systems are usually custom-designed and system problems cannot always be anticipated, which often results in some field modifications to the operating instructions.)

There are some general maintenance guidelines that can be discussed. The refractory should be inspected on a regular basis. Cracks may develop, especially in brick joints, and thermal shock damage (spalling) should be repaired with a suitable plastic of the same thermal properties. The burner should be inspected at regular intervals for signs of warpage or corrosion. Moving parts should be lubricated with graphite or a similar high-temperature lubricant. Lubricants that carbonize should not be used under any circumstances. Also dirt, mortar, carbon, or other foreign matter should be cleaned from the burner area. The pressure seals around any parts projecting through the burner or incinerator shell should be inspected. Usually, these are asbestos rope packing glands and should be fairly tight after the adjusting/retaining screws have been loosened. These seals should only be lubricated with flake or powder graphite. This should never be mixed with oil, as the oil will carbonize. The gas jets should be free of corrosion and should be cleared of any deposits on burners using gas as an auxiliary fuel.

The outer shell of the incinerator should be inspected, especially when a new lining has been installed, for signs of thermal shock, i.e., welds, especially at the outlet, should be checked for hairline cracks, which are the first signs of poor thermal design.

The auxiliary fuel piping train should be inspected in accordance with the manufacturer's instructions. Electrically operated valves and interlock switches should also be inspected frequently for conditions that might cause "shorting" (short-circuiting), e.g., dirty contacts, moisture leaks, deteriorating insulation, etc. Air supply lines and filters (to air-operated valves) should be inspected for dirt or blockages. The valves themselves are usually provided with air signal and air supply pressure gauges, and these should be checked occasionally for accuracy. If there are shutoff dampers in the ductwork to or from the incinerator, their seals should be checked frequently.

Maintenance procedures for catalytic incinerators should include catalyst cleaning every 3 months to a year. Cleaning is usually accomplished by blowing clean compressed air through the catalyst element, by vacuuming, or by washing with water or a mild detergent not containing phosphates. Iron oxide deposits can be removed by soaking with a mild organic acid followed by a water rinse.

Improving the performance of a thermal incinerator basically involves the optimization of fume combustion. Ideally, no more combustion air should be used than is required for complete combustion of the fumes and the auxiliary fuel. The auxiliary fuel should be used only in amounts required to maintain the design furnace temperature.

An incinerator operating efficiently should normally have only 1–2% O_2 and 0–1% combustibles in the outlet gases. Monitors are available that can indicate these parameters to the operator as well as provide automatic control of the incinerator when required.

The following questions often require answers. If odors are involved, has their concentration been reduced or eliminated? Has there been adequate reduction in the emissions of all types of reactive hydrocarbons? Has there been a reduction in any possible explosion or public health hazard that might exist in the manufacturing process? If so, the air pollution control system has been well designed and is being operated and maintained properly.

Recent developments with combustion equipment has centered primarily on the development of improved catalysts. Activity in the nanotechnology area has produced catalysts that have outperformed catalysts employed in the past. See text by Kunz and Theodore (R. Kunz and L. Theodore, *Nanotechnology: Environmental Implications and Solutions*, John Wiley & Sons, Hoboken, NJ, 2006) for additional details.

PROBLEMS

4.1 The Three Ts

The three Ts of combustion are commonly understood to mean (select one)

(a) Time, temperature, and type of fuel
(b) Temperature, type of fuel, and turbulence
(c) Time, temperature, and turbulence
(d) Type of fuel, turbulence, and time

Solution: As noted in Section 4.1, the "three Ts" refer to time, temperature, and turbulence. The correct answer is therefore (c).

4.2 Typical Direct Flame Incinerator Operating Range

What is a common temperature operating range for a direct flame incinerator?

(a) 400–500°F
(b) 700–900°F
(c) 1300–1500°F
(d) 2000–2400°F

Solution: Although the operating temperatures vary from application to application and from process to process, a typical value would fall within the 1300–1500°F range. The correct answer is therefore (c).

4.3 Typical Catalytic Afterburner Operating Temperature

A rule of thumb for typical operating temperature range of a catalytic afterburner

(a) 200–400°F
(b) 700–900°F

(**c**) 1200–1400°F
(**d**) 1800–2000°F

Solution: Although the operating temperatures vary with each application and each process, a typical value would fall within the 700–900°F range. The correct answer is therefore (b).

4.4 Approximate Direct Flame Incinerator Residence Time

What is an approximate residence time for gaseous material being combusted in a typical direct flame incinerator?

(**a**) 0.02 s
(**b**) 10 s
(**c**) 0.3 s
(**d**) 0.4 min

Solution: As with the answer to Problem 4.3, solutions will vary, but an approximate typical residence time is 0.3 s. The correct answer is therefore (c).

4.5 Operating Temperature Comparison

How do operating temperature ranges compare for catalytic and direct flame incinerators?

(**a**) Catalytic incinerators operate at higher combustion temperatures
(**b**) Direct flame incinerators operate at lower temperatures
(**c**) Temperature requirements are identical
(**d**) Catalytic incinerators operate at lower combustion temperatures

Solution: Catalysts increase reaction rates. Since reaction rates increase with temperature, catalytic incinerators can operate at lower temperatures with similar pollutant destruction performance. The correct answer is therefore (d).

4.6 Effect of Fine Particulates on Catalyst Life

In a catalytic combustion system, fine particulate matter in the process gas stream will cause the lifetime of the catalyst to

(**a**) Increase
(**b**) Decrease
(**c**) Remain unaffected
(**d**) None of the above

Solution: Fine particulates can become embedded in the pores of the catalyst, reducing the active area. This then reduces the performance (and lifetime) of the catalyst. The correct answer is therefore (b).

4.7 Lower Explosive Limit (LEL)

An exhaust stream with a combustible concentration lower than the LEL will

(**a**) Have an insufficient amount of oxygen to autoignite
(**b**) Be a fire hazard
(**c**) Have an insufficient amount of combustibles to autoignite
(**d**) Meet the emission limitation for lead

Solution: For a gas stream to be combustible, the concentration of the organic must be above the LEL or below the UEL. Since the concentration is below the LEL, the gas will not autoignite. The correct answer is therefore (c).

4.8 Catalytic Incineration Limitation

Catalytic incineration is unsuitable in some instances due to

(**a**) Low combustible concentrations in the gas
(**b**) Compounds such as lead and mercury, which will poison the catalyst
(**c**) Reaction of the catalyst with select inert gases
(**d**) All of the above

Solution: As discussed in Section 4.1, answers (a)–(c) limit catalytic incineration use. The correct answer is therefore (d).

4.9 Flare Application

The flare type of combustion control device is most applicable for gases with

(**a**) Periodic releases of low concentration
(**b**) Steady releases of high concentration
(**c**) Periodic releases of high concentration
(**d**) Steady releases of low concentration

Solution: Periodic releases are the norm, and the concentration of the organic pollutant must be high enough to sustain combustion. The correct answer is therefore (c).

4.10 Process Boiler Application as an Incinerator

Which of the following characteristics would be least important in using a process boiler as an incinerator?

(**a**) The boiler insulation
(**b**) Concentration of combustibles in the fume
(**c**) Oxygen concentration in the contaminated process fume
(**d**) Operating schedule of the boiler

Solution: All four answers could directly or indirectly affect incinerator performance. However, the boiler insulation would have the least effect. The correct answer is therefore (a).

4.11 Heat Recovery Methods

Which one of the following is *not* a method of heat recovery used in incineration devices for gaseous pollutants?

(**a**) Recuperative
(**b**) Distributive
(**c**) Secondary
(**d**) Regenerative

Solution: *Recuperative*, *secondary*, and *regenerative* are terms used to describe methods of heat recovery in numerous applications. The correct answer is therefore (b).

4.12 Definition of Available Heat

The available heat (HA_T) is equal to the gross heat (HV_G) minus

(**a**) Net heating value (HV_N)
(**b**) Sensible heat lost in the exit flue gas
(**c**) Latent heat of vaporization
(**d**) Heat of combustion

Solution: This is a tricky question, but the general definition of available heat (see Section 4.2) is

$$HA_T = HV - \Sigma\ \text{(sensible heat, products)} \qquad (4.18)$$

The correct answer is therefore (b).

4.13 Afterburner Volume

Calculate the volume, in cubic feet, of an afterburner whose length and diameter are 12 feet and 6 feet, respectively.

(**a**) 424
(**b**) 282
(**c**) 386
(**d**) 339

Solution: The volume is given by the cross-sectional area times the length:

$$\begin{aligned} V &= (A)(L) \\ &= \frac{\pi D^2}{4}(L) \end{aligned}$$

Substituting, one obtains

$$\begin{aligned} V &= \frac{\pi(6^2)}{4}(12) \\ &= 339.3\ \text{ft}^3 \end{aligned}$$

The correct answer is therefore (d).

4.14 Average Residence Time Calculation

Calculate the average gas residence time in a 200 ft^3 afterburner, given a gas flow rate of 24,000 acfm.

(**a**) 0.56 s
(**b**) 0.50 s
(**c**) 0.44 s
(**d**) 0.63 s

Solution: The average residence time is given by

$$\begin{aligned} t &= V/q;\ \text{consistent units} \qquad (4.1) \\ &= 200/24{,}000 \\ &= 0.00833\ \text{min} \end{aligned}$$

Converting to seconds

$$t = (0.00833)(60)$$
$$= 0.50\,\mathrm{s}$$

The correct answer is therefore (b).

4.15 Reactants/Products Ratio

The reaction equation (not balanced) for the combustion of butane is shown below:

$$C_4H_{10} + O_2 \rightarrow CO_2 + H_2O$$

Determine the mole ratio of reactants to products.

Solution: A chemical equation provides a variety of qualitative and quantitative information essential for the calculation of the quantity of reactants reacted and products formed in a chemical process. The balanced chemical equation must have the same number of atoms of each type in the reactants and products. Thus, the balanced equation for butane is

$$C_4H_{10} + \left(\frac{13}{2}\right)O_2 \rightarrow 4CO_2 + 5H_2O$$

Note that the

Number of carbons in reactants = number of carbons in products = 4
Number of oxygens in reactants = number of oxygens in products = 13
Number of hydrogens in reactants = number of hydrogens in products = 10
The mole ratio is:
Number of moles of reactants is 1 mol C_4H_{10} + 6.5 mol O_2 = 7.5 mol total
Number of moles of products is 4 mol CO_2 + 5 mol H_2O = 9 mol total
The ratio is therefore $R = 7.5/9 = 8.3$

The reader should note that although the number of moles on both sides of the equation do not balance, the masses of reactants and products (in line with the conservation law for mass) must balance.

4.16 Stoichiometric Combustion

Consider the following reaction equation:

$$C_3H_8 + 5O_2 \rightarrow 3CO_2 + 4H_2O$$

Determine the scf (standard cubic feet) of air required for stoichiometric combustion of 1.0 scf propane (C_3H_8).

Solution: Stoichiomertic air is the air required to assure complete combustion of a fuel, organic, and/or waste. For complete combustion one can assume:

1. No fuel, organic, and/or waste remains
2. No oxygen is present in the flue gas

3. Carbon has combusted to CO_2, not CO
4. Sulfur has combusted to SO_2, not SO_3

Excess air (fractional basis) is defined by

$$EA = \frac{\text{(air entering)} - \text{(stoichiometric air)}}{\text{stoichiometric air}} \tag{4.19}$$

Noting that, for an ideal gas, the number of moles is proportional to the volume. The scf of O_2 required for the complete combustion of 1 scf of propane is 5 scf. The nitrogen-to-oxygen volume (or mole) ratio in air is $\frac{79}{21}$. Therefore, the amount of N_2 in a quantity of air that contains 5.0 scf of O_2 is

$$\begin{aligned}\text{scf of } N_2 &= \left(\frac{79}{21}\right)(5) \\ &= 18.81 \text{ scf}\end{aligned}$$

The stoichiometric amount of air is then

$$\begin{aligned}\text{scf of air} &= \text{scf of } N_2 + \text{scf of } O_2 \\ &= 18.81 + 5.0 \\ &= 23.81 \text{ scf}\end{aligned}$$

Therefore amount of flue gas produced is (noting that the oxygen has reacted)

$$\begin{aligned}\text{scf of flue gas} &= \text{scf of } N_2 + \text{scf of } CO_2 + \text{scf of } H_2O \\ &= 18.81 + 3.0 + 4.0 \\ &= 25.81 \text{ scf}\end{aligned}$$

4.17 Excess Air Combustion

Benzene is incinerated at 2100°F in the presence of 50% excess air (EA). Balance the combustion reaction for this process:

$$C_6H_6 + O_2 \rightarrow CO_2 + H_2O + O_2$$

Solution: Ignoring nitrogen, the balanced equation is

$$C_6H_6 + \frac{15}{2}(EA + 1)O_2 \rightarrow 6CO_2 + 3H_2O + \frac{15}{2}(EA)O_2$$

For 50% excess air,

$$EA = 50\% = 0.5$$

so that

$$O_2 \text{ (start)} = \left(\frac{15}{2}\right)(1 + 0.5) = 11.25$$

and

$$O_2\ (\text{end}) = \left(\frac{15}{2}\right)(0.5) = 3.75$$

The balanced equation now becomes

$$C_6H_6 + 11.25O_2 \rightarrow 6CO_2 + 3H_2O + 3.75O_2$$

To include N_2,

$$N_2 = 11.25\left(\frac{79}{21}\right) = 42.32$$

The balanced reaction equation, including the inert nitrogen, is now

$$C_6H_6 + 11.25O_2 + 42.32\,N_2 \rightarrow 6CO_2 + 3H_2O + 3.75O_2 + 42.32\,N_2$$

4.18 Butanol Combustion

The offensive odor of butanol can be removed from stack gases by its complete combustion to carbon dioxide and water. It is of interest that the incomplete combustion of butanol actually results in a more serious odor pollution problem than the original one. Write the equations showing the two intermediate malodorous products formed if butanol undergoes incomplete combustion.

Solution: The malodorous products are butyraldehde (C_4H_8O) and butyric acid (C_3H_7COOH), which can be formed sequentically as follows:

$$C_4H_9OH + \frac{1}{2}O_2 \rightarrow C_4H_8O + H_2O$$

$$C_4H_8O + \frac{1}{2}O_2 \rightarrow C_3H_7COOH$$

Or, the acid can be formed directly as follows:

$$C_4H_9OH + O_2 \rightarrow C_3H_7COOH + H_2O$$

For complete combustion:

$$C_4H_9OH + 6O_2 \rightarrow 4CO_2 + 5H_2O$$

or

$$C_3H_7COOH + 5O_2 \rightarrow 4CO_2 + 4H_2O$$

4.19 Butyl Alcohol Combustion

Calculate the volume of air required to combust one pound of butyl alcohol at 60°F and 1 atm to

(**a**) Butylaldehyde
(**b**) Butyric acid
(**c**) Carbon dioxide and water

Solution: The chemical formula for butyl alcohol is C_4H_9OH; it has a molecular weight of 74. The number of moles of one pound of alcohol is

$$n = \frac{1}{74} = 0.0135 \text{ lbmol}$$

At 25°C, 1 atm, one pound of C_4H_9OH occupies

$$V = (0.0135)(379)$$
$$= 5.12 \text{ ft}^3$$

If the alcohol is converted to aldehyde and hydrogen, then

$$C_4H_9OH + (0.5)O_2 \rightarrow C_4H_8O + H_2$$

No air is therefore required. However, if the alcohol is converted to the aldehyde and water, then

$$C_4H_9OH + \frac{1}{2}O_2 \rightarrow C_4H_8O + H_2O$$

The amount of air required is therefore

$$\text{Air} = (0.5)(5.12)\left(\frac{100}{21}\right)$$
$$= 12.2 \text{ ft}^3$$

If the alcohol is converted to the acid, then

$$C_4H_9OH + O_2 \rightarrow C_3H_7COOH + H_2O$$

The amount of air required is therefore

$$\text{Air} = (1)\left(\frac{100}{21}\right)(5.12)$$
$$= 24.4 \text{ ft}^3$$

If the alcohol converted to CO_2 and H_2O, then

$$C_4H_9OH + 6O_2 \rightarrow 4CO_2 + 5H_2O$$

The amount of air is therefore

$$\text{Air} = 6\left(\frac{100}{21}\right)(4.85)$$
$$= 138.6 \text{ ft}^3$$

4.20 Ethanol Combustion

When ethanol (C_2H_5OH) is completely combusted, the products are carbon dioxide and water.

(a) Write the balanced reaction equation.
(b) If 150 lbmol/hr of water is produced, at what rate (molar) is the ethanol combusted?
(c) If 2000 kg of the ethanol is combusted, what mass of oxygen is required?

Solution: (a) The balanced reaction equation is

$$C_2H_5OH + 3O_2 \rightarrow 2CO_2 + 3H_2O$$

(b) The stoichiometric ratio of the C_2H_5OH consumed to the water produced may now be calculated.

$$\text{Ratio} = \frac{1\ \text{lbmol}\ C_2H_5OH}{3\ \text{lbmol}\ H_2O\ \text{produced}}$$

This result may be used to calculate the amount of C_2H_5OH.

$$\begin{aligned}\dot{n} &= (150\ \text{lbmol/hr}\ H_2O\ \text{produced})\left(\frac{1\ \text{lbmol}\ C_2H_5OH}{3\ \text{lbmol}\ H_2O\ \text{produced}}\right)\\ &= 50\ \text{lbmol/hr}\ C_2H_5OH\ \text{reacted}\end{aligned}$$

(c) The molar amount (in kgmol) of C_2H_5OH reacted is

$$n = \frac{2000\ \text{kg}\ C_2H_5OH}{46\ \text{kg/kgmol}\ C_2H_5OH} = 43.48\ \text{kgmol of}\ C_2H_5OH$$

Using the stoichiometric ratio of oxygen to ethanol reacted, the number of kgmol of oxygen needed is then

$$\begin{aligned}n_{O_2} &= (43.48\ \text{kgmol}\ C_2H_5OH)\left(\frac{3\ \text{kgmol}\ O_2}{1\ \text{kgmol}\ C_2H_5OH}\right)\\ &= 130.4\ \text{kgmol}\ O_2\ \text{reacted}\end{aligned}$$

The required mass of oxygen may now be calculated:

$$\begin{aligned}m_{O_2} &= (130.4\ \text{kgmol}\ O_2)(32.0\ \text{kg/kgmol})\\ &= 4174\ \text{kg}\ O_2\ \text{required}\end{aligned}$$

4.21 Limiting Reactant I

Complete combustion of carbon disulfide results in combustion products of CO_2 and SO_2 according to the reaction

$$CS_2 + O_2 \rightarrow CO_2 + SO_2$$

(**a**) Balance this reaction equation.
(**b**) If 500 lb of CS_2 is combusted with 225 lb of oxygen, which is the limiting reactant?

Data: MW of $CS_2 = 76.14$; MW of $SO_2 = 64.07$; MW of $CO_2 = 44$.

Solution:

(**a**) The balanced equation is

$$CS_2 + 3O_2 \rightarrow CO_2 + 2SO_2$$

(**b**) The initial molar amounts of each reactant is

$$\begin{aligned} n_{CS_2} &= (500\text{ lb } CS_2)\,(1\text{ lbmol } CS_2/76.14\text{ lb } CS_2) \\ &= 6.57\text{ lbmol } CS_2 \\ n_{O_2} &= (225\text{ lb } O_2)\,(1\text{ lbmol } O_2/32\text{ lb } O_2) \\ &= 7.03\text{ lbmol } O_2 \end{aligned}$$

The amount of O_2 needed to consume *all* the CS_2, i.e., the stoichiometric amount, is then

$$\begin{aligned} n_{O_2}(\text{total}) &= (6.57\text{ lbmol})(3\text{ lbmol}/1\text{ lbmol}) \\ &= 19.71\text{ lbmol } O_2 \end{aligned}$$

Therefore, O_2 is the limiting reactant since 19.7 mol of O_2 are required for complete combustion but only 7.03 mol of O_2 are available.

4.22 Limiting Reactant II

Refer to Problem 4.21. How much of each product is formed (lb)?

Solution: The limiting reactant is used to calculate the amount of products formed:

$$\begin{aligned} &(7.03\text{ lbmol } O_2)\,(1\text{ lbmol } CS_2/3\text{ lbmol } O_2) = 2.34\text{ lbmol } CS_2 \\ &(2.34\text{ lbmol } CS_2)\,(76.14\text{ lb}/1\text{ lbmol } CS_2) = 178\text{ lb } CS_2 \text{ unreacted} \\ &(7.03\text{ lbmol } O_2)\,(1\text{ lbmol } CO_2/3\text{ lbmol } O_2) = 2.34\text{ lbmol } CO_2 \\ &(2.34\text{ lbmol } CO_2)\,(44\text{ lb } CO_2/1\text{ lbmol } CO_2) = 103\text{ lb } CO_2 \text{ produced} \\ &(7.03\text{ lbmol } O_2)\,(2\text{ lbmol } SO_2/3\text{ lbmol } O_2) = 4.68\text{ lbmol } SO_2 \\ &(4.68\text{ lbmol } SO_2)\,(64.07\text{ lb}/1\text{ lbmol } SO_2) = 300\text{ lb } SO_2 \text{ produced} \end{aligned}$$

4.23 Products of Complete Combustion

Determine the products of *complete* combustion when burning a gaseous fuel of the following composition with 50% excess air: $N_2 = 0.15$, $CH_4 = 0.81$, $C_3H_8 = 0.04$. Perform the calculations assuming a volumetric flow rate of the fuel of 1.0 scfm.

Solution: Write the balanced chemical equations for complete combustion:

$$N_2 \rightarrow N_2$$
$$CH_4 + 2O_2 \rightarrow CO_2 + 2H_2O$$
$$C_3H_8 + 5O_2 \rightarrow 3CO_2 + 4H_4O$$

Determine the scfm of O_2 required from the balanced chemical equations given above:

$$\begin{aligned} CH_4\text{:} &\quad (2)\,(0.81)O_2 \\ C_3H_8\text{:} &\quad (5)\,(0.04)O_2 \\ O_2 \text{ required} &= 1.62 + 0.2 \\ &= 1.82\text{ scfm } O_2 \end{aligned}$$

Since the composition of air is 21% O_2 and 79% N_2, the theoretical air required is:

$$q_{air} = (1.82)/(0.21) = 8.67\text{ scfm}$$

For 50% excess air, one obtains

$$\begin{aligned} q_{air}\ (50\%) &= (8.67\text{ scfm air})\ (1.5) \\ &= 13.0\text{ scfm} \end{aligned}$$

The total scfm of N_2 in the flue gas is

$$\begin{aligned} q_{N_2} &= (13.0)\ (0.79) \\ &= 10.27\text{ scfm} \end{aligned}$$

Calculate the total scfm of O_2 in the flue gas:

$$\begin{aligned} O_{2,\text{flue}} &= (8.67)\ (0.5)\ (0.21) \\ &= 0.91\text{ scfm excess } O_2 \end{aligned}$$

Also calculate scfm of CO_2 and H_2O in the flue gas:

$$CH_4\text{:} \quad (0.81)(1)CO_2 + (2)(0.81)H_2O$$
$$C_3H_8\text{:} \quad (0.04)(3)CO_2 + (0.04)(4)H_2O$$

Thus,

$$CO_2 = 0.81 + 0.12 = 0.93\text{ scfm}$$
$$H_2O = 1.62 + 0.16 = 1.78\text{ scfm}$$

The total scfm of the flue gas is therefore

$$\begin{aligned} \text{Total scfm flue gas} &= 10.27\,N_2 + 0.91\text{ scfm } O_2 + 0.93\,CO_2 + 1.78\,H_2O \\ &= 13.89\text{ scfm} \end{aligned}$$

4.24 Evaluating Stoichiometric Coefficients

Balance the following:

$$C_xH_yO_zCl_wS_u + \beta O_2 \rightarrow \gamma\, CO_2 + \delta H_2O + \varepsilon HCl + \phi SO_2$$

Find all Greek stoichiometric coefficients, i.e., express them in terms of the English coefficients x, y, z, w, and u.

Solution: Proceed sequentially for each element:

$$\gamma = x$$
$$\varepsilon = w$$
$$\phi = u$$
$$\delta = \frac{y - \varepsilon}{2} = \frac{y - w}{2}$$

Perform an oxygen balance to obtain β:

$$\beta = x + \left(\frac{y - w}{4}\right) + u - \frac{z}{2}$$

Note also that

$$\delta = \frac{y}{z}$$
$$2\delta = \varepsilon \geq 0$$

4.25 Incinerator Application

C_6H_5Cl is fed into a hazardous waste incinerator at a rate of 5000 scfm (60°F, 1 atm) and is combusted in the presence of air fed at a rate of 3000 scfm (60°F, 1 atm). Both streams enter the incinerator at 70°F. Following combustion, the products are cooled from 2000°F and exit a cooler at 180°F. At what rate (lb/hr) do the products exit the cooler? The molecular weight of C_6H_5Cl is 112.5; the molecular weight of air is 29.

Solution: Convert scfm to acfm using Charles' Law for both chlorobenzene (q_1) and air (q_2):

$$q_1 = (5000\,\text{scfm})(460 + 70)/(460 + 60)$$
$$= 5096\,\text{acfm of } C_6H_5Cl$$

$$q_2 = (3000\,\text{scfm})(460 + 70)/(460 + 60)$$
$$= 3058\,\text{acfm of air}$$

Since 1 lbmol of any ideal gas occupies 379 ft^3 at 60°F and 1 atm, the molar flow rate $\dot{n}$ may be calculated by dividing the volumetric flow rates at 60°F by 379:

$$\begin{aligned}\dot{n}_1 &= 5000/379 \\ &= 13.2\ \text{lbmol/min} \\ &= 792\ \text{lbmol/hr} \\ \dot{n}_2 &= 3000/379 \\ &= 7.92\ \text{lbmol/min} \\ &= 475\ \text{lbmol/hr}\end{aligned}$$

The mass flow rate is found by multiplying the above by the molecular weight:

$$\begin{aligned}\dot{m}_1 &= (792)(112.5) \\ &= 89{,}100\ \text{lb/hr} \\ \dot{m}_2 &= (475)(29) \\ &= 13{,}800\ \text{lb/hr}\end{aligned}$$

The mass rate exiting the cooler is then

$$\begin{aligned}\dot{m}_{\text{out}} &= \dot{m}_{\text{in}} \\ &= 89{,}100 + 13{,}800 \\ &= 102{,}900\ \text{lb/hr}\end{aligned}$$

4.26 Combustion with Excess Air

A 10,000 lbmol/hr gaseous mixture of benzene and toluene (70% benzene by mole) is being burned in a thermal afterburner using 50% excess air. Assuming complete combustion, determine the amount of air required (in lbmol/hr) and the flow rate of the product gas in scfm (1 atm, at 60°F, 1 atm).

Solution: The stoichiometric combustion equations for both benzene and toluene are

$$C_6H_6 + 7.5O_2 \rightarrow 6CO_2 + 3H_2O$$
$$C_7H_8 + 9O_2 \rightarrow 7CO_2 + 4H_2O$$

Assuming a basis of 10,000 lbmol gas mixture, the molar flow rates (input) in lbmol/hr are

C_6H_6:	(0.7)(10,000) = 7000
C_7H_8:	(0.3)(10,000) = 3000
Stoichiometric O_2:	(7.5)(7000) + (9)(3000) = 79,500
Excess O_2:	(0.5)(79,500) = 39,750
Total O_2:	= 119,250
N_2:	= (79/21)(119,250) = 448,600
Air:	567,850

Molar flow rates (output), lbmol/hr are:

$$\begin{aligned} CO_2&: = (6)(7000) + (7)(3000) = 63{,}000 \\ H_2O&: = (3)(7000) + (4)(3000) = 33{,}000 \\ O_2&: = 39{,}750 \\ N_2&: = 448{,}600 \\ \text{Total}&: = 584{,}350 \end{aligned}$$

Conversion to scfm yields

$$\begin{aligned} (584{,}350)(379.0)/60 &= 3{,}691{,}200 \\ &= 3.69 \times 10^6 \text{ scfm} \end{aligned}$$

4.27 Flue Gas Composition

Refer to Problem 4.26. Determine the composition (by mole fraction) of the product gas.

Solution: The molar flow rates (output) were calculated previously:

$$\begin{aligned} CO_2 &= 63{,}000 \\ H_2O &= 33{,}000 \\ O_2 &= 39{,}750 \\ N_2 &= \underline{448{,}600} \\ \text{Total} &= 584{,}350 \end{aligned}$$

The corresponding mole fractions are

$$\begin{aligned} CO_2 &= 63{,}000/584{,}350 &= 0.108 \\ H_2O &= 33{,}000/584{,}350 &= 0.056 \\ O_2 &= 39{,}750/584{,}350 &= 0.068 \\ N_2 &= 448{,}600/584{,}350 &= \underline{0.768} \\ & & 1.000 \end{aligned}$$

4.28 Extent of Reaction Balances

In one step of the contact process for the production of sulfuric acid, sulfur dioxide is oxidized to sulfer trioxide at high pressures in the presence of a catalyst:

$$2SO_2 + O_2 \rightarrow 2SO_3$$

A mixture of 300 lbmol/min of sulfur dioxide and 400 lbmol/min of oxygen is fed to a reactor. The flow rate of the unreacted oxygen leaving the reactor is 300 lbmol/min. Determine the composition (in mol%) of the exiting gas stream by three methods:

(**a**) Molecular balances
(**b**) Atomic balances
(**c**) "Extent of reaction" balances

Solution: As noted earlier, three different types of material balance may be written when a chemical reaction is involved: the molecular balance, the atomic balance, and the "extent of reaction" balance. It is a matter of convenience which of the three types is used. Assuming a steady-state continuous reaction, the accumulation term A is zero and, for all components involved in the reaction, the molecular balance equation becomes

$$I + G = O + C \tag{4.20}$$

where I = amount or rate of input
G = amount or rate of material generated
O = amount or rate of output
C = amount or rate of material consumed

If a total material balance is performed, the appropriate form of the balanced equation must be used if the amounts of the flow rates are expressed in terms of moles, e.g., lbmol or gmol/hr since the total number of moles can change during a chemical reaction. If, however, the amounts or flow rates are given in terms of mass, e.g., kg or lb/hr, the G and C terms may be dropped, since mass cannot be lost or gained in a chemical reaction. Therefore

$$I = O \tag{4.21}$$

In general, however, when a chemical reaction is involved, it is usually more convenient to express amounts and flow rates using moles rather than mass. A material balance that is based not on the compounds (or molecules), but rather on the atoms that make up the molecules, is referred to as an *atomic balance*. Since atoms are neither created nor destroyed in a chemical reaction, the G and C terms equal zero and the balance again becomes

$$I = O$$

As an example, take once again the combination of hydrogen and oxygen to form water:

$$2H_2 + O_2 \rightarrow 2H_2O$$

As the reaction progresses, the H_2 and O_2 molecules (or moles) are consumed while H_2O molecules (or moles) are generated. On the other hand, the number of oxygen atoms(or moles of oxygen atoms) and the number of hydrogen atoms (or moles of hydrogen atoms) do not change. Care must also be taken to distinguish between molecular oxygen and atomic oxygen. If, in the preceding reaction, one starts out with 1000 lbmol of O_2 (oxygen molecules), one also starts out with 2000 lbmol of O (oxygen atoms). The extent of reaction balance gets its name from the fact that the amounts of the chemicals involved in the reaction are described in terms of how much of a particular reactant has been consumed or how much of a particular product has been generated. As an example, consider the formation of ammonia from hydrogen and nitrogen:

$$N_2 + 3H_2 \rightarrow 2NH_3$$

If one starts off with 10 mol of hydrogen and 10 mol of nitrogen, and lets z represent the amount of nitrogen consumed by the end of the reaction, the output amounts of all three components are $10 - z$ for the nitrogen, $10 - 3z$ for the hydrogen, and $2z$ for the ammonia. The following is a convenient way of representing the extent of reaction balance:

$$\begin{array}{llclll} \text{Input} \rightarrow & 10 & & 10 & 0 & \Sigma = 20 \\ & N_2 & + & 3H_2 \rightarrow & 2NH_3 & \\ \text{Output} \rightarrow & 10 - z & & 10 - z & 2z & \Sigma = 20 - 2z \end{array}$$

In this problem, the reader is asked to employ all three types of material balance for the SO_2 reaction.

A flow diagram of the process is given in Figure 4.6.

(a) Using the molecular balance approach, one obtains

$$\begin{aligned} O_2: \quad & I = O + C;\ O = 300 \\ & 400 = 300 + C \\ & C = 100\ \text{lbmol/min}\ O_2\ \text{consumed} \\ SO_2: \quad & I = O + C \\ & 300 = O + (2)(100) \\ & O = 100\ \text{lbmol/min}\ SO_2\ \text{out} \\ SO_3: \quad & G = O \\ & \left(\frac{2}{1}\right)(100) = O \\ & O = 200\ \text{lbmol/min}\ SO_3\ \text{out} \end{aligned}$$

Since the total flow rate of exiting gas is 600 lbmol/min, the mole % are:

$$\begin{aligned} O_2: \quad & (300/600)(100\%) = 50.0\% \\ SO_2: \quad & (100/600)(100\%) = 16.7\% \\ SO_3: \quad & (200/600)(100\%) = 33.3\% \end{aligned}$$

Figure 4.6 Flow diagram for Problem 4.28.

(b) Using the atomic balance approach, one obtains

$$\begin{aligned}
\text{S:}\quad & I = O \\
& \left(\tfrac{1}{1}\right)(300) = \left(\tfrac{1}{1}\right)\dot{n}_{SO_2} + \left(\tfrac{1}{1}\right)\dot{n}_{SO_3} \\
& \dot{n}_{SO_2} + \dot{n}_{SO_3} = 300 \\
\text{O:}\quad & I = O \\
& \left(\tfrac{2}{1}\right)(300) + \left(\tfrac{2}{1}\right)(400) = \left(\tfrac{2}{1}\right)(300) + \left(\tfrac{2}{1}\right)\dot{n}_{SO_2} + \left(\tfrac{3}{1}\right)\dot{n}_{SO_3} \\
& 2\dot{n}_{SO_2} + 3\dot{n}_{SO_3} = 800
\end{aligned}$$

Solving these two equations simultaneously yields the flow rates of the SO_2 and SO_3 leaving the reactor. Then, as before, one obtains

$$\begin{aligned}
\dot{n}_{SO_2} &= 100\ \text{lbmol SO}_2/\text{min out} \\
\dot{n}_{SO_3} &= 200\ \text{lbmol SO}_3/\text{min out}
\end{aligned}$$

Note: Since there are only two types of atoms, S and O, atomic balances can provide only two equations. This is not enough for a system with three unknowns. The oxygen output rate must be determined by one of the other methods.

(c) Using the extent of reaction balances approach gives

$$\begin{array}{lcccccc}
\text{Input} \rightarrow & 300 & & 400 & & 0 \\
 & 2SO_2 & + & O_2 & \rightarrow & 2SO_3 \\
\text{Output} \rightarrow & 300 - z & & 400 - \left(\tfrac{1}{2}\right)z & & 0 + z
\end{array}$$

where z is the rate (lbmol/min) of SO_2 reacted. Since the outlet O_2 flow rate is given as 300 lbmol/min, then

$$\begin{aligned}
& 300 = 400 - \left(\tfrac{1}{2}\right)z \\
& z = 200 \\
& SO_2 \quad \text{outlet flowrate} = 300 - z = 100\ \text{lbmol/min} \\
& SO_3 \quad \text{outlet flowrate} = z = 200\ \text{lbmol/min} \\
& O_2 \quad \text{outlet flowrate} = 400 - \left(\tfrac{1}{2}\right)z = 300\ \text{lbmol/min}
\end{aligned}$$

4.29 Ethylene Oxidation

Ethylene oxide is produced by the oxidation of ethylene with oxygen-enriched air:

$$C_2H_4 + \frac{1}{2}O_2 \rightarrow C_2H_4O$$

An undesired side reaction is the oxidation of ethylene to carbon dioxide:

$$C_2H_4 + 3O_2 \rightarrow 2CO_2 + 2H_2O$$

The feed stream to an ethylene oxide reactor consists of 45% (by mole) C_2H_4, 30% O_2, and 25% N_2. The amounts of ethylene oxide and carbon dioxide in the product stream are 20 gmol and 10 gmol, respectively, per 100 gmol feed. Determine the composition of the exiting gas stream.

Solution: Selecting a basis of 100 gmol feed stream, a flow diagram of the process may be generated as shown in Figure 4.7. One may apply extent of reaction balances to determine expressions for each of the components leaving the reactor in terms of the amount of C_2H_4 converted to C_2H_4O and the amount converted to CO_2:

Initial →	45		30		0		
	C_2H_4	+	$\frac{1}{2}O_2$	→	C_2H_4O		
Final →	$45 - y - z$		$30 - \left(\frac{1}{2}\right)y - \left(\frac{3}{1}\right)z$		y		
Initial →	45		30		0		0
	C_2H_4	+	$3O_2$	→	$2CO_2$	+	$2H_2O$
Final →	$45 - y - z$		$30 - \left(\frac{1}{2}\right)y - \left(\frac{3}{1}\right)z$		$2z$		$2z$

where $y =$ amount of C_2H_4 converted to C_2H_4O

$z =$ amount of C_2H_4 converted to CO_2

The amounts of all components of the product gas stream may now be calculated:

Amount $C_2H_4O = y = 20$ gmol

Amount $CO_2 = 2z = 10$ gmol $\rightarrow z = 5$

Amount $H_2O = 2z = 2(5) = 10$ gmol

Amount $C_2H_4 = 45 - y - z = 45 - 20 - 5 = 20$ gmol

$$\text{Amount } O_2 = 30 - \left(\frac{1}{2}\right)y - \left(\frac{3}{1}\right)z$$

$$= 30 - \left(\frac{1}{2}\right)20 - \left(\frac{3}{1}\right)5 = 5 \text{ gmol}$$

Amount $N_2 = 25$ gmol

Total amount of product gas = 90 gmol

Figure 4.7 Flow diagram for Problem 4.29.

The mole fractions are therefore

$$\text{Mole fraction } C_2H_4O = \frac{20}{90} = 0.2222$$

$$\text{Mole fraction } CO_2 = \frac{10}{90} = 0.1111$$

$$\text{Mole fraction } H_2O = \frac{10}{90} = 0.1111$$

$$\text{Mole fraction } C_2H_4 = \frac{20}{90} = 0.2222$$

$$\text{Mole fraction } O_2 = \frac{5}{90} = 0.0556$$

$$\text{Mole fraction } N_2 = \frac{25}{90} = \underline{0.2778}$$

$$1.000$$

4.30 Outlet Temperature Calculation

Given the following information, determine the outlet temperature of the incinerator:

$$\text{Heat input} = 18.7 \times 10^6\,\text{Btu/hr}$$
$$\text{Total gas flow} = 72{,}000/\text{lb/hr}$$
$$\text{Inlet air temperature} = 200^\circ\text{F}$$
$$\text{Average } \overline{c}_p = 0.26\,\text{Btu/lb} \cdot \text{F}$$

(**a**) 800°F
(**b**) 1400°F
(**c**) 1200°F
(**d**) 1000°F

Solution: This is a conservation of energy application that involves a sensible enthalpy calculation. Equation (4.6) is employed:

$$Q = \Delta H = \dot{m}c_p\Delta T = \dot{m}c_p(T_2 - T_1);\ T_2 = \text{outlet air temperature}$$

Substituting

$$18.7 \times 10^6 = (72{,}000)(0.26)(T_2 - 200)$$

Solving for T gives

$$T_2 = 999 + 200$$
$$= 1200^\circ F$$

The correct answer is therefore (c).

4.31 Heat Transfer Requirement

Given 10,000 scfm (60°F, 1 atm) of butanol-contaminated air heated from 300°F to 1500°F, find the heat transfer required to bring about this change of temperature. The enthalpy values of air at 300°F and 1500°F are 1870 and 10,895 Btu/lbmol, respectively.

Solution: The molar flow rate is

$$\dot{n} = 10{,}000/379$$
$$= 26.4 \text{ lbmol/min}$$

The heat load may now be calculated:

$$\dot{Q} = \dot{n}C_p(T_2 - T_1) = \dot{n}(H_2 - H_1)$$

Substituting, one obtains

$$\dot{Q} = (26.4)(10{,}895 - 1870)$$
$$= 238{,}260 \text{ Btu/min}$$
$$= 14.30 \times 10^6 \text{ Btu/hr}$$

4.32 Enthalpy of Combustion

Find the enthalpy of combustion of cyclohexane from enthalpy of formation data. Perform the calculation for both liquid and gaseous cyclohexane. The enthalpy of formation (25°C) data are listed in Table 4.2.

TABLE 4.2 Enthalpy of Formation Data

Species	ΔH_f°, cal/gmol
C_6H_{12} (l)	−37,340
C_6H_{12} (g)	−29,430
CO_2	−94,052
H_2O (l)	−68,317

Solution: To simplify the solution that follows, examine the equation

$$a\text{A} + b\text{B} \rightarrow c\text{C} + d\text{D}$$

If this reaction is assumed to occur in the standard state, the standard enthalpy of reaction ΔH° is given by

$$\Delta H^\circ = c(\Delta H_f^\circ)_C + d(\Delta H_f^\circ)_D - a(\Delta H_f^\circ)_A - b(\Delta H_f^\circ)_B \tag{4.22}$$

where $(\Delta H_f^\circ)_i$ is the standard enthalpy of formation of species i. Thus, the (standard) enthalpy of a reaction is obtained by taking the difference between the (standard) enthalpy of formation of products and reactants. If the (standard) enthalpy of reaction or formation is negative (exothermic), as is the case with most combustion reactions, then energy is liberated because of the chemical reaction. Energy is absorbed if ΔH° is positive (endothermic).

The balanced combustion equation for liquid (l) cyclohexane C_6H_{12} is

$$C_6H_{12}\ (l) + 9O_2 \rightarrow 6CO_2 + 6H_2O\ (l)$$

The enthalpy of combustion for this reaction is (noting that the enthalpy of formation for elements is zero)

$$\Delta H_C^\circ = 6(\Delta H_f^\circ)_{CO_2} + 6(\Delta H_f^\circ)_{H_2O(l)} - (\Delta H_f^\circ)_{C_6H_{12}}$$

Using the data provided gives

$$\begin{aligned}\Delta H_C^\circ &= 6(-94052) + 6(-68317) - (-37340)\\ &= -936{,}874 \text{ cal/gmol of liquid cyclohexane}\end{aligned}$$

For gaseous (g) cyclohexane

$$\begin{aligned}\Delta H_C^\circ &= 6(-94052) + 6(-68317) - (-29430)\\ &= -944{,}784 \text{ cal/gmol of gaseous cyclohexane}\end{aligned}$$

Tables of enthalpies of formation, combustion, and reaction are available in the literature (particularly thermodynamics texts/reference books) for a wide variety of compounds. It is important to note that these are valueless unless the stoichiometric equation and the states of the reactants and products are included. However, enthalpy of reaction is not always employed in engineering reaction/combustion calculations. The two other terms have been used are the gross (or higher) heating value and the net (or lower) heating value both of which were briefly reviewed in Section 4.2. These are discussed in the next problem.

4.33 Gross Heating Value

Given the gas mixture and combustion properties listed in Table 4.3, determine the gross heating value HV_G in Btu/scf.

Solution: The gross heating value (HV_G) represents the enthalpy changes or heat released when a gas is stoichiometrically combusted at 60°F, with the final (flue) products at 60°F and any water present in the liquid state. Stoichiometric combustion requires that no oxygen be present in the flue gas following combustion

TABLE 4.3 Component Gross Heating Values

Species	Mole Fraction	Gross Heating Value, Btu/scf
N_2	0.0515	0
CH_4	0.8111	1013
C_2H_6	0.0967	1792
C_3H_8	0.0351	2590
C_4H_{10}	0.0056	3370

of the hydrocarbons. The gross heating value of the gas mixture HV_G may now be calculated using the equation

$$HV_G = \sum_{i=1}^{n} x_i HV_{G,i} \tag{4.23}$$

where x_i is the mole fraction of ith component. Thus

$$\begin{aligned} HV_G &= (0.0515)(0) + (0.8111)(1013) + (0.0967)(1792) \\ &\quad + (0.0351)(2590) + (0.0056)(3370) \\ &= 1105 \text{ Btu/scf} \end{aligned}$$

As noted in Section 4.2, the *net heating value* (HV_N) is similar to HV_G except the water is in the vapor state. The net heating value is also known as the *lower heating value*, and the gross heating value is also known as the *higher heating value*.

4.34 Natural Gas Requirement

Determine the scfm of dry (at 60°F, 1atm) natural gas (available heat at 1400°F = 950 Btu/scf) required to heat 8500 scfm of a contaminated gas stream at 200°F to 1400°F. Assume that there are no heat losses and that the average heat capacity of the gas over the temperature range is 7.5 Btu/lb mol · °F:

(**a**) 120 scfm
(**b**) 213 scfm
(**c**) 1682 scfm
(**d**) 3215 scfm

Solution: The describing equation used to calculate the fuel rate: is given by

$$NG = \frac{Q}{HA_T} \tag{4.11}$$

From Equation (4.6)

$$\begin{aligned} \dot{Q} &= C_p\, \Delta T; \text{ consistent units} \\ &= (8500/379)(7.5)(1400 - 200) \\ &= 201{,}850 \text{ Btu/min} \end{aligned}$$

Therefore,

$$\begin{aligned} NG &= (201{,}850)/(950) \\ &= 212.5 \text{ scfm NG} \end{aligned}$$

The correct answer is therefore (b).

4.35 Calculations for Natural Gas Combustion

As an air pollution control engineer you have been requested to evaluate the gross heating value of a natural gas of a given composition. You are also to determine the available heat of the natural gas at a given temperature, the rate of auxiliary fuel (natural gas) required to heat a known amount of contaminated air to a given temperature, the dimensions of an afterburner treating the contaminated air stream, and the residence time. Operating and pertinent design data are provided in Table 4.4. The natural gas composition (mole or volume fraction) is identical to that provided in Problem 4.33. See also Table 4.3.

Gas velocity = 20 ft/s
Length-to-diameter ratio of the afterburner = 2.0
Temperature of dry natural gas = 60°F
Volumetric flow rate of contaminated air = 5000 scfm (60°F, 1 atm)
Molar enthalpy data are provided in Table 4.5.

It is required to heat the contaminated air from 200°F to 1200°F.

TABLE 4.4 Natural Gas Composition

Component	Mole Fraction
N_2	0.0515
CH_4	0.8111
C_2H_6	0.0967
C_3H_8	0.0351
C_4H_{10}	0.0056
Σ	1.0000

Solution: The gross heating value of this natural gas was determined in Problem 4.33:

$$\begin{aligned} HV_G &= \Sigma x_i HV_{G,i} \\ &= (0.0515)(0) + (0.8111)(1013) + (0.0967)(1792) + (0.0351)(2590) \\ &\quad + (0.0056)(3370) \\ &= 1105\ \text{Btu/scf natural gas} \end{aligned} \tag{4.23}$$

The balanced chemical combustion equations for each of the four components of the natural gas using 1 scf of natural gas as a basis is

$$0.8111CH_4 + 1.6222O_2 \rightarrow 0.8111CO_2 + 1.6222H_2O$$
$$0.0967C_2H_6 + 0.3385O_2 \rightarrow 0.1934CO_2 + 0.2901H_2O$$
$$0.0351C_3H_8 + 0.1755O_2 \rightarrow 0.1053CO_2 + 01404H_2O$$
$$0.0056C_4H_{10} + 0.0364O_2 \rightarrow 0.0224CO_2 + 0.0280H_2O$$

TABLE 4.5 Molar Enthalpies of Combustion Gases (in Btu/lbmol)

T°F	N_2	Air (MW = 28.97)	CO_2	H_2O
32	0	0	0	0
60	194.9	194.6	243.1	224.2
77	312.2	312.7	392.2	360.5
100	473.3	472.7	597.9	545.3
200	1170	1170	1527	1353
300	1868	1870	2509	2171
400	2570	2576	3537	3001
500	3277	3289	4607	3842
600	3991	4010	5714	4700
700	4713	4740	6855	5572
800	5443	5479	8026	6460
900	6182	6227	9224	7364
1000	6929	6984	10,447	8284
1200	8452	8524	12,960	10,176
1500	10,799	10,895	16,860	13,140
2000	14,840	14,970	23,630	18,380
2500	19,020	19,170	30,620	23,950
3000	23,280	23,460	37,750	29,780

Source: Kobe, K. A., and E. G. Long. "Thermochemistry for the Petroleum Industry," Petrol. Refiner 28(11): 129 (Nov. 1949), Table 9.

The number of standard cubic feet for each of the following components of combustion may now be determined.

$$\text{For } O_2\text{: } 1.6222 + 0.3385 + 0.1755 + 0.0364 = 2.172\,\text{scf/scf natural gas}$$

$$\text{For } CO_2\text{: } 0.8111 + 0.1934 + 0.1053 + 0.0224 = 1.132\,\text{scf/scf natural gas}$$

$$\text{For } H_2O\text{: } 1.6222 + 0.2901 + 0.1404 + 0.0280 = 2.081\,\text{scf/scf natural gas}$$

$$\text{For } N_2\text{: } 0.0515 + (79/21)(2.172) = 8.222\,\text{scf/scf natural gas}$$

The total cubic feet of combustion products per scf of natural gas burned are the sum of scf CO_2/scf of natural gas, scf H_2O/scf of natural gas, and scf N_2/scf of natural gas:

$$\left.\begin{array}{r}\text{Total cubic feet of combustion}\\ \text{products}\end{array}\right\} = 1.132 + 2.081 + 8.222$$

$$= 11.435 \text{ scf of products/scf of natural gas}$$

The following values of enthalpies at 60°F and 1200°F are obtained from Table 4.5: For CO_2:

$$\begin{aligned}\Delta H_{CO_2} &= H \text{ at } 1200^\circ\text{F} - H \text{ at } 60^\circ\text{F} \\ &= 12{,}960 - 243.1 \\ &= 12{,}720 \text{ Btu/lbmol}\end{aligned}$$

Since there are 379 scf per lbmol of any ideal gas and 1.132 scf CO_2/scf natural gas, one obtains

$$\begin{aligned}\Delta H_{CO_2} &= (12{,}720)\,(1.132)/(379) \\ &= 38.0 \text{ Btu/scf of natural gas}\end{aligned}$$

For N_2:

$$\begin{aligned}\Delta H_{N_2} &= (8452 - 194.9)\,(8.222)/(379) \\ &= 179.1 \text{ Btu/scf of natural gas}\end{aligned}$$

For H_2O (g):

$$\begin{aligned}\Delta H_{H_2O} &= (10{,}176 - 224.2)\,(2.081)/(379) \\ &= 54.6 \text{ Btu/scf of natural gas}\end{aligned}$$

Determine the amount of heat required of take the products of combustion from 60°F to 1200°F ($\Sigma\Delta H$). Since the water present in the combustion product is vapor, then

$$\begin{aligned}\Delta H_\lambda &= \frac{(1060\,\text{Btu/lb})(18\,\text{lb/lbmol})(2.081\text{ scf } H_2O/\text{scf natural gas})}{379\,\text{scf/lbmol natural gas}} \\ &= 104.8\,\text{Btu/scf of natural gas}\end{aligned}$$

Therefore,

$$\begin{aligned}\sum \Delta H &= \Delta H_{CO_2} + \Delta H_{N_2} + \Delta H_{H_2O} + \Delta H_\lambda \\ &= 38.0 + 179.1 + 54.6 + 104.8 \\ &= 376.5\,\text{Btu/scf of natural gas}\end{aligned}$$

The available heat (HA) of natural gas at 1200°F in Btu/scf natural gas is

$$\begin{aligned}\text{HA}_{1200} &= \text{HV}_\text{G} - \sum \Delta H \\ &= 1105 - 376.5 \\ &= 728.5\,\text{Btu/scf of natural gas}\end{aligned}$$

The enthalpy change of air going from 200°F to 1200°F is

$$\Delta H_{\text{air}} = H \text{ at } 1200°\text{F} - H \text{ at } 200°\text{F}$$
$$= 8524 - 1170$$
$$= 7354 \text{ Btu/lbmol}$$

The heat rate $\dot{Q}$ required to heat 5000 scfm of air from 200°F to 1200°F in Btu/min is

$$\dot{Q} = (5000 \text{ scfm})(7354 \text{ Btu/lbmol})/(379 \text{ scf/lbmol})$$
$$= 97{,}018 \text{ Btu/min}$$

The fuel requirement is therefore

$$\text{NG} = \dot{Q}/\text{HA} \tag{4.11}$$
$$= 97{,}018/728.5$$
$$= 133.2 \text{ scfm}$$

To size the unit, first calculate the volumetric flow rate of the products of combustion (CP) in scfm:

$$q_{\text{CP}} = 11.435 \frac{\text{scf products}}{\text{scf natural gas}}(133.2 \text{ scfm natural gas})$$
$$= 1523 \text{ scfm}$$

Also calculate the total flue gas flow rate q_{T} in scfm:

$$q_{\text{T}} = 5000 + 1523$$
$$= 6523 \text{ scfm}$$

Determine the total flue gas flow rate at 1200°F in acfm using Charles' Law:

$$q_{\text{T}} = (6523)(1200 + 460)/(60 + 460)$$
$$= 20{,}823 \text{ acfm}$$
$$= 347 \text{ acfs}$$

The diameter D and length L of the afterburner in feet are

$$D = (4q_{\text{T}}/v\pi)^{1/2}$$
$$= [(4)(347)/(20)(\pi)]^{1/2}$$
$$= 4.7 \text{ ft}$$
$$L = (2.0)(4.7)$$
$$= 9.4 \text{ ft}$$

Therefore, the residence time t for the gases in the afterburner in seconds is

$$\begin{aligned} t &= L/v \qquad (4.16)\\ &= (9.4)/(20)\\ &= 0.47\,\text{s} \end{aligned}$$

4.36 Available Heat and Fuel Flow Rate

A process gas stream (assume air) of 1.75×10^6 acfm at 32°F and 1 atm is to be incinerated at 1900°F in a process boiler that is 30 ft × 30 ft × 40 ft. Natural gas (assume reference) is to be employed with stoichiometric (primary) air. The stoichiometric combustion of natural gas produces 11.4 mol of flue gas per mole of natural gas. Find the available heat in Btu/min and calculate the required natural gas flow rate in scfm (60°F, 1 atm).

Solution: The molar flow rate and enthalpy change are first calculated. Note once again that one lbmol of an ideal gas occupies 359 ft^3 at 32°F and 1 atm:

$$\begin{aligned} \dot{n} &= \frac{(1.75 \times 10^6\ \text{ft}^3/\text{min})}{(359\ \text{ft}^3/\text{lbmol})}\\ &= 4.875 \times 10^3\ \text{lbmol/min} \end{aligned}$$

Employ Table 4.5 and linearly interpolate to obtain enthalpy values.

$$\begin{aligned} \dot{Q} = \Delta\dot{H} &= (4.875 \times 10^3)(14{,}115 - 0.0)\\ &= 6.88 \times 10^7\ \text{Btu/min} \end{aligned}$$

Use Theodore's equation to estimate the available heat:

$$\begin{aligned} \text{HA}_{1900^\circ\text{F}} &= -0.237(\text{T}) + 981 \qquad (4.10)\\ &= -0.237(1900) + 981\\ &= 530.7\ \text{Btu/scf}\,(60^\circ\text{F}) \end{aligned}$$

The natural gas flow rate is then

$$\begin{aligned} \text{NG} &= \Delta\dot{H}/\text{HA}_{1900^\circ\text{F}} = \dot{Q}/\text{HA}_{1900^\circ\text{F}} \qquad (4.11)\\ &= 6.88 \times 10^7/530.7\\ &= 1.30 \times 10^5\ \text{scfm} \end{aligned}$$

The total volumetric flow rate of the flue gas following may now be calculated:

$$q_{1900^\circ\mathrm{F}} = (1.75 \times 10^6)\left(\frac{1900+460}{32+460}\right) + (1.30 \times 10^5)(11.4)\left(\frac{1900+460}{60+460}\right)$$

$$= 1.512 \times 10^7 \text{ acfm} = 2.52 \times 10^5 \text{ acfs}$$

4.37 Required Residence Time

Refer to Problem 4.36. Does the unit meet the required residence time of 0.7s?

Solution: The volume of the system is

$$V = (30)(30)(40)$$
$$= 36{,}000 \text{ ft}^3$$

Thus, the residence time is

$$t = 36{,}000/2.52 \times 10^5$$
$$= 0.143 \text{ s}$$

The unit does *not* meet the required residence time of 0.7s.

4.38 Available Heat Calculation

A 1.75×10^6 acfm process gas stream essentially consisting of air enters a boiler at 60°F and 14.7 psia and is to be incinerated at 2000°F. The boilers dimensions are 36 m× 10.5 m× 12 m. A natural gas with a gross heating value of 1059 Btu/scf is to be used with (primary) stoichiometric air for fuel. The combustion of natural gas with stoichiometric air produces 11.5 mol of flue gas/mol of natural gas.

(**a**) Find the required natural gas flow rate in scfm (60°F, 1 atm).
(**b**) Calculate the available heat (HA) in Btu/scf.
(**c**) Does the unit meet the required residence time of 0.75 s?

Solution: This is similar to Problems 4.36 and 4.37, but with some SI units. Refer to Table 4.5.

$$\Delta H_{\mathrm{PGS}} = \mathrm{H}(2000^\circ\mathrm{F}) - \mathrm{H}(60^\circ\mathrm{F})$$
$$= 14{,}970 - 195$$
$$= 14{,}775 \text{ Btu/lbmol gas}$$

$$q_{\mathrm{PGS}} = 1.75 \times 10^6 \text{ acfm}$$

$$q_{\mathrm{PGS}} = (1.75 \times 10^6/379)(14{,}775)$$
$$= 6.82 \times 10^7 \text{ Btu/min}$$

Using Equation (4.10), one obtains

$$(HA_{2000^\circ F})_{ref} = -(0.237)(2000) + 981$$
$$= 507\ \text{Btu/scf}$$

(a)

$$NG = \dot{Q}/HA_T \quad (4.11)$$
$$= 6.82 \times 10^7/507$$
$$= 1.345 \times 10^5\ \text{scfm}$$

(b)

$$HA_T = (507)(1.345 \times 10^5)/60$$
$$= 1.137 \times 10^6\ \text{Btu/s}$$

(c)

$$q_T = q_{PGS} + q_C$$
$$q_C = (11.5)(NG) = (11.5)(1.345 \times 10^5)$$
$$= 1.548 \times 10^6\ \text{scfm}$$
$$q_{PGS} = 1.75 \times 10^6\ \text{scfm (or acfm)}$$
$$q_T = (1.548 + 1.75) \times 10^6$$
$$= 3.298 \times 10^6\ \text{scfm (60°F, 1 atm)}$$

Applying Charles' Law

$$q_T\ (\text{acfm}) = 3.298 \times 10^6 \left(\frac{2000 + 460}{60 + 460}\right)$$
$$= 15.6 \times 10^6\ \text{acfm}$$
$$= 2.60 \times 10^5\ \text{acfs}$$

The residence time is

$$t = V/q_T \quad (4.1)$$
$$= (36)(10.5)(12)(35.31)/2.60 \times 10^5$$
$$= 0.616\ \text{s}$$

Therefore, the unit does *not* meet the required residence time of 0.75 s.

4.39 Plan Review of a Direct Flame Afterburner

A regulatory agency engineer must review plans for a permit to construct a direct flame afterburner (Figure 4.8) serving a lithographer. Review is for the purpose of judging whether the proposed system, when operating as it is designed to operate, will meet emission standards. The permit application provides operating and design data. Agency experience has established design criteria that, if met in

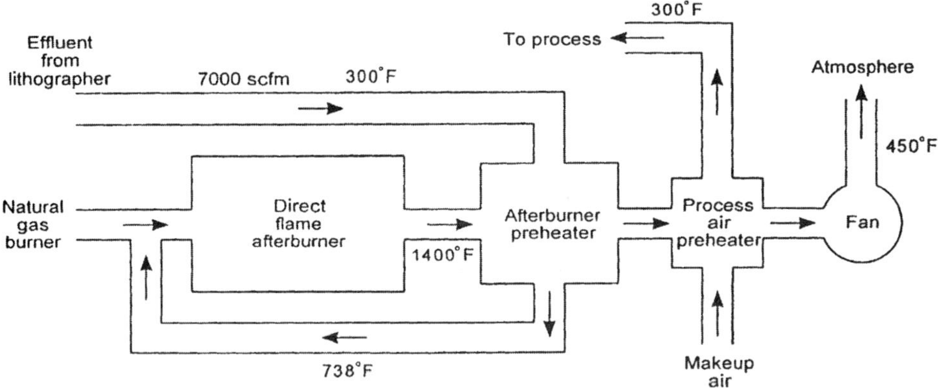

Figure 4.8 Direct flame afterburner system.

an operating system, will typically ensure compliance with standards. Operating data from the permit application are provided below.

Application = lithography
Effluent exhaust volumetric flow rate = 7000 scfm (60°F, 1 atm)
Exhaust temperature = 300°F
Hydrocarbons in effluent air to afterburner (assume hydrocarbons to be toluene) = 30 lb/hr
Afterburner entry temperature of effluent = 738°F
Afterburner heat loss = 10% in excess of calculated heat load
Afterburner dimensions = 4.2 ft in diameter, 14 ft in length

Agency Design Criteria
Afterburner temperature = 1300–1500°F
Residence time = 0.3–0.5 s
Afterburner velocity = 20–40 ft/s

Standard Data
Gross heating value of natural gas = 1059 Btu/scf of natural gas
Combustion products per cubic foot of natural gas burned = 11.5 scf/scf natural gas
Available heat of natural gas at 1400°F = 600 Btu/scf of natural gas
Molecular weight of toluene = 92
Average heat capacity of effluent gases at 738°F (above 0°F) = 7.12 Btu/(lbmol · °F)
Average heat capacity of effluent gases at 1400°F (above 0°F) = 7.38 Btu/(lbmol · °F)
Volume of air required to combust natural gas = 10.33 scf air/scf natural gas

Solution: The design temperature is already within agency criteria. To determine the fuel requirement for the afterburner, first calculate the total heat load (heating rate) $\dot{Q}$ required to raise 7000 scfm of the effluent stream from 738°F to 1400°F in Btu/min:

$$\begin{aligned} \dot{n} &= (7000\ \text{scfm})/(379\ \text{scf/lbmol}) \\ &= 18.47\ \text{lbmol/min} \end{aligned}$$

$$\begin{aligned} \dot{Q} &= \dot{n}\left[C_{P2}(T_2 - T_b) - C_{p1}(T_1 - T_b)\right] \\ &= 18.47[(7.38)(1400 - 0) - (7.12)(738 - 0)] \\ &= 93{,}780\ \text{Btu/min} \end{aligned} \tag{4.4}$$

Calculate the actual heat load required, accounting for a 10% heat loss, in Btu/min:

$$\begin{aligned} \text{Actual heat load} &= (1.1)\dot{Q} \\ &= (1.1)(93{,}780) \\ &= 103{,}200\ \text{Btu/min} \end{aligned}$$

Calculate the rate of natural gas required to supply the actual heat required to heat 7000 scfm of the effluent from 738 to 1400°F in scfm:

$$\begin{aligned} \text{NG} &= \dot{Q}/\text{HA} \\ &= (103{,}200)/(600) \\ &= 172\ \text{scfm} \end{aligned} \tag{4.11}$$

Determine the total volumetric flow rate through the afterburner q_T by first calculating the volumetric flow rate of the combustion products of the natural gas q_1 in scfm:

$$\begin{aligned} q_1 &= q_{NG}\left(11.5\ \frac{\text{scf combustion products}}{\text{scf natural gas}}\right) \qquad q_{NG} = \text{NG} \\ &= (172)(11.5) \\ &= 1978\ \text{scfm} \end{aligned}$$

Also note that the volumetric flow rate of the effluent is 7000 scfm. The volumetric flow rate of air required to combust the natural gas required q_2 in scfm is

$$\begin{aligned} q_2 &= q_{NG}\left(10.33\ \frac{\text{scf air}}{\text{scf natural gas}}\right) \\ &= (172)(10.33) \\ &= 1776\ \text{scfm} \end{aligned}$$

Calculate the total volumetric flow rate through the afterburner q_T in scfm. Since primary air is employed in the combustion of the natural gas, q_2 is not subtracted from q_T. Thus, the q_2 calculation is not required in this solution.

$$\begin{aligned} q_T &= 7000 + q_1 \\ &= 7000 + 1978 \\ &= 8978 \text{ scfm} \end{aligned}$$

Applying Charles' Law

$$\begin{aligned} &= \frac{(8978)(1400 + 460)}{(60 + 460)} \\ &= 32{,}110 \text{ acfm} \end{aligned}$$

To determine whether the afterburner velocity meets the agency criteria, first calculate the cross-sectional area of the afterburner in ft^2:

$$\begin{aligned} S &= \pi D^2/4 \\ &= (\pi)(4.2)^2/4 \\ &= 13.85 \text{ ft}^2 \end{aligned}$$

The afterburner velocity v in ft/s is then

$$\begin{aligned} v &= q_T/S \\ &= (32{,}110)/(13.85) \\ &= 2318 \text{ ft/min} \\ &= 38.6 \text{ ft/s} \end{aligned}$$

Thus, the afterburner velocity is within the agency criterion. The residence time t is then

$$\begin{aligned} t &= L/v \\ &= 14/38.6 \\ &= 0.363 \text{ s} \end{aligned} \tag{4.1}$$

This also meets the agency criterion.

4.40 Plan Review of a Catalytic Afterburner

Plans have been submitted for a catalytic afterburner. The installed afterburner is to incinerate a 3000 acfm contaminated gas stream discharged from a direct-fired paint baking oven at 350°F. The following summarizes the data taken from the plans:

Data Sheet

Exhaust flow rate from oven: 3000 acfm

Exhaust gas temperature from oven: 350°F

Solvent emission to afterburner: 0.3 lb/min
Final temperature in afterburner: 1000°F
Gross heating value of natural gas: 1100 Btu/scf
Total heat requirement: 26,884 Btu/min
Natural gas requirement: 35.0 scfm
Furnace volume: 46.0 ft^3
Exhaust flow rate from afterburner at 1000°F: 6350 acfm
Gas velocity through catalytic bed: 8.6 ft/s
Number of type A 19 × 24 × 3.75 inch catalyst elements: 4

The following additional information and rules of thumb may be required to review the plans for the catalytic afterburner:

1. Heat will be recovered from the afterburner effluent, but that process will not be considered in this problem.
2. Catalytic afterburner operating temperatures of approximately 950°F have been found sufficient to control emissions from most process ovens.
3. Preheat burners are usually designed to increase the temperature of the contaminated gases to the required catalyst discharge gas temperature without regard to the heating value of the contaminants (especially if considerable concentration variation occurs).
4. A 10% heat loss is usually a reasonable estimate for an afterburner. This may be accounted for by dividing the calculated heat load by 0.9.
5. The properties of the contaminated effluent may essentially be considered identical to those of air.
6. The natural gas is combusted using near stoichiometric (0% excess) external air.
7. The catalyst manufacturer's literature suggests a superficial gas velocity through the face surface of the catalyst element (in this case 19 inches × 24 inches) of 10 ft/s.

Two key questions are to be considered:

(a) Is the operating temperature adequate for efficiency control?
(b) Is the fuel requirement adequate to maintain the operating temperature?

Solution: (a) The plans indicate a combustion temperature of 1000°F; this is acceptable when compared to a 950°F rule-of-thumb temperature.

(b) To determine whether the fuel requirement is adequate to maintain the operating temperature, first calculate the lbmol/min of gas to be heated from 350°F to 1000°F.

$$\dot{n} = (3000\,\text{acfm})[(460 + 60)/(460 + 350)] = 1926\,\text{scfm}$$
$$1926/379 = 5.08\,\text{lbmol/min}$$

Also determine the heat requirement in Btu/min to raise the gas stream (air) temperature from 350°F to 1000°F. See Table 4.5 provided in Problem 4.35.

$$H \text{ at } 350°\text{F} = 2222\,\text{Btu/lbmol}$$
$$H \text{ at } 1000°\text{F} = 6984\,\text{Btu/lbmol}$$

$$\begin{aligned}\dot{Q} &= \dot{n}\Delta H \\ &= (5.08)(6984 - 2222) \\ &= 24{,}190\ \text{Btu/min}\end{aligned}$$

The total heat requirement Q_T in Btu/min is then

$$\begin{aligned}\dot{Q}_T &= \dot{Q}/0.9 \\ &= 24{,}190/0.9 \\ &= 26{,}870\ \text{Btu/min}\end{aligned}$$

The available heat of the natural gas in Btu/scf at 1000°F may be calculated from the following equation:

$$\begin{aligned}\text{HA}_{1000}/\text{HV}_G &= (\text{HA}_{1100}/\text{HV}_G)_{\text{ref fuel}} \qquad (4.9) \\ \text{HA}_{1000} &= 1100(745/1059) \\ &= 774\ \text{Btu/scf natural gas}\end{aligned}$$

The natural gas (NG) requirement in scfm may be calculated from the last two results:

$$\begin{aligned}\text{NG} &= \dot{Q}_T/\text{HA}_{1000} \qquad (4.11) \\ &= 26{,}888/774 \\ &= 34.7\ \text{scfm}\end{aligned}$$

4.41 Catalyst Sizing

Refer to Problem 4.40. Is the catalyst section property sized?

Solution: To determine whether the catalyst section is sized properly, first calculate the volume rate of flue products at 1000°F from the combustion of the natural gas:

$$\begin{aligned}q_c &= (11.45)(35)[(460 + 1000)/(460 + 60)] \\ &= 1125\ \text{acfm}\end{aligned}$$

Also calculate the volume of contaminated gases at 1000°F employing Charles' Law:

$$\begin{aligned}q &= (3000)[(460 + 1000)/(460 + 350)] \\ &= 5407\ \text{acfm}\end{aligned}$$

The total volumetric gas rate at 1000°F is then

$$\begin{aligned}q_T &= q_c + q \\ &= 1125 + 5407 \\ &= 6532\ \text{acfm}\end{aligned}$$

This comprises favorably with the data sheet value of 6350 acfm. Calculate the number of $19 \times 24 \times 3.75$ inch catalyst elements N required. Employ a velocity of 600 ft/min.

$$N = (6532)(144)/[(19)(24)(600)]$$
$$= 3.44$$

The catalyst section is sized properly. It is seen that four elements have been specified; this is a conservative design allowing a slightly slower gas flow through the elements. With four elements, the superficial velocity is reduced to 8.6 ft/s. This too is in agreement with the design specification.

4.42 First-order Kinetics

In order to meet recently updated pollution regulations for discharging hydrocarbons to the atmosphere, a gas stream must be reduced by 99.5% of its present hydrocarbon concentration. Owing to economic considerations, it is proposed to meet this requirement by combusting the hydrocarbons in a tubular thermal oxidizer operating at 1500°F. The gas and methane (fuel) are to be fed to the reactor at 80°F and 1 atm. Design the proposed oxidizer using kinetic principles.

Data
Flue gas flow rate (from fuel combustion) = 2500 scfm (80°F, 1 atm)
Process gas flow rate = 7200 scfm
Hydrocarbon: essentially toluene
Reaction rate constant, $k = 7.80\,\text{s}^{-1}$ at 1500°F
Average velocity = 20 ft/s;
$C_1/C_0 = 0.005/1.0$
C_1 and C_0 are the final and initial concentrations, respectively.

Solution: In terms of the actual operating conditions of the oxidizer, the volumetric flow rate is (applying Charles' Law)

$$q_a = q_s\left(\frac{T_a}{T_s}\right) = (2500 + 7200)\left(\frac{460 + 1500}{460 + 80}\right)$$
$$= 35{,}200 \text{ acfm}$$

Calculate the cross-sectional area of the oxidizer:

$$S = \left(\frac{q_a\,(\text{acfm})}{V(\text{fpm})}\right) = \left(\frac{35{,}200}{(20)(60)}\right)$$
$$= 29.33 \text{ ft}^2$$

Calculate the diameter of the oxidizer:

$$D = \left(\frac{4S}{\pi}\right)^{0.5} = \left(\frac{4 \times 29.33}{3.1416}\right)^{0.5}$$
$$= 6.11 \text{ ft}$$

Calculate the residence time required assuming first-order kinetics (see J. Reynolds, J. Jeris, and L. Theodore, *Handbook of Chemical and Environmental Engineering Calculations*, John Wiley & Sons, Hoboken, NJ, 2004 for additional details) in a plug flow oxidizer.

$$\begin{aligned} t &= -\left(\frac{1}{k}\right)\ln\left(\frac{C_1}{C_0}\right) \\ &= -\frac{1}{7.8}\ln\frac{0.005}{1.0} \\ &= 0.68 \text{ s} \end{aligned} \tag{4.24}$$

Calculate the oxidizer volume required to achieve the residence time calculated previously:

$$\begin{aligned} V &= tq_{\mathrm{a}} = (0.68)\frac{35{,}207}{60} \\ &= 399 \text{ ft}^3 \end{aligned} \tag{4.1}$$

Calculate the length of the oxidizer:

$$\begin{aligned} L &= \frac{V}{S} = \frac{399}{29.33} \\ &= 13.6 \text{ ft} \end{aligned}$$

4.43 Minimizing Incinerator Volume

A fluidized-bed incinerator is to be designed to destroy 99.99% of a unique hazardous waste. On the basis of laboratory and pilot plant studies, researchers have described the waste reaction by a first-order reversible mechanism. Their preliminary findings are given here:

$$A \underset{k'}{\overset{k}{\longleftrightarrow}} B$$

$$k = 1.0\exp(-10{,}000/T) \qquad T = \text{Rankine}$$

$$k' = 9.89\exp(-35{,}000/T) \qquad T = \text{Rankine}$$

Calculate a fluidized-bed incinerator operating temperature that will minimize the volume of the incinerator and achieve the desired degree of waste conversion (destruction). The operating temperature must be in the 1400–1700°F range.

Solution: A fluidized-bed incinerator is best described (see text referenced in previous problem) by a continuous stirred-tank reactor (CSTR) as provided in Equation (4.25).

$$t = \frac{V}{q_{\mathrm{a}}} = \frac{C_{\mathrm{A0}} - C_{\mathrm{A1}}}{-R_{\mathrm{A}}} = \frac{C_{\mathrm{A0}} - C_{\mathrm{A1}}}{kC_{\mathrm{A1}} - k'C_{\mathrm{B1}}} \tag{4.25}$$

For 99.99% destruction of the waste A, the conversion of A (X_A = moles of A reacted per mole A initially present) becomes

$$X_A = 0.9999$$

so that the outlet concentrations (state 1) are

$$C_{A1} = 0.0001 C_{A0} \qquad C_{B1} = 0.9999 C_{A0}$$

Therefore

$$\frac{V}{q_a} = \frac{C_{A0} - 0.0001 C_{A0}}{(k)(0.0001 C_{A0}) - (k')(0.9999 C_{A0})} = \frac{0.9999}{0.0001\,k - 0.9999\,k'}$$

To minimize the incinerator volume, dV/dT is set to zero. Since q_a is assumed to be constant, $d(V/q_a)/dT$ can also be set to 0

$$\frac{d(V/q_a)}{dT} = \frac{-9999[(dk/dT) - 9999(dk'/dT)]}{(k - 9999\,k')^2} = 0$$

Thus

$$\frac{dk}{dT} = \frac{10^4 dk'}{dT}$$

Substituting in the values of the reaction velocity constants, one obtains

$$k = 1.0 \exp\left(\frac{-10{,}000}{T}\right)$$

$$\frac{dk}{dT} = \left(\frac{-10{,}000}{T^2}\right) \exp\left(\frac{-10{,}000}{T}\right)$$

$$k' = 9.89 \exp\left(\frac{-35{,}000}{T}\right)$$

$$\frac{dk'}{dT} = \left(\frac{3.4615 \times 10^5}{T^2}\right) \exp\left(\frac{-35{,}000}{T}\right)$$

Therefore

$$(10{,}000) \exp\left(\frac{-10{,}000}{T}\right) = (3.4165 \times 10^9) \exp\left(\frac{-35{,}000}{T}\right)$$

Solving for T yields

$$T = 1960^\circ\mathrm{R} = 1500^\circ\mathrm{F}$$

There are simpler approaches that could have been used to solve this problem. For example, the right side of the design equation could be plotted against temperature (on which the k values depend) to determine the minimum of the function.

4.44 Length of Incinerator from Kinetic Data

It is proposed to decompose pure diethyl peroxide (A) at 225°C in a bench-scale incinerator. This pollutant will be entering the incinerator at a flow rate of 12.1 L/s. It is desired to decompose 99.995% of the diethyl peroxide. The following data are available:

$$R_A = -k_A C_A \text{ gmol/L} \cdot \text{s}; \qquad k_A = 38.3\,\text{s}^{-1} \text{ at } 225^\circ\text{C}$$

The inside diameter of the incinerator is 8.0 cm. What should the length of the incinerator be?

Solution: Assume a plug flow reactor. Use Equation (4.24) since the reactor is first-order and irreversible:

$$t = -\frac{1}{k}\ln\left(\frac{C_{A1}}{C_{A0}}\right) \tag{4.24}$$

For 99.995% destruction, one obtains

$$C_{A1} = 5 \times 10^{-5} C_{A0}$$

and

$$t = -\left(\frac{1}{38.3}\right)\ln\left(5 \times 10^{-5}\right)$$
$$= 0.259\,\text{s}$$

Calculate the incinerator volume.

$$V = (t)(q_a) = (0.259)(12.1)$$
$$= 3.13\,\text{L} = 3130\,\text{cm}^3 \tag{4.1}$$

Calculate the length of the reactor.

$$L = \frac{V}{\pi D^2/4} = \frac{3130}{(\pi)(8^2)/4}$$
$$= 62.3\,\text{cm}$$

4.45 Simplified Equation for Estimating Incinerator Temperature

Theodore and Reynolds (TAR) derived the following equation to estimate the temperature (°F) in an incinerator that combusts any waste gas/fuel with a given net heating value (NHV) and in the presence of excess air (EA, fractional basis):

$$T = 60 + \text{NHV}/[\{0.325\}\{1 + (1 + \text{EA})(7.5 \times 10^{-4})\,(\text{NHV})\}] \tag{4.26}$$

How was this equation derived? What are the key assumptions?

Solution: Some reasonable assumptions can be made to simplify a rigorous approach to calculate the temperature in an incinerator. These are detailed in this problem. When compared to the rigorous approach, a simpler (and in many instances, a more informative) set of equations result that are valid for purposes of engineering calculation.

1. The sensible enthalpy change associated with the cooling step of the feed to standard conditions is approximately zero.
2. Although the products of combustion consist of many components, the major or primary components are nitrogen, carbon dioxide, and water (vapor). The average heat capacities of these components over the temperature range 60–2000°F (the latter being a typical incinerator operating temperature) are 0.27, 0.27, and 0.52 Btu/lb·°F, respectively. The arithmetic average of these three components is 0.35. However, since this product stream consists primarily of nitrogen, the average heat capacity of the combined mixture (not including the excess air) may be assigned a value of approximately 0.325 Btu/lb · °F. For this condition

$$\Delta H_{\mathrm{P}} = m_{\mathrm{P}}(0.325)(T - T_0) \tag{4.27}$$

   where m_{p} = mass of stoichiometric air, fuel, and waste entering the incinerator per unit mass of waste-fuel mixture, or equivalently, mass of products less that of excess air per unit mass of waste–fuel mixture (lb/lb mixture). (*Note*: The energy required to raise the water resulting from combustion from the base temperature of 77°F as a liquid to boiling point at 1 atm of 212°F is equal to 1150 Btu/lb.)
3. The average heat capacity of the (excess) air is approximately 0.27 Btu/lb°F over the temperature range 60–2000°F. This value can also be rounded to 0.325.
4. Under truly adiabatic conditions, ΔH(overall) = 0. Under actual operating conditions, there will be some heat loss across the walls of the incinerator. However, if there is some preheat of the feed (usually the air), the effect of this assumption, to some extent, may balance the heat gained in the preheat (see assumption 1).
5. The reference or standard temperature is 60°F.
6. Perhaps the key assumption in this development is that associated with the stoichiometric air requirements for the combined waste-fuel mixture. The stoichiometric volumetric air requirement divided by the NHV for most hydrocarbons is approximately 0.01 ft air/Btu (or 100 Btu/ft^3 air). For example, the ratios for methane (M), benzene (B), and toluene (T) are

$$\text{M:}\quad v_{\mathrm{st}}/\text{NHV} = 9.53/913 = 0.0104\ \text{ft}^3\ \text{air/Btu}$$

$$\text{B:}\quad v_{\mathrm{st}}/\text{NHV} = 35.73/3601 = 0.0099\ \text{ft}^3\ \text{air/Btu}$$

$$\text{T:}\quad v_{\mathrm{st}}/\text{NHV} = 42.88/4284 = 0.01001\ \text{ft}^3\text{air/Btu}$$

Using the density of air at 60°F, this ratio can be converted to approximately 750 lb air/10^6 Btu or 7.5×10^{-4} lb air/Btu. Thus, for this condition the stoichiometric air requirement (m_{st}) is given by

$$m_{st} = 7.5 \times 10^{-4}\,\text{NHV} \tag{4.28}$$

It should be noted that the validity and applicability of this assumption is limited to waste–fuel mixtures consisting of pure hydrocarbons, i.e., organics with only hydrogen and carbon atoms, and to chlorinated organic mixtures where the mass fraction of the chlorine is low. However, for purposes of engineering calculations, this value for stoichiometric air will apply for almost all waste incinerator applications.

Applying the six assumptions listed above results in the following equation:

$$T = 60 + \frac{\text{NHV}}{(0.325)\,[1 + (1 + \text{EA})(7.5 \times 10^{-4})(\text{NHV})]} \tag{4.26}$$

Note: The units of T and NHV are °F and Btu/lb, respectively; EA is a dimensionless fraction representing the excess air. The reader should also note that these equations become sensitive to the value assigned to c_p (0.325 in this case). This is a function of both the temperature (T) and excess air fraction (EA), and also depends on the flue products since the heat capacities of air and CO_2 are about half that of H_2O. In addition, the 7.5×10^{-4} term "derived" earlier may vary slightly with the composition of the waste–fuel mixture combusted. The overall relationship between operating temperature and composition is therefore rather complex, and its prediction not necessarily as straightforward as shown here. (Details are available in J. Santoleri, J. Reynolds, and L. Theodore *Introduction to Hazardous Waste Incineration*, 2nd ed., John Wiley & Sons, Hoboken, NJ, 2004.)

4.46 Theoretical Flame Temperature Estimation

Estimate the theoretical flame temperature of a waste mixture containing 25% cellulose, 35% motor oil, 15% water (vapor), and 25% inerts, by mass. Assume 5% radiant heat losses. The flue gas contains 11.8% CO_2, 13% CO, and 10.4% O_2 (dry basis) by volume.

NHV of cellulose = 14,000 Btu/lb

NHV of motor oil = 25,000 Btu/lb

NHV of water = 0 Btu/lb

NHV of inerts (effective) = −1000 Btu/lb

Assume that the average heat capacity of the flue gas is 0.325 Btu/(lb · °F). Employ the Theodore and Reynolds (TAR) equation to perform this calculation.

$$T = 60 + \frac{\text{NHV}}{0.325[1 + (1 + \text{EA})(7.5 \times 10^{-4})(\text{NHV})]} \tag{4.26}$$

The EPA equation for estimating the excess air is

$$\text{EA} = \frac{0.95Y}{21 - Y} \tag{4.29}$$

where $Y = \%$ by volume O_2 (dry basis).

Solution: Determine the net heating value (NHV) for the mixture:

$$\begin{aligned}\text{NHV} &= 0.25(14{,}000\ \text{Btu/lb}) + 0.35(25{,}000\ \text{Btu/lb}) \\ &\quad + 0.15(0.0\ \text{Btu/lb}) + 0.25(-1000\ \text{Btu/lb}) \\ &= 12{,}000\ \text{Btu/lb}\end{aligned}$$

Determine the excess air employed:

$$\begin{aligned}\text{EA} &= \frac{0.95Y}{(21 - Y)} \\ &= 0.95(10.4)/(21 - 10.4) \\ &= 0.932\end{aligned} \tag{4.29}$$

Estimate the flame temperature using the Theodore–Reynolds equation:

$$\begin{aligned}T &= 60 + \frac{\text{NHV}}{0.325[1 + (1 + \text{EA})(7.5 \times 10^{-4})(\text{NHV})]} \\ &= 60 + \frac{12{,}000}{0.325[1 + (1 + 0.932)(7.5 \times 10^{-4})(12{,}000)]} \\ &= 2068^{\circ}\text{F}\end{aligned} \tag{4.29}$$

4.47 Coil-Coating Operation

A decision has been made to control the volatile organic emissions from a coil-coating operation with a combustion device. Discuss and/or outline the pros and cons of control via thermal versus catalytic incineration.

Solution: This is, to some extent, an open-ended question. The main concern is whether particulates (or other solid matter) will be present in the emission from the operation. The path of least resistance is to employ a thermal unit since the formation of any particulate would significantly reduce the effectiveness of a catalytic unit.

NOTE: Additional problems are available for all readers at www.wiley.com. Follow links for this title.

5

ABSORBERS

5.1 INTRODUCTION

Gas absorption, as applied to the control of air pollution, is concerned with the removal of one or more pollutants from a contaminated gas stream by treatment with a liquid. The necessary condition is the solubility of these pollutants in the absorbing liquid. The rate of transfer of the soluble constituents from the gas to the liquid phase is determined by diffusional processes occuring on each side of the gas–liquid interface.

Consider, for example, the process taking place when a mixture of air and sulfur dioxide is brought into contact with water. The SO_2 is soluble in water, and those molecules that come into contact with the water surface dissolve fairly rapidly. However, the SO_2 molecules are initially dispersed throughout the gas phase, and they can reach the water surface only by diffusing through the air, which is substantially insoluble in the water. When the SO_2 at the water surface has dissolved, it is distributed throughout the water phase by a second diffusional process. Consequently, the rate of absorption is determined by the rates of diffusion in both the gas and liquid phases.

Equilibrium is another extremely important factor to be considered that affects the operation of absorption systems. The rate at which the pollutant will diffuse into an absorbent liquid will depend on the departure from equilibrium that is maintained. The rate at which equilibrium is established is then essentially dependent on the rate

Air Pollution Control Equipment Calculations. By Louis Theodore

of diffusion of the pollutant through the nonabsorbed gas and through the absorbing liquid. Equilibrium concepts and relationships are considered in a later section.

The rate at which the pollutant mass is transferred from one phase to another depends also on a so-called mass transfer, or rate, coefficient which relates the quantity of mass being transferred with the driving force. As can be expected, this transfer process ceases upon the attainment of equilibrium.

Gas absorption can also be viewed on a molecular scale as a mass transfer or diffusional operation characterized by a transfer of one substance through another. The mass transfer process may be considered the result of a concentration difference driving force, the diffusing substance moving from a place of relatively high to one of relatively low concentration. The rate at which this mass is transferred depends to a great extent on the diffusional characteristics of both the diffusing substance and the medium.

The principal types of gas absorption equipment may be classified as follows:

1. Packed columns (continuous operation)
2. Plate columns (staged operation)
3. Miscellaneous

Of the three categories, the packed column is by far the most commonly used for the absorption of gaseous pollutants. The first two types of units are discussed below.

Packed columns are used for the continuous contact between liquid and gas. The countercurrent packed column (see Figure 5.1) is the most common type of unit encountered in gaseous pollutant control for the removal of the undesirable gas, vapor, or odor. This type of column has found widespread application in both the chemical and pollution control industries. The gas stream containing the pollutant moves upward through the packed bed against an absorbing or reacting liquid that is injected at the top of the packing. This results in the highest possible transfer/control efficiency. Since the pollutant concentration in the gas stream decreases as it rises through the column, there is constantly fresher liquid available for contact. This provides a maximum average driving force for the transfer process throughout the packed bed.

Liquid distribution plays an important role in the efficient operation of a packed column. A good packing from a process viewpoint can be reduced in effectiveness by poor liquid distribution across the top of its upper surface. Poor distribution reduces the effective wetted packing area and promotes liquid channeling. The final selection of the mechanism of distributing the liquid across the packing depends on the size of the column, type of packing, tendency of packing to divert liquid to column walls, and materials of construction for distribution. For stacked packing, the liquid usually has little tendency to cross distribute and thus moves down the column fairly uniformly in the cross-sectional area that it enters. In the dumped condition, most flow profiles follow a conical distribution down the column, with the apex of the cone at the liquid impingement point. For well-distributed liquid flow and reduced channeling of gas and liquid to produce efficient use of the packed bed, the impingement of the liquid onto the bed must be as uniform as possible. The liquid coming down through

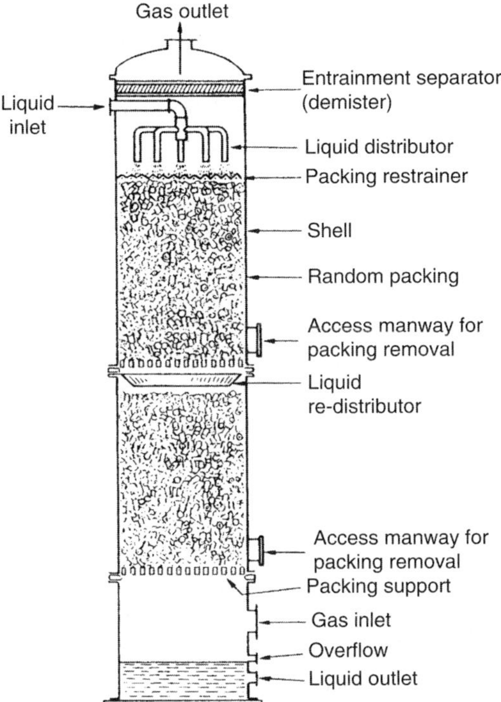

Figure 5.1 Countercurrent packed column.

the packing and on the inside wall of the column should be redistributed after a bed depth of approximately 3 column diameters for Raschig rings and 5–10 column diameters for other packing (check literature for details of packing). As a guide, Raschig rings usually have a maximum of 10–15 ft of packing per section, while other packing can use a maximum of 12–20 ft. As a general rule of thumb, however, the liquid should be redistributed every 10 ft of packed height. The redistribution brings the liquid off the wall and outer portions of the column and directs it toward the center area of the column. As noted earlier, redistribution is seldom necessary for stacked bed packings, as the liquid flows essentially in vertical streams.

Crossflow packed scrubbers are particularly successful when the process air stream requires both gas absorption and particulate removal. The crossflow scrubber operates by allowing the gas to flow horizontally across the scrubber. The scrubbing liquid is introduced at the top of the scrubber and drains vertically through the packing. Contact between the gas and liquid occurs at a right angle.

In general, crossflow scrubbers operate at lower pressure drops and liquid recycle flow rates than do countercurrent packed-bed absorbers. Vendors claim that crossflow scrubbers operate with a liquid rate and pressure drop approximately 60% less than a comparable countercurrent packed tower. The crossflow scrubber generally operates at much higher gas-to-liquid ratios than does the countercurrent packed-bed absorber.

As a result, the liquid stream in the crossflow scrubber is able to "scour" the packing media more easily than is the countercurrent packed absorber. Thus, there can be a significant reduction in plugging of the packing. Since the crossflow scrubber reduces water consumption and the size of the recirculation pump, significant savings in both operating and capital costs may be realized.

It is also easier to increase the gas flow rate through an existing crossflow absorber than a countercurrent packed-bed absorber since the crossflow scrubber can operate over a wider range of gas-to-liquid ratios. Another advantage associated with the crossflow absorber is that it can be installed horizontally in-line on an existing process. Many current crossflow scrubbers are using multiple beds with individual scrubbing sections that can remove a wide variety of pollutants. It is much more difficult and expensive to provide multiple scrubbing sections in a packed tower. The main advantage that the packed tower absorber has over the crossflow scrubber is that the packed tower can achieve very high removal efficiencies for pollutants that are difficult to absorb.

Plate columns may also be employed as absorbers, although they are used only occasionally for pollution control. These devices are essentially vertical cylinders in which the liquid and gas are contacted in stepwise fashion on plates or trays in the manner depicted schematically in Figure 5.2. The liquid enters at the top and flows downward via gravity. On the way, it flows across each plate and through a downspout (or downcomer) to the plate below. The gas passes upward through hole openings in the plate, then bubbles through the liquid to form a froth, disengages from the froth, and passes on the next plate above. The overall effect is a multiple countercurrent contacting of gas and

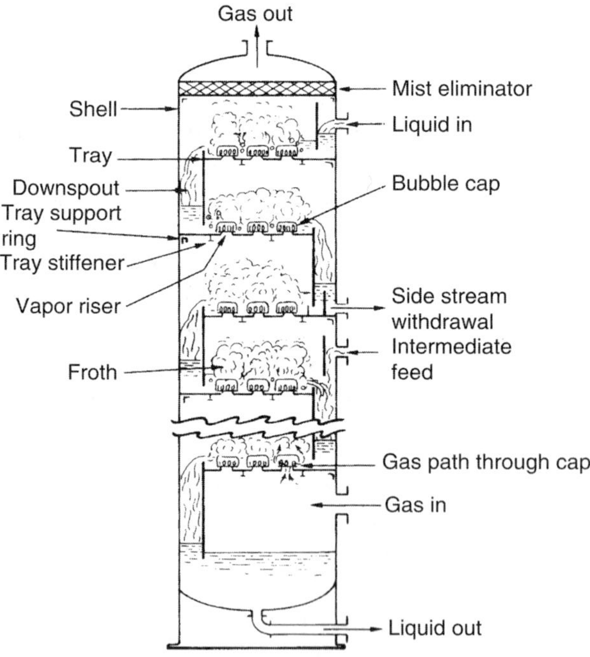

Figure 5.2 Plate column.

liquid. Each plate of the column is a stage. Since the fluids on the plate are brought into intimate contact, interphase diffusion occurs, and the fluids are then separated. The number of theoretical plates is dependent on the difficulty of the separation to be carried out and is determined solely from material balance and equilibrium considerations. The *actual* number of plates required for a given separation is greater than the *theoretical* number due to plate inefficiency. The diameter of the column, on the other hand, depends on the quantities of liquid and gas flowing through the column per unit time.

In bubble-cap plates, the vapor moves upward through *risers* into the bubble cap, out through the slots as bubbles, and into the surrounding liquid on the plates. Figure 5.2 demonstrates the vapor–liquid action for a bubble-cap plate. Although rarely used today, the bubble-cap plate design is one of the more flexible of plate designs for high and low vapor and liquid rates. On the average, plates are usually spaced approximately 24 inches apart.

5.2 DESIGN AND PERFORMANCE EQUATIONS

The equilibrium of interest in gas absorption operations is that between a relatively nonvolatile absorbing liquid (solvent) and solute gas (usually the pollutant). As described earlier, the solute is ordinarily removed from a relatively large amount of a carrier gas that does not dissolve in the absorbing liquid. The equilibrium relationship of importance is a plot (or data) of x, the mole fraction of solute in the liquid, against y^* (sometimes denoted simply as y), the mole fraction in the vapor in equilibrium with x. For cases that follow Henry's law, Henry's law constant m can be defined by the equation

$$y^* = y = mx \tag{5.1}$$

The usual operating data to be determined or estimated for isothermal systems are the liquid rate(s) and the terminal concentrations or mole fractions. An operating line that describes operating conditions in the column is obtained by a mass balance around the column (as shown in Figure 5.3)

$$\text{Total moles in} = \text{total moles out}$$

$$G_{m1} + L_{m2} = G_{m2} + L_{m1} \tag{5.2}$$

The terms G and L represent the molar flow rates of the gas and liquid, respectively.

For component A, the mole (or mass) balance becomes

$$G_{m1}\,y_{A1} + L_{m2}\,x_{A2} = G_{m2}\,y_{A2} + L_{m1}\,x_{A1} \tag{5.3}$$

Assuming $G_{m1} = G_{m2}$ and $L_{m1} = L_{m2}$ (reasonable for most air pollution control application where contaminant concentrations are usually extremely small), one may write Equation (5.3) as

$$G_m\,y_{A1} + L_m\,x_{A2} = G_m\,y_{A2} + L_m\,x_{A1} \tag{5.4}$$

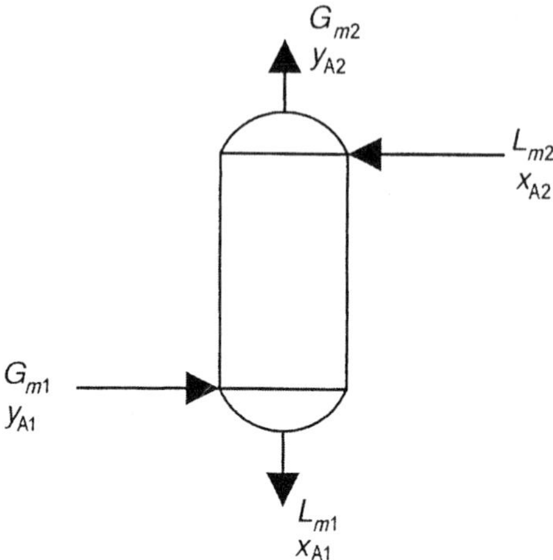

Figure 5.3 Material (mole) balance for the absorption of component A in an absorption column.

Rearranging the componential mole balance in Equation (5.4) leads to

$$\frac{L_m}{G_m} = \frac{y_{A1} - y_{A2}}{x_{A1} - x_{A2}} \tag{5.5}$$

This is the equation of a straight line known as the *operating line* since it describes operating conditions in the column. On x, y coordinates, it has a slope of L_m/G_m and passes through the points (x_{A1}, y_{A1}) and (x_{A2}, y_{A2}) as indicated in Figure 5.4.

In the design of most absorption columns, the quantity of gas to be treated G_m, the terminal concentrations y_{A1} and y_{A2}, and the composition of the entering liquid x_{A2} are ordinarily fixed by process requirements; however, the quantity of liquid solvent to be used is subject to some choice. Should this quantity already be specified, the

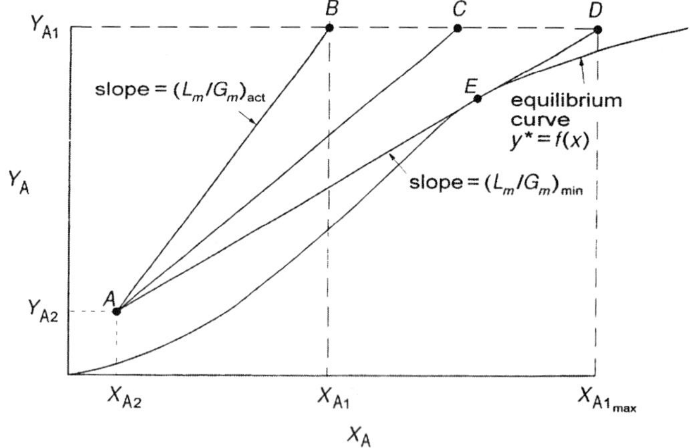

Figure 5.4 Operating and equilibrium lines.

operating line in the figure is fixed. If the quantity of solvent is unknown, the operating line is consequently unknown. However, this can be obtained through setting the determination of the minimum liquid-to-gas ratio, a topic to be discussed shortly.

With reference to Figure 5.4, the operating line must pass through point A and must terminate at the ordinate y_{A1}. If such a quantity of liquid is used to determine the operating line AB, the existing liquid will have the composition x_{A1}. If less liquid is used, the exit liquid composition will clearly be greater, as at point C, but since the driving forces (displacement of the operating line from the equilibrium line) for mass transfer are less, the absorption is more difficult. The time of contact between gas and liquid must then be greater, and the absorber must be correspondingly taller.

The minimum liquid that can be used corresponds to the operating line AD, which has the greatest slope for any line touching the equilibrium curve and is tangent to the curve at E. At point E, the diffusional concentration difference driving force is zero; the required contact time for the concentration change desired is infinite and an infinitely tall column is required. This then represents the limiting or minimum liquid-to-gas ratio.

The importance of the minimum liquid-to-gas ratio lies in the fact that column design and operation is frequently specified as some factor of the minimum liquid–gas ratio. For example, a typical situation frequently encountered is that the slope of the actual operating line, $(L_m/G_m)_{\text{act}}$, is 1.5 times the minimum, $(L_m/G_m)_{\text{min}}$.

Once all the streams entering and leaving the column and their constituents are identified, flow rates calculated, and operating conditions determined, the physical dimensions of the column can be calculated. The column must be of sufficient diameter to accommodate the gas and liquid, and of sufficient height to ensure that the required amount of mass is transfered.

Packed Column

Regarding the diameter, consider a packed column operating at a given liquid rate while the gas rate is gradually increased. After a certain point the gas rate is so high that the resistive force (drag) on the liquid is sufficient to keep the liquid from flowing freely down the column. Liquid begins to accumulate and tends to block the entire cross section for flow (so-called loading). This, of course, increases both the pressure drop and prevents the packing from mixing the gas and liquid effectively, and ultimately some liquid is even carried back up the column. This undesirable condition, known as *flooding*, occurs fairly abruptly, and the superficial gas velocity (the velocity if the column is empty) at which it occurs is called the *flooding velocity*. The calculation of column diameter is based on flooding considerations; the usual operating range is approximately 50–75% of the flooding rate.

One of the more commonly used correlations that has withstood the test of time is U.S. Stoneware's generalized pressure drop correlation, as presented in Figure 5.5. The procedure to determine the column diameter is as follows:

1. Calculate the abscissa, $(L/G)\,(\rho_G/\rho_L)^{0.5}$
2. Proceed to the flooding line and read the ordinate (design parameter)

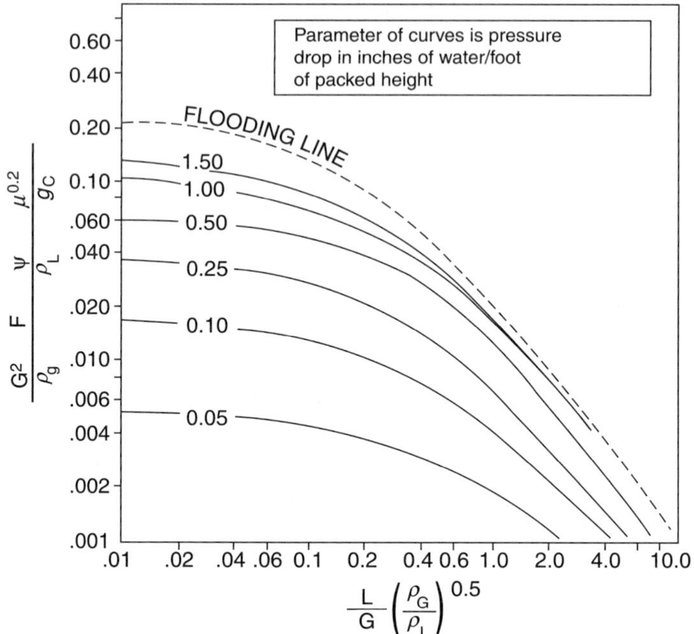

Figure 5.5 Generalized pressure drop correlation to estimate column diameter. The following units must be employed in using this figure:

$L, G = \text{lb/ft}^2 \cdot \text{s}$ ($\dot{L}$ and $\dot{G}$ are sometimes employed)

$\rho_G, \rho_L = \text{lb/ft}^3$ (ρ is often employed for ρ_G)

$F =$ packing factor, dimensionless

$\phi =$ ratio of water density to liquid density, dimensionless

$g_c = 32.2\ \text{ft} \cdot \text{lb/lb}_f \cdot \text{s}^2$

$\mu_L =$ viscosity of liquid, cP

Note that the flooding curve can also be approximately represented in equation form by:

$Y = -3.85 - 1.06X - 0.119X^2$

where $e^Y =$ vertical coordinate

$X = \ln(\text{horizontal coordinate})$.

3. Solve the ordinate term for G at flooding
4. Calculate the column cross-sectional area S for the fraction f of flooding velocity chosen for operation by the equation:

$$S = \frac{\dot{m}}{fG} \tag{5.6}$$

where $\dot{m} = \text{lb/s}$ of gas

$S = \text{area, ft}^2$

5. The diameter of the column is then determined by

$$D = 1.13(S)^{0.5} \tag{5.7}$$

Note that the proper units as designated in the correlation must be used as the plot is not dimensionless. The flooding rate should also be evaluated using (total) flows of the phases at the bottom of the column, where they are at their highest value. The pressure drop may be evaluated directly from Figure 5.5 using a revised ordinate that contains the actual, not flooding, value of G.

The column height is given by

$$Z = N_{OG}\, H_{OG} \tag{5.8}$$

where N_{OG} is the number of overall transfer units, dimensionless; H_{OG} is the height of overall transfer units, ft; and, Z is the height of packing, ft. In most air pollution design practice, the number of transfer units (N_{OG}) is obtained experimentally or calculated using any of the methods to be explained later in this section. The height of a transfer unit (H_{OG}) is also usually determined experimentally for the system under consideration. Information on many different systems using various types of packings has been compiled by the manufacturers of gas absorption equipment and should be consulted prior to design. The data are usually in the form of graphs depicting, for a specific system and packing, the H_{OG} vs. the gas rate (lb/hr · ft^2) with the liquid rate (lb/hr · ft^2) as a parameter. The packing height (Z) is then simply the product of the H_{OG} and the N_{OG}. Although there are many different approaches to determine the column height, the H_{OG}–N_{OG} approach is the most frequently used at the present time, with the H_{OG} usually being obtained from the manufacturer. A more "theoretical" approach is available and a brief discussion on this method follows.

In air pollution operations, the pollutant to be absorbed is often very dilute. For this condition

$$N_{OG} = \int \frac{dy}{(y - y^*)} \tag{5.9}$$

If the operating line and equilibrium line are both parallel and straight

$$N_{OG} = \frac{(y_1 - y_2)}{(y - y^*)} \tag{5.10}$$

If the operating line and equilibrium line are just straight

$$N_{OG} = \frac{(y_1 - y_2)}{(y - y^*)_{\mathrm{ln}}}; \quad (y - y^*)_{\mathrm{ln}} = \frac{(y - y^*)_1 - (y - y^*)_2}{\mathrm{ln}\left[\dfrac{(y - y^*)_1}{(y - y^*)_2}\right]} \tag{5.11}$$

If the operating line and/or equilibrium line is curved, the integral above should be evaluated. However, one can show that if Henry's law applies (operating line and

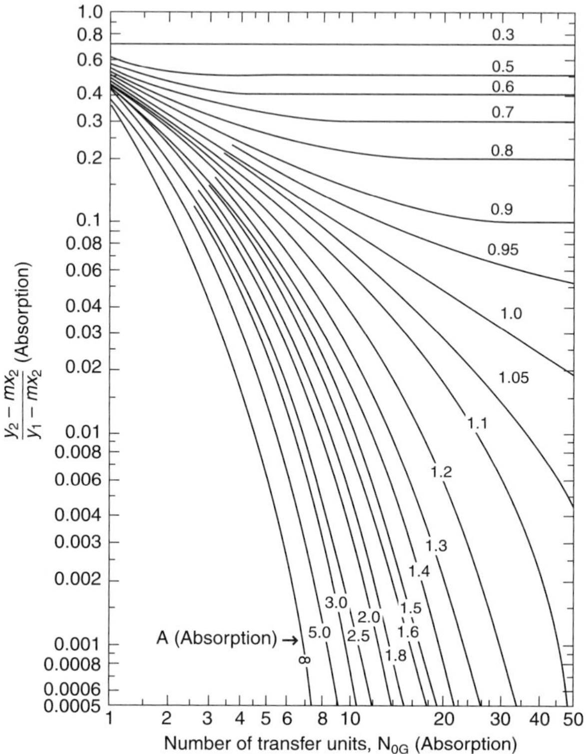

Figure 5.6 N_{OG} for an absorption column with constant absorption factor.

equilibrium line straight), the number of transfer units is given by

$$N_{OG} = \frac{\ln\left[\frac{(y_1 - mx_2)}{(y_2 - mx_2)}\left(1 - \frac{1}{A}\right) + \frac{1}{A}\right]}{1 - (1/A)}; A = L/mG \tag{5.12}$$

where the absorption factor is A (not to be confused with area) and is equal to L/mG where m is the slope of the equilibrium curve. The solution to this equation can conveniently be found graphically from Figure 5.6. Note also that the flow rates L and G are based on moles in Equation (5.12). However, if $m = 0$, Equation (5.12) reduces to

$$N_{OG} = \ln\left(\frac{y_1}{y_2}\right) \tag{5.13}$$

This equation may be applied if the gas is highly soluble or if the absorbate (pollutant) reacts with the liquid (L. Theodore: personal notes, 1976).

Qualitatively, the height of a transfer unit is a measure of the height of a contactor required to effect a standard separation, and it is a function of the gas flow rate, the liquid flow rate, the type of packing, and the chemistry of the system. As indicated above, experimental values for N_{OG} are generally available in the literature or from vendors.

Some general rules of thumb in the design of packed columns do exist. They are by no means all-inclusive in that there are other considerations that might have to be taken into account, e.g., allowable pressure drop, possible column height restrictions, etc. The rules must therefore be applied discriminately. For approximation purposes, if the gas rate is greater than about 500 acfm, a nominal packing size smaller than 1 inch would probably not be practical; similarly, at about 2000 acfm, packing sizes smaller than 2 inch would also likely be impractical. The nominal size of the packing should never exceed about 1/20th of the column diameter.

As noted earlier, the usual practice is to design so that the operational gas rate is approximately 75% of the rate that would cause flooding. If possible, column dimensions should be in readily available sizes (i.e., diameters to the nearest half-foot and heights to the nearest foot). If the column can be purchased "off the shelf" as opposed to being specially made, a substantial savings can be realized.

Plate Column

The most important design considerations for plate columns include calculation of the column diameter, type and number of plates to be used (usually sieve plates), actual plate layout and physical design, and plate spacing (which, in turn, determines the column height). To consider each of these to any great extent is beyond the scope of this book. Details are available in any standard chemical engineering unit operations or mass transfer (e.g., distillation) text. The discussion that follows, therefore, will be a relatively concise presentation of the general design techniques that will provide satisfactory results for the purpose of estimation.

The column diameter and consequently its cross section must be sufficiently large to handle the gas and liquid at velocities that will not cause flooding or excessive entrainment. The superficial gas velocity for a given type of plate at flooding is given by the relation

$$V_{\mathrm{F}} = C_{\mathrm{F}}\left(\frac{\rho_{\mathrm{L}} - \rho_{\mathrm{G}}}{\rho_{\mathrm{G}}}\right)^{0.5} \tag{5.14}$$

where V_{F} is the gas volumetric flowrate through the net column cross sectional area for gas flow, $\mathrm{ft}^3/\mathrm{s}\cdot\mathrm{ft}^2$; ρ_{L}, ρ_{G} are the liquid and gas densities, respectively, $\mathrm{lb}/\mathrm{ft}^3$; and C_{F} is an empirical coefficient that depends on the type of plate and operating conditions.

The net cross section is the difference between the column cross section and the area taken up by downcomers. In actual design, some percent of V_{F} is usually used: for non-foaming liquids 80–85% of V_{F}, and 75% or less for foaming liquids. Of course, the value is subject to a check of entrainment and pressure drop characteristics. The calculation of column diameter based on Equation (5.14) assumes that the gas flow rate is the controlling factor in its determination.

After a plate layout has been assumed, it is necessary to check the plate for its liquid-handling capacity. If the liquid-to-gas ratio is high and the column diameter large, the check will indicate whether the column will show a tendency towards flooding or gas maldistribution on the plate. If this is the case, then the liquid rate is the controlling factor in estimating column diameter and a satisfactory assumption for design purposes is a plate-handling capacity of 30 gal/min of liquid per foot of diameter. However, a well-designed single-pass, crossflow plate can ordinarily be expected to handle up to 60 gal/min of liquid per foot of diameter without excessive liquid gradient on the plate. It should also be noted that low gas rates can lead to *weeping*, a condition where the liquid flows down through the holes in the plate rather than across the plate.

The column height is determined from the product of the number of actual plates (theoretical plates divided by the overall plate efficiency) and the plate spacing chosen. The theoretical plate (or "stage," as it is sometimes called) is the theoretical unit of separation in plate column calculations. It is defined as a plate in which two dissimilar phases are brought into intimate contact with each other until equilibrium is reached and then are mechanically or otherwise separated. During the contact, various diffusing components of the mixture redistribute themselves between the phases. In an equilibrium stage the two phases are well mixed for a time sufficient to allow establishment of equilibrium between the phases leaving the stage. At equilibrium, no further net change of composition of the phases is possible for a given set of operating conditions. The number of theoretical plates can be determined graphically from the operating diagram composed of an operating line and equilibrium curve.

For cases when both the operating line and the equilibrium curve may be considered straight (dilute solutions), the number of theoretical plates may be determined directly without recourse to graphical techniques. This will frequently be the case for relatively dilute gases (as usually encountered in air pollution control) and liquid solutions where, more often than not, Henry's law is usually applicable. Since the quantity of gas absorbed is small, the total flows of liquid and gas entering and leaving the column remain essentially constant. Hence, the operating line will be substantially straight. For such cases, the Kremser–Brown–Souders equation applies for determining the number of theoretical plates N:

$$N = \frac{\log\left[\left(\dfrac{y_{N+1} - mx_0}{y_1 - mx_0}\right)\left(1 - \dfrac{1}{A}\right) + \dfrac{1}{A}\right]}{\log A} \tag{5.15}$$

Note: ln may also be employed in both the numerator and denominator. Here mx_0 is the gas composition in equilibrium with the entering liquid (m is Henry's law constant = slope of equilibrium curve). If the entering liquid contains no solute gas, then $x_0 = 0$ and Equation (5.15) can be simplified further. The solute concentrations in the gas stream, y_{N+1} and y_1 represent inlet and outlet concentrations, and L and G the total mole rates of liquid and gas flow per unit time per unit column cross-sectional area, respectively. Small variations in L and G may be roughly compensated for by using the geometric average value of each taken at the top and bottom of the column. Equation (5.15) has been plotted in Figure 5.7 for convenience and may be used for the solution to this equation.

The number of actual trays, which is based on the tray efficiency, is determined by the mechanical design used and the conditions of operation. For the case where the equilibrium curve and operating lines are straight, the overall tray efficiency E_0 can be

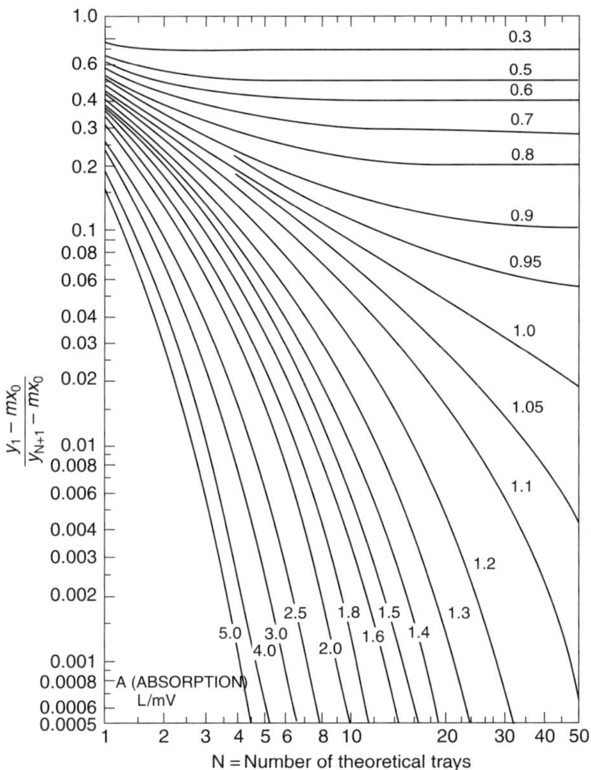

Figure 5.7 Number of theoretical stages for countercurrent absorption columns.

computed and the number of actual trays determined analytically:

$$E_0 = \frac{\text{equilibrium trays}}{\text{actual trays}}$$

$$= \frac{\log\,(1 + E_{\text{MGE}})(1/A - 1)}{\log\,(1/A)} \tag{5.16}$$

where E_{MGE} = Murphree efficiency corrected for entrainment (values available in the literature). Empirical data for standard tray designs within standard ranges of liquid and gas rates are available. These data, as shown in Figure 5.8, are accurate for bubble-cap trays and can be used as rough estimates for sieve and valve trays. After the overall efficiency of the tower is determined, the number of actual trays is calculated as:

$$N_{\text{act}} = \frac{N}{E_0} \tag{5.17}$$

The general procedure to follow in sizing a plate column is as follows:

1. Calculate the number of theoretical stages N using Figure 5.7 or Equation (5.15).

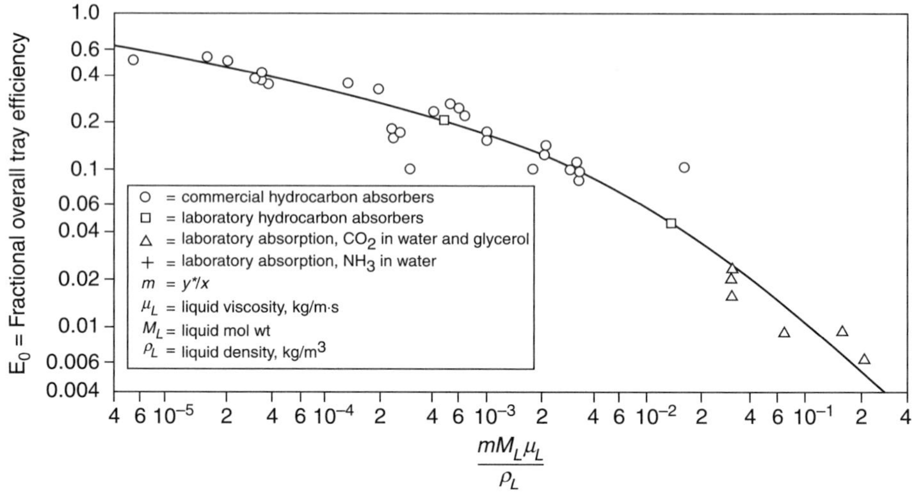

Figure 5.8 Overall tray efficiencies of bubble-cap tray absorbers.

2. Estimate the efficiency of separation E. This may be determined at the local (across plate), plate (between plates), or overall (across column) level. The overall efficiency E_0 is generally employed.
3. Calculate the actual number of plates:

$$N_{\text{act}} = \frac{N}{E_0} \tag{5.17}$$

4. Obtain the height between plates h. This is usually in the 12–36 inch range. Most towers use a 24-inch plate spacing.
5. The tower height Z is then

$$Z = (N_{\text{act}})(h) \tag{5.18}$$

6. The diameter may be calculated directly from Equation (5.14).
7. The plate or overall pressure drop is difficult to quantify. It is usually in the 2–6-inch H_2O per plate range for most columns with the lower and upper values applying to small and large diameters, respectively. This pressure drop value represents the approximate height of the liquid on the plates.

Strippers

Absorbers are also used for stripping to regenerate the liquid. During this operation, the pollutant (absorbate) is transferred from the liquid to the gas (the stripping medium). The calculations essentially remain the same for both packed and plate towers, except the height of a packed tower is given by

$$Z = (H_{\text{OL}})(N_{\text{OL}}) \tag{5.19}$$

Summary of Key Equations

The key equations for stripping calculations, including a summary of earlier material, is presented below.

For packed tower absorption

$$N_{\mathrm{OG}} = \frac{\ln\left[\dfrac{(y_1 - mx_2)}{(y_2 - mx_2)}\left(1 - \dfrac{1}{A}\right) + \dfrac{1}{A}\right]}{1 - \dfrac{1}{A}} \tag{5.20}$$

For stripping

$$N_{\mathrm{OL}} = \frac{\ln\left[\dfrac{(x_2 - y_1/m)}{(x_1 - y_1/m)}(1 - A) + A\right]}{1 - A} \tag{5.21}$$

where the subscripts 1 and 2 refer to bottom and top conditions. In addition, $A = L/mG$ and $S = 1.0/A$.

To use Figure 5.6 for stripping calculations, replace the y coordinate, x coordinate, and parameter by $[x_1 - (y_1/m)]/[x_2 - (y_2/m)]$, N_{OL}, and S, respectively, where

$$S = 1.0/A = mG/L \tag{5.22}$$

For plate tower absorption

$$N = \frac{\log\left[\left(\dfrac{y_{N+1} - mx_0}{y_1 - mx_0}\right)\left(1 - \dfrac{1}{A}\right) + \dfrac{1}{A}\right]}{\log A} \tag{5.23}$$

Note: ln may also be employed in both the numerator and denominator. If A approaches unity, Equation (5.23) becomes

$$N = \frac{y_{N+1} - y_1}{y_1 - mx_0} \tag{5.24}$$

or

$$\frac{y_{N+1} - y_1}{y_{N+1} - mx_0} = \frac{N}{N+1} \tag{5.25}$$

Note that the subscripts 1 and N refer to the top and bottom of the column, respectively. For stripping in plate towers

$$N = \frac{\log\left[\left(\dfrac{x_0 - y_{N+1}/m}{x_N - y_{N+1}/m}\right)\left(1 - \dfrac{1}{S}\right) + \dfrac{1}{S}\right]}{\log S} \tag{5.26}$$

or

$$\frac{x_0 - x_N}{x_0 - (y_{N+1}/m)} = \frac{S^{N+1} - S}{S^{N+1} - 1} \tag{5.27}$$

If S is approximate 1.0, one may use either of the following equations:

$$N = \frac{x_0 - x_N}{x_N - (y_{N+1}/m)} \tag{5.28}$$

$$\frac{x_0 - x_N}{x_0 - (y_{N+1}/m)} = \frac{N}{N+1} \tag{5.29}$$

To use Figure 5.7 for stripping, replace the y coordinate and the parameter A by $[x_N - (y_{N+1}/m)]/x_0 - (y_{N+1}/m)]$ and S, respectively.

5.3 OPERATION AND MAINTENANCE, AND IMPROVING PERFORMANCE

Since packed towers are used primarily for gaseous pollutant control, this device will be emphasized in the presentation below, although much of this material will also be applicable to plate and other towers.

Normal preventive maintenance requires only periodic checks of the fan, pumps, chemical feed system, piping, duct, and liquid distributor. The normally irrigated packing may never have to be cleaned during the life of the absorber as long as the absorber has been properly designed and operated. However, should the absorber be run dry or the entering air stream contain unexpectedly high solids loading, a heavy formation of solids, crystallized salts, or other foreign matter may accumulate on the packing, and this must be removed. In most cases, removal can be accomplished by recirculating, for a short period of time, a chemical solution into which the solids will dissolve or react. Sometimes, chemical treatment will not remove the buildup and it may be necessary to use high-pressure water, hot water, or atmospheric steam. The absorber manufacturer should be consulted to verify the resistance of the internals and shell to the cleaning process. Prior to using any chemical, hot water, or steam, in a very few instances, such as CaF_2 deposition or extremely heavy solids deposition, it may be necessary to remove the packing medium from the absorber for cleaning.

The normally unirrigated entrainment separator must be periodically flushed with sprays to prevent buildup and eventual plugging. The intervals between routine washings must be determined by experience, as the collection of solid materials is a function of specific operating conditions.

A maintenance checklist is suggested to ensure proper operation and unexpected problems with the absorber system. The periodic time intervals indicated for maintenance will vary depending on the specific system's operating conditions and the equipment manufacturer's recommendation.

There are options available that should be considered when purchasing a packed absorber (or any absorber) in order to reduce maintenance costs and allow the equipment to be more operationally reliable between scheduled maintenance shutdowns. Any of these options generally increase the initial cost of the equipment, and for this reason are seldom considered. However, some of these items can pay for themselves in a short time by reducing maintenance costs.

Another source of periodic maintenance relates to the strainers or filters in the recycle piping and the recycle pumps. To prevent total system shutdown during cleaning or maintenance of these items, dual strainers or pumps may be considered, valved separately, so that the system can be maintained in a fully operational mode during strainer cleaning or pump maintenance checks.

Proper instrumentation used for the purpose of providing operational data is useful in determining areas where problems or maintenance may be required. The type of instrumentation will depend totally on the system requirements. Areas of specific interest would be pressure drop across the absorber, both the wetted bed and entrainment separator, low liquid recycle flow and/or pressure, liquid makeup rates to the system, and liquid temperatures.

The absorber can be designed to provide future additional packing height requirements should greater future gas absorption capability be required. This can be accomplished in one of two ways: the shell of the absorber can be flanged to allow a future stub section to be inserted, or the tower can be initially designed with a void space above the packing between the liquid distributors so that future packing heights can be added without shell modification. Although the initial cost of this may seem high, it could preclude the need to field-modify the absorber or to purchase a new absorber for greater efficiency, when and if the air pollution laws are tightened.

Recent developments with absorbers have centered primarily on improvement of

1. Packing types
2. Plate manufacture
3. Liquid distributor manufacture (the author favors one spray).

However, the reader should note that these units are still designed essentially as they were nearly 100 years ago.

PROBLEMS

5.1 Effect of Pressure on the Absorption Process

If the pressure of the system increases, the amount of absorbate retained in solution

(**a**) Increases
(**b**) Decreases
(**c**) Remains unaffected
(**d**) Is doubled

Solution: In line with physical chemistry principles, as the pressure is increased, more absorbate is driven from the gas phase into the liquid. The correct answer is therefore (a).

5.2 Effect of Temperature on the Absorption Process

During gas absorption, as the temperature of the system increases, the amount of pollutant absorbed generally

(**a**) Increases
(**b**) Decreases
(**c**) Remains constant
(**d**) Varies with time

Solution: In line with physical chemistry principles more absorbate is driven from the liquid to the gas as the temperature decreases. The correct answer is therefore (b).

5.3 Absorber Equipment

Which of the equipment listed below is generally not classified as an absorber?

(**a**) Packed tower
(**b**) Spray tower
(**c**) Plate column
(**d**) Indirect condenser

Solution: As described in Section 5.1, packed towers, spray towers, and plate columns are classified as absorbers. The correct answer is therefore (d).

5.4 Absorber Flow Classification

The type of packed tower in which the average concentration gradient is maximized (the most dilute pollutant in the gas phase is contacted with the purest liquid absorbent) is the

(**a**) Concurrent
(**b**) Countercurrent
(**c**) Cross-current
(**d**) Cyclonic

Solution: During countercurrent flow, the most dilute pollutant is contacted with the purest liquid absorbent at the top of the column. The net effect is to produce the highest average concentration difference driving force across the entire length of the column. The correct answer is therefore (b).

5.5 Henry's Law Limitation

Henry's law applies if the equilibrium diagram of a system is a straight line. For most systems, the equilibrium diagram is a straight line if

(**a**) The solution is very concentrated
(**b**) The solution is dilute
(**c**) Water is the solute
(**d**) The partial pressure of the absorbate is large

Solution: For most systems, an equilibrium plot of y (or p) vs. x is concave μp for higher values of x. It is linear at lower values. The correct answer is therefore (b).

5.6 Effect of Heavy Inlet Particulate Load

Which of the following absorbers would be the worst choice if the exhaust stream contained a heavy particulate load?

(**a**) Venturi
(**b**) Spray tower

(**c**) Cross-flow packed tower
(**d**) Countercurrent packed tower

Solution: A heavy particulate load would severely impact any absorber that contained packing. Therefore, only (c) and (d) are eligible answers. However, the ability to retain the particulate is further enhanced with countercurrent flow, i.e., the likelihood of washing out the particulates is reduced. The correct answer is therefore (d).

5.7 Countercurrent Packed Tower/Spray Tower Comparison

Countercurrent packed flow absorbers are more efficient than spray towers because of the packed tower's increased

(**a**) Pressure drop used to collect the gases
(**b**) Diameter
(**c**) Tower height
(**d**) Gas–liquid interfacial area

Solution: Packing provides for more intimate mixing between the gas and liquid phases, producing more interfacial area for control. The correct answer is therefore (d).

5.8 Gas versus Liquid Dispersal Methods

In gas absorption, the two methods of contacting the gas and liquid streams are gas dispersal and liquid dispersal. An absorber that uses the gas dispersal method is the

(**a**) Spray tower
(**b**) Packed tower
(**c**) Plate tower
(**d**) Moving-bed absorber

Solution: In a plate tower, the rising gas/vapor stream is forced to "bubble" through the liquid. The plate openings disperse the gas as it "bubbles" through liquid. The correct answer is therefore (c).

5.9 Limiting Efficiency of a Venturi Scrubber

One of the main disadvantages that limits the efficiency of a venturi-type absorber for removing gases is the

(**a**) Short residence time
(**b**) High L/G ratio
(**c**) Low L/G ratio
(**d**) Small interface area

Solution: Since the flow rate of gases traveling through the venturi (see Chapter 11) throat is usually above 200 ft/s, the residence time for mass transfer to occur is minimal. The correct answer is therefore (a).

5.10 Packing Application

Raschig rings or berl saddles would be used

(**a**) Around the throat of a venturi scrubber
(**b**) In a cyclonic separator before the demister

(**c**) At the top of a cyclonic spray scrubber
(**d**) In a cross flow packed scrubber

Solution: Answers (a)–(c) do not pertain to packing material. The correct answer is therefore (d).

5.11 Plastic Packing Advantage

Discuss why plastic packing is often employed in absorber applications.

Solution: Thermoplastic materials of construction for packings are the most widely used for gas absorption in air pollution control. Care must be taken to choose the proper thermoplastic material for the corrosive environment and temperature. Thermoplastic packings are available in polypropylene, polyethylene, polyvinyl chloride (PVC), Noryl, Kynar, Udel polysulfone, and Tefzel, to name the most common. If the proper material is chosen, it can be virtually inert to a corrosive environment. The smooth surface of plastic packings reduces fouling since any solids that may be deposited do not cling, as they would on a rough surface, and are more easily washed away. Lightweight plastic allows a more open support plate to be used, reducing pressure drop and solids deposition. This lighter weight also reduces the shell cost and is less expensive to install and support. Plastic packings are easily installed by simply dumping the packing into an empty column. When installing the packing, care should be exercised to prevent bridging of the packing.

5.12 Packing Characteristics

The packing is the heart of the performance of packed column. Its proper selection entails an understanding of packing operational characteristics. List the main points to be considered in choosing the column packing.

Solution:

1. *Durability and Corrosion Resistance*. The packing should be chemically inert to the fluids being processed.
2. *Free (void) Space per Unit Volume of Packed Space*. This controls the liquid holdup in the column as well as the pressure drop across it.
3. *Wetted Surface Area per Unit Volume of Packed Space*. This affects the pressure drop across the column.
4. *Packing Resistance to the Flow of Gas*. This affects the pressure drop across the column.
5. *Packing Stability and Structural Strength*. This permits easy handling and installation.
6. *Weight per Unit Volume of Packed Space*.
7. *Cost per Unit Area of Effective Surface*.

5.13 Absorber Advantages

List at least three advantages of using an absorber for gaseous control.

Solution:

1. High efficiency
2. Moderate/low pressure drop
3. Moderate/low capital cost
4. Moderate/low operating cost

5.14 Absorber Disadvantages

List at least three disadvantages of using an absorber for gaseous control.

Solution:

1. Vapor plume
2. Contaminated liquid
3. Some operating problems
4. Distribution problems

5.15 Absorber Types

List the principal types of gas absorption equipment.

Solution:

1. Packed columns (continuous operation)
2. Plate columns (staged operation)
3. Miscellaneous

Of the three categories, the packed column is by far the most commonly used for the absorption of gaseous pollutants. It might also be mentioned again at this time that the exhaust (cleaned gas) from an absorption air pollution control system is usually released to the atmosphere through a stack. To prevent condensation of the vapor plume in and around the stack, the temperature of this exhaust gas should be above its dew point. A general rule of thumb is to ensure that the exhaust gas stream temperature is approximately 50°F above its dew point.

5.16 Packed-Column Characteristics

Packed columns are characterized by a number of features to which their widespread popularity may be attributed. List some of these features.

Solution:

1. *Minimum Internal Problems*. The packed column needs only a packing support and liquid distributor about every 10 ft along its height.
2. *Versatility*. The packing material can be changed by simply dumping it and replacing it with a type giving better efficiency, lower pressure drop, or higher capacity. The depth of packing can also be easily changed if the operating efficiency turns out to be less than anticipated, or if the feed or product specifications change.

3. *Corrosive-Fluids Handling*. Ceramic packing is common and often preferable to metal or plastic because of its corrosion resistance. It may also be preferred at the base or inlet of a tower when handling hot combustion gases, as in a hazardous waste incineration facility.
4. *Low Pressure Drop*. Unless operated at very high liquid rates, where the liquid becomes the continuous phase as its dropwise films thicken and merge, the pressure drop per linear foot of packed height is relatively low.
5. *Low Investment(s)*. When plastic packings are satisfactory or when the columns are less than 3–4 ft in diameter, cost is relatively low.

5.17 Solvent Selection

List 10 factors that should be considered when choosing a solvent for a gas absorption column that is to be used as an emission control device.

Solution: Solvent selection for use in an absorption column for gaseous pollutant removal should be based on the following criteria:

1. *Solubility of the Gas in the Solvent*. High solubility is desirable as it reduces the amount of solvent needed. Generally, a polar gas will dissolve best in a polar solvent and a nonpolar gas will dissolve best in a nonpolar solvent.
2. *Vapor Pressure*. A solvent with a low vapor pressure is preferred to minimize loss of solvent.
3. *Corrosivity*. Corrosive solvents may damage the equipment. A solvent with low corrosivity will extend equipment life.
4. *Cost*. In general, the less expensive, the better, However, an inexpensive solvent is not always the best choice if it is too costly to dispose of and/or recycle after it has been used.
5. *Viscosity*. Solvents with low viscosity offer benefits such as better adsorption rates, better heat transfer properties, lower pressure drops, lower pumping costs, and improved flooding characteristics.
6. *Reactivity*. Solvents that react with the contaminant gas to produce an unreactive product are undesirable because the scrubbing solution may be recirculated. Reactive solvents should produce few unwanted side reactions with the gases that are to be absorbed.
7. *Low Freezing Point*. Resistance to freezing lessens the chance of solid formation and clogging of the column. See criterion 2 (above) on boiling point (vapor pressure).
8. *Availability*. If the solvent is "exotic," it generally has a higher cost and may not be readily available for long-term continuous use. Water is often the natural choice according to these criteria.
9. *Flammability*. Lower flammability or nonflammable solvents decrease safety problems.
10. *Toxicity*. A solvent with low toxicity is desirable.

5.18 Fouling

Discuss fouling in packed towers.

Solution: The most common maintenance problem associated with packing media is fouling or plugging. Plugging can result from deposition of undissolved solids in the liquid, deposition of insoluble dust in the inlet air stream, precipitation of dissolved salts in the liquid that exceed their solubility, and precipitation of an insoluble salt that is formed by the reaction of a soluble salt in solution with an absorbed gas. Of those mentioned, the most prevalent and the most difficult to prevent is the precipitation of an insoluble salt formed by reaction. Typically, this will occur because of the Ca^{2+} or Mg^{2+} in hard water reacting with the absorbed CO_2 from the air to form $CaCO_3$ and $MgCO_3$ scaling. In most cases, this will occur with the use of an alkaline scrubbing liquid. This scaling can also be removed by periodic flushing with an acidic solution such as HCl. Scaling can also be reduced by using a packing having a large free (void) volume. The other plugging problems mentioned can be easily reduced by increasing the liquid flow through the column, adding a strainer in the recycled liquid piping, or increasing the water makeup and overflow rates.

5.19 Liquid Distributors in Packed Columns

Discuss the use of liquid distributors in packed columns.

Solution: Although the type and size of packing are important in designing the physical absorber size and packing depth to achieve a specified removal efficiency of the gaseous pollutants, an equally important factor is the method of liquid distribution over the packing. If the liquid is not distributed evenly across the packing, or if portions of the packing are not wetted, channeling will occur and reduce performance. Again, bear in mind that it is the liquid not the packing, that absorbs the gases. Also, packing that is not wetted will tend to plug much *faster* because there is no liquid continually washing away any solids deposition.

There are four basic types of liquid distributors: spray, weir, perforated pipe, and open splash plate. The spray type using spray nozzles provides the most effective means of evenly distributing the liquid over the packing. It also provides an area above the packed section for additional contact between the air stream and the spray droplets for gas absorption. A properly designed spray distributor should provide overlapping sprays, and the top of the packing should be at a specified distance below the sprays for optimum effectiveness. If the packing level is too high, some areas near the tower wall will not be wetted; conversely, if the packing level is too low, too great a portion of the liquid is sprayed on the vessel wall and becomes ineffective. Full cone-type spray nozzles are suggested and should be as nonclogging as possible. Spray nozzle plugging is the major disadvantage. However, with the proper nozzle selection and by installing a strainer in the recycle piping which is capable of removing solids smaller than the nozzle orifice size, nozzle plugging can be drastically reduced, if not eliminated.

5.20 Crossflow Absorbers

Discuss some of the features of crossflow absorbers.

Solution: The crossflow design is unique and has become a commonly used option for the removal of highly or moderately soluble gases, or where rapid chemical reactions will occur. In this mode, the air stream flows horizontally through the packing while the liquid flows vertically downward. The crossflow mode has some attractive features over the countercurrent design, while affording lower pressure drop. Lower pressure drop results in the crossflow unit where the liquid is at right angles to the airflow, resulting in less pressure drop when compared to the countercurrent mode, where the gas and liquid flow in opposite directions. In addition, the crossflow has greater solids-handling capacity without causing fouling. This can be achieved by increasing the liquid flow over the first section of packing to flush off any collected solid to prevent buildup, and by adding a front cocurrent spray to prevent blinding at the front packing support plate. Care must be taken to prevent gas from bypassing the packed bed. This bypassing occurs primarily in the liquid distribution zone and in the sump; however, it can be eliminated by the proper use of internal baffles.

5.21 Packed versus Plate Column

Briefly describe the differences between a packed column and a plate column.

Solution: Of the various types of gas absorption devices mentioned, packed columns and plate columns are the most commonly used in industrial practice. Although packed columns are used more often in air pollution control, both have their special areas of usefulness, and the relative advantages and disadvantages of each are worth considering. In general:

1. The pressure drop of the gas passing through the packed column is smaller.
2. The plate column can treat an extremely low liquid feed and permits a higher gas feed than the packed column. It can also be designed to handle liquid rates that would ordinarily flood the packed column.
3. If the liquid deposits a sediment, the plate column is more advisable. By fitting the column with manholes, the plate column can be cleaned of accumulated sediment that would clog many packing materials and warrant necessary costly removal and refilling of the column. Packed columns are also susceptible to plugging if the gas contains particulate contaminants.
4. In mass transfer processes accompanied by considerable heat effects, cooling or heating the liquid is accomplished more easily in the plate column. A system of pipes immersed in the liquid can be placed between the plates and heat can be removed or supplied through the pipe wall directly to the area in which the process is taking place. The solution of the same problem for a packed column leads to the division of this process into a number of sections, with the cooling or heating of the liquid taking place between these sections.
5. The total weight of the plate column is usually less than that for the packed column designed for the same capacity.

6. A well-installed plate column avoids serious channeling difficulties, ensuring good, continuous contact between the gas and liquid throughout the column.
7. In highly corrosive atmospheres, the packed column is simpler and cheaper to construct.
8. The liquid holdup in the packed column is considerably less than in the plate column.
9. Temperature changes are apt to do more damage to a packed column than to a plate column.
10. Plate columns are advantageous for absorption processes with an accompanying chemical reaction (particularly when it is not very rapid). The process is favored by a long residence time of the liquid in the column and by easier control of the reaction.
11. Packed column are preferred for liquids with high foaming tendencies.
12. The relative merits of the plate column and packed column for a specified purpose are properly determined only by comparison of the actual cost figures resulting from a detailed design analysis for each type. Most conditions being equal, packed columns in the smaller sizes (diameters up to approximately 2–3 ft) are on the average less expensive. In the larger sizes, plate columns tend to be the more economical.

5.22 Henry's Law

Given Henry's law constant and the partial pressure of H_2S, determine the maximum mole fraction of H_2S that can be dissolved in solution at the given conditions. Data are provided below:

Partial pressure of $H_2S = 0.01$ atm
Total pressure $= 1.0$ atm
Temperature $= 60^\circ$F
Henry's law constant, $H_{H_2S} = 483$ atm/mol fraction (1 atm, 60°F)

Solution: An important equilibrium phase relationship (see Chapter 3 for more details) is that between liquid and vapor. Raoult's and Henry's laws theoretically describe liquid–vapor behavior and under certain conditions are applicable in practice. Raoult's law is sometimes useful for mixtures of components of similar structure. It states that the partial pressure of any component in the vapor is equal to the product of the vapor pressure of the pure component and the mole fraction of that component in the liquid, i.e.,

$$p_i = p'x_i \tag{5.30}$$

where $p_i =$ partial pressure of component i in the vapor; $p' =$ vapor pressure of pure i at the same temperature; $x_i =$ mole fraction of component i in the liquid. This expression may be applied to all components. If the gas phase is ideal, then

$$p_i = y_iP \tag{5.31}$$

and the first equation can be written as follows:

$$y_i = (p'/P)x_i \tag{5.32}$$

where y_i = mole fraction component i in the vapor; P = total system pressure. Thus, the mole fraction of water vapor in a gas that is saturated, i.e., in equilibrium contact with pure water ($x = 1.0$), is given simply by the ratio of the vapor pressure of water at the system temperature divided by the system pressure. These equations find application in distillation, absorption, and stripping calculations.

Unfortunately, relatively few mixtures follow Raoult's law. Henry's law is a more empirical relationship used for representing data on many systems. Here

$$p_i = H_i x_i \tag{5.33}$$

where H_i is Henry's law constant for component i (in units of pressure). If the gas behaves ideally, Equation (5.33) may also be written as

$$y_i = m_i x_i \tag{5.34}$$

where m_i is a constant (dimensionless). See Section 5.2 for additional details. This is a more convenient form of Henry's law. The constant m_i (or H_i) has been determined experimentally for a large number of compounds and is usually valid at low concentrations. In word equation form, Henry's law states that the partial pressure of a solute in equilibrium in a solution is proportional to its liquid mole fraction. The law is exact in the limit as the concentration approaches zero.

The most appropriate form of Henry's law for this problem is Equation (5.33). The maximum mole fraction of H_2S that can be dissolved in solution in the problem statement can now be calculated:

$$\begin{aligned} x_i &= p_i/H_{H_2S} \\ &= 0.01/483 \\ &= 0.0000207 \\ &= 20.7\,\text{ppm} \end{aligned} \tag{5.33}$$

Henry's law may be assumed to apply for most dilute solutions. This law finds widespread use in absorber calculations since the concentration of the solute in some process gas streams is often dilute. This greatly simplifies the study and design of absorbers. One should note, however, that Henry's law constant is a strong function of temperature.

5.23 Absorption Equilibria

Experimental data are provided for an absorption system to be used for scrubbing ammonia (NH_3) from air with water. The water rate is 300 lb/min and the gas rate is 250 lb/min at 72°F. The applicable equilibrium data for the ammonia-air system is shown in Table 5.1(a) (R. H. Perry and D. W. Green, eds, *Perry's Chemical Engineers' Handbook*, 7th ed. McGraw-Hill, New York,

TABLE 5.1(a) Ammonia Equilibrium Data I

Equilibrium Partial Pressure (mm Hg)	NH_3 Concentration (lb NH_3/100 lb H_2O)
3.4	0.5
7.4	1.0
9.1	1.2
12.0	1.6
15.3	2.0
19.4	2.5
23.5	3.0

1996). The air to be scrubbed has 1.5% (mass basis) NH_3 at 72°F and 1 atm pressure and is to be vented with 95% of the ammonia removed. The inlet scrubber water is ammonia-free.

(a) Plot the equilibrium data in mole fraction units.
(b) Perform the material balance and plot the operating line on the equilibrium plot.

Solution: (a) Employing the data provided in the problem statement, convert the equilibrium partial pressure data and the liquid concentration data to mole fractions as shown in Table 5.1(b). Plotting the mole fraction values on the graph in Figure 5.9 results in a straight line. The slope of the equilibrium line is approximately 1.0.
(b) Convert the liquid and gas rates to lbmol/min:

$$L_M = L = 300/18$$
$$= 16.67 \text{ lbmol/min}$$

$$G_M = G = 250/29$$
$$= 8.62 \text{ lbmol/min}$$

TABLE 5.1(b) Ammonia Equilibrium Data II

Gas Mole Fraction y	Liquid Mole Fraction x
0.00447	0.0053
0.00973	0.0106
0.0120	0.0127
0.0158	0.0169
0.0201	0.0212
0.0255	0.0265
0.0309	0.0318

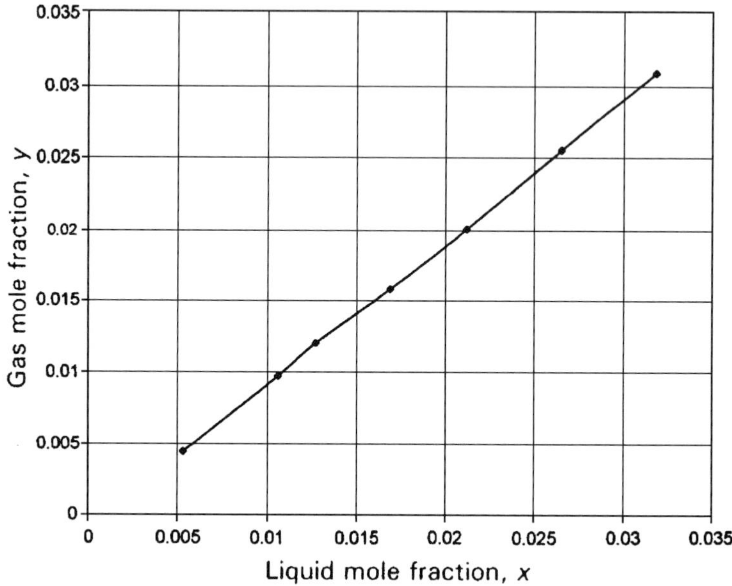

Figure 5.9 Equilibrium line for Problem 5.23.

Determine the inlet and outlet mole fractions for the gas, y_1 and y_2, respectively:

$$\begin{aligned} y_1 &= \frac{1.5/17}{1.5/17 + 98.5/29} \\ &= 0.0253 \end{aligned}$$

$$\begin{aligned} y_2 &= \frac{(0.05)(1.5/17)}{(0.05)(1.5/17) + (98.5/29)} \\ &= 0.0013 \end{aligned}$$

The inlet liquid mole fraction x_2 is given as zero, and the describing equation for x_1, the outlet liquid mole fraction, is

$$\begin{aligned} x_1 &= (G/L)(y_1 - y_2) + x_2 \\ &= (8.62/16.67)(0.0253 - 0.0013) + 0 \qquad (5.51) \\ &= 0.0124 \end{aligned}$$

One may now use the inlet and outlet mole fractions to plot the operating line on the graph in Figure 5.10.

The slope of the operating line is

$$\begin{aligned} \text{Slope} &= \frac{0.0253 - 0.0013}{0.0124 - 0.0} \\ &= 1.935 \end{aligned}$$

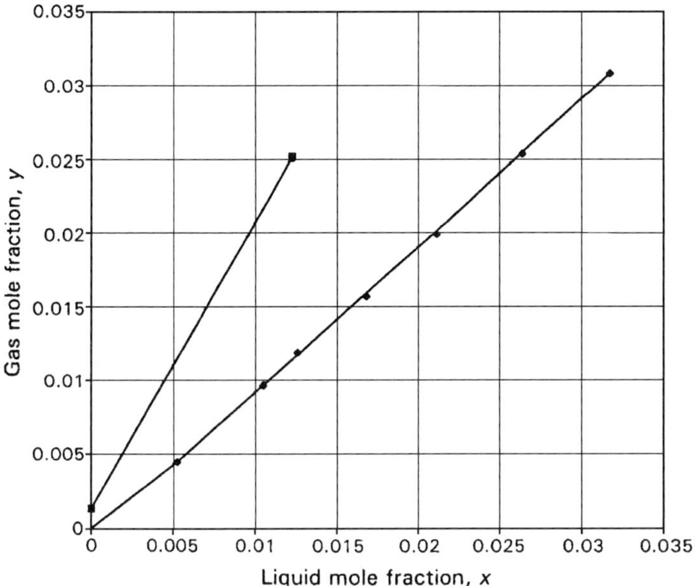

Figure 5.10 Operating line for Problem 5.23.

5.24 Column Mole Balance

Given the following information for a packed countercurrent gas scrubber, determine the liquid flow rate in lbmol/hr · ft^2.

Gas flow rate = 18 lbmol/hr · ft^2
The mole fractions of pollutant in inlet and outlet gas are 0.08 and 0.002, respectively
The mole fractions of pollutant in inlet and outlet liquid are 0.001 and 0.05, respectively.

(**a**) 28.7
(**b**) 36.0
(**c**) 40.0
(**d**) 57.3

Solution: Applying a componential mole balance (dropping the subscripts) on the pollutant gives

$$L(x_{\text{out}} - x_{\text{in}}) = G(y_{\text{out}} - y_{\text{in}}) \tag{5.4}$$

Substituting, one obtains

$$L(0.05 - 0.001) = 18(0.08 - 0.002)$$
$$L = 18(0.078)/0.049$$
$$= 28.7 \text{ lbmol/hr} \cdot \text{ft}^2$$

The correct answer is therefore (a).

5.25 Liquid Requirement

The EPA has conducted investigations on the exhaust streams from Valco Chemical Company. To Valco's dismay, their hydrocarbon emissions are too high and must be reduced in order to continue operation. At present conditions, the exhaust stream contains 1.0% benzene at a total flow of 40,000 ft^3/hr. It has been determined that if the exhaust stream is brought down to 0.01% benzene, the company can continue operation. An absorber that is currently out of commission will be used to absorb the benzene, employing a light wash oil as the absorbent. The pump on the liquid feed has a maximum liquid pumping rate of 50 ft^3/hr. Does this pump have enough capacity to do the job if the outlet benzene concentration in the wash oil cannot exceed 5.0%? Operating data are provided below:

Gas temperature = 100°F
Liquid temperature = 60°F
Light oil molecular weight (MW) = 156 lb/lbmol
Light oil density = 55.1 lb/ft^3
Actual liquid flow rate = $(1.5)L_{m,\min}$

Solution: Find the gas molar flow rate G_m.

$$G_m = (40{,}000\ \text{ft}^3/\text{hr})(\text{lbmol}/397\ \text{ft}^3)[(460+60)/(460+100)]$$
$$= 98\ \text{lbmol/hr}$$

Calculate the minimum liquid rate $L_{m,\min}$ by applying Equation (5.5):

$$\frac{L_m}{G_m} = \left[\frac{(y_{\text{in}} - y_{\text{out}})}{(x_{\text{out}} - x_{\text{in}})}\right] \tag{5.5}$$

$$L_{m,\min} = 98\left(\frac{0.01 - 0.0001}{0.05 - 0}\right)$$
$$= 19.4\ \text{lbmol/hr}$$

The actual operating liquid molar flow rate is

$$L_m = 1.5(19.4)$$
$$= 29.1\ \text{lbmol/hr}$$

The liquid mass flow rate is

$$L = 29.1(156\ \text{lb/lbmol})$$
$$= 4540\ \text{lb/hr}$$

The liquid volumetric flow rate is then

$$q_{\text{L}} = 4540\ \text{ft}^3/55.1\ \text{lb}$$
$$= 82.4\ \text{ft}^3/\text{hr}$$

The pump will therefore not do the job! Finally, the reader should consider whether there is another way to perform the calculation.

5.26 N_{OG} Calculation for a Highly Soluble Gas

The number of overall gas transfer units N_{OG} in a packed tower is given by

$$N_{OG} = \ln\left(\frac{y_1}{y_2}\right) \tag{5.13}$$

if the gas is highly soluble. Calculate N_{OG} if $y_1 = 200$ ppm and $y_2 = 0.5$ ppm.

(**a**) 3.91
(**b**) 4.26
(**c**) 5.99
(**d**) 5.30

Solution: Substituting in Equation (5.13) yields

$$\begin{aligned} N_{OG} &= \ln(200/0.5) \\ &= \ln(400) \\ &= 5.99 \end{aligned}$$

The correct answer is therefore (c).

5.27 Inlet Concentration for a Very Soluble Gas

Assuming that N_{OG} for a very soluble gas is 4.6, calculate y_2, given $y_1 = 100$ ppm:

(**a**) 1.0
(**b**) 2.0
(**c**) 5.0
(**d**) 10.0

Solution: The solution to this problem is the reverse of that to Problem 5.26 since y_2 is unknown. Applying Equation (5.13) once again,

$$\begin{aligned} 4.6 &= \ln(100/y_2) \\ y_2 &= 1.0 \end{aligned}$$

The correct answer is therefore (a).

5.28 N_{OG} Calculations

(**a**) Find the number of overall gas transfer units (N_{OG}) in a packed tower required to remove 90% of a gas in an inlet air stream containing 10 mole percent (mol%) of the pollution gas using pure water at a rate 20% greater than the minimum rate. Assume $m = 1.485$.
(**b**) How many N_{OG} values would be required if instead of pure water, water containing 0.1, 0.3, 0.5, and 0.65 mol% (mole percent) of the gas were used instead?
(**c**) What mole percent of the gas in water would make such a separation impossible?

Solution:

(**a**) Calculations can be performed on a mole fraction or solute-free mole fraction basis. Since $y_1 = 0.1$, the solute-free mole fraction is

$$\begin{aligned} Y_1 &= y_1/(1 - y_1) \\ &= 0.1/(1 - 0.1) \\ &= 0.1111 \end{aligned} \tag{5.35}$$

In addition,

$$\begin{aligned} y_2 &= (0.1)(0.1111) \\ &= 0.0111 \end{aligned}$$

and

$$\begin{aligned} Y_2 &= 0.0111/(1 - 0.0111) \\ &= 0.0112 \end{aligned}$$

For the minimum rate, one obtains

$$\begin{aligned} X_1^* &= Y_1/1.485 \\ &= 0.1111/1.485 \\ &= 0.0748 \end{aligned}$$

The minimum liquid-to-gas ratio is then

$$\begin{aligned} \left(\frac{L}{G}\right)_{\text{min}} &= \frac{0.1111 - 0.0112}{X_1^* - X_0} \\ &= \frac{0.1111 - 0.0112}{0.0748 - 0} \\ &= 1.34 \end{aligned}$$

The actual rate is

$$\begin{aligned} \left(\frac{L}{G}\right)_{\text{act}} &= (1.2)(1.34) \\ &= 1.60 \end{aligned}$$

Since

$$\begin{aligned} A &= \frac{L}{mG} \\ &= 1.60/1.485 \\ &= 1.08; \quad 1/A = 0.928 \end{aligned}$$

and

$$\frac{y_2}{y_1} \simeq \frac{Y_2}{Y_1}$$

$$= 0.111; \quad \frac{y_1}{y_2} = 9.01$$

one may employ either Equation (5.12) or Figure 5.6. From Figure 5.6

$$N_{OG} \simeq 6.0$$

From Equation (5.12)

$$N_{OG} = \frac{\ln\left[\left(\frac{y_1 - mx_2}{y_2 - mx_2}\right)\left(1 - \frac{1}{A}\right) + \frac{1}{A}\right]}{1 - (1/A)}$$

$$= \frac{\ln\left[\left(\frac{0.1}{0.0111}\right)(0.072) + 0.928\right]}{0.072}$$

$$= 6.32$$

(**b**) For this condition

$$x_2 = X_2 = 0.001$$

Once again

$$\left(\frac{L}{G}\right)_{\min} = \frac{0.1111 - 0.0111}{0.0748 - 0.001}$$

$$= 1.35$$

and

$$\left(\frac{L}{G}\right)_{\text{act}} = (1.2)(1.35)$$

$$= 1.63$$

If Equation (5.12) is used, then

$$A = 1.63/1.485$$
$$= 1.098; \ 1/A = 0.911$$

and

$$\frac{y_2 - mx_2}{y_1 - mx_2} = \frac{0.011 - 0.001485}{0.1 - 0.001485}$$

$$= 0.096$$

The N_{OG} value from Equation (5.12) is

$$N_{OG} = 6.82$$

The corresponding values for $x_2 = 0.003$, 0.05, and 0.065 (employing the same procedure above) are 7.45, 9.66, and 13.5, respectively. The reader is left with the exercise of repeating the calculations [(a)–(c)] for the number of theoretical stages or plates (N) for a plate tower. (This is addressed in Problem 5.42.)

(**c**) The separation becomes impossible when

$$y_2 - mx_2 = 0$$

$$0.011 - 1.485 \quad x_2 = 0$$

$$y_2 = X_2 = 0.00741 = 0.741\%$$

and (of course)

$$N_{OG} = \infty$$

5.29 Outlet Concentration from a Spray Tower

A waste incinerator emits 300 ppm HCl with peak values of 600 ppm. The airflow is a constant 5000 acfm at 75°F and 1 atm. Only sketchy information was submitted with the scrubber permit application to the state agency for a spray tower. You are requested to determine if the spray unit is satisfactory.

Data
Emission limit = 30 ppm HCl
Maximum gas velocity allowed through the tower = 3 ft/s
Number of sprays = 6
Diameter of the tower = 10 ft

The number of transfer units N_{OG} to meet the regulations, assuming the worst-case scenario, is

$$N_{OG} = \ln(y_1/y_2) \tag{5.13}$$

where y_1 = inlet mole fraction
y_2 = outlet mole fraction

As a rule of thumb in a spray tower, the N_{OG} of the first or top spray is 0.7. Each lower spray will have approximately 60% of the N_{OG} of the spray above it. The absorption that occurs in the inlet duct adds no height but has an N_{OG} of 0.5.

Solution: First, check the gas velocity:

$$v = \frac{q}{S} = \frac{q}{\pi D^2/4}$$

$$= \frac{5000/60}{\pi(10)^2/4}$$

$$= 1.06 \text{ ft/s} < \text{maximum of } 3 \text{ ft/s}$$

The number of transfer units required is obtained from Equation (5.13).

$$N_{OG} = \ln(600/30)$$
$$= 3.00$$

For a tower with five spray sections, see Table 5.2.

TABLE 5.2 Spray Tower Calculation

Spray Section	N_{OG}
Top	0.70
Second	0.42
Third	0.25
Fourth	0.15
Fifth	0.09
Sixth (inlet)	0.50
	$\Sigma = 2.11$

Therefore

$$2.11 = \ln(600/y_2)$$
$$y_2 = 72.74 \text{ ppm}$$

Since the outlet HCl concentration exceeds 30 ppm, the outlet concentration does not meet the specification!

5.30 Spray Efficiency Calculation

A steel pickling operation is "stuck" with a scrubber designed by Dr. Theodore. The spray tower is currently unable to meet the HCl emission limit of 25 ppm. You are requested to determine under what individual spray efficiency parameter this design would comply with the limit, given an inlet concentration of 500 ppm.

Solution: This involves a trial-and-error calculation. The required N_{OG} value is

$$N_{OG} = \ln\left(\frac{500}{y_2}\right)$$
$$= \ln\left(\frac{500}{25}\right)$$
$$= 3.0$$

Calculations for 90% and 80% of each individual spray efficiencies are provided below, based on the usual 70% efficiency for the top spray:

TABLE 5.3 Individual Spray Efficiencies

	90%	80%
Top	0.7	0.7
2	0.63	0.56
3	0.567	0.448
4	0.5103	0.3584
5	0.4592	0.2867
Inlet	0.4133	0.2293
Total	3.2%	2.5%

An individual spray efficiency between 85 to 90% will do the job.

5.31 Packing Height

Doyle Unlimited, a Daniel F. Rodenci Corporation, has submitted design plans to Theodore Consultants for a packed ammonia scrubber on an air stream containing NH_3. The operating and design data provided by Doyle Unlimited, Inc. are given below. Theodore Consultants remember reviewing plans for a nearly identical scrubber for Doyle Unlimited, Inc. in 2005. After consulting old files, the consultants find all the conditions were identical except for the gas flow rate. What recommendation should be made?

Tower diameter = 3.57 ft
Packed height of column = 8 ft
Gas and liquid temperature = 75°F inlet
Operating pressure = 1.0 atm
Ammonia-free liquid flow rate (mass flux or mass velocity) = 1000 $\text{lb}/(\text{ft}^2\cdot\text{hr})$
Gas flow rate = 1575 acfm
Inlet NH_3 gas concentration = 2.0 mol%
Air density = $0.0743\ \text{lb}/\text{ft}^3$
Molecular weight of air = 29
Molecular weight of water = 18
Henry's law constant $m = 0.972$
Figure 5.11; packing type A is used
Colburn chart (Figure 5.12) applies
Emission regulation = 0.1% NH_3 (by mole or volume)
Packing height = $H_{OG}N_{OG}$

Solution: Calculate the cross-sectional area of the tower S (note that S is now employed for the area) in ft^2:

$$S = \pi D^2/4$$

$$= (\pi)(3.57)^2/(4)$$

$$= 10.0\ \text{ft}^2$$

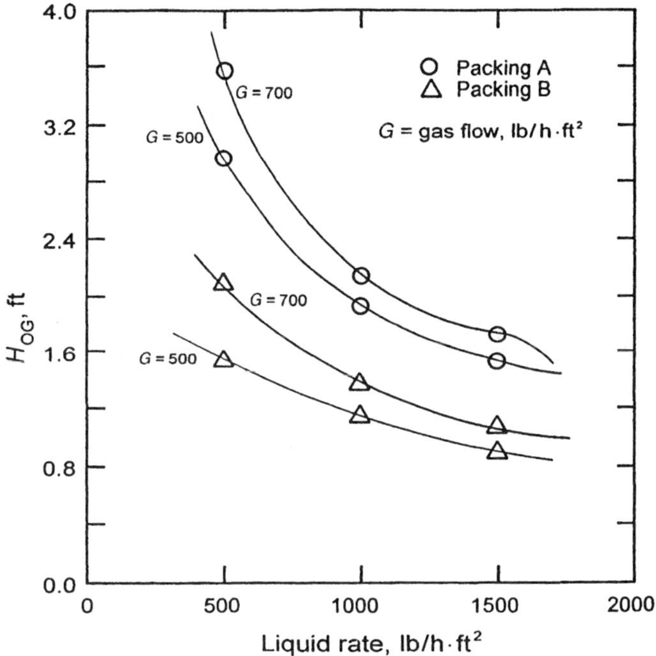

Figure 5.11 H_{OG} values for different types of packing.

Calculate the gas molar flux (molar flow rate per unit cross section) and liquid molar flux in lbmol/(ft^2 · hr):

$$
\begin{aligned}
G_m &= q\rho/S(MW)_G \\
&= (1575)(0.0743)/[(10.0)(29)] \\
&= 0.404\ \text{lbmol}/(\text{ft}^2 \cdot \text{min}) \\
&= 24.2\ \text{lbmol}/(\text{ft}^2 \cdot \text{hr}) \\
L_m &= L/(MW)_L \\
&= (1000)/(18) \\
&= 55.6\ \text{lbmol}/(\text{ft}^2 \cdot \text{hr})
\end{aligned}
$$

The value mG_m/L_m is therefore

$$
\begin{aligned}
mG_m/L_m &= (0.972)(24.2/55.6) \\
&= 0.423
\end{aligned}
$$

The absorption factor A is defined as

$$
\begin{aligned}
A &= L_m/(mG_m) \\
&= 1/0.423 = 2.364
\end{aligned}
$$

The value of $(y_1 - mx_2)/(y_2 - mx_2)$ is

$$\frac{y_1 - mx_2}{y_2 - mx_2} = \frac{0.02 - (0.972)(0)}{0.001 - (0.972)(0)}$$

$$= 20.0$$

N_{OG} is calculated from Colburn's equation (or read from Colburn's chart, Figure 5.12—a modified version of Figure 5.6):

$$N_{OG} = \frac{\ln [(y_1 - mx_2)/(y_2 - mx_2)(1 - \{1/A\}) + \{1/A\}]}{1 - \{1/A\}}$$

$$= \frac{\ln (20.0)(1 - \{1/2.364\}) + \{1/2.364\}}{1 - \{1/2.364\}}$$

$$= 4.30$$

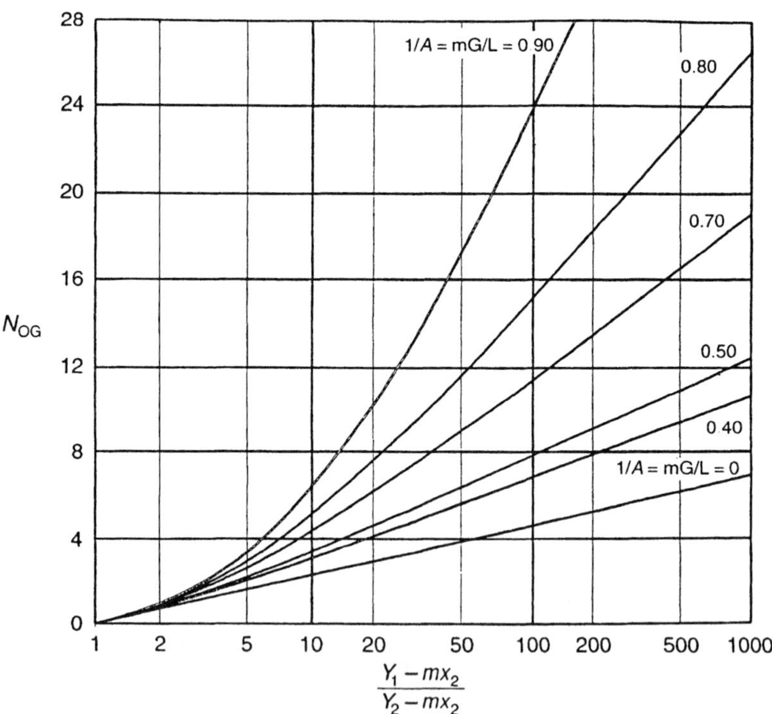

Figure 5.12 Colburn chart.

To calculate the height of an overall gas transfer unit, H_{OG}, first calculate the gas mass velocity G in $lb/ft^2 \cdot hr$:

$$\begin{aligned} G &= q\rho/S \\ &= (1575)(0.0743)/10.0 \\ &= 11.7\,lb/(ft^2 \cdot min) \\ &= 702\ lb/(ft^2 \cdot hr) \end{aligned}$$

From Figure 5.11, one obtains (for packing A)

$$H_{OG} = 2.2\ ft$$

The required packed column height Z, in feet is

$$\begin{aligned} Z &= N_{OG}\,H_{OG} \\ &= (4.3)(2.2) \\ &= 9.46\,ft \end{aligned}$$

The application should be rejected.

5.32 Packing Height for a Reactive Gas

Determine the packing height of a packed countercurrent absorber required to reduce the Cl_2 concentration in a gas by 99% assuming that a dilute NaOH solution is employed. The following information is given:

Liquid rate = $1000\,lb/hr \cdot ft^2$ (essentially water)
Gas rate = $750\,lb/hr \cdot ft^2$ (essentially air)
Mole fraction of Cl_2 in inlet gas = 0.00043
$H_{OG} = 1.63$ ft

(**a**) 1.63 ft
(**b**) 3.75 ft
(**c**) 7.50 ft
(**d**) 8.71 ft

Solution: It can be assumed that Cl_2 will react with the dilute NaOH solution. When the absorbate (pollutant) reacts with the liquid, it can be assumed (L. Theodore: personal notes, 1979) that $m = 0$ so that Equation (5.13) applies. For a 99% reduction:

$$\begin{aligned} N_{OG} &= \ln(100/1) \\ &= 4.6 \end{aligned}$$

The packing height is therefore

$$\begin{aligned} Z &= H_{OG} N_{OG} \\ &= (1.63)(4.6) \\ &= 7.5 \text{ ft} \end{aligned}$$

The correct answer is therefore (c).

5.33 Packing Height for a Nonreactive Gas

Determine the packing height (in feet) of a countercurrent scrubber required to reduce an inlet ammonia concentration by 90%, given the following information:

Liquid flow rate (water) $=1000 \text{ lbmol/hr} \cdot \text{ft}^2$
Gas flow rate (air) $=700 \text{ lbmol/hr} \cdot \text{ft}^2$
Mole fraction of NH_3 in inlet gas $= 0.023$
Mole fraction of NH_3 in outlet liquid $= 0.015$
Assume no NH_3 in inlet water stream
Slope of equilibrium line $= 0.93$
$H_{OG} = 1.5 \text{ ft}$

(**a**) 4.4
(**b**) 5.2
(**c**) 6.1
(**d**) 8.1

Solution: First check whether the material balance is satisfied:

$$\begin{aligned} 1000(0.015 - 0) &= 700(0.023 - 0.0023); \quad y_{out} = (0.023)(1 - 0.9) \\ 15 &= 14.49 \end{aligned}$$

The material balance is satisfied. Now apply Coburn's equation to calculate N_{OG}:

$$N_{OG} = \frac{\ln\left[\frac{(y_1 - mx_2)}{(y_2 - mx_2)}\left(1 - \frac{1}{A}\right) + \frac{1}{A}\right]}{1 - (1/A)} \tag{5.12}$$

$$\begin{aligned} A &= L/mG \\ &= 1000/(0.93)(700) \\ &= 1.536 \end{aligned}$$

Substituting, one obtains

$$\begin{aligned} N_{OG} &= \frac{\ln\left[\frac{(0.015 - 0)}{(0.0015 - 0)}\left(1 - \frac{1}{1.536}\right) + \frac{1}{1.536}\right]}{1 - 1/1.536} \\ &= 4.07 \end{aligned}$$

Therefore

$$\begin{aligned} Z &= H_{OG}N_{OG}; \quad H_{OG} = 1.5\,\text{ft} \\ &= (1.5)(4.07) \\ &= 6.12\,\text{ft} \end{aligned}$$

The correct answer is therefore (c).

5.34 Minimum Diameter Operating Condition

The *minimum tower* diameter for a *countercurrent* packed tower may be determined by the gas flow rate that causes a condition in the tower called

(**a**) Stripping
(**b**) Spraying
(**c**) Flooding
(**d**) Vapor saturation

Solution: There are numerous diameter/pressure drop calculational procedures available in the literature. Most of the diameter calculations center on a condition defined as *flooding*. The correct answer is therefore (c).

5.35 Pressure Drop Correlation

The generalized flooding and pressure drop correlation for packed towers requires calculation of the term $(L/G)(\rho_G/\rho_L)^{0.5}$. Given $(L/G) = 1.0$ and $(\rho_G/\rho_L) = 0.002$, determine the value of this term:

(**a**) 0.036
(**b**) 0.045
(**c**) 0.027
(**d**) 0.067

Solution: Substituting gives

$$\begin{aligned} (L/G)(\rho_G/\rho_L)^{0.5} &= 1.0(0.002)^{0.5} \\ &= 0.045 \end{aligned}$$

The correct answer is therefore (b).

5.36 Tower Height and Diameter

A packed column is used to absorb a toxic pollutant from a gas stream. From the data given below, calculate the height of packing and column diameter. The unit operates at 50% of the flooding gas mass velocity, the actual liquid flow rate is 40% more than the minimum, and 95% of the pollutant is to be collected. Employ the generalized correlation provided in Figure 5.5 to estimate the column diameter.

Gas mass flow rate = 3500 lb/hr
Pollutant concentration in inlet gas stream = 1.1 mol%
Scrubbing liquid = pure water
Packing type = l-inch Raschig rings; packing factor F = 160

H_{OG} of the column = 2.5 ft
Henry's law constant $m = 0.98$
Density of gas (air) = 0.075 lb/ft^3
Density of water = 62.4 lb/ft^3
Viscosity of water = 1.8 cP

Solution: To calculate the number of overall gas transfer units N_{OG}, first calculate the equilibrium outlet concentration x_1^* at $y_1 = 0.011$:

$$x_1^* = y_1/m \quad (5.1)$$
$$= 0.011/0.98$$
$$= 0.0112$$

Determine y_2 for 95% removal:

$$y_2 = \frac{(0.05)y_1}{(1 - y_1) + (0.05)y_1}$$
$$= \frac{(0.05)(0.011)}{(1 - 0.011) + (0.05)(0.011)}$$
$$= 5.56 \times 10^{-4}$$

The minimum ratio of molar liquid flow rate to molar gas flow rate, $(L_m/G_m)_{\min}$, is determined by a material balance:

$$(L_m/G_m)_{\min} = \frac{y_1 - y_2}{x_1^* - x_2} \quad (5.5)$$
$$= \frac{0.011 - 5.56 \times 10^{-4}}{0.0112 - 0}$$
$$= 0.933$$

The actual ratio of molar liquid flow rate to molar gas flow rate L_m/G_m is

$$L_m/G_m = (1.40)(L_m/G_m)_{\min}$$
$$= (1.40)(0.933)$$
$$= 1.306$$

In addition

$$(mG_m)/L_m = (0.98)/(1.306) = 0.7504$$

The absorption factor A is defined (see Equation 5.12) as

$$A = L_m/(mG_m) = 1/0.7504 = 1.333$$

The value of $(y_1 - mx_2)(y_2 - mx_2)$ is

$$\frac{y_1 - mx_2}{y_2 - mx_2} = \frac{0.011 - (0.98)(0)}{5.56 \times 10^{-4} - (0.98)(0)} = 19.78$$

N_{OG} is then calculated from Colburn's equation (or read from Colburn's chart—Figure 5.6:

$$\begin{aligned} N_{OG} &= \frac{\ln[(y_1 - mx_2)/(y_2 - mx_2)(1 - \{1/A\}) + \{1 + A\}]}{1 - \{1/A\}} \\ &= \frac{\ln[(19.78)(1 - \{1/1.333\}) + \{1/1.3333\}]}{1 - \{1/1.333\}} \\ &= 6.96 \end{aligned}$$

The packing height Z is then

$$\begin{aligned} Z &= N_{OG}H_{OG} \\ &= (6.96)(2.5) \\ &= 17.4 \text{ ft} \end{aligned}$$

To determine the diameter of the packed column, the ordinate of Figure 5.5 is first calculated:

$$\begin{aligned} \left(\frac{L}{G}\right)\left(\frac{\rho_G}{\rho_L}\right)^{0.5} &= \left(\frac{L_m}{G_m}\right)\left(\frac{18}{29}\right)\left(\frac{\rho_G}{\rho_L}\right)^{0.5} = (1.306)\left(\frac{18}{29}\right)\left(\frac{0.075}{62.4}\right)^{0.5} \\ &= 0.0281 \end{aligned}$$

The value of the abcissa at the flooding line is determined from the same figure:

$$\frac{G^2 F \psi \mu_L^{0.2}}{\rho_L \rho g_c} = 0.21; \quad \rho = \rho_G$$

The flooding gas mass velocity G_f in $\text{lb}/(\text{ft}^2 \cdot \text{s})$ is

$$\begin{aligned} G_f &= \left(\frac{0.21 \rho_L \rho g_c}{F \psi \mu_L^{0.2}}\right)^{1/2} = \left(\frac{(0.21)(62.4)(0.075)(32.2)}{(160)(1)(1.8)^{0.2}}\right)^{1/2} \\ &= 0.419 \text{ lb}/(\text{ft}^2 \cdot \text{s}) \end{aligned}$$

The actual gas mass velocity G_{act} in $lb/(ft^2 \cdot s)$ is

$$G_{act} = (0.5)(0.419)$$
$$= 0.2095\ lb/(ft^2 \cdot s)$$
$$= 754\ lb/(ft^2 \cdot hr)$$

Calculate the diameter of the column in feet:

$$D = [(4m)/(G_{act}\pi)]^{1/2}$$
$$= [(4)(3500)/(754)(\pi)]^{1/2}$$
$$= 2.43\ ft$$

The column height (packing) and diameter are 17.4 and 2.43 ft, respectively.

5.37 Estimation of Liquid Rate and Inlet Concentration

A packed tower is employed on an air stream containing NH_3. The operating and design data are given below. Use Figure 5.11 to estimate the liquid rate in $lb/hr \cdot ft^2$ and Figure 5.6 (or the corresponding equation) to calculate the inlet NH_3 gas composition (mol%):

Packed column height = 8.5 ft
Packing type A (see Figure 5.11)
$H_{OG} = 2.6$ ft
Outlet NH_3 gas composition = 0.3 mol%
Henry's law constant = 0.95
Molecular weight of air = 29 lb/lbmol
Molecular weight of water = 18 lb/lbmol
Gas flow rate = 2000 acfm
Air density = $0.0743\ lb/ft^3$
Tower diameter = 4.04 ft

Solution: First calculate the tower area:

$$S_{tower} = \pi D^2/4$$
$$= (0.785)(4.04)^2$$
$$= 12.82\ ft^2$$

The gas mass flow rate is

$$G = q\rho/S$$
$$= (2000)(0.073)(160)/12.82$$
$$= 695\ lb/hr \cdot ft^2$$

For $H_{OG} = 2.6$ ft and $G = 700\ \text{lb/hr}\cdot\text{ft}^2$, Figure 5.11 gives

$$L = 800\ \text{lb/hr}\cdot\text{ft}^2$$

Apply Equation (5.8) to calculate N_{OG}:

$$\begin{aligned} N_{OG} &= Z/H_{OG} \\ &= 8.5/2.6 \\ &= 3.3 \end{aligned}$$

Use Figure 5.6 to estimate y_2:

$$G_m = 695/29 = 24.0\ \text{lbmol/hr}\cdot\text{ft}^2$$

$$\begin{aligned} L_m &= 800/18 = 44.4\ \text{lbmol/hr}\cdot\text{ft}^2 \\ 1/A &= mG_m/L_m \\ &= (0.95)(24.0/44.4) \\ &= 0.51 \end{aligned}$$

Apply Equation (5.12). By trial and error, one obtains

$$\frac{y_1 - mx_2}{y_2 - mx_2} \approx 9.0$$

Since $x_2 = 0$, it follows that $y_1/y_2 = 9.0$. Thus

$$\begin{aligned} y_1 &= (9)(0.003) \\ &= 0.027 = 2.7\%\ \text{inlet NH}_3\ \text{concentration} \end{aligned}$$

One may apply the Colburn equation, which provides more accurate results.

5.38 Sizing of a Packed Tower with No Data

Qualitatively outline how one can size (diameter, height) a packed tower to achieve a given degree of separation without any information on the physical and chemical properties of a gas to be cleaned.

Solution: To calculate the height, one needs both the height of a gas transfer unit H_{OG} and the number of gas transfer units N_{OG}. Since equilibrium data are not available, assume that m (slope of equilibrium curve) approaches zero. This is not an unreasonable assumption for most solvents that preferentially absorb (or react with) the solute. For this condition:

$$N_{OG} = \ln(y_1/y_2) \tag{5.13}$$

where y_1 and y_2 represent inlet and outlet concentrations, respectively. Since it is reasonable to assume the scrubbing medium to be water or a solvent that effectively has the physical and chemical properties of water, H_{OG} can be assigned values usually encountered for water systems. These are given in Table 5.4. For plastic packing, the liquid and gas flow rates are both typically in the

TABLE 5.4 Packing Diameter versus H_{OG}

Packing Diameter, inches	Plastic Packing H_{OG}, feet	Ceramic Packing H_{OG}, feet
1.0	1.0	2.0
1.5	1.25	2.5
2.0	1.5	3.0
3.0	2.25	4.5
3.5	2.75	5.5

range of 1500–2000 lb/(hr · ft^2) of cross-sectional area. (*Note*: The "flow rates" are more properly termed "fluxes"; a *flux* is a flow rate per unit cross-sectional area with units of lb/hr · ft^2 in this problem. However, it is common practice to use the term *rate*.) For ceramic packing, the range of flow rates is 500–1000 lb/hr · ft^2. For difficult-to-absorb gases, the gas flow rate is usually lower and the liquid flow rate higher. Superficial gas velocities (velocity of the gas if the column is empty) are in the 3–6-ft/s range. The height Z is then calculated from

$$Z = (H_{OG})(N_{OG})(\mathrm{SF}) \tag{5.36}$$

where SF is a safety factor, the value of which can range from 1.25 to 1.5. Pressure drops can vary from 0.15 to 0.40 inch H_2O/ft packing. Packing size increases with increasing tower diameter. (*Note*: This problem and design procedure were originally developed by Dr. Louis Theodore in 1985 and later published in 1988. These materials have appeared elsewhere without properly acknowledging the author, and without permission from the original publisher.) The reader is left the exercise of verifying the chart in Table 5.5 for plastic packing.

TABLE 5.5 Packing Height, Z (ft), as a Function of Efficiency and Plastic Packing Size

	Plastic Packing Size, inches				
Removal Efficiency, %	1.0	1.5	2.0	3.0	3.5
63.2	1.0	1.25	1.5	2.25	2.75
77.7	1.5	1.9	2.25	3.4	4.1
86.5	2.0	2.5	3.0	4.5	5.5
90	2.3	3.0	3.45	5.25	6.25
95	3.0	3.75	4.5	6.75	8.2
98	3.9	4.9	5.9	8.8	10.75
99	4.6	5.75	6.9	10.4	12.7
99.5	5.3	6.6	8.0	11.9	14.6
99.9	6.9	8.6	10.4	15.5	19.0
99.99	9.2	11.5	13.8	20.7	25.3

5.39 Packed Tower Absorber Design with No Data

A 1600 acfm gas stream is to be treated in a packed tower containing ceramic packing. The gas stream contains 100 ppm of a toxic pollutant that is to be reduced to 1 ppm. Estimate the tower's cross-sectional area, diameter, height, pressure drop, and packing size. Use the procedure outlined in Problem 5.38.

Solution: Key information from Problem 5.38 for ceramic packing is provided in Table 5.6. The equation for the cross-sectional area of the tower S in terms of the gas volumetric flow rate q in acfs is

$$S(\text{ft}^2) = q(\text{acfs})/4$$

An equation to estimate the tower packing pressure drop ΔP in terms of Z is

$$\Delta P(\text{in H}_2\text{O}) = (0.2)Z; \quad z = \text{ft} \tag{5.37}$$

The following packing size(s) is (are) recommended:

For $D \approx 3$ ft, use 1-inch packing
For $D < 3$ ft, use $<$1-inch packing
For $D > 3$ ft, use $>$1-inch packing

As a rule, recommended packing size increases with increasing diameter.

For the problem at hand

$$S = 1600/4 = 400 \text{ ft}^2$$

The diameter D is

$$D = (4S/\pi)^{0.5}$$
$$= [(4)(400)/\pi]^{0.5}$$
$$= 22.6 \text{ ft}$$

TABLE 5.6 Packing Height, Z(ft), as a Function of Efficiency and Ceramic Packing Size

	Ceramic Packing Size, inches				
Removal Efficiency, %	1.0	1.5	2	3	3.5
63.2	2.0	2.5	3.0	4.5	5.5
77.7	3.0	3.7	4.5	6.75	8.25
86.5	4.0	5.0	6.0	9.0	11.0
90	4.6	5.75	6.9	10.4	12.7
95	6.0	7.5	9.0	13.5	16.5
98	7.8	9.8	11.7	17.6	21.5
99	9.2	11.5	13.8	20.7	25.3
99.5	10.6	13.25	15.9	23.8	29.1
99.9	13.8	17.25	20.7	31.1	38.0
99.99	18.4	23.0	27.6	41.4	50.7

For a tower this large, the 3.5-inch packing should be used. Additional information is available from L. Theodore, "Engineering Calculations: Sizing Packed-Tower Absorber without Data," *Chem. Eng. Progress*, p. 18, May 2005.

5.40 Two Absorbers to Replace One

The calculations for an absorber indicate that it would be excessively tall, and the five schemes in the diagrams in Figure 5.13 are being considered as a means of using two shorter absorbers. Make freehand sketches of operating lines, one for each scheme showing the relation between the operating lines for the two absorbers on the left-hand side of the figure and the equilibrium curve. Mark the concentrations on the figure for each diagram. No calculations are required. Assume dilute solutions.

Solution: Operating lines for each scheme are provided on the right-hand side of Figures 5.13—1, 2, 3, 4, and 5.

5.41 Design of a Packed Tower Air Stripper

An atmospheric packed tower air stripper is used to clean contaminated groundwater with a concentration of 100 ppm trichloroethylene (TCE). The stripper was designed such that the packing height is 13 ft, the diameter is 5 ft, and the height of a transfer unit (HTU) is 3.25 ft. Assume that Henry's law applies with a constant (H) of 324 atm at 68°F. Also, at these conditions the molar density of water is 3.47 lbmol/ft^3 and the air–water mole ratio (G/L) is related to the air–water volume ratio (G''/L'') through $G''/L'' = 130\, G/L$, where the units of G'' and L'' are $\text{ft}^3/(\text{s}\cdot\text{ft}^2)$.

(**a**) If the stripping factor (R) used in the design is 5.0, what is the removal efficiency?

(**b**) If the air blower produces a maximum air flow (q) of 106 acfm, what is the maximum water flow (in gpm) that can be treated by the stripper?

As described earlier, the height of packed tower can be calculated by

$$Z = (N_{\text{OG}})(H_{\text{OG}}) = (\text{NTU})(\text{HTU}) \tag{5.38}$$

In addition, the following equation has been developed for calculation of the number of transfer units (NTUs) for an air–water stripping system and is based on the stripping factor R and the inlet/outlet concentrations:

$$\text{NTU} = \frac{R}{R-1}\ln\left(\frac{(C_{\text{in}}/C_{\text{out}})(R-1)+1}{R}\right) \tag{5.39}$$

where C_{in} = inlet contaminant concentration, ppm
C_{out} = outlet contaminant concentration, ppm
R = stripping factor

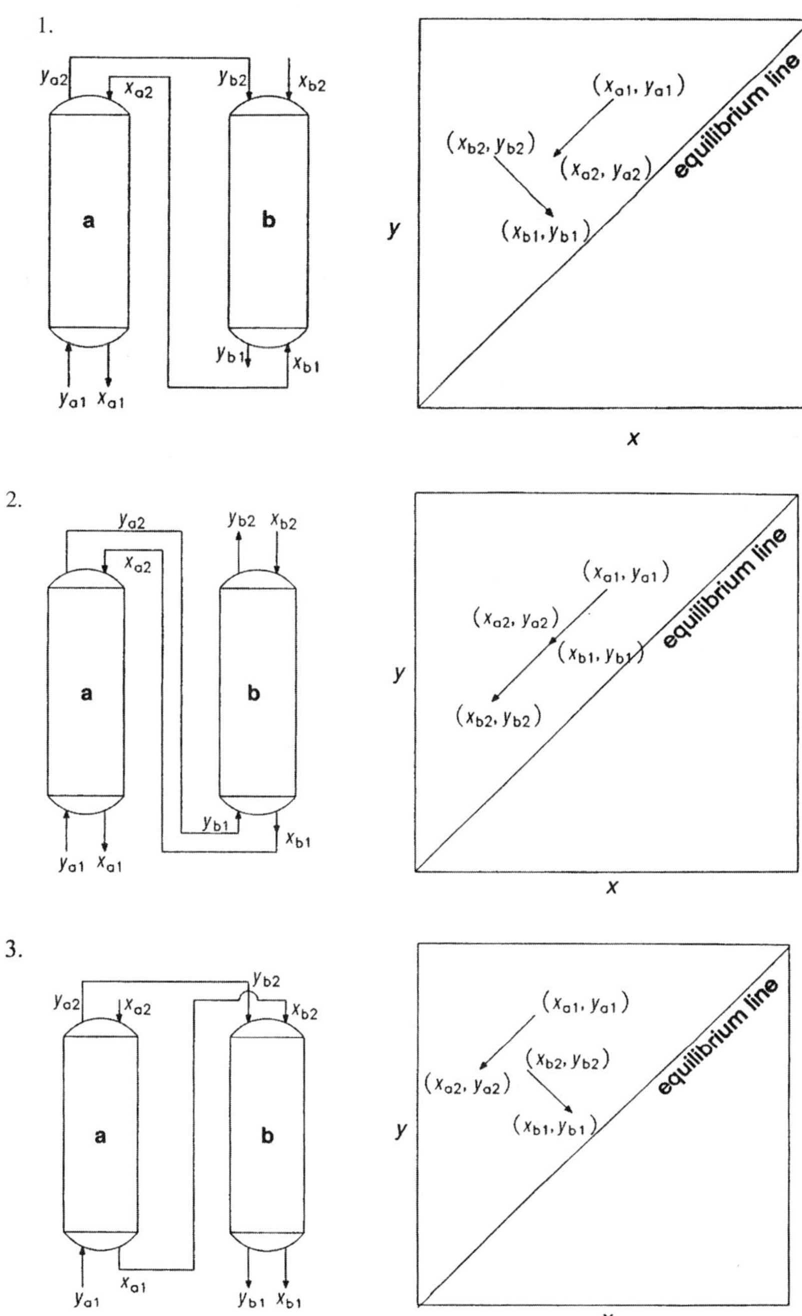

Figure 5.13 Answers to Problem 5.40.

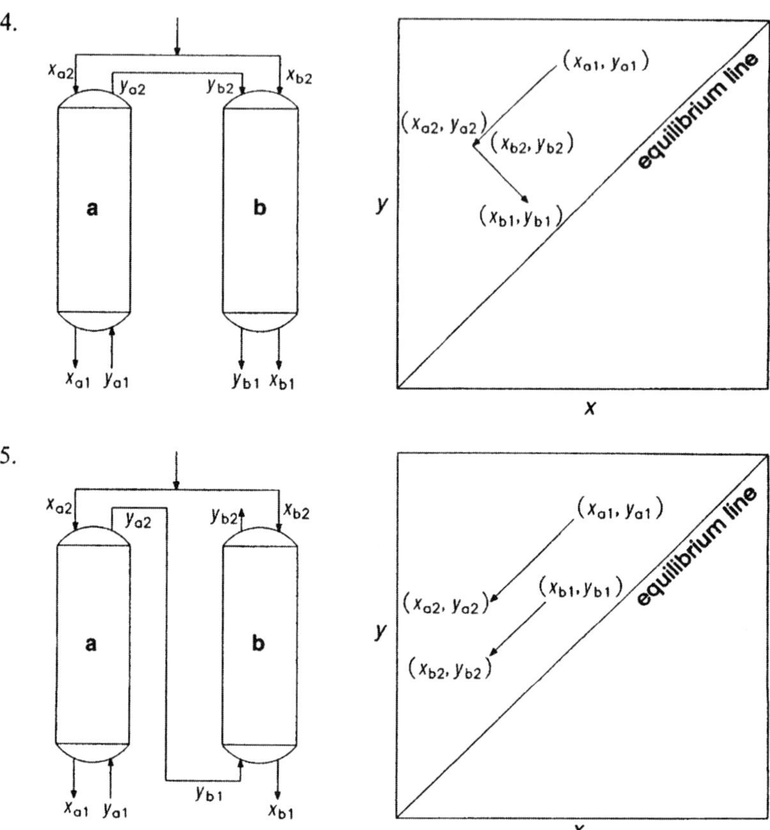

Figure 5.13 *(Continued)*.

For the purposes of this problem

$$R = \frac{H}{L}\frac{G}{P} \tag{5.40}$$

where H = Henry's law constant, atm
P = system pressure, atm
G = gas (air) loading rate (or flux) lbmol/(s · ft^2)
L = liquid loading rate (or flux) lbmol/(s · ft^2)

Solution: (a) The number of transfer units (NTUs) is first calculated:

$$\begin{aligned} Z &= (N_{OG})(H_{OG}) = (\text{NTU})(\text{HTU}) \\ \text{NTU} &= Z/\text{HTU} \\ &= 13/3.25 \\ &= 4 \end{aligned}$$

Rearranging Equation (5.39), one obtains

$$\begin{aligned} C_{\text{out}} &= \frac{C_{\text{in}}(R-1)}{R\exp[(\text{NTU})(R-1)/R]-1} \\ &= \frac{(100)(5.0-1)}{(5.0)\exp[(4.0)(5.0-1)/5.0]-1} \\ &= 3.3 \end{aligned}$$

The removal efficiency (RE) is then

$$\begin{aligned} \text{RE} &= [(C_{\text{in}}-C_{\text{out}})/C_{\text{out}}]100\% \\ &= [(100-3.3)/100]100\% \\ &= 96.7\% \end{aligned}$$

(b) The air–water mole ratio G/L is

$$\begin{aligned} G/L &= (P)(R)/H \\ &= (1\text{ atm})(5)/(324\text{ atm}) \\ &= 0.0154 \end{aligned}$$

In addition

$$\begin{aligned} G''/L'' &= 130\ G/L \\ &= 130(0.0154) \\ &= 2 \end{aligned}$$

Since the tower cross-sectional area S in ft^2, is

$$\begin{aligned} \text{Area} = S &= \pi D^2/4 \\ &= \pi(5\text{ ft})^2/4 \\ &= 19.63\text{ ft}^2 \end{aligned}$$

the air volumetric loading rate G'' in ft^3/(min · ft^2) is then

$$\begin{aligned} G'' &= (106\text{ ft}^3/\text{min})/(19.63\text{ ft}^2) \\ &= 5.4\text{ ft}^3/(\text{min}\cdot\text{ft})^2 \end{aligned}$$

Also, the water volumetric loading rate L'' in $\text{ft}^3/(\text{min}\cdot\text{ft}^2)$ is

$$\begin{aligned} L'' &= 2G'' \\ &= 2[5.4\ \text{ft}^3/(\text{min}\cdot\text{ft}^2)] \\ &= 10.8\ \text{ft}^3/(\text{min}\cdot\text{ft}^2) \end{aligned}$$

This can be converted to gallons per minute (gpm).

$$\begin{aligned} L &= \frac{[10.8\ \text{ft}^3/(\text{min}\cdot\text{ft}^2)]\,(3.47\ \text{lbmol}/\text{ft}^3)\,(18\ \text{lb}/\text{lbmol})\,(19.63\ \text{ft}^2)}{8.33\ \text{lb}/\text{gal}} \\ &= 1590\ \text{gpm} \end{aligned}$$

Once the volatile organic compounds (VOCs) have been recovered from a process wastewater or groundwater stream, the off-gas (air/VOC mixture) usually needs to be treated. This entails the installation of other equipment to handle the disposal of the VOCs. Some typical methods include flaring, carbon adsorption, and incineration. Flaring (see Chapter 4) is potentially hazardous because of the oxygen that is allowed to enter the flare header. Carbon adsorption (see Chapter 6) can be an efficient means of recovering the VOCs from the off-gas, but it can generate large quantities of solid hazardous waste. A utility boiler (incineration) is probably the best alternative since VOC destruction is typically more than 99%, and it is safer and more inexpensive. In addition, catalytic incinerators can also achieve high VOC destruction efficiencies.

5.42 Number of Theoretical Plates Calculation

Refer to Problem 5.28. Repeat calculations (a)–(c) as they apply to a plate column. In effect, determine the number of theoretical plates N. Employ Equation (5.23).

Solution: Equation (5.23) applies for a plate column:

$$N = \frac{\log\left[\left(\dfrac{y_{N+1} - mx_0}{y_1 - mx_0}\right)\left(1 - \dfrac{1}{A}\right) + \dfrac{1}{A}\right]}{\log A} \tag{5.23}$$

once again

$$\begin{aligned} m &= 1.485 \\ (L/V)_{\text{act}} &= 1.2\,(L/V)_{\text{min}} \end{aligned}$$

Results are provided in Table 5.7. The maximum value of x is given (as before) by

$$\begin{aligned} x &= y/m \\ &= 0.011/1.485 \\ &= 0.00741 = 0.741\% \end{aligned}$$

If this calculation is performed on a solute-free basis, then

$$\begin{aligned} X &= Y/m \\ &= 0.0111/1.485 \\ &= 0.00747 = 0.747\% \end{aligned}$$

TABLE 5.7 Number of Plates For Problem 5.42

X_2	$(L/V)_{min}$	$(L/V)_{act}$	A	N_p
0	1.384	1.661	1.119	5.48
0.001	1.4031	1.6837	1.1339	5.87
0.003	1.4438	1.7326	1.1667	6.93
0.005	1.4872	1.7846	1.2018	8.82
0.0065	1.5215	1.8258	1.2295	12.21

5.43 Raschig Rings versus Tellerettes Economic Comparison

A small packed absorption tower is being designed to remove ammonia from air by scrubbing with water at 68°F and 1 atm. It was decided to use either 1-inch Raschig rings or 1-inch Tellerettes as the packing in the tower. The value of N_{OG} is 6.0 for either packing, but the H_{OG} for the Raschig rings is 2.5 ft and that for Tellerettes is 1.0 ft. Each 1-inch Raschig ring costs \$57.56/ft^3 and each Tellerette costs \$26.40/ft^3. The installed cost of the tower shell is \$18.30 per foot of tower height. Determine which of the packings will be the more economical for the system. Assume a tower diameter of 20 inches for both cases.

Solution: For a Raschig ring tower, the required packing height is

$$\begin{aligned} Z &= N_{OG}H_{OG} \\ &= (6.0)\ (2.5) = 15.0 \text{ ft} \end{aligned} \tag{5.8}$$

The tower volume is (packed volume)

$$\begin{aligned} V_R &= \frac{\pi D^2}{4} Z \\ &= \frac{\pi}{4}\left(\frac{20}{12}\right)^2 (15.0) \\ &= 32.7 \text{ ft}^3 \end{aligned}$$

The packing cost is therefore

$$\begin{aligned} \mathrm{RC} &= (7.56/\mathrm{ft}^3)\,(32.7\ \mathrm{ft}^3) \\ &= \$247 \end{aligned}$$

The tower shell cost is

$$\begin{aligned} \mathrm{RS} &= (\$18.30/\mathrm{ft})\,(15.0\ \mathrm{ft}) \\ &= \$274.50 \end{aligned}$$

The total tower cost is

$$\begin{aligned} \mathrm{TRC} &= 247 + 274 \\ &= \$521 \end{aligned}$$

For a Tellerette packed tower the required packing height is

$$\begin{aligned} Z &= N_{\mathrm{OG}} H_{\mathrm{OG}} \\ &= 6.0 \times 1.0 = 6.0\ \mathrm{ft} \end{aligned} \tag{5.8}$$

The tower volume is (packed volume)

$$\begin{aligned} V_{\mathrm{T}} &= \frac{\pi D^2}{4} Z \\ &= \frac{\pi}{4}\left(\frac{20}{12}\right)^2 (6.0) \\ &= 13.1\ \mathrm{ft}^3 \end{aligned}$$

The packing cost is

$$\begin{aligned} \mathrm{TC} &= (\$26.40/\mathrm{ft}^3)\,(13.1\ \mathrm{ft}^3) \\ &= \$346 \end{aligned}$$

The tower shell cost is

$$\begin{aligned} \mathrm{TS} &= (\$18.30/\mathrm{ft})\,(6.0\,\mathrm{ft}) \\ &= \$110 \end{aligned}$$

The total tower cost is

$$\begin{aligned} \mathrm{TTC} &= \$110 + \$346 \\ &= \$456 \end{aligned}$$

The cost of a Raschig ring tower is 1.14 (14%) times the cost of a Tellerette packed tower in this service. In addition, the pressure drop will be significantly lower with the Tellerettes.

5.44 Recovery Versus Operating Costs: Breakeven Calculation

A plant emits 50,000 acfm of air containing an organic gas at a concentration of 2.0 grains/ft^3. The gas is worth \$0.01/lb. A theoretical equation has been previously developed relating the efficiency of gas absorption E to the pressure drop ΔP (lb_f/ft^2):

$$E = \frac{\Delta P}{\Delta P + 90}$$

The fan is 55% efficient (overall) and electric power costs \$0.03/(kw) (hr).

(**a**) At what E is the cost of power equal to the value of the recovered (absorbed) gas?

(**b**) What is ΔP, in inches of H_2O, at the condition in (a)?

Solution: The value of the recovered organic is

$$\begin{aligned} R &= (50{,}000\ \text{ft}^3/\text{min})(2.0\ \text{gr/ft}^3)(\text{lb}/7000\ \text{gr})(\$0.01/\text{lb})(E) \\ &= 0.143E,\ \$/\text{min} \end{aligned}$$

The cost of power (electricity) is:

$$\begin{aligned} P &= (50{,}000\ \text{ft}^3/\text{min})[\Delta P(\text{lb}_f/\text{ft}^2)](\text{kW}/44{,}200\ \text{ft}\cdot\text{lb/min}) \\ &\quad \times (1/0.55)(\$0.03/\text{kW}\cdot\text{hr})(\text{hr}/60\ \text{min}) \\ &= 0.001\Delta P,\ \$/\text{min} \end{aligned}$$

Setting $R = P$, and noting that

$$E = \frac{\Delta P}{\Delta P + 90}$$

leads to

$$0.143E = 0.143\frac{\Delta P}{\Delta P + 90} = 0.001\Delta P$$

Solving for ΔP gives

$$\begin{aligned} \Delta P &= 53\ \text{lb}_f/\text{ft}^2 \\ &= 10.2\ \text{in}\ H_2O \end{aligned}$$

and

$$\begin{aligned} E &= \frac{53}{53 + 90} \\ &= 0.37 = 37\% \end{aligned}$$

5.45 Recovery Versus Operating Costs: Maximizing Profit

Refer to Problem 5.44. At what efficiency will maximum profit be realized?

Solution: The profit (PR) in \$/min is

$$\begin{aligned} PR &= R - P \\ &= (0.143)\left(\frac{\Delta P}{\Delta P + 90}\right) - 0.001\Delta P \end{aligned}$$

Either solve by trial and error or take the derivative of PR with respect to either ΔP or E and set the result equal to zero.

$$\frac{d(PR)}{d(\Delta P)} = (0.143)\left[\frac{90}{(\Delta P + 90)^2}\right] - 0.001 = 0$$

Solving for ΔP gives

$$\begin{aligned} \Delta P &= 23.45\,\text{lb/ft}^2 \\ &= 4.53 \text{ in } H_2O \end{aligned}$$

The value of ΔP represents the operating pressure drop that produces the maximum profit.

5.46 Economic Equation Applicability

On the basis of the results of Problems 5.44 and 5.45, comment on the applicability of the (E, ΔP) equation.

Solution: The results seem reasonable, as can be seen by plotting PR versus either ΔP or E. This plot is provided in Figure 5.14.

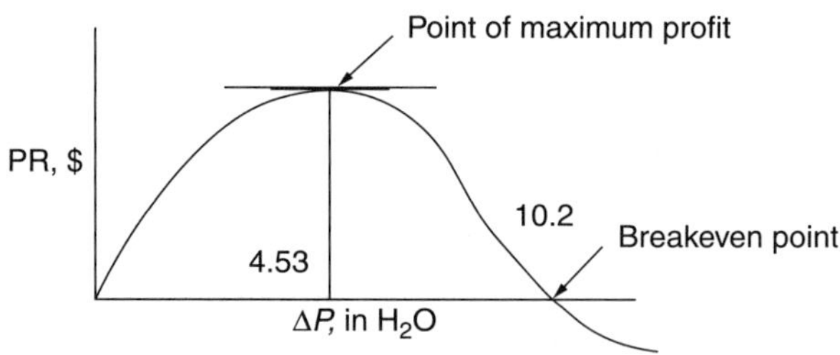

Figure 5.14 Profit versus pressure drop.

5.47 Enthalpy of Solution Effects

In an attempt to quantify the effect of enthalpy of solution effects on the absorption of HCl into scrubbing water in an absorber, Pallechi Consultants reviewed the literature (personal notes: L. Theodore, 2006) and obtained the following rough estimates of this effect. The data provided temperature increases as a function of increasing HCl concentration (mass percent basis) in water:

$$0 - 1.5\% = 10^\circ\text{C}$$
$$0 - 3.0\% = 15^\circ\text{C}$$
$$0 - 5.0\% = 20^\circ\text{C}$$

Apply the above data and estimate the discharge temperature increase for the following three HCl scenarios:

Scenario 1: 0.0% inlet to 1.5% outlet
Scenario 2: 0.5% inlet to 3.0% outlet
Scenario 3: 0.5% inlet to 5.0% outlet

Solution: Since enthalpy is a point function, it is reasonable to assume that the temperature effects are additive. Therefore, the temperature increases are

Scenario 1:

$$\begin{aligned}\Delta T &= \Delta T_{1.5} - \Delta T_{0.0}\\ &= 10 - 0\\ &= 10^\circ\text{C}\end{aligned}$$

Scenario 2:

$$\begin{aligned}\Delta T &= \Delta T_{3.0} - \Delta T_{1.5}\\ &= 15 - 10\\ &= 5^\circ\text{C}\end{aligned}$$

Scenario 3:

$$\begin{aligned}\Delta T &= \Delta T_{5.0} - \Delta T_{1.5}\\ &= 20 - 10\\ &= 10^\circ\text{C}\end{aligned}$$

The reader should note that this is an effect that often should be reflected in engineering applications since any increase in the temperature of the scrubbing liquid adversely affects the equilibrium, reducing the equilibrium capacity of the liquid.

5.48 Absorber Failure to Meet Performance Guarantee

Consider the absorber system shown in Figure 5.15. Corenza Engineers designed the unit to operate with a maximum discharge concentration of 50 ppm. Once the unit was installed and running, the unit operated with a discharge of 60 ppm. Rather than purchase a new unit, what options are available to bring the unit into compliance with the specified design concentration?

Solution: This is obviously an open-ended question. On the basis of the material presented earlier in Section 5.1 and the solutions to several of the problems in this chapter, one may employ any one or a combination of the following suggestions (L. Theodore: personal notes, 1986):

1. Increase the temperature
2. Decrease the pressure

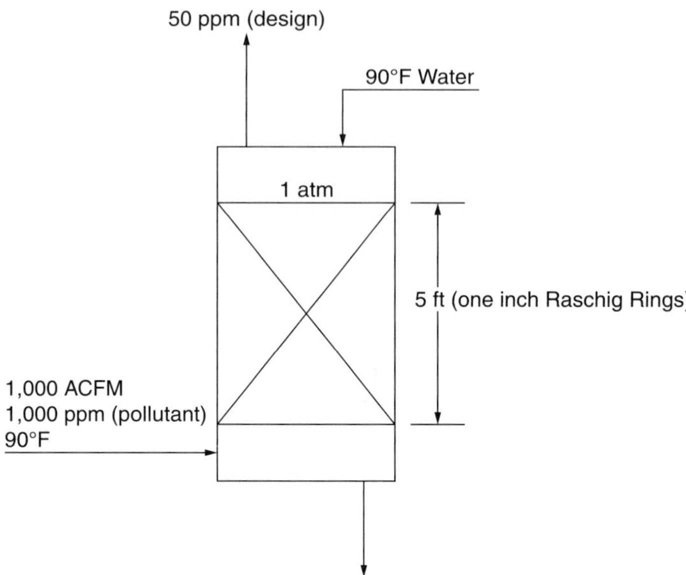

Figure 5.15 Absorber failure to meet design performance.

3. Increase the height of the packing
4. Place sprays at the inlet before the packing
5. Increase the liquid flow rate, but check the pressure drop increase and any potential effect on the fan
6. Change the liquid, but check the pressure drop increase and any potential effect on the fan
7. Change packing size and/or type, but check the pressure drop increase and any potential effect on the fan
8. Process modification: reduce inlet pollutant concentration
9. Process modification: reduce gas flow rate
10. Design a new system
11. Add more packing on top of existing packing (if possible), but check the pressure drop increase and any potential effect on the fan
12. Fire the design engineer
13. Take the regulatory official/inspector out to lunch
14. Shut down the plant

Any other suggestions?

NOTE: Additional problems are available for all readers at www.wiley.com. Follow links for this title.

6

ADSORBERS

6.1 INTRODUCTION

During adsorption, one or more gaseous components are removed from an effluent gas stream by adhering to the surface of a solid. The gas molecules being removed are referred to as the *adsorbate*, while the solid doing the adsorbing is called the *adsorbent*. Adsorbents are highly porous particles and adsorption occurs primarily on the internal surface of the particles.

The attractive forces that hold the gas to the surface of the solid are the same that cause vapors to condense (*van der Waals* forces). All gas–solid interfaces exhibit this attraction, some more than others. Adsorption systems use materials that are highly attracted to each other to separate these gases from the nonadsorbing components of an air stream. For air pollution control purposes, adsorption is not a final control process. The contaminant gas is merely stored on the surface of the adsorbent. After it becomes saturated with adsorbate, the adsorbent must either be disposed of and replaced, or the vapors must be desorbed. Desorbed vapors are highly concentrated and may be recovered more easily and more economically than before the adsorption step.

Traditionally, adsorption has been used for air purification and solvent recovery. Air purification processes are those in which the contaminant is often present in trace

Air Pollution Control Equipment Calculations. By Louis Theodore

quantities (less than 1.0 ppm) but can be highly odorous and toxic. Systems used for air purification are usually small *thin-bed* adsorbers. When the bed becomes saturated with contaminant, it is taken out and replaced. Solvent recovery processes require much larger systems and are designed to control organic emissions whose concentrations are usually greater than 1000 ppm. This is the point where the recovery value of the solvent could justify the expense of the large adsorption–desorption system. Currently, adsorption is used as a method of recovering valuable organic vapors from the (flue) gases at all concentration levels. This is due to present regulations limiting volatile organic emission and the higher costs of solvents.

Adsorption Forces—Physical and Chemical

The adsorption process is classified as either physical or chemical. The basic difference between physical and chemical adsorption is the manner in which the gas molecule is bonded to the adsorbent. In physical adsorption the gas molecule is bonded to the solid surface by weak forces of intermolecular cohesion. The chemical nature of the adsorbed gas remains unchanged; therefore, physical adsorption is a readily reversible process. In chemical adsorption a much stronger bond is formed between the gas molecule and adsorbent. A sharing or exchange of electrons takes place—as happens in a chemical bond. Chemical adsorption is not easily reversible.

The forces active in physical adsorption are electrostatic in nature. These forces are present in all states of matter: gas, liquid, and solid. They are the same forces of attraction that cause gases to condense and real gases to deviate from ideal behavior. Physical adsorption is sometimes referred to as *van der Waals* adsorption. The electrostatic effect that produces the van der Waals forces depends on the polarity of both the gas and solid molecules. Molecules in any state are either polar or non polar, depending on their chemical structure. Polar substances are those that exhibit a separation of positive and negative charges within the compound. This separation of positive and negative charges is referred to as a *permanent dipole*. Water is a prime example of a polar substance. Nonpolar substances have both their positive and negative charges in one center, so they have no permanent dipole. Most organic compounds, because of their symmetry, are nonpolar.

Physical, or van der Walls adsorption can occur from three different effects: an orientation effect, a dispersion effect, or induction effect. For polar molecules, attraction to each other occurs because of the orientation effect. The orientation effect describes the attraction that occurs between the dipoles of two polar molecules. The negative area of one is attracted to the positive area of the other. An example of this type of adsorption would be the removal of water vapor (polar) from an exhaust stream by using silica gel (polar).

Chemical adsorption, or chemisorption, results from the chemical interaction between a gas and a solid. The gas is held to the surface of the adsorbate by the formation of a chemical bond. Adsorbents used in chemisorption can be either pure substances or chemicals deposited on an inert carrier material. One example is using pure iron oxide chips to adsorb H_2S. Another example is using activated carbon which has been impregnated with sulfur to remove mercury vapor.

All known adsorption processes are *exothermic*, whether adsorption occurs from chemical or physical forces. In adsorption, molecules are transferred from the gas to the surface of a solid. The fast-moving gas molecules lose their kinetic energy of motion to the adsorbent in the form of heat. In chemisorption, the heat of adsorption is comparable to the heat evolved from a chemical reaction, usually over 10 kcal/gmol. The heat given off by physical adsorption is much lower, approximately 100 cal/gmol, which is comparable to the heat of condensation (or vaporization).

Adsorbent Materials

Several materials are used efficiently as adsorbing agents. The most common adsorbents used industrially are activated carbon, silica gel, activated alumina (alumina oxide), and zeolites (molecular sieves). Adsorbents are characterized by their chemical nature, extent of their surface area, pore distribution, and particle size. In physical adsorption the most important characteristic in distinguishing between adsorbents is their surface polarity. As discussed previously, the surface polarity determines the type of vapors for which a particular adsorbent will have the greatest affinity. Of the above adsorbents, activated carbon is the primary nonpolar adsorbent. It is possible to manufacture other adsorbing material having nonpolar surfaces, but since their surface areas are much less than that of activated carbon, they are not used commercially. Polar adsorbents will preferentially adsorb any water vapor that may be present in a gas stream. Since moisture is present in most pollutant airstreams, the use of polar adsorbents is severely limited for an air pollution system. Therefore, the emphasis is placed on the use of activated carbon in further discussion, although some of the information is applicable to polar adsorption systems.

Activated Carbon. Activated carbon can be produced from a variety of feedstocks such as wood, coal, coconut, nutshells, and petroleum-based products. The activation process takes place in two steps. First, the feedstock is carbonized. Carbonization involves heating the material (usually in the absence of air) to a temperature high enough (600°C) to drive off all volatile material. Thus, carbon is essentially all that is left. To increase the surface area the carbon is then "activated" by using steam, air, or carbon dioxide at higher temperatures. These gases attack the carbon and increase the pore structure. The temperatures involved, the amount of oxygen present, and the type of feedstock all greatly affect the adsorption qualities of the carbon. Manufacturers vary these parameters to produce activated carbon suitable for specific purposes. In sales literature, the activity and retentivity of carbons are based on their ability to adsorb a standard solvent, such as carbon tetrachloride (CCl_4).

Because of its nonpolar surface, activated carbon is used to control emission of organic solvents, odors, toxic gases, and gasoline vapors. Carbons used in gas phase adsorption systems are manufactured in granular form, usually ranging from 4×6 to 4×20 mesh in size. (A 4×6 mesh is one that will pass the carbon through a 4-wire-per-inch Tyler mesh screen, but will be captured by a 6-wire-per-inch screen.) The bulk density of the packed bed can range from 0.5 to 0.08 g/cm^3 (from 30 to 5 lb/ft^3) depending on the internal porosity of the carbon. (The *bulk density* was defined earlier as the mass of carbon divided by the volume occupied by both the carbon and void spaces between the carbon particles.) The surface area of the carbon can range from 600 to 1600 m^2/g (2.9×10^6 to 7.8×10^6 ft^2/lb); this is equivalent to having the surface the area of two to five football fields in one gram of carbon.

Silica Gel. Silica gels are made from sodium silicate. Sodium silicate is mixed with sulfuric acid, resulting in a jelly-like precipitate from which the "gel" name comes. The precipitate is then dried and roasted. Depending on the processes used in manufacturing the gel, different grades of varying activity can be produced. Silica gels have surface areas of approximately 3.7×10^6 ft^2/lb (750 m^2/g). Silica gels are used primarily to remove moisture from exhaust streams, but are ineffective at temperatures above 500°F (260°C).

Molecular Sieves. Unlike the other adsorbents, which are amorphous (not crystalline) in nature, molecular sieves have a crystalline structure. The pores, therefore, are relatively uniform in diameter. Molecular sieves can be used to capture or separate gases on the basis of molecular size and shape. An example of this are refining processes, which sometimes use molecular sieves to separate straight-chained paraffins from branched and cyclic compounds. However, the main use of molecular sieves is in the removal of moisture from exhaust streams. The surface area of molecular sieves range from 2.9×10^6 to 3.4×10^6 ft^2/lb (from 600 to 700 m^2/g).

Aluminum Oxide (Activated Alumina). Aluminum oxides are manufactured by thermally activated alumina or bauxite. This is accomplished by heating the alumina in an inert atmosphere to produce a porous aluminum oxide pellet. Aluminum oxides are not commonly used in air pollution applications. They are used primarily for drying of gases, especially under high pressure, and as support material in catalytic reactions. A prime example is the impregnating of the alumina with platinum or palladium for use in catalytic incineration. The surface area of activated alumina can range from 0.98×10^6 to 1.5×10^6 ft^2/lb (from 200 to 300 m^2/g).

Pore Size Distribution

The physical properties of the adsorbent affect the adsorption capacity, rate, and pressure drop across the adsorber bed. Table 6.1 summarizes these properties for the above adsorbents. Since adsorption occurs at the gas–solid interface, the surface area available to the vapor molecules determines the effectiveness of the adsorbent. Generally, the larger the surface area, the higher the adsorbent's capacity. However, the surface area must be available in certain pore sizes if it is to be effective as a vapor adsorber. The pores in activated carbon are generally classified as micropores, transitional pores, or macropores.

Micropores are openings whose radii are 200 nm (20 Å) or less. Pores larger than 2000 nm (200 Å) are macropores. Transitional pores are those with radii between 200 and 2000 nm (between 0.2 and 2.0 μm).

TABLE 6.1 Physical Properties of Major Types of Adsorbents

Adsorbent	Internal Porosity, %	Surface Area, m^2/g	Pore Volume, cm^3	Bulk Density, g/cm^3	Mean Pore Diameter, nm
Activated carbon	55–75	600–1600	0.80–1.20	0.35–0.50	150–200[a]
Activated alumina	30–40	200–300	0.29–037	0.90–1.00	180–200
Molecular sieves	40–55	600–700	0.27–038	0.80	30–90
Silica gel	70	750	0.40	0.70	220

[a]The 150–200 nm average is for the micropores only, since 95% of the surface area is associated with them.

Most gaseous air pollutant molecules are in the 40–90 nm size range. If a large portion of an adsorbent's surface area is in pores smaller than 40 nm, many contaminant molecules will be unable to reach these active sites. Figure 6.1 illustrates molecule movement in the pores. In addition, the larger pore sizes (macro- and transitional) contribute little to molecule capture. The vapor pressure of the contaminant in these larger areas is too low to be effectively removed. These larger pores serve mainly as passageways to the smaller micropore areas where the adsorption forces are strongest. Adsorption forces are strongest in pores that are not more than approximately twice the size of the contaminant molecule. These strong adsorption forces result from the overlapping attraction of the closely spaced walls.

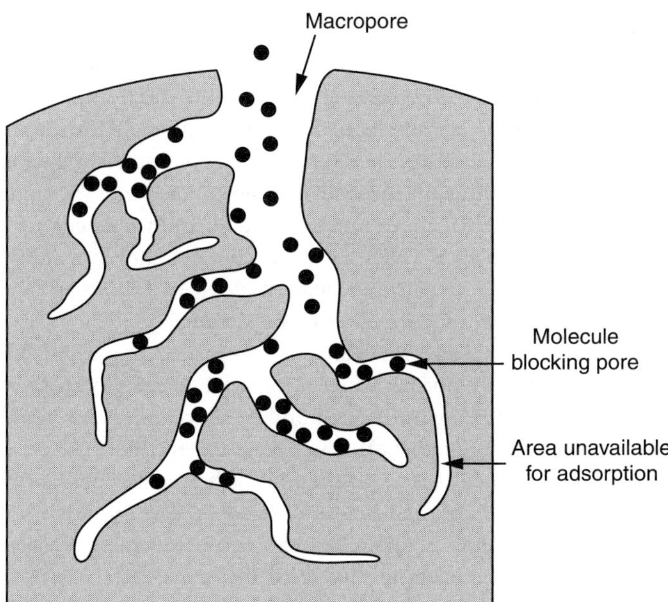

Figure 6.1 Molecular screening in pores of activated carbon.

Another phenomenon, *capillary condensation*, usually occurs only in the micropores. Capillary condensation occurs when multilayers of adsorbed contaminant molecules build up on both sides of the pore wall, totally packing the pore and condensing on it. The amount of contaminant that is removed increases since additional molecules condense on the surface previously occupied by the liquid molecules. However, desorption is often not as complete if capillary condensation occurs, since the forces that hold a liquid together are much stronger than the physical adsorption forces.

Air pollution control applications involve contaminant vapors at low partial pressures. Therefore, the micropore structure of an adsorbent plays an important role in determining the overall efficiency. Another reason for the wide use of activated carbon is that 90–95% of its surface area is in the micropore size range.

Adsorption control systems can be classified as either regenerable or nonregenerable. Nonregenerable systems are normally used to control exhaust streams with low pollutant concentrations, below 1.0 ppm. Generally, these pollutants are highly odorous or toxic to some degree. When these systems become saturated and start emitting pollutants (breakthrough), the bed is taken off stream and replaced with a fresh bed. The used carbon can then be sent back to the manufacturer for reactivation. Regenerable systems are used for higher pollutant concentrations such as in solvent recovery operations. Once the bed reaches the breakthrough point in a regenerable system, the pollutant vapors are directed to a second bed while the first has the vapors desorbed. Details on both classes of system follows.

Nonregenerable Adsorption Systems

Nonregnerable adsorption systems are manufactured in a variety of configurations. Bed areas are sized to control the airflow through them at 20–60 ft/min (6.0–18 m/min). They usually consist of a thin adsorbent bed depth, ranging in thickness from 0.5 to 4.0 inches (1.25–10.0 cm). These thin beds have a low pressure drop, normally below 0.025 in H_2O (62 Pa) depending on the bed thickness, gas velocity, and particle size of the adsorbent. Service time for these units can range from 6 months for heavy contaminant concentrations to up to 2 years for trace concentrations or intermittent operations. They are used mainly as air purification devices for small airflows in the office, laboratory exhaust, and other small exhaust streams.

The shapes of these thin-bed adsorbers are flat, cylindrical, or pleated. The granules of activated carbon are retained by porous support material, usually perforated sheetmetal. The adsorber system usually consists of a number of *retainers* or *panels* placed in one frame. The panels are similar to home air filters except that instead of containing steel wool they contain activated carbon as the filter. The pleated cell is one continuous retainer of activated carbon, rather than individual panels. The cylindrical canisters (see Figure 6.2) are usually small units designed to handle low flow rates of approximately 25 cfm (0.12 m/s). Cylindrical canisters are made of the same materials as the panel and pleated absorbers except their shape is round rather than square. Panel and pleated beds are dimensionally about the same size, normally 2 ft^2 (0.6 m^2), with the carbon

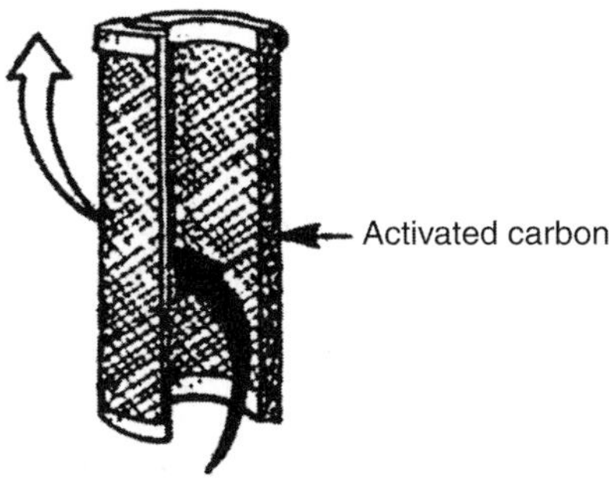

Figure 6.2 Canister adsorber.

depth ranging from 8 in to 2 ft (0.2–0.6 m). Flat-panel beds are sized to handle higher exhaust flow rates of approximately 2000 cfm (4.7 m/s).

In addition to thin-bed systems, thick-bed nonregenerable systems are also available. One system that can be used is essentially just a 55-gal drum. The bottom of the drum is filled with gravel to support a bed of activated carbon weighing approximately 330 lb (150 kg). A typical unit is shown in Figure 6.3. These units are used to treat small flow rates (100 cfm or 0.5 m/s) from laboratory hoods, chemical storage tank vents, and chemical reactors.

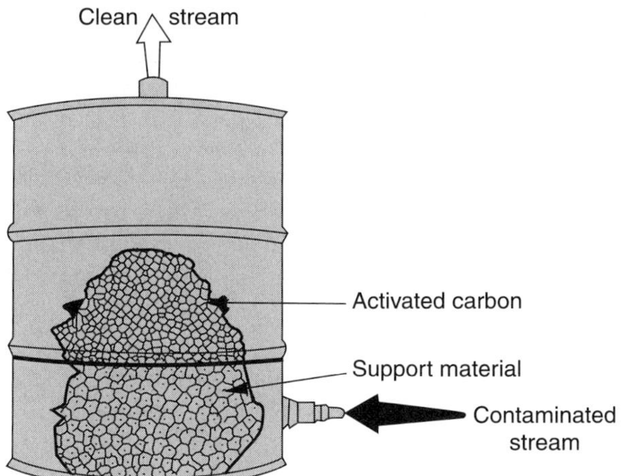

Figure 6.3 Thick-bed canister adsorber.

Regenerable Adsorption Systems

A large regenerable adsorption system can be categorized as a fixed, moving, or fluidized bed. The name refers to the manner in which the vapor stream and adsorbent are brought into contact. The choice of a particular system depends on the pollutants to be controlled and the recovery requirements. The most common adsorption system for controlling air pollutants is the fixed carbon bed. These systems are used to control a variety of organic vapors and are usually regenerated by direct steaming of the bed. The organic compounds may be recovered by condensing the exhaust from the regeneration step and separating out the water and solvent.

Fixed-bed adsorption systems generally involve multiple beds. One or more beds treat the process exhaust while the other beds are either being regenerated or cooled. A typical three-bed adsorption system is shown in Figure 6.4. The solvent-laden air stream is first pretreated to remove any solid or liquid particles that could blind the carbon bed and decrease efficiency. The solvent-laden air stream then usually passes down through the fixed carbon bed. Upward flow through the bed is usually avoided (unless flow rates are low <500 cfm to eliminate the risk of entraining carbon particles in the exhaust stream).

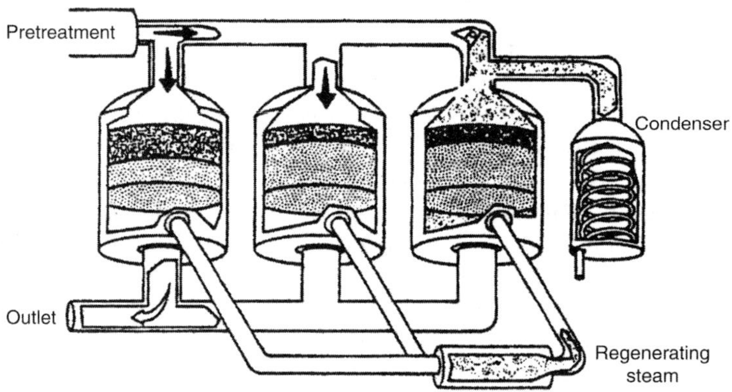

Figure 6.4 Three-bed system.

After a predetermined length of time, referred to as the *cycle time*, the solvent-laden air stream is directed to the second adsorber by a series of valves. Often, steam is then injected into the first bed to remove the adsorbed vapors. The steam and desorbed vapors are then usually sent to a recovery system. If the solvents are immiscible in water, they can be separated by condensing the exhaust and decanting off the solvent. If the solvents are miscible in water, distillation may be required.

Before the first adsorber is returned to service, cooling and drying of the carbon is usually provided. This will ensure against immediate breakthrough occurring from the "hot, wet" carbon bed. This can be accomplished by venting the solvent-laden air stream through the hot, wet adsorber, and then to the on-line adsorber to maintain a high removal efficiency.

Regenerable fixed carbon beds are usually between 1 and 6 ft (0.3 and 1.8 m) thick. The maximum adsorbent depth is primarily based on pressure drop considerations. Superficial gas velocities through the adsorber range from 20 to 100 ft/min (6.0–30.0 m/min); the maximum permissible velocity is approximately 90 ft/min. Pressure drops normally range from 3 to 15 in H_2O (from 750 to 3730 Pa) depending on the gas velocity, bed depth, and carbon particle size.

The two types of fixed-bed adsorbers can be distinguished by bed orientation in relation to airflow. The first is referred to as a *vertical flow adsorber*. The bed length is vertical, as is the direction of airflow. The gas stream usually flows downward.

For larger flow rates, *horizontal flow adsorbers* are used. Structurally, they are more suited to handle larger gas volumes. In horizontal flow units, the bed length is horizontal as is the direction of the incoming stream. The stream flows across the bed and then down through the bed. This system is illustrated in Figure 6.5. Adsorbers of this type are manufactured as a package system capable of handling flow rates of up to 10,000 cfm. Larger units must be engineered and fabricated for a specific application.

With a fixed-bed adsorber, it is necessary to periodically remove an adsorber from service for regeneration, drying, and cooling. This process can be done automatically if the bed is arranged to move in a prescribed way. Adsorbers where the bed moves from the polluted fluid to the regeneration fluid to the drying bed and cooling fluid are called *moving-bed adsorbers*. Although there are several possible arrangements, the most common type of moving-bed adsorber has a cylindrical bed that slowly rotates about its axis with an angular velocity while the frame and partitions are stationary. There are three sections of this moving-bed adsorber: an adsorption section, a regeneration section, and a drying–cooling section. As the bed revolves past the radial partitions, an element of the bed will pass from the adsorption section to the regeneration section, another element will pass from the regeneration section to the drying and cooling section, and a third element will pass into the adsorption section. In this way, the bed becomes continuously saturated, regenerated, and then dried and cooled in preparation for another adsorption cycle.

A fluidized-bed adsorption system operates in the same manner as a tray tower absorber (see Chapter 5). Instead of liquid flowing down the column from tray to

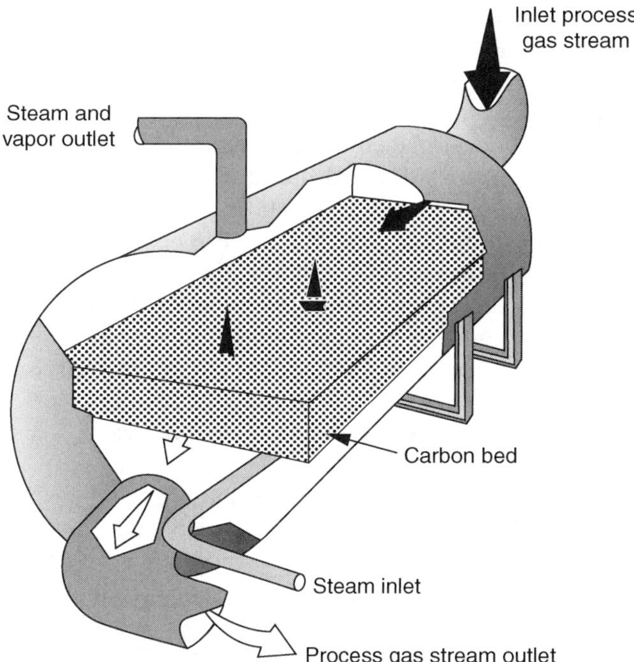

Figure 6.5 Horizontal bed absorber.

tray, granular activated carbon is used. The solvent-laden air stream is introduced at the middle of the tower. Then it passes up through the tower, fluidizing the activated carbon in a series of trays. The carbon then flows down through a vessel from tray to tray until it reaches the desorption section. Regeneration is accomplished in the bottom half of the vessel and the activated carbon is air-conveyed back to the top of the tower. As with the moving bed, the fluidized bed also provides continuous operation and more efficient utilization of the adsorbent. The need for multiple vessels is eliminated, which can greatly reduce the cost of the system. Gas velocities around 200 ft/min (1 m/s) are needed to fluidize the bed. This allows for use of a much smaller vessel for comparable airflow and helps to achieve uniform gas distribution. The main disadvantage with fluidized-bed adsorption systems is the high attrition (wear) losses of the granular activated carbon. There are also operational problems, as with virtually all systems that involve the movement of solids.

6.2 DESIGN AND PERFORMANCE EQUATIONS

Most available data on adsorption systems are determined at equilibrium conditions. *Adsorption equilibrium* is the set of conditions at which the number of molecules arriving on the surface of the adsorbent equals the number of molecules that are leaving. The adsorbent bed is then said to be "saturated with vapors" and can remove no more vapors from the exhaust stream. Thus, equilibrium determines the maximum amount of vapor that may be adsorbed at a given set of operating conditions. Although a number of variables affect adsorption, the two most important ones in determining equilibrium for a given system are temperature and pressure. Three types of equilibrium graphs are used to describe adsorption systems: isotherm at constant temperature, isobar at constant pressure, and isostere at constant amount of vapors adsorbed.

The most common and useful adsorption equilibrium data is the adsorption *isotherm*. The isotherm is a plot of the adsorbent capacity vs. the partial pressure of the adsorbate at a constant temperature, T. Adsorbent capacity is usually given in weight percent expressed as grams of adsorbate per 100 g of adsorbent. Figure 6.6 shows a typical example of an adsorption isotherm for carbon tetrachloride on activated carbon. Graphs of this type are used to estimate the size of adsorption systems, as described later. Attempts have been made to develop generalized equations that can predict adsorption equilibrium from physical data. This is very difficult because adsorption isotherms take many shapes depending on the forces involved. Isotherms may be concave upward, concave downward, or "S"-shaped. To date, most of the theories agree with data only for specific adsorbate systems and are valid over limited concentration ranges.

Two additional adsorption equilibrium relationships are the isostere and the isobar. The isostere is a plot of the natural log of the partial pressure p vs. $1/T$ at a constant amount of vapor adsorbed. Adsorption isostere lines are usually straight for most adsorbate–adsorbent systems. The isostere is important in that the slope of the isostere (approximately) corresponds to the heat of adsorption. The isobar is a plot of the amount of vapors adsorbed vs. T at a constant partial pressure. However, in the design of a

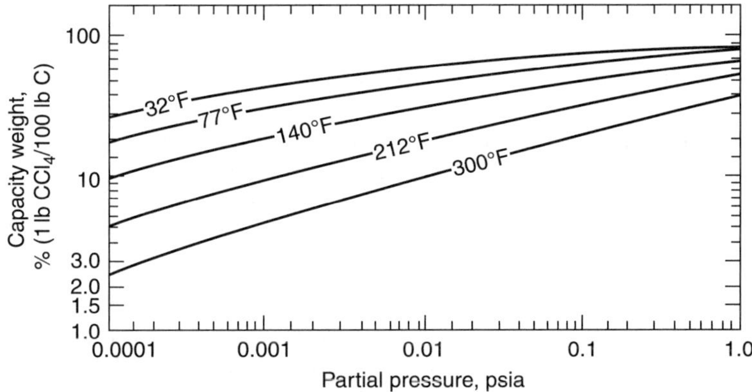

Figure 6.6 Adsorption isotherms for carbon tetrachloride on activated carbon.

pollution control system, the adsorption isotherm is by far the most commonly used equilibrium relationship.

The movement of vapors through an adsorbent bed is often referred to as a *dynamic process*. The term dynamic refers to motion, both in the movement of gas through the adsorbent bed and the change in vapor concentration(s) as it moves through the bed. There are a variety of configurations in which the contaminant air stream and adsorbent are brought into contact. The most common configuration is to pass the air stream down through a fixed volume or bed of adsorbent. Figure 6.7 illustrates how adsorption (mass transfer) occurs as vapors pass down through the bed.

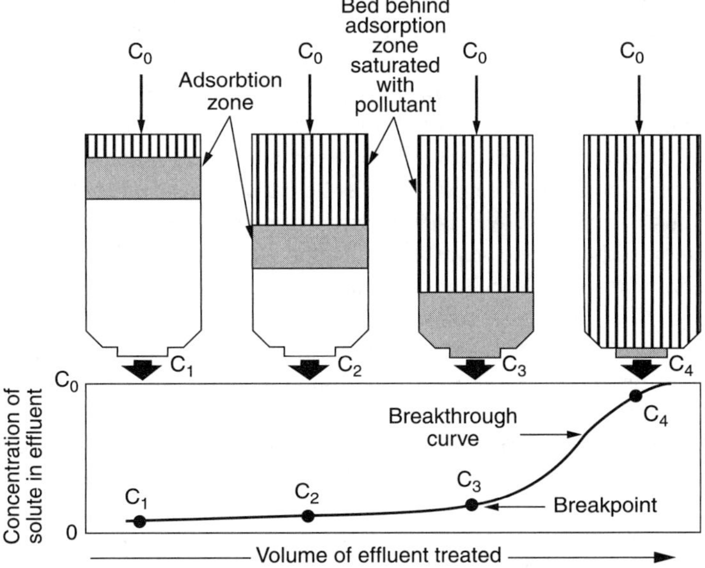

Figure 6.7 Adsorption wavefront.

The gas stream containing the pollutant at an initial concentration C_0 is passed down through a (deep) bed of adsorbent material that is free of any contaminant. Most of the contaminant is readily adsorbed by the top portion of the bed. The small amount of contaminant that is left is easily adsorbed in the remaining section of the bed. The effluent from the bottom of the bed is essentially pollutant-free, denoted at C_1.

After a period of time, the top layer of the adsorbent bed becomes saturated with contaminant. The majority of adsorption (approximately 95%) now occurs in a narrow portion of the bed directly below this saturation section. The narrow zone of adsorption is referred to as the *mass transfer zone* (MTZ). As additional contaminant vapors pass through the bed, the saturated section of the bed becomes larger and the MTZ moves further down the length of the adsorbent. The actual length of the MTZ remains fairly constant as it travels through the adsorbent bed. Additional absorption occurs as the vapors pass through the "unused" portion of the bed. The effluent concentration at C_2 is essentially still zero, since there is still an unsaturated section of the bed.

Finally, when the lower portion of the MTZ reaches the bottom of the bed, the concentration of contaminant in the effluent suddenly begins to rise. This is referred to as the *breakthrough point* (or *breakpoint*)—where untreated vapors are being exhausted from the adsorber. If the contaminated air stream is not switched to a fresh bed, the concentration of contaminant in the outlet will quickly rise until it approaches the initial concentration, illustrated at point C_4. It should be noted that in most air pollution control systems even trace amounts of contaminants in the effluent stream are undesirable.

To achieve continuous operation, adsorbers must be either replaced or cycled from adsorption to desorption before breakthrough occurs. In *desorption* or *regeneration*, the contaminant vapors are removed from the used bed in preparation for the next cycle. Most commercial adsorption systems are the regenerable type.

In regard to regenerable adsorption systems, three important terms are used to describe the *equilibrium capacity* (CAP) of the adsorbent bed. Other capacity terms are measured in mass of vapor per mass of adsorbent. First, the *breakthrough capacity* (BRK) is defined as the capacity of the bed at which unreacted vapors begin to be exhausted. The *saturation capacity* (SAT) is the maximum amount of vapor that can be adsorbed per unit mass of carbon (this is the capacity read from the adsorption isotherm) and thus is equal to CAP. The *working capacity* or *working charge* (WC) is the actual amount adsorbed in the adsorber. Thus, the working capacity is a certain fraction of the saturation capacity. Working capacities can range from 0.1 to 0.7 times the saturation capacity. (*Note*: A smaller capacity increases the amount of carbon required.) This fraction is set by the designer for individual systems by often balancing the cost of carbon and adsorber operation vs. preventing breakthrough while allowing for an adequate cycle time.

Another factor in determining the working capacity is that it is uneconomical to desorb all the vapors from the adsorber bed. The small amount of residual vapors left on the bed is referred to as the *HEEL*. This HEEL accounts for a large portion of the difference between the saturation capacity and the working capacity. In some cases, the working capacity can be estimated by assuming it to be equal to the saturation capacity minus the HEEL.

The following equations and procedures may be used to estimate some of the terms described above for an adsorber bed of height Z:

$$\text{BRK} = [(0.5)(\text{CAP})(\text{MTZ}) + (\text{CAP})(Z - \text{MTZ})]/Z \tag{6.1}$$

$$\text{WC} = [(0.5)(\text{CAP})(\text{MTZ})/Z] + [(\text{CAP})(Z - \text{MTZ})]/Z - \text{HEEL} \tag{6.2}$$

As discussed, the working charge can sometimes be taken to be some fraction of the saturated (equilibrium) capacity of the adsorbent (CAP):

$$\text{WC} = f(\text{CAP}); \quad 0 \leq f \leq 1.0 \tag{6.3}$$

For multicomponent adsorption, the maximum working charge may be calculated from the following (L. Theodore: personal notes, 1985)

$$\text{WC} = 1.0 \Big/ \sum_{i=1}^{n} (w_i/\text{CAP}_i) \tag{6.4}$$

where n = number of components
w_i = mass fraction of i in n components (not including carrier gas)
CAP_i = equilibrium capacity of component i

For a two-component (A, B) system, Equation (6.4) reduces to

$$\text{WC} = \frac{(\text{CAP}_\text{A})(\text{CAP}_\text{B})}{(w_\text{A})(\text{CAP}_\text{B}) + (w_\text{B})(\text{CAP}_\text{A})} \tag{6.5}$$

Note once again that the acronym SAT has also been used to represent the equilibrium capacity, i.e., CAP.

After determining the service time and/or working charge necessary for a particular application, there are other factors to be considered. These include:

1. The adsorbent particle size
2. The physical adsorbent bed depth
3. The gas velocity
4. The temperature of the inlet gas stream to the adsorbent
5. The contaminant concentration to be adsorbed
6. The contaminant concentration not to be adsorbed, including moisture
7. The removal efficiency
8. Possible decomposition or polymerization of the contaminant on the adsorbent
9. The frequency of operation(s)
10. Regeneration conditions
11. The system pressure
12. The system pressure drop

The design of fixed-bed adsorption systems also requires the capability of estimating pressure drop through the bed. Ergun derived a correlation to estimate the pressure drop for the flow of a fluid through a bed of packed solids when it alone fills the void in the bed. This correlation is given by the relationship [S. Ergun, *Chem. Eng. Progress* (1952)]:

$$\frac{\Delta P g_c d_p \varepsilon^3}{(Z)(2)\rho v^2(1-\varepsilon)} = \frac{75(1-\varepsilon)}{\text{Re}} + 0.875 \tag{6.6}$$

where ΔP = pressure drop of gas, lb_f/ft^2
Z = depth of packing, ft
g_c = conversion constant, 4.18 (10^8) $\text{ft}\cdot\text{lb}/\text{lb}_f\cdot\text{hr}^2$
d_p = effective particle diameter (the diameter of a sphere of the same surface/volume ratio as the packing in place), ft
ε = fractional void volume in a dry-packed bed, ft^3 voids/ft^3 packed volume
ρ = gas density, lb/ft^3
v = gas velocity, ft/hr
Re = $d_p\, v\rho/\mu$
μ = gas viscosity, $\text{lb}/\text{ft}\cdot\text{hr}$

Pressure drop information on different mesh carbon sizes is presented in Figure 6.8 (EPA chart). These data are also often used in pressure drop calculations.

Adsorbent regeneration is an integral part of the design procedure. Periodic replacement or regeneration of the adsorbent bed is mandatory in order to maintain continuous

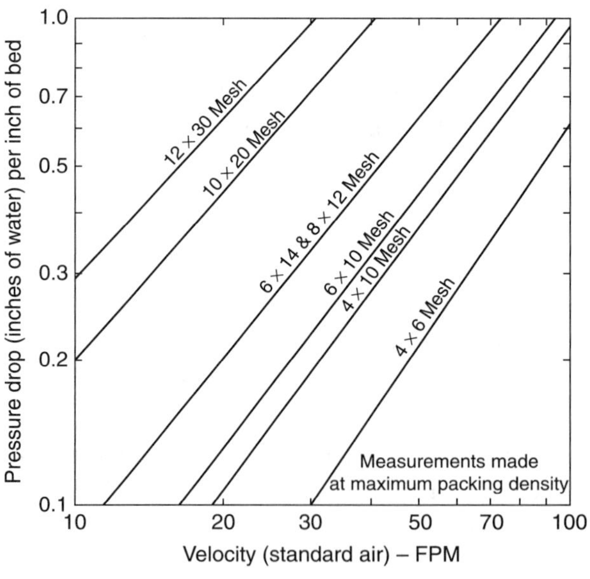

Figure 6.8 Activated carbon pressure along curves.

operation. When the adsorbate concentration is high and/or the cycle time is short (less than 12 hr), replacement of the adsorbent is not feasible and in situ regeneration is required. Regeneration is accomplished by reversing the adsorption process, usually by increasing the temperature or decreasing the pressure. Commercially, four methods are used in regeneration:

1. *Thermal Swing*. The bed is heated so that the adsorption capacity is reduced to a lower level. The adsorbate leaves the surface of the carbon and is removed from the vessel by a stream of purge gas. Cooling must be provided before the subsequent adsorption cycle begins. Steam regeneration is an example of thermal swing regeneration.
2. *Pressure Swing*. The pressure is lowered at a constant temperature to reduce the adsorbent capacity.
3. *Inert Purge Gas Stripping*. The stripping action is caused by an inert gas that reduces the partial pressure or the contaminant in the gas phase, reversing the concentration gradient. Molecules migrate from the surface into the gas stream.
4. *Displacement Cycle*. The adsorbates are displaced by some preferentially adsorbed material. This method is usually a last resort for situations in which the adsorbate is both valuable and is heat-sensitive, and for which pressure swing regeneration is ineffective.

Because it is simple and relatively inexpensive, steam stripping is one of the more commonly used desorption techniques. Several additional advantages to using steam for desorption are:

1. At high pressures, the steam temperature (>212°F, 100°C) is high enough to desorb most solvents of interest without damaging the carbon or the desorbed vapors.
2. Steam readily condenses in the adsorber bed releasing its (the steam's) heat of condensation, aiding in desorption.
3. Many organic compounds can easily be separated and recovered from the effluent stream by condensation or distillation.
4. Residual moisture in the bed is easily removed by a stream of cool dry air (either pure or process effluent).
5. Steam is a more concentrated source of heat than hot air so it is very effective in raising the temperature of the adsorber bed very quickly.

The amount of steam required for regeneration depends on the adsorbate loading of the bed. The longer a carbon bed is steamed, the more adsorbate will be desorbed, thus decreasing the HEEL. It is usually not cost-effective to try to desorb all of the adsorbed vapors from the bed. Acceptable working capacities can be achieved by using less steam and leaving a small portion of adsorbate in the bed.

During the initial heating period no vapors are desorbed. This arises because a fixed amount of steam is first required to raise the temperature of the cold bed to the desorption temperature. After this initial period, a substantial amount of adsorbate vapor is released until a plateau is reached. The plateau represents the optimum steam requirement, usually in the range of 0.25–0.35 lb of steam/lb of carbon, or approximately 2–3 lb steam/lb adsorbed vapor. In these systems, steam is usually supplied at or slightly above atmospheric pressure.

This section concludes with a rather simplified overall design procedure for a system adsorbing an organic that consists of two horizontal units (one on/one off) that are regenerated with steam. (*Note*: This design procedure was originally developed by Dr. Louis Theodore in 1985 and later published in 1988. This procedure has appeared elsewhere in the literature without properly acknowledging the author.)

1. Select adsorbent type and size.
2. Select cycle time; estimate regeneration time; set adsorption time equal to regeneration time; set cycle time to twice the regeneration time; generally, try to minimize regeneration time.
3. Set velocity; v is usually 80 ft/min, but can be increased to 100 ft/min.
4. Set steam/solvent ratio.
5. Calculate (or obtain) WC for the above.
6. Calculate amount of solvent adsorbed during $\frac{1}{2}$ cycle time, M_s.
7. Calculate adsorbent required, M_{AC}.

$$M_{AC} = M_s/\mathrm{WC} \tag{6.7}$$

8. Calculate adsorbent volume requirement:

$$V_{AC} = M_{AC}/\rho_B; \quad \rho_B = \text{adsorbent bulk density} \tag{6.8}$$

9. Calculate the face area of the bed:

$$A_{AC} = q/v; \quad q = \text{gas flowrate} \tag{6.9}$$

10. Calculate the bed height:

$$H = (V/A)_{AC} \tag{6.10}$$

11. Set L/D (length-to-diameter) ratio.
12. Calculate L and D from $A = LD$; constraints: for $L < 30$ ft, $D < 10$ ft; L/D of 3–4 acceptable if $v < 30$ ft/min.
13. Design (structurally) to handle if filled with water.
14. Design vertically if $q < 2500$ actual cubic feet per min (acfm).

6.3 OPERATION AND MAINTENANCE, AND IMPROVING PERFORMANCE

Utilizing adsorption as an air pollution control measure on sources emitting volatile organic hydrocarbons has proved to be extremely effective if a proper design procedure is applied and rigid operating procedures are established and followed. In terms of operation and maintenances, it is always desirable to check over the process being controlled to ensure that normal operation is being experienced. Of course, the blower must be running with the fan turning in the correct direction. Three-phase motors that are running backward may be corrected by changing two of the phase wires to the motor. Initial startups will probably require a system balance of the exhaust ducts. If multiple processes are being exhausted into the same system, adjustments of individual slide dampers may be required to obtain the correct airflow in each branch duct. Airflow switches are inexpensive and should be placed in each branch to sense airflow and allow the process to operate when the airflow is adequate. Airflows below 100 ft/min through the carbon beds will provide adequate retention time for solvents in the air stream to be adsorbed on the activated carbon. Excessive flows will reduce carbon efficiencies and allow volatiles to escape into the atmosphere. Excessive airflows are also detrimental to process operations where unnecessary solvents are evaporated and lost from process tanks and delivered to the carbon beds, posing additional loads on the system.

Additional details are available from L. Theodore, "Engineering Calculations: Adsorber Sizing Made Easy," *Chem. Eng. Progress*, pp. 16–17 (March 2005).

Optimizing performance of air pollution control equipment such as carbon adsorbers involves consideration and monitoring of the following factors:

1. Operation of process controls to minimize solvent emissions
2. Quality of solvent/air inlet stream
3. Characteristics of the inlet stream, such as concentration, temperature, and flow
4. Duration of the adsorption cycle
5. Quality and quantity of available steam for regeneration
6. Duration of steam–stripping cycle
7. Saturation and retentivity of the carbon
8. Quantity and quality of cooling water
9. Effectiveness of the water/solvent separator
10. Quality of reclaimed solvent
11. Quality of the wastewater
12. Quality of the exhaust stream from the adsorber bed

More recent development with adsorbers has centered primarily on improvement of adsorbents. Activity in the nanotechnology area has produced adsorbents such as carbon nanotubes, which have outperformed adsorbents employed in the past. See text by Kunz and Theodore (R. Kunz and L. Theodore, *Nanotechnology: Environmental Implications and Solutions*, John Wiley & Sons, Hoboken, NJ, 2006) for additional details.

PROBLEMS

6.1 Principal Adsorbent

The adsorbent most often used to control organic vapors is (select one)

(**a**) Molecular sieves
(**b**) Activated carbon
(**c**) Silica gel
(**d**) Zeolites

Solution: All four have been used for adsorption purposes, with (d) being a class of (a). However, it is activated carbon that is almost always used for organics. The correct answer is therefore (b).

6.2 Commercial Adsorbent

Which of the following materials is not a commercial adsorbent?

(**a**) Silica gel
(**b**) Activated carbon
(**c**) Methane
(**d**) Molecular sieve

Solution: Answers (a), (b), and (d) are commercial adsorbents. The correct answer is therefore (c).

6.3 Adsorption Process

The adsorption process is almost always

(**a**) Exothermic
(**b**) Endothermic
(**c**) Neutral
(**d**) Hydroscopic

Solution: The process is always exothermic since energy is liberated on adsorption during either physical or chemical adsorption. The correct answer is therefore (a).

6.4 Effect of Temperature

During gas adsorption, as the temperature of the system increases, the amount of material adsorbed

(**a**) Increases
(**b**) Decreases
(**c**) Remains constant
(**d**) Is doubled

Solution: Increasing the temperature during the adsorption process displaces the adsorbate from the adsorbent. Therefore, the amount adsorbed decreases. The correct answer is therefore (b).

6.5 Increasing the Amount Adsorbed

For adsorption processes, the amount of pollutant retained by the collecting media increases as

(**a**) The temperature of the system increases
(**b**) The liquid-to-gas flow ratio is increased
(**c**) The molecular weight of the pollutant decreases
(**d**) The pressure of the system increases

Solution: During the regenerative step, the pressure is often decreased to remove (displace) the pollutant from the collecting media. The opposite is true when the pressure is increased. The correct answer is therefore (d).

6.6 Chemical Versus Physical Adsorption

When compared to the heat of adsorption for physical adsorption, that for chemical adsorption is

(**a**) Always greater
(**b**) Always less
(**c**) About equal
(**d**) Dependent on the carbon pore size

Solution: The heat liberated during physical adsorption usually may be assumed to be approximately equal to the enthalpy (heat) of condensation. The heat liberated during chemical adsorption usually may be assumed to be approximately equal to the enthalpy (heat) of reaction. Further, the heat of reaction is usually several orders of magnitude greater than the heat of condensation. The correct answer is therefore (a).

6.7 Physical Versus Chemical Adsorption Characteristics

Provide a summary of the characteristics of physical vs. chemical adsorption in tabular form.
Solution: The answer is provided in Table 6.2.

TABLE 6.2 Summary of Characteristics of Chemisorption and Physical Adsorption

Chemisorption	Physical Adsorption
Releases high heat, 10,000 cal/mol	Releases low energy, 100 cal/mol
Forms a chemical compound	Dipolar interaction
Desorption is difficult	Desorption is easy
Impossible absorbate recovery	Easy absorbate recovery

6.8 Representing Equilibrium Data

The representation of adsorption data at a constant temperature that indicates the amount adsorbed versus the partial pressure of the adsorbate is called an

(**a**) Isobar
(**b**) Isostere
(**c**) Isotherm
(**d**) Isokinetic

Solution: Adsorption equilibrium data at a constant temperature are referred to collectively as an *isotherm*. The correct answer is therefore (c).

6.9 Process Options

Of the following vapors, which one would be most readily adsorbed using activated carbon?

(**a**) H_2O at 140°F
(**b**) CH_4 at 70°F
(**c**) C_4H_{10} at 140°F
(**d**) C_4H_{10} at 70°F

Solution: The general rule of thumb is that organics are more easily adsorbed on activated carbon. Furthermore, the higher the molecular weight, the more easily it can be captured (because of the larger molecular diameter). Finally, increasing the temperature decreases the adsorption capability. The correct answer is therefore (d).

6.10 Saturation Capacity

The saturation capacity of an adsorbent can be obtained by

(**a**) Subtracting the heel from the breakthrough capacity
(**b**) Adding the breakthrough and working capacity
(**c**) Reading a graph of the applicable isotherm
(**d**) Subtracting the heel from the working capacity

Solution: Saturation capacity represents the equilibrium capacity of the adsorbent for a particular system and specific operating conditions. It therefore represents the equilibrium value. The correct answer is therefore (c).

6.11 Mass Transfer Zone

In adsorption, if the mass transfer zone (MTZ) is larger than the bed depth, then

(**a**) The bed will become blinded with water
(**b**) The pressure drop will be too great
(**c**) The system is well designed
(**d**) Untreated vapors will be exhausted

Solution: The mass transfer zone represents the length of the bed across which the concentration gradient exists. Therefore, all the vapors will not be captured since the bed at the discharge location contains residual vapors. Note that "untreated vapors" refers to discharge under normal operating conditions. The correct answer is therefore (d).

6.12 Thin Beds

A thin-bed adsorption system is employed mainly for

(**a**) High power consumption per cfm
(**b**) Low concentration of adsorbate in the gas
(**c**) High pressure drop
(**d**) High collection efficiency

Solution: Answers (a), (c), and (d) are incorrect. A thin bed cannot be used if the concentration is high, as this would result in the bed becoming saturated in a short period of time. However, a low concentration is acceptable. The correct answer is therefore (b).

6.13 Advantages Versus Disadvantages

Describe three advantages and three disadvantages of using an adsorber for gaseous control.

Solution: Three advantages are:

1. Moderate capital costs
2. Moderate operating costs
3. Ability to regenerate at high recovery efficiencies

Three disadvantages are:

1. Pollutant is not permanently destroyed
2. Carbon has to be periodically replaced
3. Plugging and/or short-circuiting can occur

6.14 Design and Process Factors

As noted earlier, activated carbon adsorption equipment is commonly used for the control of gaseous pollutants. Aside from the size of such a piece of equipment, list and briefly discuss at least five design or process factors that affect its effectiveness in controlling gaseous pollutants.

Solution:

1. Temperature
2. Pressure
3. Molecular weight
4. Partial pressure of other gases
5. Reactivity, e.g., decomposition and polymerization
6. Size and/or surface area of adsorbent
7. Design considerations, including height, velocity, pressure drop, concentration (in and out), and regeneration frequency

6.15 Carbon Dioxide Adsorption

As described earlier, the relation between the amount of substance adsorbed by an adsorbent and the equilibrium partial pressure or concentration at constant temperature is called the *adsorption isotherm*. The adsorption isotherm is the most important and by far the most often used of the various equilibria data that can be measured. To represent the variation of the amount of adsorption per unit area or unit mass with pressure, Freundlich proposed the equation

$$Y = kp^{(1/n)} \tag{6.11}$$

where Y = weight or volume of gas (or vapor) adsorbed per unit area or unit mass of adsorbent

p = equilibrium partial pressure of adsorbed gas

k, n = empirical constants dependent on the nature of the solid and adsorbate and on the temperature

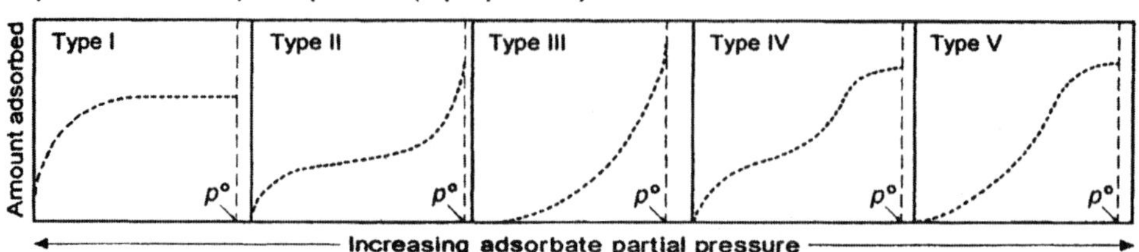

Figure 6.9 Types of adsorption isotherms.

Equation (6.11) may also be written as follows. Taking logarithms of both sides one obtains Equation (6.12)

$$\log Y = \log k + (1/n)\log p \tag{6.12}$$

If log Y is now plotted against log p, a straight line should result with a slope equal to $1/n$ and a log Y intercept equal to log k. Although the requirements of the equation are often met satisfactorily at low pressures, at higher pressures experimental points tend to deviate from a straight line, indicating that this equation does not have general applicability in reproducing adsorption of gases (or vapors) by solids. A much better equation for type I isotherms (see Figure 6.9) was deduced by Langmuir from theoretical considerations:

$$Y = \frac{ap}{1 + bp} \tag{6.13}$$

This equation is the Langmuir adsorption isotherm. The constants a and b are characteristic of the system under consideration and are evaluated from experimental data. Their magnitude also depends on the temperature. At any one temperature the validity of the Langmuir adsorption equation can be verified most conveniently by first dividing both sides of this equation by p and then taking reciprocals. The result is

$$\frac{p}{Y} = \frac{1}{a} + \frac{b}{a}p \tag{6.14}$$

Since a and b are constants, a plot of p/Y vs. p should yield a straight line with slope equal to b/a and an ordinate intercept equal to $1/a$.

The carbon dioxide adsorption data on Columbia (Columbia is a registered trade-mark of Union Carbide Corporation) activated carbon are presented in Table 6.3 for a temperature of 50°C. Determine the constants of both the Freundlich equation and the Langmuir equation.

Solution: For the Freundlich equation, Table 6.4 can be generated using the data in the problem statement A plot of log Y vs. log p yields the equation (see Figure 6.10)

$$Y = 30p^{0.7}$$

TABLE 6.3 Carbon Dioxide Equilibrium Data

Equilibrium Capacity, cm^3/g	Partial Pressure CO_2, atm
30	1
51	2
67	3
81	4
93	5
104	6

TABLE 6.4 Freundlich Calculations

Equilibrium Capacity, cm^3/g	Partial Pressure CO_2, atm	log Y	log p
30	1	1.477	0.000
51	2	1.708	0.301
67	3	1.826	0.477
81	4	1.909	0.602
93	5	1.969	0.699
104	6	2.017	0.778

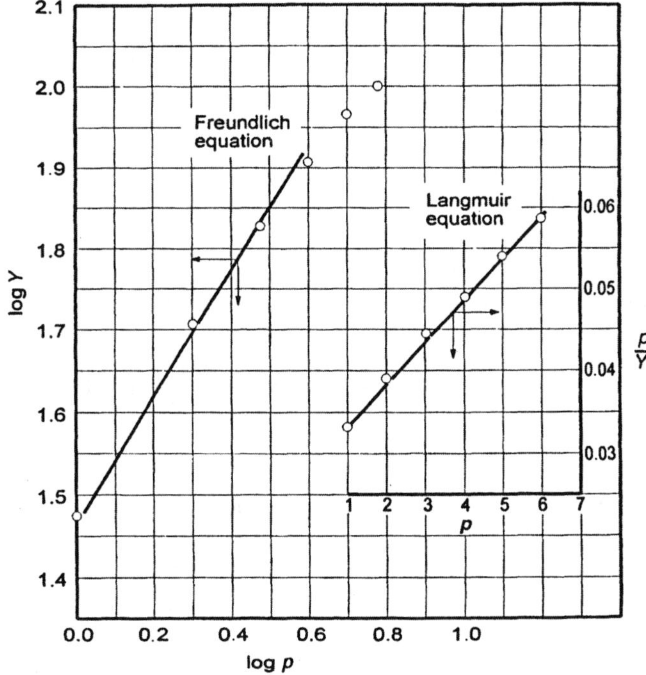

Figure 6.10 Plots of log Y vs. log p and p/Y vs. p.

TABLE 6.5 Langmuir Calculations

p/Y	p
0.063	1
0.039	2
0.045	3
0.049	4
0.054	5
0.058	6

For the Langmuir equation, (see Table 6.5)

$$\frac{p}{Y} = \frac{1}{a} + \frac{b}{a}p$$

A plot of p/Y vs. p yields the equation (see Figure 6.10).

$$\frac{p}{Y} = \frac{3.57p}{(1 + 0.186)p}$$

The Langmuir equation appears to fit the data better. Regressing the data for both equations and generating regression coefficients are left as exercises for the reader.

6.16 Breakthrough and Working Capacities

Calculate both the breakthrough capacity and the working capacity of an adsorption bed given the saturation (equilibrium) capacity (SAT), mass transfer zone (MTZ), and HEEL data provided below:

Depth of adsorption bed $Z = 3$ ft
SAT = 39% = 0.39; lb solvent/lb adsorbent
MTZ = 4 inches
HEEL = 2.5% = 0.025; lb solvent/lb adsorbent

Solution: The working charge (WC) may be estimated from

$$\text{WC} = \text{SAT}\left(\frac{Z - \text{MTZ}}{Z}\right) + (0.5)\text{SAT}\left(\frac{\text{MTZ}}{Z}\right) - \text{HEEL} \tag{6.2}$$

where SAT (or CAP) is the equilibrium capacity and HEEL is the residual adsorbate present in the bed following regeneration. (The breakthrough capacity is given in the above equation without the inclusion of the HEEL term.) Regarding the problem, first calculate the breakthrough capacity (BRK):

$$\text{BRK} = \frac{(0.5)(0.39)(4) + (0.39)(36 - 4)}{36}$$

$$= 0.368 = 36.8\% \tag{6.1}$$

Calculate the working capacity (WC):

$$\begin{aligned} \text{WC} &= 36.8 - 2.5 \\ &= 34.3\% = 0.343 \end{aligned}$$

As noted earlier, the working charge may also be taken to be some fraction *f* of the saturated (equilibrium) capacity of the adsorbent:

$$\text{WC} = (f)(\text{SAT}) \quad 0 \leq f \leq 1.0 \tag{6.3}$$

Note once again that the notation CAP is often employed in place of SAT (see Problem 6.18).

6.17 Carbon Tetrachloride Saturation Capacity

Determine the saturation capacity of the carbon bed during the adsorption phase for carbon tetrachloride (CCl_4) at the following operating conditions:

Airflow rate = 12,000 cfm at 77°F
Concentration of CCl_4 air (inlet) = 410 ppmv
System at atmospheric pressure
Steam regeneration at 212°F used

Refer to Figure 6.6. The saturation capacity is

(a) 40%
(b) 30%
(c) 20%
(d) 18%

Solution: The only pertinent data provided are conditions 1–3, i.e., the operating temperature and pressure and CCl_4 partial pressure during the adsorption step. At 410 ppmv, the partial pressure is given by

$$\begin{aligned} p &= (410/10^6)(14.7); \quad P = 1\,\text{atm} = 14.7\,\text{psia} \\ &= 6.03 \times 10^{-3}\,\text{psia} \end{aligned}$$

Referring to Figure 6.6, the saturation capacity at this partial pressure at a temperature of 77°F is approximately 40%. The correct answer is therefore (a).

6.18 Working Charge Calculation

The saturated carbon capacity in a 3-ft adsorption bed is 39%. The MTZ and HEEL were determined to be 4 inches and 2.5%, respectively. Calculate the working charge in percent.

(a) 41.5
(b) 36.5
(c) 34.3
(d) 32.0

Solution: First note that answer (a) is not possible since it is greater than the CAP specified. The describing equation is

$$\mathrm{WC} = \mathrm{CAP}\left(\frac{Z - \mathrm{MTZ}}{Z}\right) + (0.5)\mathrm{CAP}\left(\frac{\mathrm{MTZ}}{Z}\right) - \mathrm{HEEL};\ \text{consistent units} \quad (6.2)$$

Substituting yields

$$\begin{aligned}\mathrm{WC} &= \left(\frac{36-4}{36}\right)39 + (0.5)\left(\frac{4}{36}\right)39 - 2.5\\ &= 34.7 + 2.2 - 2.5\\ &= 34.4\end{aligned}$$

The correct answer is therefore (c).

6.19 MTZ Calculation

The following data sheet was extracted from a company's files.
Test performed: measurement of length of MTZ, 20°C
Material tested: carbon "B"
Test number: B–17
Conditions:

Carbon bed depth = 62.0 cm (2.0 ft)
Saturation capacity = 26% at 20°C
Temperature of test = 20°C
Breakthrough capacity = 24.9% (calculated from test)

Calculate the MTZ in centimeters and inches. Comment on the result.

Solutions: Apply Equation (6.1).

$$\mathrm{BRK} = \frac{0.5(\mathrm{SAT})(\mathrm{MTZ}) + (\mathrm{SAT})(Z - \mathrm{MTZ})}{Z} \quad (6.1)$$

$$Z = 62\ \mathrm{cm}$$

$$\mathrm{SAT} = 0.26$$

Capacity at breakthrough, BRK = 0.249. Substituting into Equation (6.1), one obtains

$$0.249 = \left[\frac{(0.5)(0.26)(\mathrm{MTZ}) + 0.26(62 - \mathrm{MTZ})}{62}\right]$$

Solving yields

$$\begin{aligned}\mathrm{MTZ} &= 5.25\ \mathrm{cm}\\ &= 2.08\ \text{inches}\end{aligned}$$

For activated carbon, MTZs of 2–4 inches are to be expected. For tall beds, the MTZ effect may be neglected.

6.20 Adsorbent Breakthrough Calculation

The R&D group at a local adsorbent manufacturer plans to apply one of its adsorbents for gaseous emission control to water systems. The granulated activated carbon (GAC) adsorbent, JB26 is the adsorbent of interest. Some data have been collected on the adsorption isotherm for a few solutes, but no extensive tests have been conducted as of yet. However, a major client is very interested in the new adsorbent and would like to know approximately how long one of its 65-ft^3 units could operate with JB26 before breakthrough would occur.

The following information was given to estimate how many days a 56,000-gal/day unit could run until breakthrough. From limited testing, the isotherm of interest is described by

$$Y_T = 0.002C^{3.11} \tag{6.15}$$

where Y_T = lb adsorbate/lb adsorbent
C = adsorbate concentration, mg/L

Currently, the client's unit operates 30 days until regeneration; the client would like to double that time if possible so as to limit downtime and increase profits. The density of JB26 is 42 lb/ft^3, and it will treat a stream with an inlet concentration (C_i) of 3.5 mg/L. In addition, the breakthrough concentration has been set at 0.5 mg/L. In the design of the GAC it is important to be able to measure or predict the approximate time until an adsorbent will reach its maximum capacity for adsorption. The point at which this occurs is referred to as the *breakthrough adsorption capacity* Y_B, and corresponds to an adsorbate (solute) concentration at breakthrough C_B. Once breakthrough occurs, undesired solute will pass through the bed without being adsorbed, contaminating product quality. The breakthrough adsorption capacity typically ranges within 25–50% of the theoretical capacity Y_T, which is determined from the adsorption isotherm, evaluated at the initial solute concentration in solution C_i. The time to breakthrough is then given by the following equation:

$$t_B = \frac{Y_B M_{AC}}{8.34q[C_i - (C_B/2)]} \tag{6.16}$$

where t_B = time to breakthrough, days
Y_B = adsorption capacity at breakthrough, lb adsorbate/lb adsorbent
M_{AC} = mass of carbon in column, lb
q = volumetric flow rate of solution, Mgal/day (millions of gallons per day)
C_i = adsorbate feed concentration, mg/L
C_B = adsorbate concentration at breakthrough, mg/L

Solution: The theoretical adsorption capacity Y_T is

$$Y_T = 0.002C^{3.11} \tag{6.15}$$

$$= 0.002(3.5)^{3.11}$$

$$= 0.09842 \text{ lb adsorbate/lb adsorbent}$$

Assume the actual value is 50% of the theoretical value (see comment above). Thus,

$$Y_B = (0.5)(0.09842) = 0.04921 \text{ lb adsorbate/lb adsorbent}$$

The mass of carbon in the unit is then

$$M_{AC} = (65 \text{ ft}^3)(42 \text{ lb/ft}^3)$$

$$= 2730 \text{ lb carbon}$$

The breakthrough time can now be calculated from Equation (6.16).

$$t_B = \frac{(0.04921)(2730)}{(8.34)(0.56)[3.5 - (0.5/2)]}$$

$$= 8.85 \text{ days}$$

6.21 Activated Carbon Requirement

How many pounds of carbon are needed for an adsorption system to remove 120 lb/hr of carbon tetrachloride from an air stream (77°F, 1 atm) if the partial pressure of the CCl_4 in the air stream is 0.005 psia and the operation is on a 4-hr cycle (one on–one off)? Assume that the working charge is estimated by doubling the amount of carbon needed at the saturation capacity. Refer to Figure 6.6 for adsorption equilibrium data.

(**a**) 1000
(**b**) 1600
(**c**) 2400
(**d**) 3200

Solution: At 77°F and a partial pressure of 0.005 psia, the saturation capacity (from Figure 6.6) is approximately 40%. The carbon required for each adsorber for a 2-hr period is then

$$M_{AC} = \frac{(120 \text{ lb/hr})(2 \text{ hr})}{0.4 \text{ lb/lb AC}}$$

$$= 600 \text{ lb}$$

(where subscript AC denotes activated carbon). Since there are two adsorbers, the answer should be doubled. In addition, the amount calculated from the WC is also to be doubled. Thus, the total carbon required is 2400 lb. The correct answer is therefore (c).

6.22 Adsorber Column Height

Determine the required height of adsorbent for an adsorber that treats a degreaser ventilation stream contaminated with trichloroethylene (TCE) given the following operating and design data:

Volumetric flow rate of contaminated air stream = 10,000 scfm
Standard conditions = 60°F, 1 atm
Operating temperature = 70°F
Operating pressure = 20 psia
Adsorbent = activated carbon
Bulk density of activated carbon, $\rho_B = 36\ \text{lb/ft}^3$
Working capacity of activated carbon = 28 lb TCE/100 lb carbon
Inlet concentration of TCE = 2000 ppm (ppm by volume)
Molecular weight of TCE = 131.5

The adsorption column cycle is set at 4 hr in the adsorption mode, 2 hr in heating and desorbing, 1 hr in cooling, and 1 hr in standby. The adsorber recovers 99.5% by weight of TCE. A horizontal cylinder unit with an inside diameter of 6 ft and length of 15 ft is used. Employ the design procedure provided in Section 6.2.

Solution: The actual volumetric flow rate of the contaminated gas stream q in acfh (actual cubic feet per hour) is obtained using Charles' Law:

$$\begin{aligned} q &= 10{,}000\left(\frac{70+460}{60+460}\right)\left(\frac{14.7}{20}\right) \\ &= 7491\ \text{acfm} \\ &= 4.5 \times 10^5\ \text{acfh} \end{aligned}$$

The volumetric flow rate of TCE in acfh is

$$\begin{aligned} q_{\text{TCE}} &= (y_{\text{TCE}})(q_a) \\ &= (2000 \times 10^{-6})(4.5 \times 10^5) \\ &= 900\ \text{acfh} \end{aligned}$$

The mass flow rate of TCE, $\dot{m}$, in lb/hr can be calculated employing the ideal gas law.

$$\begin{aligned} \dot{m} = \dot{m}_{\text{TCE}} &= q_{\text{TCE}}\frac{P(\text{MW})}{\text{RT}} \\ &= (900)\frac{(131.5)(20)}{(10.73)(70+460)} \\ &= 416.2\ \text{lb/hr} \end{aligned}$$

The mass of TCE adsorbed-during the 4-hr period is

$$\begin{aligned}\text{TCE adsorbed} = m_{\text{TCE}} &= (\dot{m})(0.995)(4) \\ &= (416.2)(0.995)(4) \\ &= 1656\,\text{lb}\end{aligned}$$

The volume of activated carbon, V_{AC} required is

$$\begin{aligned}V_{\text{AC}} &= \frac{m_{\text{TCE}}}{(\text{WC})(\rho_{\text{B}})} \\ &= \frac{\text{TCE adsorbed}}{(28\,\text{lb TCE adsorbed}/100\,\text{lb carbon})(\text{bulk density})} \\ &= \frac{1656}{(28/100)(36)} \\ &= 164\ \text{ft}^3\end{aligned}$$

The height of the adsorbent Z is

$$\begin{aligned}Z &= \frac{\text{activated carbon volume}}{\text{cross-sectional area}} \\ &= \frac{164}{(6)(15)} \\ &= 1.82\ \text{ft}\end{aligned}$$

6.23 Calculation of Bed Volume and Height

Theodore Consultants has been selected by Fitzmaurice Chemicals Inc., to design an adsorber which treats a degreaser ventilation stream contaminated with trichloroethylene (TCE). Fitzmaurice Chemicals has provided basic operating data to Theodore Consultants that requires the use of activated carbon as an adsorbent. Determine the volume of activated carbon required to treat the gas and the height of the adsorption column. Operating data are provided below:

Flow rate of contaminated air stream = 12,000 scfm
Standard conditions = 1atm, 65°F
Operating conditions = 25 psia and 75°F
Bulk density of activated carbon = 38 lb/ft^3
Working capacity of activated carbon = 25 lb TCE per 100 lb carbon
Inlet concentration of TCE = 2000 ppmv

The adsorption column cycle is set at 5 hr in the adsorption mode, 2 hr in heating–desorbing mode, 2 hr in cooling, and 1 hr in standby. The adsorber recovers 96% of TCE by weight. A horizontal unit with a 5 ft × 20 ft cross-sectional area is recommended.

Solution: First apply Charles' law to obtain the actual gas flow rate:

$$\begin{aligned} q_a &= q_s\left(\frac{T_a}{T_s}\right)\left(\frac{P_a}{P_s}\right) \\ &= 12{,}000\left(\frac{75+460}{65+460}\right)\left(\frac{14.7}{25}\right) \\ &= 7190.4\ \text{acfm} \\ &= 4.3\times 10^5\ \text{acfh} \end{aligned}$$

The TCE flow rate is

$$\begin{aligned} q_{TCE} &= y_{TCE}\, q_a \\ &= \left(\frac{2000}{10^6}\right)(4.3\times 10^5) \\ &= 860\ \text{acfh} \end{aligned}$$

The TCE mass flow rate is

$$\begin{aligned} \dot{m}_{TCE} &= q_{TCE}\,(\rho_{TCE}) \\ &= 860\frac{(25)(131.5)}{(10.73)(535)} \\ &= 492.5\ \text{lb/hr} \end{aligned}$$

For a 5-hr period, one obtains

$$\begin{aligned} m_{TCE} &= (492.5)(0.96)(5) \\ &= 2364\ \text{lb} \\ V_{AC} &= \frac{m_{TCE}}{(\text{WC})(\rho_B)} \\ &= \frac{2364}{(0.25)(38)} \\ &= 248.8\ \text{ft}^3 \\ Z &= \frac{248.8}{(5)(20)} \\ &= 2.49\ \text{ft} = 2.5\ \text{ft} \end{aligned}$$

6.24 Performance of a Two-Bed Carbon Adsorption System

A two-bed carbon adsorption system is being used to control odors emitting from a drum-filling operation. The material being drummed is a high-purity grade of pyridine (C_5H_5N) that has a human odor detection level of 100 ppm. It has been reported that an odor has been detected from outside the drumming area when the equipment is in service, i.e., when drums are being filled. Discussions with operating personnel have indicated that the adsorption system is the source of the odor. You are requested to determine if the adsorption equipment/emission is the source of the odor or if the equipment is capable of containing/controlling the pyridine emission. Design and actual operating data are provided below.

1. The adsorption units are twin horizontal units with face dimensions for flow of 5×12 ft. Each unit contains new 4×6 mesh activated carbon B that was installed one month ago. The measured bed height is 12 inches.
2. The carbon manufacturer maintains that the breakthrough capacity of the carbon is 0.49 lb pyridine/lb carbon and that the carbon has a bulk density of 25 lb/ft^3.
3. Laboratory tests performed by plant personnel indicate that the carbon contains a HEEL of approximately 0.03 lb pyridine/lb carbon when regenerated with 4.0 lb of steam/lb pyridine at 10 psig.
4. The ventilation blower for the drum filling station has a flow of 5000 acfm at 25°C and 14.7 psia and contains a pyridine concentration of 2000 ppm (plant hygienist data).
5. The drum filling operation operates on a 24-hr/day basis and the adsorption units are operated on an 8-hr adsorption, 5-hr regeneration cycle, with 3 hr for cooling and stand-by. The steam used during the 5-hr regeneration cycle was determined to be 2725 lb (mass flowmeter).
6. The adsorption unit was designed based on a pressure drop through the bed following the relationship

$$\Delta P = 0.37Z(v/100)^{1.56} \tag{6.17}$$

   where Z is the bed depth in inches, v is the velocity in ft/min, and the pressure drop is in inches of water. The measured operational pressure drop is 3.3 inches of water.
7. The fractional fan efficiency is 0.58.

To evaluate the adsorber's performance, please determine the following:

1. The mass of pyridine to be captured in the adsorption period
2. The working capacity of the carbon B
3. The mass and volume of carbon that should be used in each unit
4. The required bed height

5. The design pressure drop through the bed using the required bed height for full capture
6. The horsepower requirement for this process
7. The required steam to regenerate the bed to the HEEL level of 0.03lb pyridine/lb carbon

Solution: Calculate the mole fraction of pyridine (P) in the gas stream.

$$y_P = 2000/10^6$$
$$= 0.0020$$

Calculate the volumetric flow rate of P in acfm.

$$q_P = y_P q$$
$$= (0.0020)(5000)$$
$$= 10.0 \text{ acfm}$$

Determine the density of the P vapor at the operating conditions.

$$\rho_P = \frac{P(MW)}{RT} = \frac{(14.7)(79)}{(10.73)(537)}$$
$$= 0.2015 \text{ lb/ft}^3$$

The mass of P collected during the adsorption period is then

$$m_P = (10)(0.2015)(8)(60)$$
$$= 967.2 \text{ lb}$$

Estimate the working capacity (WC) of carbon B for this system.

$$WC = BC - HEEL$$
$$= 0.49 - 0.03$$
$$= 0.46 \text{ lb P/lb carbon B}$$

Calculate the mass of carbon that should be used for each unit.

$$M_{AC} = m_P/WC$$
$$= 967.2/0.46$$
$$= 2103 \text{ lb carbon B}$$

The volume of activated carbon V_{AC} is

$$V_{AC} = M_{AC}/\rho_B$$
$$= 2103/25$$
$$= 84.1 \text{ ft}^3$$

The cross-sectional area A of the carbon that is presently available for flow is

$$A = (5)(12) = 60\ \text{ft}^2$$

Since the bed height is 12 inches or 1.0 ft, the actual volume currently employed is 60 ft^3. Because this is below the required 84 ft^3, the odor problem is present; the equipment is not capable of controlling the emission to eliminate the odor.

The "required" height of the adsorbent in the unit is

$$\begin{aligned} Z &= V_{\text{AC}}/A \\ &= 84.1/60 \\ &= 1.40\ \text{ft} \\ &= 16.8\ \text{inches} \\ &\approx 17.0\ \text{inches} \end{aligned}$$

Estimate the pressure drop across the required adsorbent in inches of H_2O using Equation (6.17).

$$\begin{aligned} \Delta P &= 0.37Z\left(\frac{v}{100}\right)^{1.56} = 0.37Z\left(\frac{q/A}{100}\right)^{1.56} \\ &= (0.37)(17)\left(\frac{5000/60}{100}\right)^{1.56} \\ &= 4.73\ \text{in}\ \text{H}_2\text{O} \end{aligned}$$

The total pressure drop across the bed in $\text{lb}_\text{f}/\text{ft}^2$ is then

$$\begin{aligned} \Delta P_{\text{total}} &= (4.73)(5.2) \\ &= 24.6\ \text{lb}_\text{f}/\text{ft}^2 \end{aligned}$$

while the HP requirement is

$$\begin{aligned} \text{HP} &= \frac{(24.6)(5000)}{(60)(550)(0.58)} \\ &= 6.43 \end{aligned}$$

Finally, the steam requirement for regeneration is.

$$\begin{aligned} m_{\text{steam}} &= (4.0)(967.2) \\ &= 3869\ \text{lb steam during regeneration} \end{aligned}$$

The actual operating steam rate (2725 lb for 5 hr) is also below the required value.

6.25 Adsorber Capacity

The operating temperature and pressure for a column are 16°C and 1.2 atm, respectively. An acetone (molecular weight = 58) gas stream enters the column at a concentration of 100 ppm by volume. Activated carbon is used to remove the pollutant; its bulk density is 0.575 g/cm^3. The activated carbon in the column has a saturation capacity and HEEL of 55% and 4.5%, respectively, and the mass transfer zone is 0.15 m. The column is horizontal with the dimensions of 2.5 m in height, 30.75 m in length, and 6 m in diameter. The column cycles every 5 hr, assume that to be the time in the adsorption mode. The adsorber recovers 98% by weight of acetone. Assuming that the standard conditions are 21°C, and 1 atm, find the volumetric flow rate of the gas stream containing the acetone in scfm, am^3/min, sm^3/min, acfm.

Solution: This problem involves the use of SI units. First calculate BRK and WC.

$$\text{BRK} = \frac{(0.5)(\text{CAP})(\text{MTZ}) + \text{CAP}(Z - \text{MTZ})}{Z} \qquad (6.1)$$

$$= \frac{(0.5)(0.55)(0.15) + 0.55(2.5 - 0.15)}{2.5}$$

$$= 0.5335 = 53.35\%$$

$$\text{WC} = 53.35 - 4.5$$

$$= 48.85\% = 0.485$$

The volume and mass of the activated carbon are

$$V_{\text{AC}} = (30.75)(2.5)(6)$$

$$= 461.25\ \text{m}^3$$

$$M_{\text{AC}} = (461.25)(10^6\ \text{cm}^3/\text{m}^3)(0.575\ \text{g/cm}^3)(1\ \text{kg}/1000\ \text{g})$$

$$= 119{,}350\ \text{kg}$$

For acetone (converting to English units)

$$\dot{m}_{\text{A}}(5\ \text{hr}) = 119{,}350/5$$

$$= 23{,}870\ \text{kg A/hr}$$

$$\dot{m}_{\text{A}}(\text{inlet}) = 23{,}870/0.98 = 24{,}350\ \text{kg A/hr}$$

$$= (24{,}350)(2.2)$$

$$= 53{,}585\ \text{lb A/hr}$$

$$\rho_{\text{A}} = \text{P(MW)/RT}$$

$$= (1.2)(58)/(0.73)(16 + 273)(1.8)$$

$$= 0.1833\ \text{lb/ft}^3$$

$$\begin{aligned} q_A &= \dot{m}_A/\rho_A = 53{,}585/0.1833 \\ &= 292{,}335\ \text{ft}^3/\text{hr} \\ &= 4872\ \text{ft}^3/\text{min} \\ q_{\text{total}} &= 4872(10^6/10^2) \\ &= 4.872 \times 10^7\ \text{ft}^3/\text{min} \end{aligned}$$

Applying Charles' Law,

$$\begin{aligned} q_{\text{total}} &= 4.872 \times 10^7[(21 + 273)/(16 + 273)]/(1.2/1.0) \\ &= 5.95 \times 10^7\ \text{scfm} \end{aligned}$$

The conversion back to SI units is left as an exercise for the reader.

6.26 Sizing a Carbon Adsorber

A printing company must reduce and recover the amount of toluene it emits from its Rotograve printing operation. The company submits some preliminary information on installing a carbon adsorption system. You, the primary consultant, are given the following information:

Airflow = 20,000 acfm (77°F, 1 atm)
Adsorption capacity for toluene = 0.175 lb toluene/lb activated carbon
Operation at 10% of LEL (lower explosivity limit) for toluene in the exit air from printer
LEL for toluene = 1.2%
Toluene molecular weight = 92.1
Carbon bulk density (4 × 6 mesh) = 30 lb/ft^3
Working charge = 60% of saturation capacity
Regeneration just under one hour; assume 1.0 hr
Maximum velocity through adsorber = 100 fpm (ft/min)

Determine the minimum size of adsorber you would recommend for a 1 × 1 system. Calculations should include the pertinent dimensions of the adsorber, the amount of carbon, the depth of the bed and an estimate of the pressure drop. Also calculate the fan horsepower if the blower/motor efficiency is 58%.

Solution: Initially, base the calculations on 1 hr regeneration time so that 1 hr of adsorption is available. Key calculations and results are provided below for the toluene (TOL) and activated carbon (AC). Design for operation at 10% of LEL.

$$\begin{aligned} q_{\text{TOL}} &= (20{,}000)(0.10)(0.012) \\ &= 24\ \text{acfm} \\ \dot{m}_{\text{TOL}} &= \frac{(24)(492/537)(92.1)(60)}{359} \\ &= 338\ \text{lb/hr} \end{aligned}$$

$$p_{\text{TOL}} = (24/20{,}000)(14.7)$$
$$= (0.0012)(14.7)$$
$$= 0.01764 \text{ psia}$$

$$\text{SAT} = 17.5\% = 0.175\,\text{lb}_{\text{TOL}}/\text{lb}_{\text{AC}}$$

$$\text{WC} = (0.175)(0.60)$$
$$= 0.105\,\text{lb}_{\text{TOL}}/\text{lb}_{\text{AC}}$$
$$= 10.5\,\text{lb}_{\text{TOL}}/100\,\text{lb}_{\text{AC}}$$

$$M_{\text{AC}} = (338/0.105)(1.0)$$
$$= 3220\,\text{lb}_{\text{AC}} \text{ for one bed}$$
$$= 6440\,\text{lb}_{\text{AC}} \text{ for both beds}$$

$$V_{\text{AC}} = 3220/30 = 107\,\text{ft}^3$$

$$A_{\text{AC}} = 20{,}000/100 = 200\,\text{ft}^2$$

$$Z = H = 107/200$$
$$= 0.535\,\text{ft} = 6.4\,\text{in}$$

Suggest a horizontal 10-ft diameter × 20 ft–long design.
Refer to Figure 6.6. At 100 ft/min, $\Delta P = 0.625$ in H_2O/in bed. Therefore

$$\Delta P_{\text{total}} = (0.625)(6.4)$$
$$= 4.0\,\text{in } H_2O$$
$$\text{HP} = \frac{(20{,}000)(4.0)(5.2)}{(0.58)(33{,}000)}$$
$$= 22\,\text{HP}$$

Note: This represents a marginal design since H is only slightly higher than 0.5 ft.

6.27 Breakthrough Time Calculation

A degreaser ventilation stream contaminated with trichloroethylene (TCE) is treated through the use of a horizontal carbon bed adsorber. The adsorber is normally designed to operate at a gas flow of 8000 scfm (60°F, 1 atm), and the concentration of TCE at the adsorber inlet is 1500 ppmv. The capture efficiency of the adsorber is 99% ($E_c = 0.99$) under normal design conditions. Design parameters are as follows:

Actual conditions: 25psia, 90°F
SAT = 35%
Z = depth of bed = 2.5 ft
L = length of adsorber = 2.5 ft

D = diameter of adsorber = 8 ft
MTZ = 5 in
HEEL = 2.0%
Bulk density of carbon bed = 35 lb/ft^3

Determine the time before breakthrough occurs.

Solution: The calculations are provided below.

$$q = 8000\left(\frac{14.7}{25}\right)\left(\frac{90+460}{60+460}\right)$$
$$= 4975 \text{ acfm}$$

$$q_{\text{TCE}} = (1500 \times 10^{-6})(4975)$$
$$= 7.46 \text{ acfm}$$

$$\dot{m}_{\text{TCE}} = \frac{P(\text{MW})q_{\text{TCE}}}{RT} = \frac{(25)(131.5)(7.46)}{(10.73)(90+460)}$$
$$= 4.16 \text{ lb/min}$$

$$\text{BRK} = \frac{0.5(C_s)(\text{MTZ}) + (C_s)(Z - \text{MTZ})}{Z} \tag{6.1}$$

$$= \frac{0.5(0.35)\left(\frac{5}{12}\right) + (0.35)\left[2.5 - \left(\frac{5}{12}\right)\right]}{2.5}$$

$$= 0.32$$

$$\text{WC} = \text{BC} - \text{HEEL} \pm \text{SF (safety factor)}$$
$$= 0.32 - 0.02 - 0$$
$$= 0.30 = 30\%$$

$$V_{\text{AC}} = (25)(8)(2.5)$$
$$= 500 \text{ ft}^3$$

$$M_{\text{AC}} = (500)(35)$$
$$= 17{,}500 \text{ lb}$$

$$t \text{ (to saturation)} = \frac{(\text{WC})(M_{AC})}{\dot{m}_{\text{TCE}}E_c} \tag{6.18}$$

$$= \frac{(0.30)(17{,}500)}{(4.16)(0.99)}$$

$$= 1275 \text{ min} \approx 21 \text{ hr}$$

6.28 Transient Operation

Refer to Problem 6.27. Recalculate the time before breakthrough occurs, based on the following transient condition. The adsorber system is on line for one hour at the previous normal design conditions when the inlet concentration of

TCE rises to an average value of 2500 ppmv due to a malfunction in the degreaser process; the efficiency also drops to 97.5% during this time. Assume that the SAT remains the same.

Solution: For transient conditions

$$\begin{aligned} m_{\text{TCE}} \text{ (in carbon)} &= (0.30)(500)(35) \\ &= 5250\,\text{lb}_{\text{TCE}}, \text{ maximum} \\ m_{\text{TCE}} \text{ (flow, first hour)} &= (4.16)(60)(0.99) \\ &= 247.1\,\text{lb}_{\text{TCE}} \text{ captured} \end{aligned}$$

The remaining capacity of the bed is now available for adsorption after the first hour.

$$\begin{aligned} m_{\text{TCE}} \text{ (after first hour)} &= 5250 - 247.1 \\ &= 5003\,\text{lb}_{\text{TCE}} \end{aligned}$$

$$t \text{ (to transient saturation)} = t_{\text{ts}} = \frac{5003}{\dot{m}_{\text{TCE}\cdot\text{transient}}(0.975)}$$

$$\begin{aligned} \dot{m}_{\text{TCE}\cdot\text{transient}} &= (4.16)(2500/1500) \\ &= 6.93\,\text{lb/min} \end{aligned}$$

$$\begin{aligned} t_{\text{tp}} &= \frac{5003}{(6.93)(0.975)} \\ &= 740\,\text{min} = 12.34\,\text{hr} \end{aligned}$$

The time to breakthrough, following the transient period, is

$$\begin{aligned} t_{\text{B}} &= t + t_{\text{tp}} \\ &= 60 + 740.1 \\ &= 800.1\text{ min} = 13.34\,\text{hr} \end{aligned}$$

Thus, the time to breakthrough has been reduced from 21 to 13.3 hr.

6.29 Ethyl Acetate Application

A solvent recovery plant is to recover 1800 lb/hr of ethyl acetate (EA) vapor from a mixture with air at a concentration of 1.8 lb vapor/1000 ft^3 air at 60°F, 1 atm pressure. The adsorbent will be activated carbon, 4 × 10 mesh (average particle diameter 0.0091 ft) apparent density of individual particles 42 lb/ft^3, and apparent density of the packed bed 28 lb/ft^3. The carbon is capable of adsorbing 0.25 lb vapor/lb carbon including the HEEL up to the breakpoint.

Determine the amount of carbon required and choose suitable dimensions for the carbon beds on what you consider a good design. Estimate the pressure drop by using the EPA chart (Figure 6.8). Using this pressure drop, calculate the horsepower requirement for the system.

Solution: In some respects, this is an open-ended problem since design values need to be set. Select a 1×1 system, operating 2 hr on and 2 hr off. Thus, select as a basis 2 hr of operation.

$$\begin{aligned} M_{\mathrm{EA}} &= (2)(1800) \\ &= 3600\ \mathrm{lb} \\ M_{\mathrm{AC}} &= 3600/0.25 \\ &= 14{,}400\ \mathrm{lb};\ 28{,}800 \text{ total for two units} \\ V_{\mathrm{AC}} &= 14{,}400/28 \\ &= 514.3\ \mathrm{ft}^3 \end{aligned}$$

The volumetric flow rate is

$$\begin{aligned} q &= [(1800)/(1.8/1000)](1/60) \\ &= 16{,}700\ \mathrm{acfm} \end{aligned}$$

A horizontal unit should definitely be used. Select

$$v = 80\ \mathrm{ft/min}$$

so that

$$\begin{aligned} A &= 16{,}700/80 \\ &= 209\ \mathrm{ft}^3 = (L)(D) \\ H &= Z = 514.3/209 \\ &= 2.5\ \mathrm{ft} \end{aligned}$$

If $L/D = 2$, then $D = 10.2$ ft and $L = 20.4$ ft. If $L/D = 3$, then $D = 8.3$ ft, $L = 25.0$ ft. Select $D = 9.5$ ft so that

$$\begin{aligned} L &= A/D \\ L &= 209/9.5 \\ &= 22\ \mathrm{ft} \end{aligned}$$

For this design

$$\begin{aligned} L/D &= 22/9.5 = 2.3 \\ H/D &= 2.5/9.5 = 0.263 \end{aligned}$$

Select a 4×10 mesh carbon. From Figure 6.8, one obtains

$$\Delta P/L = 0.72\ \mathrm{in}\ H_2O/\mathrm{ft\ bed}$$

Thus

$$\begin{aligned} \Delta P &= (0.72)(12)(2.5) \\ &= 21.6\ \mathrm{in}\ H_2O \end{aligned}$$

Finally (for 70% efficiency)

$$\begin{aligned} \text{HP} &= (q)(\Delta P)/E_f \\ &= (16{,}700)(21.6)(5.21)/(60)(550)(0.7) \\ &= 81.2 \end{aligned}$$

6.30 Acetone Recovery System Design

You have been asked to design a system to recover a 1.3% by volume acetone mixture in air. The air stream flow rate is 4.32×10^7 acfd (actual ft^3/day) at 20°C and 1 atm. Your boss has given you plenty of latitude but suggests (and of course you want to anyway) working within the operating conditions and design procedure suggested by the eminent one, Dr. L. Theodore. The most immediate adsorbent equilibrium data available indicate that the equilibrium capacity for acetone in the 4.0–8.0% relative saturation range is 0.35. It has further been suggested to employ 4×6 mesh carbon as the adsorbent and to operate with a 2-hr regeneration period. The average particle diameter can be assumed to be 0.0091 ft. The apparent and bulk density for all types of carbon particles available are 45 lb/ft^3 and 26 lb/ft^3, respectively. Assume $v = 80$ fpm and WC = (0.8) (CAP) and provide horsepower ($E_f = 0.65$) requirements (use the EPA chart, Figure 6.8).

Solution: First calculate the relative saturation (RS) of the acetone on the gas stream:

$$\begin{aligned} \text{RS} &= yP/p' = p/p' \\ &= (0.013)(760)/170 \\ &= 0.0581 = 5.81\% \end{aligned}$$

The equilibrium capacity is therefore 0.35

$$\text{CAP} = 0.35$$

Following the usual design procedure

$$\begin{aligned} q &= 4.32 \times 10^7/1440 \\ &= 30{,}000 \text{ acfm} \end{aligned}$$

For $v = 80$ ft/min

$$\begin{aligned} A_{AC} &= 30{,}000/80 \\ &= 375 \text{ ft}^2 \end{aligned}$$

$$\begin{aligned} \dot{m}_A &= (0.013)(4.32 \times 10^7)(50)(273/293)/359 \\ &= 84{,}450 \text{ lb/day} \end{aligned}$$

In a 2-hr period

$$\begin{aligned} m_A &= (84{,}450)(2/24) \\ &= 7045 \text{ lb} \end{aligned}$$

The working charge is

$$\begin{aligned} \text{WC} &= (0.35)(0.8) \\ &= 0.28 \end{aligned}$$

Therefore

$$\begin{aligned} M_{\text{AC}} &= 7045/0.28 \\ &= 25{,}160\ \text{lb} \end{aligned}$$

$$\begin{aligned} V_{\text{AC}} &= 25{,}160/26 \\ &= 968\ \text{ft}^3 \\ H &= V_{\text{AC}}/A_{\text{AC}} = 968/375 \\ &= 2.58\ \text{ft} = 31\ \text{in} \end{aligned}$$

From Figure 6.8

$$\begin{aligned} \Delta P/L &= 0.44\ \text{in}\ H_2O/\text{in bed} \\ \Delta P &= (0.44)(31) \\ &= 13.6\ \text{in}\ H_2O \end{aligned}$$

Finally

$$\begin{aligned} \text{HP} &= (30{,}000)(13.6)/(33{,}000)(0.65) \\ &= 19.0 \end{aligned}$$

6.31 Multicomponent Adsorption

Provide an equation to estimate the saturation capacity, (SAT) for a multi-component mixture.

Solution: For multicomponent adsorption the saturation capacity may be calculated from the following (personal notes, L. Theodore):

$$\text{SAT} = \frac{1.0}{\sum_{i=1}^{n} (w_i/\text{SAT}_i)} \tag{6.4}$$

where n = number of components
w_i = mass fraction of i in n components (not including carrier gas)
SAT_i = equilibrium capacity of component i

For a two-component (A, B) system, Equation (6.4) reduces to

$$\text{SAT} = \frac{(\text{SAT})_{\text{A}}(\text{SAT})_{\text{B}}}{w_{\text{A}}(\text{SAT})_{\text{B}} + w_{\text{B}}(\text{SAT})_{\text{A}}} \tag{6.5}$$

Refer to Section 6.2 for additional details.

6.32 Multicomponent Organic Vapor Application

An air stream containing organic vapors is to be cleaned with an adsorber using activated carbon as the adsorbent. The organic vapor concentrations in the air stream are provided in Table 6.6. Calculate the "theoretical" working capacity and the "actual" capacity given a HEEL of 0.025 (fractional basic) and a packing factor of 0.03 (negative).

TABLE 6.6 Equilibrium Data for Organic Vapor

w_i (Solute-Free Basis)	Mass Fraction, w_i (Solute-Free Basis)	Equilibrium Capacity, lb/lb
Methane	0.67	0.39
Toluene	0.05	0.08
Propane	0.25	0.40
Diphenyl	0.02	0.11
Benzyl alcohol	0.01	0.16

Solution: The working capacity is calculated from

$$\mathrm{WC} = \frac{1.0}{\sum_{i=1}^{n} (w_i/\mathrm{CAP})} \tag{6.4}$$

Substituting, one obtains

$$\begin{aligned}\mathrm{WC} &= \frac{1.0}{(0.67/0.39) + (0.05/0.08) + (0.25/0.40) + (0.02/0.11) + (0.01/0.16)} \\ &= \frac{1.0}{1.718 + 0.625 + 0.625 + 0.182 + 0.063} \\ &= 0.3113 = 31.13\%\end{aligned}$$

This represents the "theoretical" or maximum working charge. The actual WC, including the HEEL effect, is

$$\begin{aligned}\mathrm{WC} &= 0.3113 - 0.025 \\ &= 0.2863 = 28.63\%\end{aligned}$$

Including the packing factor leads to

$$\begin{aligned}\mathrm{WC} &= 0.3113 - 0.025 - 0.03 \\ &= 0.2563 = 25.63\%\end{aligned}$$

6.33 Molecular Sieve Application

A 1000-acfm gas stream at 40 psi and 75°F is processed by a 1×1 molecular sieve adsorber system to remove CO and CO_2. The concentrations of CO and CO_2 are 5% and 1% (by volume), respectively. It is estimated that the system will remove 99.5% CO and 98% CO_2. Size the unit and determine the total mass of molecular sieve required.

Data:

$CAP_{CO} = 0.4$ lb CO/lb MS
$CAP_{CO_2} = 0.3$ lb CO_2/lb MS
Bulk density $= 36$ lb/ft^3
Velocity $= 60$ fpm
Adsorption time $= 4$ hr
Actual working charge $= (0.5)$ ideal working charge

Solution: For this system, the key calculations for both CO and CO_2 are:

	CO	CO_2
$q_i = 1000(y_i)$; ft^3/min	50	10
$V_{i,4\text{hr}} = (q_i)(60)(4)$; ft^3	12,000	2400
$\rho_i = (40)(\text{MW})/(10.73)(536)$; lb/ft^3	0.195	0.306
$m_i = (V_i)(\rho_i)$; lb	2340	734
$m_{i,\text{rec}} = (\%\ \text{rec}/100)\, m_i$; lb	2328.3	719.3
$w_i = m_i/(m_i + m_j)$; mass fraction	0.761	0.239

Employ Equation (6.5) to calculate the ideal working charge.

$$\text{WC} = \frac{(\text{CAP})_A(\text{CAP})_B}{(w_A\text{CAP}_B + w_B\text{CAP}_A)} \tag{6.5}$$

$$= 0.37$$

$$\text{WC}_{\text{act}} = (0.5)(0.37) = 0.185$$

The actual WC is used in the design calculations:

$$\text{M}_{\text{MS}} = (2328.3 + 719.3)/0.185$$
$$= 16{,}473.6 \text{ lb MS}$$
$$A = 1000/60$$
$$= 16.7 \text{ ft}^2$$
$$V_{\text{MS}} = 16{,}473.6/36$$
$$= 457.6 \text{ ft}^3$$
$$Z = H = 457.6/16.7$$
$$= 27.4 \text{ ft}$$

A vertical unit is recommended because of the low gas flow rate:

$$D = (4A/\pi)^{0.5}$$
$$= 4.6 \text{ ft}$$

The total mass of sieve is

$$M_{\text{MS,total}} = (2)(16{,}473.6) = 32{,}947 \text{ lb}$$

6.34 Benzene and Pyridine Application

A vertical 10-ft-diameter vessel is used to adsorb 40 ppm benzene and 30 ppm pyridine from air stream at 1 atm and 70°F. The superficial velocity through the vessel is 64 ft/min. The adsorbent is 4×6 mesh activated carbon having a bulk density 30 lb/ft^3. Inlet concentrations indicate that the equilibrium capacities for benzene and pyridine are 0.12 and 0.19, respectively. The working charge can be assumed equal to 80% of the ideal value.

Determine the pressure drop and the amount of adsorbent in the bed if the bed is in operation 5 days a week, 24 hr a day. Use the EPA chart (Figure 6.8) to estimate the pressure drop.

Solution: Preliminary calculations are provided below:

$$A = (\pi/4)D^2 = (0.7854)(10)^2$$
$$= 78.54 \text{ ft}^2$$
$$q = (64)(78.54)$$
$$= 5027 \text{ ft}^3/\text{min}$$

Applying the ideal gas law yields

$$\rho_{\text{B}} = \frac{(1)(78.11)}{(530)(0.7302)}; \quad \text{MW}_{\text{B}} = 78.11$$
$$= 0.20 \text{ lb/ ft}^3$$
$$\rho_{\text{P}} = 0.20 \text{ lb/ ft}^3; \quad \text{MW}_{\text{P}} = 79.1$$

The volume and mass of each pollutant are now calculated:

$$V_{\text{B}} = (5027)(60)(24)(5)(40/10^6)$$
$$= 1448 \text{ ft}^3$$
$$m_{\text{B}} = (1448)(0.2)$$
$$= 290 \text{ lb}$$

$$V_P = (5027)(60)(24)(5)(30/10^6)$$
$$= 1086 \text{ ft}^3$$
$$m_P = (1086)(0.2)$$
$$= 217 \text{ lb}$$

The total mass adsorbed is

$$m_T = m_B + m_P = 290 + 217$$
$$= 507 \text{ lb}$$

Calculate the two mass fractions:

$$w_B = 290/507$$
$$= 0.57$$
$$w_P = 217/507$$
$$= 0.428$$

Noting that $CAP_B = 0.12$ and $CAP_P = 0.19$ and applying Equation (6.5), one obtains

$$WC_I = \frac{(CAP_B)(CAP_P)}{(w_B)(CAP_P) + (w_P)(CAP_B)}$$
$$= \frac{(0.12)(0.19)}{(0.571)(0.19) + (0.428)(0.12)}$$
$$= 0.143$$

The actual WC is

$$WC = (0.143)(0.8)$$
$$= 0.114$$

The mass and volume of carbon are therefore

$$M_{AC} = 507/0.114$$
$$= 4447 \text{ lb}$$
$$V_{AC} = 4447/30$$
$$= 148 \text{ ft}^3$$
$$H = Z = 148/78.54$$
$$= 1.89 \text{ ft}$$
$$= 22.6 \text{ in} \approx 23 \text{ in}$$

From Figure 6.8, one obtains

$$\Delta P = 0.31 \text{ in } H_2O/\text{in bed}$$
$$\Delta P_{bed} = (0.31)(23)$$
$$= 7.1 \text{ in } H_2O$$

The reader should note that one could also perform the calculation by obtaining the individual working charges for each compound:

$$WC_B = (0.8)(0.12)$$
$$= 0.096$$
$$WC_P = (0.8)(0.19)$$
$$= 0.152$$

The carbon required is then calculated for both benzene and pyridine. The sum represents the total carbon needed to do the job.

6.35 Landfill Emissions

Gas emissions are being collected from a landfill and must be treated before being released into the environment. There are several options for treatment. As one of the project engineers, you have been given the task to look at the use of a horizontal activated carbon adsorber to collect the methane in the gas stream (assume 95% removal). Perform the following calculations:

1. Mass of CH_4 collected per operating period
2. Mass of activated carbon needed
3. Depth of AC bed

The following data are provided:

Flow rate = 11,000 acfh
Operating pressure of the adsorber = 1 atm
Operating temperature of the adsorber = 70°F
Time between regeneration = 24 hr
Gas stream contains (by mole fraction): N_2 0.10
CH_4 0.50
CO_2 0.40
Bulk density of activated carbon = 30 lb/ft^3
Width of AC bed = 15 ft
Length of AC bed = 20 ft
CAP = 0.39 lb CH_4/lb AC
HEEL = 0.05 lb CH_4/lb AC

Solution: The mass flow rate of the methane is (applying the ideal gas low)

$$\dot{m}_M = (0.5)(11{,}000)(14.7)(16)/(10.73)(530)$$
$$= 227 \text{ lb/hr}$$

For the mass of methane collected in one day,

$$\begin{aligned} m_{\mathrm{M}} &= 0.95(227)(24) \\ &= 5176\,\mathrm{lb} \end{aligned}$$

A key assumption that should be made here is that only the methane contributes to WC:

$$\begin{aligned} \mathrm{WC} &= 0.39 - 0.05 \\ &= 0.34 \end{aligned}$$

The mass of carbon required is

$$\begin{aligned} M_{\mathrm{AC}} &= 5176/0.34 \\ &= 15{,}224\,\mathrm{lb} \end{aligned}$$

To complete the calculations,

$$\begin{aligned} V_{\mathrm{AC}} &= 15{,}224/30 \\ &= 508\,\mathrm{ft}^3 \\ H = Z &= 508/(15)(20) \\ &= 1.7\ \mathrm{ft} \\ &= 20.4\,\mathrm{in} \end{aligned}$$

6.36 Vinyl Chloride Monomer Application

It is desired to recover vinyl chloride monomer (VCM), MW = 62.5, from a side stream at a PVC plant with a flow rate of 1000 acfm at 75°F and 1.0 atm. The stream contains 750 ppm VCM, and 99% recovery is desired. Using a 1 × 1, 4 × 10 mesh activated carbon system, determine a suitable bed diameter, height, pressure drop through the bed, and horsepower requirements ($E_{\mathrm{f}} = 0.55$). Assume a bulk density of 30 lb/ft^3, 8 hr of adsorption time, and a superficial velocity of 50 ft/min. After regeneration, the activated carbon is at equilibrium with the VCM at a partial pressure of 0.0001 psia. Neglect MTZ effects. Equilibrium data at 75°F and 1.0 atm are available in Table 6.7.

Solution:

$$\begin{aligned} p_{\mathrm{VCM}} &= (750 \times 10^{-6})(14.7\ \mathrm{psia}) \\ &= 0.0110\ \mathrm{psia} \end{aligned}$$

By linear interpolation (see Table 6.7)

Equilibrium capacity = 8.15 lb VCM/100 lb AC
HELL (at 0.0001 psia) = 1.0 lb VCM/100 lb AC

TABLE 6.7 Vinyl Chloride Monomer Equilibrium Data

Equilibrium Capacity, lb VCM/100 lb Carbon	Partial Pressure VCM, psia
1.0	0.0001
1.7	0.0005
3.0	0.001
6.0	0.005
8.0	0.01
14.0	0.05

Therefore

$$\begin{aligned}
\text{WC} &= 8.15 - 1.0 \\
&= 7.15\,\text{lb VCM}/100\,\text{lb AC} \\
m_{\text{VCM}} &= \left[\frac{(1000)(60)}{(379)(535/520)}\right](8)(750 \times 10^{-6})(62.5)(0.99) \\
&= 57.1\,\text{lb VCM} \\
M_{\text{AC}} &= 57.1/(7.15/100) \\
&= 799\,\text{lb AC} \\
V_{\text{AC}} &= 799/30 \\
&= 26.6\ \text{ft}^3 \\
A &= 1000/50 \\
&= 20\ \text{ft}^2 \\
D &= (4A/\pi)^{0.5} \\
&= 5.05\ \text{ft} = 5\,\text{ft};\quad A(D = 5\,\text{ft}) = 19.6\ \text{ft}^2 \\
H &= 26.6/19.6 \\
&= 1.36\,\text{ft} = 16.3\ \text{in}
\end{aligned}$$

Pressure drop is obtained from EPA chart Figure 6.8, 4×10 mesh at 50 ft/min:

$$\begin{aligned}
\Delta P &= 0.38\,\text{in}\,\text{H}_2\text{O}/\text{in bed} \\
\Delta P_{\text{T}} &= (0.38)(16.3) \\
&= 6.2\,\text{in}\,\text{H}_2\text{O} \\
&= 0.22\ \text{psi} \\
\text{HP} &= (1000)(6.2)/(6356)(0.55) \\
&= 2.0
\end{aligned}$$

6.37 Molecular Sieve Regeneration

Use the graph in Figure 6.11 to solve the following problem. Estimate the pounds of CO_2 that can be adsorbed by 100 lb of Davison 4 Å molecular sieve from a

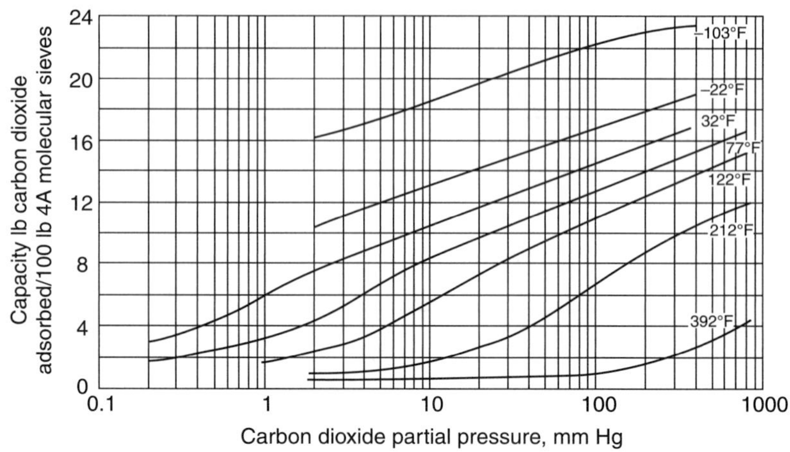

Figure 6.11 Vapor–solid equlibrium isotherms.

discharge gas mixture at 77°F and 40 psia containing 10,000 ppmv (ppm by volume) CO_2.

Solution: Calculate the mole fraction of CO_2 in the discharge gas mixture:

$$\begin{aligned} y_{CO_2} &= \text{ppm}/10^6 \\ &= 10{,}000/10^6 \\ &= 0.01 \end{aligned}$$

Also determine the partial pressure of CO_2 in psia and mm Hg:

$$\begin{aligned} p_{CO_2} &= y_{CO_2}P \\ &= (0.01)(40) \\ &= 0.4 \text{ psia} \\ &= (0.4)(760/14.7) \\ &= 20.7 \text{ mm Hg} \end{aligned}$$

Estimate the adsorbent capacity, SAT, at 77°F:

$$\text{SAT} = 9.8 \text{ lb } CO_2/100 \text{ lb sieve (from Figure 6.11)}$$

6.38 Refer to Problem 6.37. What percentage of this adsorbed vapor would be recovered by passing superheated steam at a temperature of 392°F through the adsorbent until the partial pressure of the CO_2 in the stream leaving is reduced to 1.0 mm Hg?

Solution: Estimate the adsorbent capacity at 392°F and 1.0 mm Hg. Note that this represents the HEEL:

$$\text{HEEL} = 0.8 \text{ lb } CO_2/100 \text{ lb sieve}$$

The amount of CO_2 recovered is therefore

$$\begin{aligned} CO_2 \text{ recovered} &= 9.8 - 0.8 \\ &= 9.0 \text{ lb } CO_2/100 \text{ lb sieve} \end{aligned}$$

While the percent recovery is

$$\begin{aligned} \% \text{ recovery} &= (9.0/9.8)100 \\ &= 91.8\% \end{aligned}$$

Note that this represents the percent recovery relative to the HEEL.

6.39 Refer to Problem 6.37. What is the residual CO_2 partial pressure in a gas mixture at 77°F in contact with the freshly stripped sieve in Problem 6.38.
Solution: Estimate the partial pressure of CO_2 in mm Hg in equilibrium at 77°F with sieve containing 0.8 lb CO_2/100 lb sieve (the HEEL):

$$p_{CO_2} \approx 0.05 \text{ mm Hg}$$

The equilibrium CO_2 partial pressure may be converted to ppm:

$$\begin{aligned} \text{ppm} &= (p_{CO_2}/P)10^6 \\ &= (0.05)(14.7/760)(10^6)/40 \\ &= 24.1 \end{aligned}$$

This represents the discharge ppm following regeneration.

6.40 Steam Requirement for Benzene Adsorption

Benzene, at a concentration of 0.15% by volume, is to be recovered from a 25°C, 1 atm air stream of 30,000 acfm by using a 4 × 10 mesh carbon adsorber. The capacity of the carbon is 0.26 lb benzene/lb carbon at saturation. The mass transfer zone (MTZ) at an operating gas velocity of 70 ft/min is 4 inches. The density of carbon is 28 lb/ft^3 and, from experience, the best steam–solvent mass ratio for desorption is 2.75 to 1. The pressure drop equation for 4 × 10 mesh packing is

$$\frac{\Delta P}{L} = 0.00165\, v^{1.39} \tag{6.19}$$

where the pressure drop is in inches H_2O, L is the bed depth in inches, and v is the gas velocity in ft/min.

For an adsorption bed depth of 1 ft, determine the following:

(**a**) The working charge if the HEEL is neglected
(**b**) Cycle time
(**c**) Pressure drop
(**d**) Steam required for desorption

Solution:

(**a**) Use the following equation for the MTZ to calculate the bed capacity, which can be assumed equal to WC:

$$\mathrm{MTZ} = 2Z[(1 - (\mathrm{WC/CAP})] \tag{6.20}$$

Solve this equation for WC:

$$4.0 = (2)(12)[1 - (\mathrm{WC}/0.26)]$$
$$\mathrm{WC} = 0.217\ \mathrm{lbB/lbAC}$$

(**b**) Key calculations follow:

$$A = 30{,}000/70$$
$$= 428.6\ \mathrm{ft}^2$$
$$M_{\mathrm{AC}} = (428.6)(1.0)(28)$$
$$= 12{,}000\ \mathrm{lb\ AC}$$
$$m_{\mathrm{B}} = (12{,}000)(0.217)$$
$$= 2604\ \mathrm{lb\ B}$$
$$\dot{m}_{\mathrm{B}} = (30{,}000)(0.0015)$$
$$= 45\ \mathrm{ft^3B/\min}$$
$$= 45(78/359)(273/298)$$
$$= 8.96\ \mathrm{lb\ B/min}$$

The time is then

$$t = 2604/8.96$$
$$= 290\ \mathrm{min} = 4\ \mathrm{hr}\ 50\ \mathrm{min}$$

The cycle time is dictated by the number of adsorbers. For a 1×1 system

$$t_{\mathrm{C}} = (2)(290)$$
$$= 580\ \mathrm{min} = 9\ \mathrm{hr}\ 40\ \mathrm{min}$$

(**c**) Employ the equation provided for the pressure drop:

$$\Delta P = 0.00165(70)^{1.39}(12)$$
$$= 7.27\ \mathrm{in}\ H_2O$$

(**d**) The steam requirement is simply

$$\mathrm{SR} = (2.75)(2604)$$
$$= 7161\ \mathrm{lb\ steam/cycle}$$

6.41 Steam Requirement for Bed Regeneration I

Benzene is to be recovered from a dilute mixture with air by adsorption on a 6×10 mesh activated carbon ($\rho_B = 30\ \text{lb/ft}^3$). The gas enters the adsorber at a rate of $6500\ \text{ft}^3/\text{hr}$ (60°F, 1.0 atm) and contains 3.8% by volume benzene. A two-bed unit (one on, one off) is used, adsorbing at 80°F and 1.0 atm, where approximately 95% of benzene is removed. Experience indicates that a superficial gas velocity through the adsorber of 25 ft/min is satisfactory. The adsorption time is 4 hr. Regeneration is to be accomplished with 150 psia saturated steam, and 15% excess carbon is used in the beds. During operation, the activated carbon will retain 0.3 lb benzene per lb carbon at 80°F. Determine the following:

(a) The amount of carbon needed in each bed.
(b) The gas flow area of each bed.
(c) The dimensions the adsorber.
(d) The pressure drop across the bed (use EPA chart in Figure 6.8).
(e) The heat required to heat the vessel to the 150 psia steam temperature is 90,000 Btu and the heat of desorption (with sensible heat) is 40 Btu/lb benzene. The heat capacity of carbon is approximately 0.25 Btu/lb · F. How much steam must be theoretically supplied for regeneration of the bed?

Solution: **(a)** Key calculations are provided below.

$$
\begin{aligned}
q &= 6500(540/520) \\
&= 6750\ \text{acfh} \\
q_B &= (0.038)(6750) \\
&= 256.5\ \text{ft}^3/\text{hr} \\
\dot{m}_B &= (256.5)(78/379)(520/540) \\
&= 50.83\ \text{lb/hr} \\
\dot{m}_B/4\ \text{hr} = m_B &= (50.83)(0.95)(4) \\
&= 193.2\ \text{lbB} \\
M_{AC} &= (193.2/0.3)1.15 \\
&= 740.5\ \text{lb AC}
\end{aligned}
$$

(b) The face area of the bed may now be calculated:

$$
\begin{aligned}
V_{AC} &= 740.5/30 = 24.68\ \text{ft}^3 \\
A_{AC} &= 6750/(25)(60) \\
&= 4.5\ \text{ft}^2
\end{aligned}
$$

(**c**) The dimensions of the bed are (assuming a vertical column)

$$\begin{aligned} H &= 24.68/4.5 \\ &= 5.5 \text{ ft} \end{aligned}$$

$$\begin{aligned} D &= (4.5/0.785)^{1/2} \\ &= 2.4 \text{ ft} \end{aligned}$$

(**d**) From Figure 6.8, one obtains

$$\begin{aligned} \Delta P &= 0.145 \text{ in } H_2O/\text{in bed} \\ \Delta P_{\text{total}} &= (0.145)(5.5)(12) \\ &= 9.57 \text{ in } H_2O \end{aligned}$$

(**e**) To heat the bed

$$\begin{aligned} Q_{\text{b}} &= (0.25)(740.5)(359 - 80) \\ &= 51{,}650 \text{ Btu} \end{aligned}$$

To desorb

$$\begin{aligned} Q_{\text{d}} &= (40)(193.2) \\ &= 7{,}727 \text{ Btu} \\ Q_{\text{total}} &= 90{,}000 + 51{,}650 + 7{,}727 \\ &= 149{,}400 \text{ Btu} \end{aligned}$$

The steam requirement is

$$\begin{aligned} &= 149{,}400/862 \\ SR &= 173 \text{ lb steam} \end{aligned}$$

for each 4 hr cycle. In actual practice, more steam would be required.

6.42 Stream Requirement for Bed Regeneration II

A process N_2 stream contains 3.8 vol% toluene is to be recovered by adsorption. The gas flow rate is 350 lb/min. A 4 × 10 mesh granular activated carbon, two-bed unit, adsorbing at 80°F is to be used. The bed depth is to be 2 ft and the adsorption time 1 hr. Regeneration is to be accomplished with 25 psia steam. No bed cooling is required. Bed switching is to be done at the onset of break-through, and 10% excess carbon is used in the beds. The carbon will retain 0.27 lb toluene/lb carbon at 80°F. The bulk density is 30 lb/ft^3

(**a**) How much carbon is required for each bed?
(**b**) What is the gas flow area if the bed height is 2 ft?
(**c**) Estimate the pressure drop across the beds (use EPA chart).

(**d**) During regeneration the vessel heat requirement is 100,000 Btu, the carbon must by heated from 80°F to the 25 psia steam temperature, (240°F) and the heat of desorption is 50 Btu/lb of toluene. If the heat capacity of the carbon is 0.25 Btu/lb°F, how much steam must be supplied for the regeneration of the bed?

Solution:

(**a**) The average molecular weight of the gas stream is first calculated:

$$\begin{aligned} \text{MW} &= (28)(1 - 0.038) + (92)(0.038) \\ &= 30.43 \end{aligned}$$

The gas mass and molar flow rate are

$$\begin{aligned} \dot{m} &= 350\,\text{lb/min} \\ &= 21{,}000\,\text{lb/hr} \\ \dot{n} &= 21{,}000/30.43 \\ &= 690.1\,\text{lbmol/hr} \end{aligned}$$

The mass flow rate of toluene is

$$\begin{aligned} \dot{m}_{\text{T}} &= (690.1)(0.038)(92) \\ &= 2412.6\,\text{lb/hr} \\ m_{\text{T}} &= 2412.6\,\text{lb T in } 1.0\,\text{hr} \end{aligned}$$

The mass of AC is therefore

$$\begin{aligned} M_{\text{AC}} &= (2412.6/0.27)(1.1) \\ &= 9829.1\,\text{lb} \end{aligned}$$

(**b**) To size the bed

$$\begin{aligned} V_{\text{AC}} &= 9829/30 \\ &= 327.6\ \text{ft}^3 \end{aligned}$$

Since $H = 2$ ft, it follows that

$$\begin{aligned} A &= 327.6/2 \\ &= 163.8\ \text{ft}^2 \end{aligned}$$

(**c**) For operation at 1.0 atm:

$$\begin{aligned} \rho &= (1.0)(30.43)/(0.7302)(540) \\ &= 0.07725\,\text{lb/ft}^3 \end{aligned}$$

The volumetric flowrate is

$$q = 350/0.0772$$
$$= 4533.7 \text{ ft}^3/\text{min}$$

and

$$v = 4533.7/163.8$$
$$= 27.7 \text{ ft/min}$$

From Figure 6.8, one obtains

$$\Delta P \approx 1.0 \text{ in } H_2O/\text{in bed}$$

and

$$\Delta P_{\text{total}} = (2.0)(12)(1.0)$$
$$= 24 \text{ in } H_2O$$

From standard steam tables, at 25 psia, $T_{\text{SAT}} = 240°\text{F}$ and

$$H_l = 208.41 \text{ Btu/lb}$$
$$H_v = 1160.4 \text{ Btu/lb}$$

(d) To heat the bed

$$Q_b = (0.25)(9829)(240 - 80)$$
$$= 393{,}160 \text{ Btu}$$

To desorb

$$Q_d = (50)(2412.6)$$
$$= 120{,}630 \text{ Btu}$$

Total heat is

$$Q_{\text{total}} = 100{,}000 + 393{,}160 + 120{,}630$$
$$= 613{,}800 \text{ Btu}$$

The steam requirement is therefore

$$M_{\text{steam}} = 613{,}800/(1160 - 208.4)$$
$$= 645 \text{ lb}$$

during each cycle.

6.43 Dust Particle Resistivity

An electrostatic precipitator (see Chapter 10 for more details) is collecting dust particles whose average surface area is equivalent to that of spheres of 1.0 μm radius. The resistivity of the dust is too high for satisfactory operation and it is desired to reduce the resistivity by adding a gas, known as a *conditioning agent*, which will be adsorbed on the surface of the dust particle. The following facts have been determined by experiment:

1. The conditioning agent has a molecular weight of 60.6, and a molecular diameter of 1 Å.
2. The adsorption of a monolayer will bring the resistivity down to a satisfactory level.
3. The area void fraction in monolayer packing of the conditioning agent on the dust particle is 0.09307.
4. The particle has a specific gravity of 3.0.
5. If CAP is the ratio by weight of adsorbate to adsorbent, and y, the mole fraction of the conditioning agent in the gas stream, then

$$\text{CAP} = 0.064\ (1 - e^{10^4(y)}) \tag{6.21}$$

Find the minimum value of y that will permit satisfactory precipitation.

Solution: Key calculations are provided below.

$$\text{Volume of one particle} = \frac{4}{3}\pi r_{\text{p}}^3; \quad r_{\text{p}} = 1.0\ \mu\text{m}$$
$$= 4.189 \times 10^{-12}\text{cm}^3$$

$$\text{Surface area of one particle} = 4\pi r_{\text{p}}^2 = 4\pi(10^{-4})^2$$
$$= 1.257 \times 10^{-7}\ \text{cm}^2$$

$$\text{Density of particle} = (3.0)(0.9982) = 2.995\ \text{g/cm}^3$$

$$\text{Mass of one particle} = (4.189 \times 10^{-12})(2.995)$$
$$= 1.254 \times 10^{-11}\ \text{g}$$

$$\text{Total projected area of one molecule} = \pi r_{\text{m}}^2/(1 - \varepsilon);\ 1\ \text{Å} = 10^{-8}\ \text{cm}$$
$$= \pi(0.5 \times 10^{-8})^2/(1 - 0.09307)$$
$$= 8.659 \times 10^{-17}\ \text{cm}^2$$

$$\text{Number of gas molecules/particle} = 1.257 \times 10^{-7}/8.659 \times 10^{-17}$$
$$= 1.452 \times 10^{9}$$

$$\text{Weight of monolayer deposition} = \left(\frac{1.452 \times 10^{9}\ \text{molecules}}{6.023 \times 10^{23}\ \text{molecules/mol}}\right) \times (60.6\,\text{g})$$
$$= 1.460 \times 10^{-13}\ \text{g}$$

Therefore

$$\text{CAP} = \frac{1.460 \times 10^{-13}}{1.254 \times 10^{-11}}$$
$$= 0.01165$$

Equation (6.21) can now be solved for the conditioning agent mole fraction:

$$\text{CAP} = 0.064(1 - \text{e}^{-(10^{4})(y)})$$
$$y = 2.01 \times 10^{-5}$$
$$= 20.1\ \text{ppm}$$

If the void area correction is not included in the analysis, one obtains

$$y = 22\ \text{ppm}$$

6.44 Cost of Retrofitting a Plant

The 40,000 acfm drier effluent of SKBP Chemicals, Inc. contains 10,000 parts per million (by weight) of benzene. The EPA has required that they reduce the benzene content of their effluent air streams to no more than 100 parts per million. Determine the cost (or profit) of retrofitting the plant with an activated carbon absorption unit if the recovered benzene can be sold, FOB plant site, for \$0.016/lb (net). Specifications and various unit costs are as follows:

1. Operating temperature is 130°F, 330 operating days/year.
2. Three absorbers will be used, 1-hr adsorption period followed by 2-hr steam-stripping period.
3. Capital cost, including multiple beds, steam boiler, fan, solvent separator, initial carbon charges, installation, and interest rate is \$12.90/acfm.
4. 4 × 10 mesh activated carbon, 70 ft/min effluent velocity through bed, 20 inches H_2O pressure drop, 33.3 inch bed depth.
5. Solvent loading for 1% breakthrough is 0.241 lb solvent per lb of carbon working charge.
6. 3.0 lb of steam (saturated vapor at 220°F, produced from boiler inlet water at 70°F) is used in bed regeneration per lb of adsorbed solvent.

7. Steam generation cost is \$0.90 per million Btu (enthalpy value of 1115.4 Btu/lb).
8. Fan operating cost is \$140/horsepower · year, 60% fan efficiency.
9. 8-year straight line depreciation of capital equipment with salvage value of 10% at the end of 8 years.
10. Maintenance charges are 4000 years.
11. Carbon regeneration costs are \$0.16 per lb of carbon per year, including makeup carbon.
12. The activated carbon bulk density is 23.4 lb/ft^3.

Solution: The actual volumetric flow rate is calculated using Charles' Law:

$$q = 35{,}250\left(\frac{460 + 130}{460 + 60}\right) = 40{,}000\,\text{acfm}$$

The various charges and cost are listed below:
Depreciation charges:

(40,000 acfm/8 year) (12.90/acfm) (1.0 − 0.1) = 58,050/year

Value of benzene recovered (neglecting discharge):

$$\left(\frac{10{,}000 - 100}{10^6}\right)(40{,}000)\left(\frac{492}{590}\right)\left(\frac{78}{359}\right)$$

$$= (72.5\,\text{lb/min})(330\,\text{days/year})(24\,\text{hr/day})(60\text{min/hr}) = \$551{,}000$$

Steam consumption cost:

$$(72.5\,\text{lb/min})\,(3\,\text{lb steam/lb})(1153.4 - 38.05)(\text{Btu/lb}) = 243{,}000\,\text{Btu/min}$$

$$(243\,\text{Btu/min})\,(\$0.90/10^6\,\text{Btu})\,(330)\,(24)\,(60) = \$103{,}900$$

Fan operating costs:

$$\text{HP} = \frac{(40{,}000)\,(20)\,(5.2)}{(33{,}000)\,(0.6)} = 210$$

$$(210\,\text{HP})(\$140/\text{HP} \cdot \text{year}) = 29{,}400/\text{year}$$

Maintenance:

\$4000/year

Carbon regeneration costs:

Area = 40,000/70
Height = 33.3/12

$$M_{\text{AC}} = \left(\frac{40{,}000}{70}\right)\left(\frac{33.3}{12}\right)23.4 = 37{,}110\,\text{lb/bed}$$

For three beds

$$(3)(37{,}110)(1.60) = \$178{,}100/\text{year}$$

Summary of annual costs:

Capital	58,050
Steam	103,900
Blower	29,400
Regeneration	178,100
Maintenance	4000
	$373,450
Income from benzene sale:	
	$551,000

Profit:

$$507{,}500 - 373{,}450 = \$134{,}000/\text{year}$$

Note: The economic analysis is very sensitive to the C_6H_6 value. Furthermore, this analysis does include the value and money. In effect, the $516,000 ($40{,}000 \times 12.90$) is money that must be shelled out initially, and cannot be invested elsewhere. If money is worth 10% then $516,000 is also lost and should be deducted from the above mentioned profit.

6.45 Adsorber not in Compliance

Consider the adsorber system in Figure 6.12. It is designed to operate with a maximum discharge concentration of 50 ppm. Once the unit is installed and running, it operates with a discharge of 60 ppm. Rather than purchase a new unit, what options are available to bring the unit into compliance with the specified design concentration?

Solution: This is the first of two open-ended problems that close out this chapter. In a very real sense this is a takeoff of a similar problem in Chapter 5 ("Absorbers"). The following suggestions and options (if possible) are recommended for adsorbers:

1. Reduce the inlet temperature.
2. Increase the system pressure.
3. Change to finer-mesh carbon with a higher CAP, but check the effects on both the pressure drop and the fan.
4. Increase the amount of carbon in the bed, but check the effects on both the pressure drop and fan.
5. Employ a higher-temperature steam to reduce the HEEL.
6. Make sure that the top of the bed is level.
7. Process modification: reduce gas flow rate.
8. Process modification: reduce inlet concentration of pollutant.

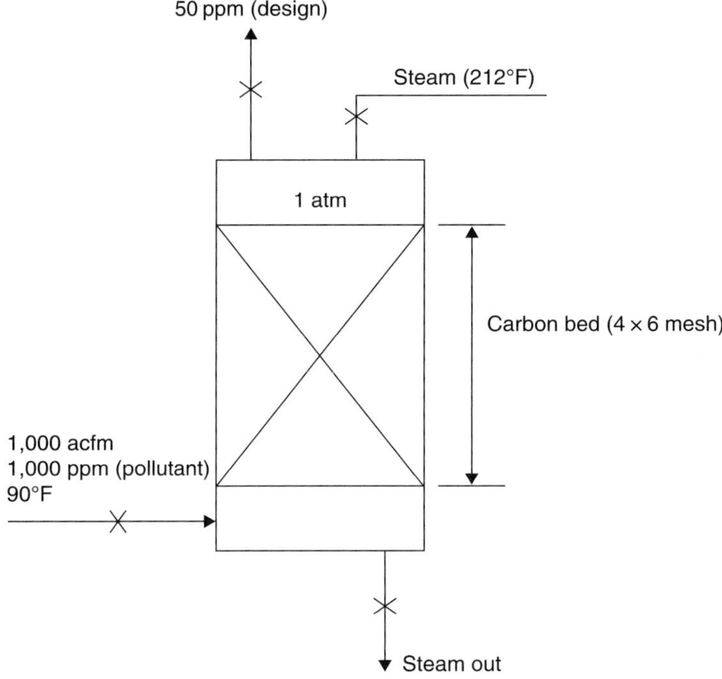

Figure 6.12 Adsorber not in compliance.

9. Consider changing the type of adsorbent.
10. Eliminate the process and/or shut down the plant.
11. Take the regulatory official out to lunch.

6.46 Accident and Emergency Management

On the basis of the air pollution control equipment course recently completed under the direction of the internationally recognized authority Dr. Theodore, a young engineer is now aware that during the operation of an adsorber, a fire and/or explosion can develop, particularly when treating ketones. Since two of the adsorbers at her plant site have been retrofitted to recover/control a dilute MIBK air stream from a process vent, she is concerned about a possible accident and/or explosion. Outline what steps or procedures she should take or follow, as a competent and responsible environmental engineer, to help reduce this hazard.

Solution: This, too, is an open-ended question. Here are some suggestions:

1. Thermocouples are a good indicator of a fire; however, they should be placed in the gas, not in the adsorbent.
2. Carbon monoxide and carbon dioxide are also good indicators of a fire; a gas chromatograph is satisfactory.

3. Ketones are particularly bad actors; be careful on startup, begin with the regenerative step.
4. If a fire develops, steam (if available) the unit to death.
5. Be sure that concentration levels are below the LEL.
6. Operating above the LEL can be dangerous, particularly if the concentration drops.
7. Keep ignition sources away.

NOTE: Additional problems are available for all readers at www.wiley.com. Follow links for this title.

7

FUNDAMENTALS: PARTICULATES

[*Note*: The material presented in this chapter primarily reviews fluid–particle dynamics, particle size distribution, and size–efficiency calculations. The reader is referred to Chapter 3 ("Fundamentals: Gases") for additional material that is applicable to particulate control equipment; topics of importance include conversion of units, physical and chemical properties, and the ideal gas law.]

7.1 INTRODUCTION

The author has employed several definitions for particulates: a small, discrete mass of solid or liquid matter; a fine liquid or a solid particle that is found in the air or emissions; any solid or liquid matter that is dispersed in a gas; or, insoluble solid matter dispersed in a liquid so as to produce a heterogeneous mixture. Further, particulate matter 10 (PM10) is defined as particulate matter with a diameter less than or equal to 10 micrometers (μm) while particulate matter 2.5 (PM2.5) is particulate matter with a diameter less than or equal to 2.5 micrometers (μm).

Air Pollution Control Equipment Calculations. By Louis Theodore

Particulates may be classified into two broad categories: (1) natural and (2) human-made (engineered or synthetic). Natural sources include (but are not limited to):

1. Windblown dust
2. Volcanic ash and gases
3. Smoke and fly ash from forest fires
4. Pollens and other aeroallergens

These sources contribute to background values over which control activities can have little, if any, effect. Human-made sources cover a wide spectrum of chemical and physical activities and are contributors to urban air pollution. Particulates in the United States pour out from over 200 million vehicles, from the refuse of nearly 300 million people, the generation of billions of kilowatts of electricity, and the production of innumerable products demanded by everyday living.

Particulates may also be classified by their origin.

1. Primary—emitted directly from a process
2. Secondary—subsequently formed as a result of a chemical reaction

Particulates may be classified as solid or liquid matter whose effective diameter (to be defined shortly) is larger than a molecule but smaller than approximately 1000 μm. Particulates dispersed in a gaseous medium are collectively termed an *aerosol*. The terms "smoke," "fog," "haze," and "dust" are commonly used to describe particular types of aerosols, depending on the size, shape, and characteristic behavior of the dispersed particles. Aerosols are rather difficult to classify on a scientific basis in terms of their fundamental properties such as settling rate under the influence of external forces, optical activity, ability to absorb electric charge, particle size and structure, surface-to-volume ratio, reaction activity, physiological action, and so on. In general, particle size and settling rate have been the most characteristic properties emphasized by engineers and scientists. For example, particles larger than 100 μm may be excluded from the category of dispersions because they settle too rapidly. On the other hand, particles on the order of 1 μm or less settle so slowly that, for all practical purposes, they are regarded as permanent suspensions. Despite possible advantages of scientific classification schemes, the use of popular descriptive terms such as smoke, dust, and mist, which are traditionally and essentially based on the mode of formation, appears to be a satisfactory and convenient method of classification. In addition, this approach is so well established and understood that it undoubtedly would be difficult to change.

When a liquid or solid substance is emitted to the air as particulate matter, its properties and effects may be changed. As a substance is broken up into smaller and smaller particles, more of its surface area is exposed to the air. Under these circumstances, the substance—whatever its chemical composition—tends to physically or chemically combine with other particulates or gases in the atmosphere. The resulting combinations are frequently unpredictable. Very small aerosol particles (measuring 0.001–0.1 μm, or

1–100 nm) can act as condensation nuclei to facilitate the condensation of water vapor, thus promoting the formation of fog and ground mist. Particles less than 2 or 3 μm in size—about half (by weight) of the particles suspended in urban air—can penetrate into the mucus, and attract and convey harmful chemicals such as sulfur dioxide. By virtue of the increased surface area of the small aerosol particles, and as a result of the adsorption of gas molecules or other such properties that are able to facilitate chemical reactions, aerosols tend to exhibit greatly enhanced surface activity, a phenomenon well documented in the nanotechnology literature.

The range of sizes of particles formed in a process is largely dependent on the types of the particle formation mechanisms present. It is sometimes possible to estimate the general size range simply by recognizing which of these are important in the process being evaluated. The most important particle formation mechanisms in air pollution sources include the following.

1. Physical attrition/mechanical dispersion
2. Combustion particle burnout
3. Homogeneous nucleation
4. Heterogeneous nucleation
5. Droplet evaporation

Particle size is the single most important characteristic that affects the behavior of a particle. The range in sizes of particles observed in practice is remarkable. Some of the droplets collected in the demisters of wet scrubbers and the solid particles collected in large-diameter cyclones are as large as raindrops. However, some of the particles created in high-temperature incinerators and metallurgical processes can consist of a few molecules clustered together. These particles cannot be seen by sensitive light microscopes because they are extremely small in size; these sizes approach those of individual gas molecules (which range from 0.2 to 1.0 nm). In fact, particles composed of a few molecules clustered together can exist in a stable form. Some of these industrial processes generate particles in the range of 10–100 nm. However, particles in this size range can grow and agglomerate to yield particles in the 100+-nm range.

7.2 PARTICLE COLLECTION MECHANISMS

The overall collection/removal process for particulates in a fluid essentially consists of four steps (L. Theodore: personal notes, 1976).

1. An external force (or forces) must be applied that enables the particle to develop a velocity that will displace and/or direct it to a collection or retrieval section or area.
2. The particle should be retained at this area with strong enough forces so that it is not reentrained.
3. As collected/recovered particles accumulate, they are subsequently removed.
4. The ultimate disposition of the particles completes the process.

Obviously, the first is the most important step. The particle collection mechanisms discussed below are generally applicable when the fluid is air; however, they may also apply if the fluid is water. In most commercial particulate systems, it is necessary to remove the particles from the gas stream and collect them in layers, dust cakes, or other forms. In the case of electrostatic precipitators (see Chapter 10), the dust is collected on a layer on a vertical collection plate. In the case of fabric filters (see Chapter 12), the particles collect as dust cakes on vertical bags. This is the initial collection step, and most particulate devices use one or a combination of collection mechanisms to accomplish this objective.

The forces listed below are basically the "tools" that can be used for particulate/recovery collection:

1. Gravity settling
2. Centrifugal action
3. Inertial impaction and interception
4. Electrostatic attraction
5. Themophoresis and diffusiophoresis
6. Brownian motion

Note that all of these collection mechanism forces are strongly dependent on particle size. Each mechanism is briefly described below.

As indicated above, all particulate devices collect or recover particles by a variety of mechanisms involving an applied force—the simplest of which is *gravity*. Large particles moving slowly (relatively speaking) in a gas stream can settle out and be collected or recovered. This mechanism is responsible for particle capture in the simplest of all devices—the settling chamber. The distance (d) that the particle travels while settling (traveling) with velocity (v) in time (t) is given by

$$d = vt; \quad \text{consistent units} \tag{7.1}$$

Details of gravity forces can be found in Chapter 8.

Centrifugal force is another collection mechanism used for particle capture. The shape or curvature of the collector causes the gas stream to rotate in a spiral motion. Large particles move toward the outside of the wall by virtue of their momentum. The particles lose kinetic energy there and are separated from the gas stream. Particles are then acted on by gravitational forces and are collected. Thus, centrifugal and gravitational forces are both responsible for particle collection in a cyclone. See Chapter 9 for more details.

Inertial impaction occurs when an object (e.g., a fiber or liquid droplet), placed in the path of a particulate-laden gas stream, causes the gas to diverge and flow around it. Larger particles, however, tend to continue in a straight path because of their inertia; they may impinge on the obstacle and be collected (as in Figure 7.1). Since the trajectories of particle centers can be calculated, it is possible to theoretically determine the probability of collision.

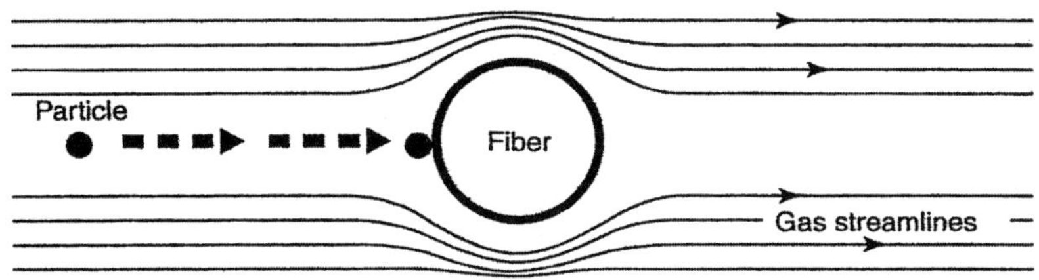

Figure 7.1 Impaction.

Direct interception also depends on inertia and is merely a secondary form of impaction. Note that the trajectory of a particle's center can be calculated; however, even though the center may bypass the target object, a collision might occur since the particle has finite size (Figure 7.2). A collision occurs due to direct interception if the particle's center misses the target object by some dimension less than the particle's radius. Direct interception is, therefore, not a separate principle, but only an extension of inertial impaction.

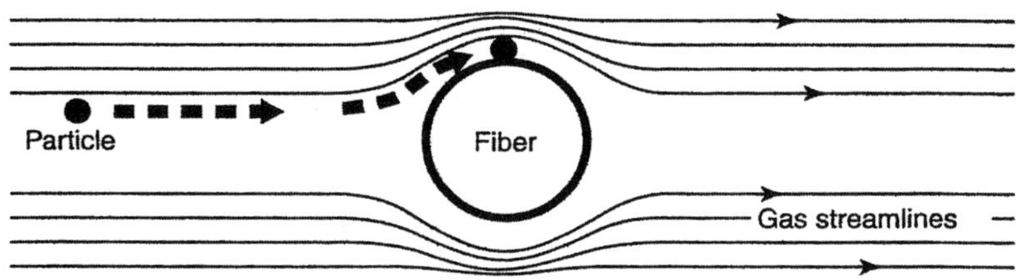

Figure 7.2 Direct interception.

Inertial impaction is analogous to a small car riding down an interstate highway at 65 mph (mi/hr) and approaching a merge lane where a slow-moving large truck has entered. If the car is not able to get into the passing lane to go around the merged truck, there could be an "impaction" incident. The faster the car is going, the more probable the impaction. The larger the car, the more difficulty it will have going around the truck. Conversely, if the truck enters the interstate road at the same speed as the car, the likelihood of impaction is reduced.

Another primary particle collection mechanism involves electrostatic forces. The particles can be naturally charged, or, as in most cases involving *electrostatic attraction*, be charged by subjecting the particle to a strong electric field. The charged particles migrate to an oppositely charged collection surface. This is the collection mechanism responsible for particle capture in both electrostatic precipitators and charged droplet scrubbers (see Appendix). In an electrostatic precipitator, particle collection occurs as a result of electrostatic forces only. The electrostatic force F_E experienced by a charged particle in an electric field is given by

$$F_E = qE_p; \quad \text{consistent units} \tag{7.2}$$

where q = particle charge (not to be confused with volume flowrate)
E_p = the collection field intensity (electric field)

Thermophoresis and *diffusiophoresis* are two relatively weak forces that can affect particle collection. *Thermophoresis* is particle movement caused by thermal differences on two sides of the particle. The gas molecule kinetic energies on the hot side of the particle are higher. Therefore, collisions with the particle on this side transfer more energy than the molecular collisions on the cold side. Accordingly, the particle is deflected toward the cold area. *Diffusiophoresis* is caused by an imbalance in the kinetic energies being transmitted to the particles by the surrounding molecules. When there is a strong difference in the concentration of molecules between two sides of the particle, there is a difference in the number of molecular collisions. The particle moves toward the area of lower concentration.

Very small particles deflect slightly when they are struck by gas molecules. The deflection is caused by the transfer of kinetic energy from the rapidly moving gas molecule to the small particle. As the particle size, mass, and density increase, the extent of the particle movement decreases. Thus, deflection begins to be effective as a capture mechanism for particles less than approximately 1.0 μm, and it is significant for particles less than 0.5 μm. Because of the bombardment by fluid molecules, particles suspended in a fluid will be subjected to a random and chaotic molecular motion known as *Brownian motion*. This movement arises in addition to any net motion in a given direction due to the action of other external forces such as gravity. Thus, this effect becomes the predominant collection mechanism for particles in the 0.1–1.0 μm range. Additional details are provided in the next section.

Other factors affecting collection mechanisms include:

1. Nonspherical particles
2. Wall effects
3. Multiparticle effects
4. Multidimensional flow

Details are available in the literature (L. Theodore, *Nanotechnology: Basic Calculations for Engineers and Scientists*, John Wiley & Sons, Hoboken, NJ, 2007).

7.3 FLUID–PARTICLE DYNAMICS

Most industrial techniques used for the separation of particles from gases involve the relative motion of the two phases under the action of various external forces. The collection methods for particulate pollutants are based on the movement of solid particles (or liquid droplets) through a gas. The final objective is their removal in order to comply with applicable standards and regulations and/or their recovery for economic reasons. In order to accomplish this, the particle is subjected to external forces—forces large enough to separate the particle from the gas stream during its residence time in the control unit.

Whenever a difference in velocity exists between a particle and its surrounding fluid, the fluid will exert a resistive force on the particle. Either the fluid (gas) may be at rest with the particle moving through it, or the particle may be at rest with the gas flowing past it. It is generally immaterial which phase (solid or gas) is assumed to be at rest; it is the *relative* velocity between the two that is important. The resistive force exerted on the particle by the gas is called the *drag*.

In treating fluid flow through pipes, a *friction factor* term is used in many engineering calculations. An analogous factor, called the *drag coefficient*, is employed in drag force calculations for flow past particles.

Consider a fluid flowing past a stationary solid sphere. If F_D is the drag force and ρ is the density of the gas, the drag coefficient C_D is defined as

$$C_D = \frac{F_D/A_p}{\rho v^2/2g_c} \tag{7.3}$$

where A_p is given by $\pi d_p^2/4$. In the following analysis, it is assumed that

1. The particle is a rigid sphere (with a diameter d_p) surrounded by gas in an infinite medium (no wall or multiparticle effects).
2. The particle or fluid is not accelerating.

From dimensional analysis, one can show that the drag coefficient is solely a function of the particle Reynolds number Re:

$$C_D = C_D(\text{Re}) \tag{7.4}$$

where

$$\text{Re} = \frac{d_p v \rho}{\mu} \tag{7.5}$$

and v is the relative velocity, μ is the fluid (gas) viscosity and ρ is the fluid (gas) density. The quantitative use of the equation of particle motion (to be developed shortly) requires numerical and/or graphical values of the drag coefficient. Graphical values are presented in Figure 7.3.

The drag force F_D exerted on a particle by a gas at low Reynolds numbers is given in equation form by

$$F_D = 6\pi\mu v a/g_c = 3\pi\mu v d_p/g_c \tag{7.6}$$

Equation (7.6) is known as *Stokes' law* and can be derived theoretically (L. Theodore: personal notes, 1960). However, keep in mind that Stokes' equation is valid only for very low Reynolds numbers—up to $\text{Re} \approx 0.1$; at $\text{Re} = 1$, it predicts a value for the drag force that is nearly 10% too low. In practical applications, Stokes' law is generally assumed applicable up to a Reynolds number of 2.0. By rearranging Stokes' law in the

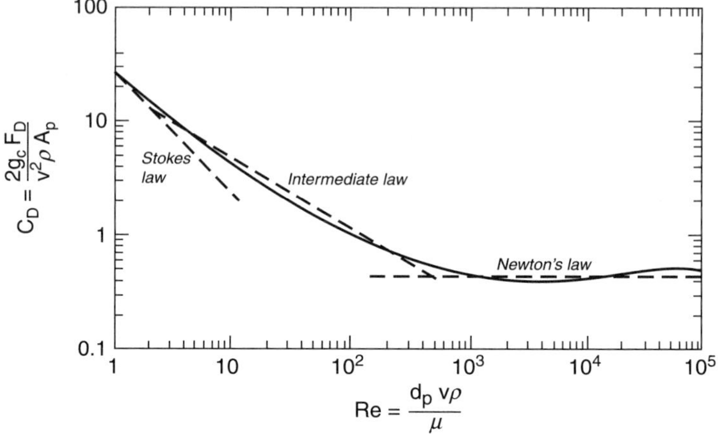

Figure 7.3 Drag coefficient for spheres.

form of Equation (7.3), the drag coefficient becomes

$$C_{\mathrm{D}} = \frac{6\pi\mu va/\pi a^2}{\rho v^2/2} \tag{7.7}$$

where a equals the particle radius. Hence, for "creeping flow" around a particle, one obtains

$$C_{\mathrm{D}} = 24/\mathrm{Re} \tag{7.8}$$

This is the straight-line portion of the log–log plot of C_{D} vs. Re (Figure 7.3). For higher values of the Reynolds number, it is almost impossible to perform purely theoretical calculations. However, several investigators have managed to estimate, with a considerable amount of effort, the drag and/or drag coefficient at higher Reynolds numbers.

In addition to the analytical equation [Equation (7.8)], one may use

$$C_{\mathrm{D}} = 18.5/\mathrm{Re}^{0.6}; \quad 2 < \mathrm{Re} < 500 \tag{7.9}$$

for the "intermediate" range (i.e., between the Stokes' law range and the Newton's law range, discussed later). This indicates a lesser dependence than Stokes' law on Re; it is less accurate than Stokes' law for Re < 2. At higher Re the drag coefficient is approximately constant. This is the Newton's law range, for which

$$C_{\mathrm{D}} \approx 0.44; \quad 500 < \mathrm{Re} < 200{,}000 \tag{7.10}$$

In this region the drag force—see Equation (7.3)—on the sphere is proportional to the square of the gas velocity. (Note that Newton's law for the drag force is not to be confused with Newton's law of viscosity or Newton's laws of motion.)

A simple two-coefficient model of the form

$$C_D = \alpha \text{ Re}^{-\beta} \tag{7.11}$$

can be used over the three Reynolds number ranges given in Equations (7.8)–(7.10). The numerical values of α and β are given in Table 7.1.

TABLE 7.1 Drag Model Coefficients

Re Range	α	β
Stokes range	24.0	1.0
Intermediate range	18.5	0.6
Newton range	0.44	0.0

Using the model in Equation (7.3) with Equation (7.11), the drag force becomes

$$F_D = \frac{\alpha\pi(d_p v)^{2-\beta}\mu^{\beta}\rho^{1-\beta}}{8g_c} \tag{7.12}$$

This equation reduces to

$$F_D = 3\pi\mu v d_p/g_c \tag{7.13}$$

for the Stokes' law range (Re < 2),

$$F_D = 2.31\pi(d_p v)^{1.4}\mu^{0.6}\rho^{0.4}/g_c \tag{7.14}$$

for the intermediate range (2 < Re < 500), and

$$F_D = 0.055\pi(d_p v)^2\rho/g_c \tag{7.15}$$

for the Newton's law range (500 < Re < 200,000). This two-coefficient, three-Reynolds-number range model will be used for drag force calculations in this and subsequent chapters. A review of this model and these equations indicates that they are fairly consistent with the experimental values found in the literature.

Another empirical drag coefficient model is given by Equation (7.16).

$$\begin{aligned}\log C_D = {} & 1.35237 - 0.60810(\log \text{Re}) - 0.22961(\log \text{Re})^2 \\ & + 0.098938(\log \text{Re})^3 + 0.041528(\log \text{Re})^4 \\ & - 0.32717(\log \text{Re})^5 + 0.007329(\log \text{Re})^6 \\ & - 0.0005568(\log \text{Re})^7\end{aligned} \tag{7.16}$$

This is an empirical equation (L. Theodore: personal notes, 1974) that was obtained by the use of a statistical fitting technique. This correlation gives excellent results over the entire range of Reynolds numbers. An advantage of using this correlation is that it is not partitioned for application to only a specific Reynolds number range.

Still another empirical equation is (S. Barnea and I. Mizraki, PhD thesis, Haifa University, Haifa, 1972):

$$C_D = [0.63 + (4.80/\sqrt{\text{Re}})]^2 \tag{7.17}$$

This correlation is also valid over the entire spectrum of Reynolds numbers. Its agreement with literature values is generally good. However, in the range of $30 < \text{Re} < 10{,}000$, there is considerable deviation. For $\text{Re} < 30$ or $\text{Re} > 10{,}000$, the agreement is excellent.

Consider now a solid spherical particle located in a gas stream and moving in one direction with a velocity v relative to the gas. The net or resultant force experienced by the particle is given by the summation of all the forces acting on the particle. These forces include drag, buoyancy, and one or more external forces (such as gravity, centrifugal, and electrostatic). In order to simplify the presentation, the direction of particle movement relative to the gas is always assumed to be positive. Newton's law of motion is then

$$F_R = F - F_B - F_D \tag{7.18}$$

where F_R is the resultant or net force, F is the external force, F_B is the buoyant force, and F_D is the drag force. The net force results in acceleration of the particle, given by

$$F_R = \frac{m}{g_c}\left(\frac{dv}{dt}\right) \tag{7.19}$$

where m is the mass of the particle ($\pi d_p^3 \rho_p/6$); and ρ_p is the particle density. The external force per unit mass is denoted as f. The external force F on the particle is then

$$F = mf \tag{7.20}$$

Unless the particle is in a vacuum, it will experience a buoyant force in conjunction with the external force(s); this is given by

$$F_B = m_f f \tag{7.21}$$

where m_f is the mass of gas (fluid) displaced by the particle.

The equation of motion now becomes

$$\begin{aligned} \frac{(dv/dt)}{g_c} &= f - \left(\frac{m_f}{m}\right)f - \left(\frac{F_D}{m}\right) \\ &= f[1 - (m_f/m)] - (F_D/m) \\ &= f[(m - m_f)/m] - (F_D/m) \end{aligned} \tag{7.22}$$

This equation may also be written as

$$(dv/dt)/g_c = \left(fm_{eq}/m\right) - (F_D/m) \tag{7.23}$$

where $m_{eq} = (m - m_f)$, or

$$\begin{aligned}(dv/dt)/g_c &= f[1 - (\rho/\rho_p)] - (F_D/m) \\ &= f[(\rho_p - \rho)/\rho_p] - (F_D/m)\end{aligned} \tag{7.24}$$

For gases, $\rho_p >>> \rho$, so that the bracketed term in Equations (7.22) and (7.24) reduce to unity.

The particle may be acted on by one or more external forces. If the external force is gravity, then

$$f_g = g/g_c$$

and

$$F_g = m(g/g_c)$$

The describing equation for the particle motion then becomes

$$(dv/dt)/g_c = (g/g_c) - (F_D/m) \tag{7.25}$$

If the particle experiences an electrostatic force F_E, then

$$F_E = mf_E$$

so that

$$(dv/dt)/g_c = f_E - (F_D/m) \tag{7.26}$$

where f_E is the electrostatic force per unit mass of particle. If the external force is from a centrifugal field, then

$$f_C = r\omega^2/g_c = v_\phi^2/g_c r$$

where r is the radius of the path of the particle, f_c is the centrifugal force per unit mass of particle, ω is the angular velocity, and v_ϕ is the tangential velocity at that point. The centrifugal force F_C is then

$$F_C = mf_c$$

The describing equation becomes

$$(dv/dt)/g_c = (r\omega^2/g_c) - (F_D/m)$$

or

$$(dv/dt)/g_c = (v_\phi^2/g_c r) - (F_D/m) \tag{7.27}$$

The reader is reminded of the use of g_c. Any term or group of terms in these equations may be indiscriminately multiplied or divided by this conversion constant.

If a particle is initially at rest in a stationary gas and is then set in motion by the application of a constant external force or forces, the resulting motion occurs in two stages. The first period involves acceleration, during which time the particle velocity increases from zero to some maximum velocity. The second stage occurs when the particle achieves this maximum velocity and remains constant. During the second stage, the particle is not accelerating. The LHS (left-hand side) of those above equations containing the term dv/dt are, therefore, zero. The final, constant, and maximum velocity attained is defined as the *terminal settling velocity* of the particle. Most particles can be shown to reach their terminal settling velocity almost instantaneously (L. Theodore: personal notes, 1966).

Consider the equations examined above under terminal settling conditions. Since

$$dv/dt = 0$$

the general equation for particle motion becomes

$$0 = f - (F_D/m)$$

or

$$f = F_D/m \tag{7.28}$$

The units of f in this equation are ft/s^2. The general equation for the terminal settling velocity is obtained by direct substitution of Equation (7.12) into Equation (7.28) and solving for v. Thus

$$f = 3\alpha v^2 \mu^\beta \rho / 4 d_p (d_p v \rho)^\beta$$

so that

$$v = [4 f d_p^{1+\beta} \rho_p / 3\alpha \mu^\beta \rho^{1-\beta}]^{1/(2-\beta)} \tag{7.29}$$

For the Stokes' law range, Equation (7.29) becomes (see Table 7.1)

$$v = f d_p^2 \rho_p \Big/ 18\mu \tag{7.30}$$

For the intermediate range

$$v = 0.153 f^{0.71} d_p^{1.14} \rho_p^{0.71} \Big/ \mu^{0.43} \rho^{0.29} \tag{7.31}$$

Finally, for Newton's law range, one obtains

$$v = 1.74(f d_p \rho_p / \rho)^{0.5} \tag{7.32}$$

Keep in mind that f denotes the external force per unit mass of particle. One consistent set of units for the equations above is ft/s^2 for f, ft for d_p, lb/ft^3 for ρ, lb/ft · s for μ, and ft/s for v.

Ordinarily, determining the settling velocity of a particle of known diameter would require a trial-and-error calculation since the particle's Reynolds number is unknown and one cannot select the proper describing drag force equation. This iterative calculation can be circumvented by rearrangement of the drag force equations and solving for the settling velocity directly.

Both sides of Equations (7.30) and (7.32) can be multiplied by

$$(d_p\rho/\mu)$$

A dimensionless constant (K) is defined as

$$K = d_p\left(f\rho_p\rho/\mu^2\right)^{1/3} \tag{7.33}$$

Equations (7.30) and (7.32) can now be rewritten, respectively, as

$$\mathrm{Re} = K^3/18; \quad \mathrm{Re} \leq 2.0 \tag{7.34}$$

and

$$\mathrm{Re} = 1.74\,K^{1.5}; \quad \mathrm{Re} \geq 500 \tag{7.35}$$

Since K is not a function of the settling velocity, the choice of drag force equations may now be based on calculated K values. These new K range limits are given as follows:

$$\begin{aligned} K &< 3.3 \quad \text{Stokes' law range} \\ 43.6 > K &> 3.3 \quad \text{Intermediate range} \\ 2360 > K &> 43.6 \quad \text{Newton's law range} \end{aligned}$$

Interestingly, Larocca [*Chem. Eng.* p. 73 (1987)] and Theodore (personal notes, 1978), using the same approach employed above, defined a dimensionless term W that would enable one to calculate the diameter of a particle if the terminal velocity were known (or given). This particular approach has found application in catalytic reactor particle size calculations. The term W, which does not depend on the particle diameter, is given by

$$W = \frac{v^3\rho^2}{g\mu\rho_p} \tag{7.36}$$

The two key values of W that are employed in a manner similar to that for K are 0.2222 and 1514, i.e., for

$W < 0.2222$; Stokes' law range
$0.2222 < W < 1514$; Intermediate law range
$1514 < W$; Newton's law range

Another important consideration involves the Cunningham correction factor, CCF [C. E. Cunningham, *Proc. Roy. Soc. London Sec. A* **83**: 357 (1910)]. At very low values of the Reynolds number, when particles approach sizes comparable to the mean free path of the fluid molecules, the medium can no longer be regarded as continuous. For this condition, particles can fall between the molecules at a faster rate than predicted by the aerodynamic theories that led to the previous standard drag coefficients. To allow for this "slip," Cunningham introduced a multiplying correction factor to Stokes' law. This can alter the equations used to describe flow in the laminar regime. This effect is treated in Problem 7.20.

7.4 PARTICLE SIZING AND MEASUREMENT METHODS

Unfortunately, the term "measurement methods" has come to mean different things to engineers and scientists in the instrumentation field. Traditionally, this term has referred to sampling techniques employed for source characterization purposes. Source characterization includes either the determination of particle size distribution, concentration and flow rate, or chemical composition and other properties, or both. The equipment and procedures employed to measure particle data in a stream (usually air) represents the sampling part of the overall procedure. The analysis of the data is also important, usually providing the aforementioned information on particle size distribution, concentration, and flow rate.

An accurate quantitative analysis of the discharge of particulates from a process must be determined prior to the design and/or selection of recovery/control equipment. If the unit is properly engineered, utilizing the emission data as input to the device, most particulates can be successfully controlled by one or a combination of the methods to be discussed later in this book.

There are many techniques available for measuring the particle size distribution of particulates. The wide size range covered, from nanometers to millimeters, cannot always be analyzed using a single measurement principle. Added to this are the usual constraints of capital costs versus operating costs, speed of operation, degree of skill required, and most important, the end-use requirement.

Physical sizing is traditionally one of the most common methods available for classifying (or sizing) particles. This has been achieved by dry or wet screening with sieves, microscopic analysis, electric-grating techniques, and light-scattering methods.

More realistic determinations of particle behavior in any environment must consider the size, shape, and density. The technique best able to accomplish this is aerodynamic sizing. Only by knowing the aerodynamic size of particles is it possible to determine how they will behave in an air stream and the kind(s) of control equipment required to capture them.

Consider, for example, a ping-pong ball and a golf ball. On close examination they will appear almost equal in size; however, if both were tossed into a moving air stream, they would behave quite differently. Even though the size and shape are similar, the density is quite different, and the behavior of the two objects is far from being similar.

Since the density is quite different, the behavior of the objects is far from being similar aerodynamically. This is the primary fallacy in physical sizing.

A common method of specifying large particle sizes is to designate the screen mesh that has an aperture corresponding to the particle diameter. Since various screen scales are in use, confusion may result unless the screen scale involved is specified. The screen mesh generally refers to the number of screen openings per unit of length or area. The aperture for a given mesh will depend on the wire employed. The Tyler and the US Standard Screen Scales (Table 7.2) are the most widely used in the United States. The screens are generally constructed of wire mesh cloth, with the diameters of the wire and the spacing of the wires being specified. These screens form the bottoms of metal pans about 8 inches in diameter and 2 inches high, whose sides are so fashioned that the bottom of one sieve nests snugly on the top of the next.

The clear space between the individual wires of the screen is termed the screen aperture. As indicated above, the term *mesh* is applied to the number of apertures per linear inch; for example, a 10-mesh screen will have 10 openings per inch, and the aperture will be 0.1 inch minus the diameter of the wire.

Particle size itself is difficult to define in terms that accurately represent the types of particles of interest. This difficulty stems from the fact that particles exist in a wide variety of shapes, not just spheres. In the case of spherical particles, the definition of particle size is easy: it is simply the diameter. For the irregularly shaped particles, there are a variety of ways to define the size. The most convenient of these is to directly relate it to how the particle behaves in a fluid such as air, and the particle size definition that is most

TABLE 7.2 Tyler and US Standard Screen Scales

Tyler Mesh	Aperture Microns	US Mesh	Aperture Microns
400	37	400	37
325	43	325	44
270	53	270	53
250	61	230	62
200	74	200	74
170	88	170	88
150	104	140	105
100	147	100	149
65	208	70	210
48	295	50	297
35	417	40	420
28	589	30	590
20	833	20	840
14	1168	16	1190
10	1651	12	1680
8	2362		
6	3327		
4	4699		
3	6680		

useful for evaluating particle motion in a fluid is termed the *aerodynamic diameter*. It takes into account the particle density as shown in Equation (7.37)

$$d_{pa} = d_p \sqrt{\rho_p C_c} \tag{7.37}$$

where d_{pa} = aerodynamic diameter, μm
d_p = actual diameter, μm
ρ_p = particle density, g/cm^3
C_c = Cunningham correction factor (discussed later), dimensionless

Note that particles that appear to be different in physical size and shape can have the same aerodynamic diameter. Conversely, some particles that appear to be visually similar can have somewhat different aerodynamic diameters. Thus, the term *aerodynamic diameter* is useful in describing all particles, including fibers and particle clusters. Although it is not a true size because nonspherical particles require more than one dimension to characterize their size, the aerodynamic diameter provides a simple means of categorizing the sizes of particles with a single dimension and in a way that relates to how particles move in a fluid.

If one were able to design an ideal particle measuring device, the device should be able to:

1. Measure the exact size of each particle
2. Report data instantaneously without averaging data over some specified time interval
3. Determine the composition of each particle including shape, density, chemical nature, and so on

It would be an extremely difficult task to produce such an instrument. At this time there are devices that incorporate only one or two of these ideal functions. Various sizing techniques are listed below. Details are available in the literature (L. Theodore, *Nanotechnology: Basic Calculations For Engineers and Scientists*, John Wiley & Sons, Hoboken, NJ, 2007).

1. Microscopy
2. Optical counters
3. Electrical aerosol analyzers
4. Bahco microparticle classifier
5. Impactors
6. Photon correlation spectroscopy (PCS)

7.5 PARTICLE SIZE DISTRIBUTION

Particulate emissions from both human-made (synthetic) and natural sources do not consist of particles of any one size. Instead, they are composed of particles over a relatively wide size range. It is often necessary to describe this size range. The author has defined particle size analysis as a determination of the particle size distribution of a sample, and particle size distribution (PSD) as an equation, graph, or table that quantifies, usually by percent, the various sizes of particulates found in a sample.

Particle size distributions are often characterized by a "mean" particle diameter. Although numerous "means" have been defined in the literature, the most common are the arithmetic mean and the geometric mean. The arithmetic mean diameter is simply the sum of the diameters of each of the particles divided by the number of particles measured. The geometric mean diameter is the *n*th root of the product of the *n* number of particles in the sample. In addition to the arithmetic and geometric means, a particle size distribution may also be characterized by the "median" diameter. The median diameter is that diameter for which 50% of the particles are larger in size and 50% are smaller in size. Another important characteristic is the measure of central tendency. It is sometimes referred to as the *dispersion* or *variability*. The most common term employed is the *standard deviation*. These terms are discussed later in this chapter.

There are various methods for expressing the results of particle size measurements; the most common method is by plotting either (1) a frequency distribution curve or (2) a cumulative distribution curve. These topics are considered in this section.

Ultimately, the decision of how to represent particle size distribution information is left to the user and/or practitioner. Considerations include

1. Choosing an approach
2. Checking on the reasonableness of the choice
3. Ability to draw the proper conclusions from the approach selected

A typical particulate size analysis method of representation employed in the past is provided in Table 7.3. These numbers mean that 40% of the particles by weight are greater than 5 nm in size, 27% are less the 5 nm but greater than 2.5 nm, 20% are less than 2.5 nm but greater than 1.5 nm, and the remainder (13%) are less than 1.5 nm. Another form of representing data is provided in Table 7.4.

TABLE 7.3 Early PSD Representation

—	>5.0	nm	40%
<5	>2.5	nm	27%
<2.5	>1.5	nm	20%
<1.5	—	nm	13%
—	—	—	100%

One basic way of summarizing data is by the computation of a central value. The most commonly used central value statistic is the aforementioned arithmetic average, or the mean. This statistic is particularly useful when applied to a set of data having a fairly symmetrical distribution. The mean is an important statistic in that it summarizes all the data in the set and because each piece of data is taken into account in its

TABLE 7.4 Size Ranges in Arithmetic Increments

Size Range, μm	Percent in Size Range, %
0–2	10
2–4	15
4–6	30
6–8	30
8–10	10
>10	5

computation. The formula for computing the mean is

$$\overline{X} = \frac{X_1 + X_2 + X_3 + \cdots + X_n}{n} = \frac{\sum_{i=1}^{n} X_i}{n} \tag{7.38}$$

where $\overline{X}$ = arithmetic mean
X_i = any individual measurement
n = total number of observations
$X_1, X_2, X_3 \ldots$ = measurements 1, 2, and 3, respectively

The most commonly used measure of dispersion, or variability, of sets of data is the *standard deviation*, *s*. Its defining formula is given by expression

$$s = \sqrt{\frac{\sum (X_i - \overline{X})^2}{n - 1}} \tag{7.39}$$

where s = standard deviation (always positive)
X_i = value of the ith data point
$\overline{X}$ = mean of the data sample
n = number of observations

(Note: The above term s represents the *sample* standard deviation. In statistical circles, the symbol σ represents another, but near similar, standard deviation. Unfortunately, σ has been employed in the air pollution literature to represent the *sample* standard deviation. For this reason, the symbol σ is used in this text to represent the *sample* standard deviation. (Refer to S. Shaefer and L. Theodore, "Probability and Statistics Applications for Environmental Science," CRC Press, Boca Raton, FL, 2007 for additional details.)

Frequency distribution curves are usually plotted on regular coordinate (linear) paper. The curve describes the amount of material (particles) contained within each size range. When gasborne particles produced in a given operation are measured, the data generally have a tendency to show a preferential (maximum amount of) particle size. A plot of percent mass versus particle size (d_p) on a linear scale gives a curve with a peak at the preferential size. Such a curve is shown in Figure 7.4. This figure shows a *normal* probability distribution (see also later discussion) that is symmetric about the preferential size. This curve is rarely encountered for particulates consisting of very fine

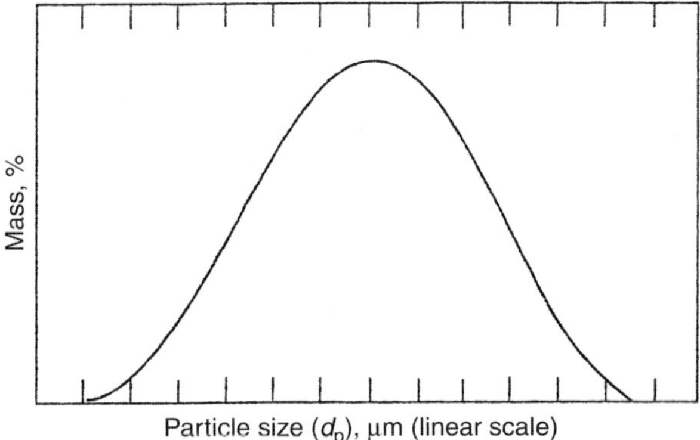

Figure 7.4 Particle size distribution.

particles. However, this curve may occasionally be approached for particles such as fumes formed by vapor phase reaction and condensation, or for tar and acid mists.

Particle size data can also be plotted as a cumulative plot. Particle size for each size range is plotted on the ordinate. The cumulative percent by weight (frequency) is plotted on the abscissa. The cumulative percent by weight can be given as cumulative percent less than stated particle size (%LTTS) or cumulative percent larger than stated particle size (%GTSS). The cumulative percent by weight can be plotted on either a linear percentage or a probability percentage scale. The particle size range (ordinate) is usually a logarithmic scale.

Frequently, the cumulative distribution is plotted on special coordinate paper called *log-probability* paper. The particle size of each size range is plotted on a logarithmic ordinate. The percent by weight larger than d_p is plotted on the probability scale as the abscissa. If the distribution is lognormal (see later discussion), the distribution curve plots out as a straight line. It should be noted that one can just as easily plot percent mass less than d_p (%LTSS) on the abscissa. In fact, %LTSS is preferred because the author's initials are carved in the acronym.

When measuring particulate emissions from industrial and/or human-made sources, the graph of the particle size distribution often displays the logarithmic variation of the normal distribution. The normal distribution has a fundamental defect related to its use in particle sizing analysis; implicit in the statement that a random variable is normally distributed is the concept that the values of particle size are at equal distances from the central tendency or preferential size. Suppose that the mean particle size or central tendency of the distribution were 20 μm. It would be equally probable to find either a 15- or 25-μm particle. One might also find a particle the size of 50 μm in the distribution. If the distribution were normal, it would be equally likely to find a particle the size of minus 10 μm (not physically possible). However, if it can be assumed that the logarithm of the particle size is randomly distributed, then this problem is avoided. The ratios of particle size about a central tendency are equally probable, and the ratios are then bounded on the lower end by zero.

The usefulness of the lognormal distribution is more evident when the frequency distribution curve is characteristically *skewed*. The data can then be plotted as a cumulative plot on log-probability paper. If the distribution follows the *lognormal* relationship, then the plot will result in a straight line. The linearity of the relationship allows one to describe the distribution statistically with a minimum of two parameters: the geometric mean, d_{gm} and the geometric standard deviation σ_{gm}.

The geometric mean value of a lognormal distribution can be read directly from a plot similar to that represented in Figure 7.5. The geometric mean size is the 50% size on the plot.

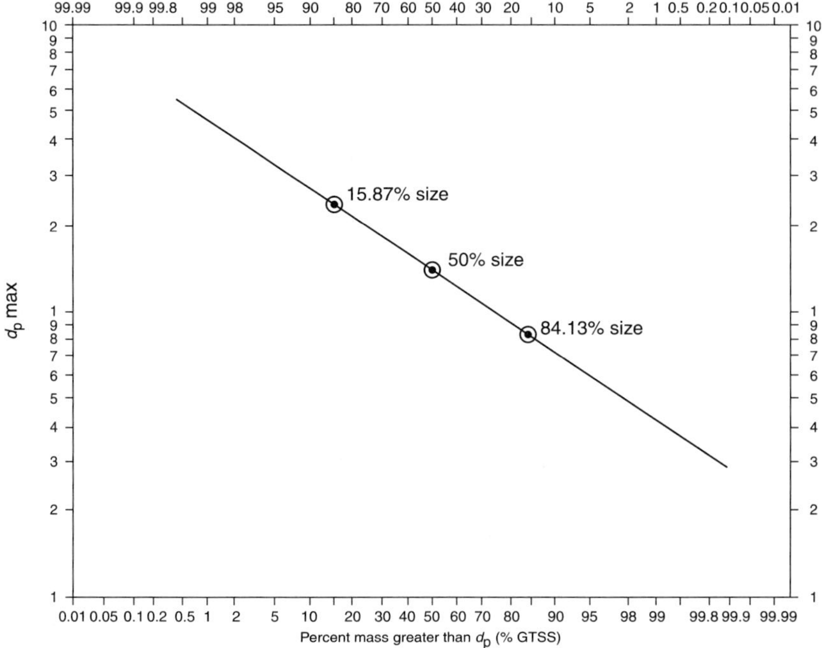

Figure 7.5 Geometric mean and standard deviation from a lognormal distribution plot.

As discussed previously, the *geometric standard deviation* is a good measure of the dispersion or spread of a distribution. The geometric standard deviation is the root mean square deviation about the mean value and can be read directly from a plot such as shown in Figure 7.5. For a lognormal distribution (which plots d_p maximum versus percent mass larger than d_p), the geometric standard deviation is given by

$$\sigma = \sigma_{gm} = \frac{50\% \text{ size}}{84.13\% \text{ size}}$$

or

$$\sigma = \sigma_{gm} = \frac{15.87\% \text{ size}}{50\% \text{ size}} \tag{7.40}$$

All one must do is determine the 50% size and the 84.13% size from the plot and divide to determine the geometric standard deviation. Therefore, knowing any two lognormal combination of these parameters (15.87%, 50%, 84.13%, or σ_{gm}) describes the entire range of lognormal particle size distributions.

It should also be noted that particle size distributions resulting from complex particle formation mechanisms or several simultaneous formation mechanisms may not be lognormal. In these cases, plots of the data on log-probability paper will not yield a straight line. In order to characterize this type of particle size data, it may be necessary to treat the data differently, e.g., as two separate lognormal distributions.

7.6 COLLECTION EFFICIENCY

Efficiency is the other characteristic quantity that warrants further discussion. The efficiency of a particulate control device is usually expressed as the percentage of mass collected by the unit compared with that entering the unit. It may be calculated on a particle number basis:

$$E_N = \left(\frac{\text{particles collected}}{\text{particles entering}}\right)100 \tag{7.41}$$

or on a total mass basis:

$$E = \left(\frac{(\text{inlet loading}) - (\text{outlet loading})}{\text{inlet loading}}\right) \times 100 \tag{7.42}$$

It is extremely important to distinguish between the two. Larger particles, which possess greater mass and are more easily removed in a control device, will contribute more significantly to the efficiency calculated on a mass or weight basis. Thus, if one considers a volume of aerosol which contains 100 1-μm particles and 100 100-μm particles, and if the efficiency of separation is 90% for 1-μm particles and 99% for 100-μm particles, then on a particle count basis 90 1-μm and 99 100-μm particles will be removed out of a total of 200. This gives a particle count efficiency of

$$E_N = \left(\frac{189}{200}\right)100 = 94.5\%$$

On a weight basis, however, if a 1-μm particle has unit mass, a 100-μm particle has 10^6 mass unit. The weight efficiency is then given by

$$E = \left(\frac{90(1) + 99(10^6)}{100(1) + 100(10^6)}\right)100 = 99\%$$

Any expression of the efficiency of a particulate removal device is therefore of little value without a careful description of the size spectrum of particles involved.

It is interesting to note certain observations. The smaller particles, equivalent in mass to a considerably smaller number of large particles, have a much greater impact on visibility, health, and water droplet nucleation than the larger particles. When large tonnages (significant mass discharges) are involved, the high mass efficiencies often reported for particle collection may lead to overly optimistic conclusions about emissions. The small weight percentages of particles that pass through the collector can still represent large *numbers* of particles escaping to the atmosphere. It would then seem that tonnage collection figures and weight removal efficiencies may not really be that adequate to delineate the entire particulate emission problem.

As noted earlier, the capture efficiency is a strong function of particle size. Combining this size–efficiency material with the particle size distribution information in Section 7.3 allows one to obtain and/or calculate the overall efficiency of a unit. In practice, this is treated analytically by methods other than the somewhat simple example given above. A more detailed presentation is given in the "Problems" section of this chapter.

There is one simplified procedure that is noteworthy since the standard detailed method of analysis may not be warranted. Furthermore, the detailed calculations may be inaccurate since they can involve a large number of values read from a graph. Sundberg (R. Sundberg, "The Prediction of Overall Collection Efficiency," *Proc. APCA Annual Meeting*, 1973, pp. 73–298) suggests that if the particle size distribution of the dust *and* the size–efficiency of the collector can be adequately approximated by lognormal functions, the collector's overall efficiency can be determined from d_{p50}, σ_g, plus d'_{p50}, and σ'_g (the corresponding size–efficiency values). These four parameters are obtained from the two lognormal curves, or the equivalent. The overall efficiency can be expressed in integral form as

$$E = \int_{-\infty}^{+\infty} \Phi_1(d'_p)d[\Phi_2(d_p)] \tag{7.43}$$

where $\Phi_1(d'_p)$ = lognormal function for the fractional efficiency
$\Phi_2(d_p)$ = lognormal function for the particle size distribution

The solution of Equation (7.43) is a simple relation between the overall collection efficiency and the error function (erf) or table of areas under the standard normal curve, and given by

$$E = \operatorname{erf}\left[\frac{\ln\left(d_{p50}/d'_{p50}\right)}{\left\{\left(\ln\sigma_g\right)^2+\left(\ln\sigma'_g\right)^2\right\}^{0.5}}\right]; \text{ fractional basis} \tag{7.44}$$

The overall collection efficiency can thus be computed by evaluating Equation (7.44) and reading the efficiency from the error function table (a portion of which is presented in Table 7.5).

Finally, it should be noted that the potential for the agglomeration of particles provides hope for higher collection efficiencies, particularly for fine particulates. Recently, a unique potential technology—the *indigo bipolar agglomerator*—has been

TABLE 7.5 Error Function

Areas under the Standard Normal Curve from $-\infty$ to d_p

$$\text{erf}(d_p) = \frac{1}{2\pi}\int_{-\infty}^{d_p} e^{-(1/2)f^2}\,df$$

d_p	0	2	4	6	8
0.0	0.5000	0.5080	0.5160	0.5239	0.5319
0.1	0.5398	0.5478	0.5557	0.5636	0.5714
0.2	0.5793	0.5871	0.5948	0.6026	0.6103
0.3	0.6179	0.6255	0.6331	0.6406	0.6480
0.4	0.6554	0.6628	0.6700	0.6772	0.6844
0.5	0.6915	0.6985	0.7054	0.7123	0.7190
0.6	0.7258	0.7324	0.7389	0.7454	0.7518
0.7	0.7580	0.7642	0.7704	0.7764	0.7823
0.8	0.7881	0.7939	0.7996	0.8051	0.8106
0.9	0.8159	0.8212	0.8264	0.8315	0.8365
1.0	0.8413	0.8461	0.8508	0.8554	0.8599
1.1	0.8643	0.8686	0.8729	0.8770	0.8810
1.2	0.8849	0.8888	0.8925	0.8962	0.8997
1.3	0.9032	0.9066	0.9099	0.9131	0.9162
1.4	0.9192	0.9222	0.9251	0.9279	0.9306
1.5	0.9332	0.9357	0.9382	0.9406	0.9429
1.6	0.9452	0.9474	0.9495	0.9515	0.9535
1.7	0.9554	0.9573	0.9591	0.9608	0.9625
1.8	0.9641	0.9656	0.9671	0.9686	0.9699
1.9	0.9713	0.9726	0.9738	0.9750	0.9761
2.0	0.9772	0.9783	0.9793	0.9803	0.9812
2.1	0.9821	0.9830	0.9838	0.9846	0.9854
2.2	0.9861	0.9868	0.9875	0.9881	0.9887
2.3	0.9893	0.9898	0.9904	0.9909	0.9913
2.4	0.9918	0.9922	0.9927	0.9931	0.9934
2.5	0.9938	0.9941	0.9945	0.9948	0.9951
2.6	0.9953	0.9956	0.9959	0.9961	0.9963
2.7	0.9965	0.9967	0.9969	0.9971	0.9973
2.8	0.9974	0.9976	0.9977	0.9979	0.9980
2.9	0.9981	0.9982	0.9984	0.9985	0.9986

applied successfully at several installations. This new device performs the following functions:

1. Treats the dust prior to entering a control device.
2. Uses electrostatic forces to attach the fine particles to the larger, easily collected particles.
3. The bipolar charger charges half the dust positively and the other half negatively.

4. Fine negative particles are mixed with large positive particles and large negative particles are mixed with fine positive particles.
5. When a fine particle comes close to an oppositely charged large particle, it is attracted and attaches to the large particle.

The unit consists of or is characterized by

1. Grounded plates
2. Discharge electrodes
3. Alternating arrangement
4. High-voltage energization
5. Low pressure drop

It features

1. No collection problems
2. No hoppers
3. No rappers
4. High velocity
5. Compact size
6. Low power consumption

The end result is that finer particles ($<5\ \mu m$) attach to larger particles ($>10\ \mu m$) that are more readily collected in an electrostatic precipitator (ESP) or other equipment. The benefits include

1. Reduced mass emissions
2. Reduced visible emissions
3. Reduced PM2.5 emissions
4. Greater compliance safety margin
5. Increased choice of coal supply (for coal fired utility boilers)
6. Higher tolerance for (a) boiler upsets and (b) lost electrical sections of ESP

Some key operating features are

1. No moving parts ensures high reliability and low maintenance
2. Low power consumption provides low operating cost (5 kW/100 MW generation)
3. High gas velocity ensures no dust buildup on electrodes or bottom of duct
4. Small size and prefabrication reduce outage time for installation to a few days
5. Low pressure loss (<1.0 in H_2O)

PROBLEMS

7.1 Particle Velocity

A particle migrates with a velocity of 8 ft/min. What distance (inches) will the particle travel in 0.5 seconds?

(**a**) 0.4
(**b**) 0.6
(**c**) 0.8
(**d**) None of the above

Solution: Apply Equation (7.1), being careful to employ consistent units:

$$\begin{aligned} d &= vt \qquad (7.1)\\ &= (8\ \text{ft/min})(0.5\ \text{s})(\text{min}/60\ \text{s})(12\ \text{in/ft})\\ &= 0.8\ \text{in}\end{aligned}$$

The correct answer is therefore (c).

7.2H Particle "Circularity"

The cross section of a particle has the shape of a square. Estimate its "circularity."

Solution: Apply Equation (7.57) (see Problem 7.29) and represent the particle's "circularity" by its equivalent diameter (ED):

$$\text{ED} = 4(A)/P \qquad (7.45)$$

where A = area
P = perimeter

For a square of side L

$$A = L^2$$

and

$$P = 4L$$

Therefore

$$\begin{aligned}\text{ED} &= (4)L^2/4L\\ &= L\end{aligned}$$

7.3 Particle Reynolds Number

Calculate the following Reynolds numbers:

(**a**) 10-μm particle moving at 50 ft/s
(**b**) 100-μm particle moving at 500 ft/s
(**c**) 100-μm particle moving at 50 ft/s
(**d**) 1-μm particle moving at 50 ft/s

Solutions: The Re values above cannot be calculated since the state of the medium through which the particles are moving has *not* been specified. If ambient air (60°F, 1 atm) conditions are assumed, then $\rho = 0.0765\ \text{lb/ft}^3$ and $\mu = 1.14 \times 10^{-5}\ \text{lb/ft}\cdot\text{s}$. Thus

(a) $$\begin{aligned} \text{Re} &= d_p v \rho/\mu \\ &= (10)(3.281 \times 10^{-6})(50)(0.0765)/(1.14 \times 10^{-5}) \\ &= 11.0; \text{ intermediate regime} \end{aligned}$$

(b) $$\begin{aligned} \text{Re} &= (11.0)(500/50)(100/10) \\ &= 1110; \text{ Newton regime} \end{aligned}$$

(c) $$\begin{aligned} \text{Re} &= (11.0)(100/10) \\ &= 110; \text{ intermediate regime} \end{aligned}$$

(d) $$\begin{aligned} \text{Re} &= 11.0(1/10) \\ &= 1.1; \text{ Stokes regime} \end{aligned}$$

7.4 Micrometer Magnitude

How many 1-μm particles could fit across a 1-inch space?

Solution: The solution to this problem can be obtained by applying the appropriate conversion factors, as demonstrated in Chapter 3. However, if one proceeds to the Appendix, it is noted that there are $(39.37)^{-1}$ m in an inch. Since there are 10^6 μm in a meter, there are

$$(39.37)^{-1} \times 10^6\ \mu\text{m/inch}$$

or

$$2.54 \times 10^4\ \mu\text{m/inch}$$

or

$$25{,}400\ \mu\text{m/inch}$$

Thus, 25,400 μm particles could fit across a 1-inch gap.

7.5 Increase in Cube Surface Area

Consider a cube with side A of 1.0 m. Assuming that the same mass of cube is converted to cubes with sides of 1.0 μm, calculate the increase in surface area of the smaller cubes.

Solution: The volume of a cube is given by

$$V = A^3$$

Therefore

$$\begin{aligned} V(1.0\,\text{m})) &= (1)^3 \\ &= 1.0\,\text{m}^3 \\ &= 1.0 \times 10^9 (\mu\text{m})^3 \end{aligned}$$

The volume of the 1.0-μm-sided cube is

$$V\,(1.0\,\mu\text{m}) = 1.0\,(\mu\text{m})^3$$

Since that volume remains the same, the number of smaller cubes is

$$\begin{aligned} N &= 10^9/1.0 \\ &= 10^9 \end{aligned}$$

The surface area of a 1.0 m sided cube is

$$\begin{aligned} \text{SA}(1.0\,\text{m}) &= 6A^2 \\ &= 6\,(\text{m})^2 \\ &= 6 \times 10^6\,(\mu\text{m})^2 \end{aligned}$$

and

$$\begin{aligned} \text{SA}(1.0\,\mu\text{m}) &= 6A^2 \\ &= 6\,(\mu\text{m})^2 \end{aligned}$$

Since there are 10^9 of the 1.0-μm cubes, their total surface area is $6 \times 10^9\,(\mu\text{m})^2$. Note that the reduction in size has increased the surface area by a factor of 1000 due to a decrease in the side length of the cube.

7.6 Sphere, Cube, Rectangular Parallelepiped, and Cylinder

Calculate the area and volume of the following shaped particles:

(a) Sphere of radius R;
(b) Cube of side A;
(c) Rectangular parallelepiped of sides A and B;
(d) Cylinder of radius R and height H.

Solution

(**a**) For a sphere of radius R:

$$\text{Volume} = \frac{4}{3}\pi R^3$$

$$\text{Area} = 4\pi R^2$$

(**b**) For a cube of side A:

$$\text{Volume} = A^3$$

$$\text{Area} = 6A^2$$

(**c**) For a rectangular parallelepiped of depth C:

$$\text{Volume} = ABC$$
$$\text{Area} = 2(AB + AC + BC)$$

(**d**) For a cylinder of radial R and height H:

$$\text{Volume} = \pi R^2 H$$
$$\text{Area} = 2\pi RH$$

7.7 Parallelogram, Triangle, and Trapezoid

Calculate the area and perimeter of the following shaped particles:

(**a**) Parallelogram of height H, side A, and base B.

(**b**) Triangle of height H, base B, and left side A. The angle between side A and base B is ϕ.

(**c**) Trapezoid of height H and parallel sides A and B. The lower left angle is ϕ, and the lower right angle is θ.

Solution

(**a**) For the parallelogram:

$$\text{Area} = (H)\,(B) = (A)\,(B)\sin\phi$$
$$\text{Perimeter} = 2A + 2B$$

(**b**) For the triangle:

$$\text{Area} = \frac{1}{2}(B)\,(H) = \frac{1}{2}(A)\,(B)\sin\phi$$
$$\text{Perimeter} = A + B + C$$

(**c**) For the trapezoid:

$$\text{Area} = \frac{1}{2}H(A + B)$$
$$\text{Perimeter} = A + B + H\left(\frac{1}{\sin\phi} + \frac{1}{\sin\theta}\right)$$
$$= A + B + H(\cos\phi + \cos\theta)$$

7.8 Area-to-Volume Ratios

Develop the area-to-volume ratio equations for the shaped particles from Problem 7.6.

Solution: The following ratios are obtained for each of the four particles:

(**a**) Area/volume $= 3/R$
(**b**) Area/volume $= 6/A$
(**c**) Area/volume $= 2[(1/A) + (1/B) + (1/C)]$
(**d**) Area/volume $= 2/R$

7.9 Dust Explosions

Qualitatively describe dust explosions.

Solution: Seven key factors are required for a dust explosion to take place (L. Theodore: personal notes, 1990).

1. Air (oxygen)
2. Fuel source (dust)
3. Mixing
4. Dry state
5. Minimum concentration
6. Ignition source
7. Enclosure

All seven factors need to be present for an explosion to occur. Thus, eliminating only one prevents the explosion.

An *explosion* is defined as an occurrence where energy is released over a sufficiently short period of time and in an enclosed (usually) volume to generate a pressure wave of finite amplitude traveling away from the source of the explosion. This energy may have been originally stored in the system as chemical, nuclear, electrical, or pressure energy. However, the release is not considered to be explosive unless it is rapid and concentrated enough to produce a pressure wave that can be heard. Many substances that oxidize slowly in a massive state will oxidize extremely fast or possibly even explode when dispersed as fine particles in air. Dust explosions are often caused by the unstable burning or oxidation of combustible particles, brought about by their relatively large specific surfaces.

7.10 Brownian Motion/Molecular Diffusion

Qualitatively describe Brownian motion.

Solution: Brownian motion (also referred to as *molecular diffusion*) becomes the dominant collection mechanism for particles less than 500 nm (0.5 μm) and is especially significant as the particles become smaller. As discussed earlier, very small particles deflect slightly when gas molecules impact on them. The transfer of kinetic energy from the fast moving gas molecules to the small particle causes this deflection, which is termed *Brownian motion*. *Diffusivity* provides a measure of the extent to which molecular collisions cause these very small particles to move in a random manner across the direction of gas flow. The diffusion coefficient in the equation below represents the diffusivity (D) of a particle at given gas stream conditions.

$$D = \frac{CKT}{3\pi d_{pa}\mu}; \quad \text{consistent units} \tag{7.46}$$

where D = diffusion coefficient (diffusivity)
C = Cunningham correction factor (dimensionless)
K = Boltzmann constant
T = absolute temperature
d_{pa} = particle aerodynamic diameter
μ = gas viscosity

Small particles obtain a high diffusion coefficient because the diffusion coefficient is inversely proportional to particle size. Thus, small particles in a fluid are subject to a random displacement known as the aforementioned Brownian motion. This occurs in addition to the net motion in a given direction owing to the action of any of the external forces mentioned earlier. This "chaotic" motion is represented by the statistical average displacement of particles in a given period of time. By definition, the average displacement of all particles over time is then zero. However, the actual path or trajectory taken by a given particle consists of an extremely large number of irregular and zig-zag jumps.

7.11 Inertial Impaction versus Molecular Diffusion

Discuss the difference between particles captured by inertial impaction and molecular diffusion.

Solution: The reader is referred to the body of this chapter for the answer. It should be noted that these two collection mechanisms play important roles with both venturi scrubbers and baghouses.

7.12 Cunningham Correction Factor

The Cunningham correction factor (CCF) is used to

(**a**) Correct the stack gas to standard conditions
(**b**) Correct the drag coefficient for fluid flow in the laminar regime
(**c**) Determine the settling velocity of a particle in the turbulent regime
(**d**) Determine the aerodynamic drag force on a particle.

Solution: Answers (a) and (c) never apply. Answer (d) applies only if the particle is in the laminar Stokes regime. Answer (b) always applies. The correct answer is therefore (b).

7.13 Cunningham Correction Factor Calculations

Discuss the role that the Cunningham correction factor has on calculations involving

(**a**) Stokes' law
(**b**) Intermediate law
(**c**) Newton's law
(**d**) Overall collection efficiency

Solution: As indicated in Problem 7.12, the Cunningham correction factor applies only in the Stokes regime. The correct answer is therefore (a).

7.14 Cunningham Correction Factor Effect on Particle Size

Briefly describe the effect that the Cunningham correction factor (CCF) has on a particle size distribution whose minimum particle size is 2.0 μm.

Solution: Generally, the CCF affects particle motion for particle sizes below 2 μm. Since the minimum particle size is 2.0 μm, the CCF effect can be neglected.

7.15 Effect of Cunningham Correction Factor on Collection Efficiency and Pressure Drop

Briefly describe the effect of the Cunningham correction factor on

1. Collection efficiency
2. Pressure drop

Solution: Neither of the two effects come into play. However, one could argue that the CCF indirectly affects the collection efficiency performance because of its effect on small-particle behavior. These efficiency calculations increase because of the reduced drag force that the particle experiences during the capture process.

7.16 The Drag Force

Qualitatively and quantitatively define the drag force.

Solution: Whenever a difference in velocity exists between a particle and its surrounding fluid, the fluid will exert a resistive force on the particle. Either the fluid (gas) may be at rest with the particle moving through it or the particle may be at rest with the gas flowing past it. It is generally immaterial which phase (solid or gas) is assumed to be at rest; it is the *relative velocity* between the two that is important. This resistive force exerted on the particle by the gas is called the *drag*. See Section 7.3 for more details. In treating fluid flow through pipes, a *friction factor* term is used in many engineering calculations. An analogous factor, called the *drag coefficient*, is employed in drag force calculations for flow past particles. The drag force expression, including the drag coefficient, can be obtained by rewriting Equation (7.3):

$$F_{\mathrm{D}} = \left(\frac{\rho v^2}{2\,g_{\mathrm{c}}}\right) A_{\mathrm{p}} C_{\mathrm{D}} \tag{7.47}$$

For a sphere

$$A_{\mathrm{p}} = \frac{\pi d_{\mathrm{p}}^2}{4}$$

Substituting gives

$$F_{\mathrm{D}} = \frac{\pi d_{\mathrm{p}}^2 \rho v^2 C_{\mathrm{D}}}{8\,g_{\mathrm{c}}}; \quad \text{consistent units} \tag{7.48}$$

where C_{D} = drag coefficient, dimensionless
v = particle velocity
d_{p} = particle diameter
ρ = density of the (air) fluid

As indicated above, the drag force arises when a particle moves through a fluid. The particle clears or displaces the fluid immediately in front of it, imparting momentum to the fluid. The drag force produced is equal to the momentum (mv) per unit time imparted to the fluid by the particle. Since the moving particle has a velocity (v), a portion of the particle's velocity is transferred by momentum to the fluid as fluid velocity v_{α}. The amount of energy imparted from v to v_{α} is related to a friction factor that has been termed the *drag coefficient*, (C_{D}).

7.17 Settling Particle Reynolds Number

A 72.7 μm-diameter particle moving at its terminal settling velocity has a drag coefficient in 70°F air determined to be 12 for the Stokes law regime and 12 for the intermediate or transition regime. What is the Reynolds number?

(**a**) 18.5
(**b**) 2.0
(**c**) 500
(**d**) 24.0

Solution: Employ the equation for the drag force. For Stokes' law [see Equation (7.8)]

$$C_D = 24/\mathrm{Re} \tag{7.8}$$

For

$$C_D = 12$$

One obtains

$$\mathrm{Re} = 2.0$$

For the intermediate law regime, Equation (7.9) applies:

$$C_D = 18.5/\mathrm{Re}^{0.6} \tag{7.9}$$

For

$$C_D = 12$$

One obtains (rounding to two significant figures)

$$\mathrm{Re} = 2.0$$

The correct answer is therefore (b).

7.18 Particle Settling Velocity Equations

Obtain the various equations describing the terminal velocity of spherical particles settling under the influence of gravity.

Solution: The appropriate terminal particle settling velocity equation is derived by settling F_G equal to F_D (see Section 7.3, Equations 7.30–7.32). For the three regimes one obtains (with g replacing f)

$$v_t = \frac{g\rho_p d_p^2}{18\mu}; \quad \text{laminar regime} \tag{7.49}$$

$$v_t = \frac{0.153\, g^{0.71} d_p^{1.14} e_p^{0.71}}{\mu^{0.43} e^{0.29}}; \quad \text{transition regime} \tag{7.50}$$

$$v_t = 1.74\left(\frac{g d_p e_p}{e}\right)^{0.5}; \quad \text{turbulent regime} \tag{7.51}$$

Once again, the term g can be replaced by another external force (centrifugal, electrostatic, etc.), provided consistent units are employed.

7.19 Settling Velocity Application

Calculate the settling velocity of a particle settling by gravity in a gas stream. Assume the following information:

$$\rho_p = 0.899\ \text{g/cm}^3$$
$$\rho = 0.0012\ \text{g/cm}^3$$
$$\mu_{(\text{air})} = 1.82 \times 10^{-4}\ \text{g/cm} \cdot \text{s}$$
$$g = 980\ \text{cm/s}^2$$
$$d_p = 45\ \mu\text{m}$$

Solution: Calculate the K parameter to determine the proper flow regime. Employ Equation (7.33):

$$K = d_p(g\rho_p\rho/\mu^2)^{1/3} \tag{7.33}$$

$$= 45 \times 10^{-4}\text{cm}\left[\frac{(980\ \text{cm/s}^2)(0.899\ \text{g/cm}^3)(0.0012\ \text{g/cm}^3)}{(1.82 \times 10^{-4}\ \text{g/cm} \cdot \text{s})^2}\right]^{1/3}$$

$$= 1.43$$

Therefore, the flow regime is laminar; that is, Stokes' law applies. The settling velocity is calculated using the equation for Stokes' law:

$$v_t = \frac{g\rho_p d_p^2}{18\mu}; \quad g_c = 1.0\ \text{(metric units)} \tag{7.49}$$

$$= \frac{(980\ \text{cm/s}^2)(0.899\ \text{g/cm}^3)(45 \times 10^{-4}\text{cm})^2}{18(1.82 \times 10^{-4}\ \text{g/cm} \cdot \text{s})}$$

$$= 5.45\ \text{cm/s}$$

7.20 The Cunningham Correction Factor

Quantitatively describe the Cunningham correction factor (CCF).

Solution: As discussed in Section 7.3, at very low values of the Reynolds number, when particles approach sizes comparable to the mean free path of the fluid molecules, the medium can no longer be regarded as continuous. To allow for this "slip," Cunningham introduced a multiplying correction factor to Stokes' law that took the form

$$v = \frac{f d_p^2 \rho_p}{18\mu}\left(\frac{1 + 2A\lambda}{d_p}\right) \tag{7.52}$$

or

$$v = \frac{f d_p^2 \rho_p}{18\mu}(C); \quad C = 1 + \frac{2A\lambda}{d_p} \tag{7.53}$$

where λ = mean free path of the fluid molecules, consistent units
$A = 1.257 + 0.40 \exp(-1.10\, d_p/2\lambda)$
C = Cunningham correction factor, dimensionless

The modified Stokes' law equation, which is usually referred to as the *Stokes–Cunningham equation*, is then

$$F_D = \frac{3\pi\mu d_p v}{C g_c} \tag{7.54}$$

As shown in, Problem 7.21, the correction factor should definitely be included in the drag force term when dealing with submicron and nanosize particles.

A word of interpretation is in order for λ, the mean free path of the fluid molecules. (Note that the value of λ is required in the calculation for C.) Based on the kinetic theory of gases, λ is given by

$$\lambda = \frac{\mu}{0.499\rho\sqrt{8RT/P(\text{MW})}}; \quad \text{consistent units} \tag{7.55}$$

where μ = gas viscosity
ρ = gas density
R = ideal gas constant
T = absolute temperature of gas
P = absolute measure of gas
MW = molecular weight of gas

For many gases, λ is approximately 0.1 μm or 100 nm. For air at 70°F and 1 atm, $\lambda = 653$ nm. As noted above, this correction is of importance only for particles smaller than 1.0 μm or 1000 nm. Finally, the CCF may be estimated from the following equation:

$$C = 1.0 + \frac{6.21 \times 10^{-4}}{d_p} T \tag{7.56}$$

where T = temperature, K
d_p = particle diameter, μm

7.21 Cunningham Correction Factor Values for Air at Atmospheric Pressure

Calculate the Cunningham correction factor (CCF) for particle size variation from 1.0 to 10^4 nm at temperatures of 70°F, 212°F, and 500°F. Include a sample calculation for a particle diameter of 400 nm (0.4 μm) at 70°F, 1 atm.

Solution: Employ the equations presented in Problem 7.20. The calculated results are provided in Table 7.6, along with a sample calculation.

TABLE 7.6 Cunningham Correction Factors

d_p nm	d_p μm	C (70°F)	C (212°F)	C (500°F)
1.0	0.001	216.966	274.0	405.32
10	0.01	22.218	27.92	39.90
100	0.1	2.867	3.61	5.14
250	0.25	1.682	1.952	2.528
500	0.5	1.330	1.446	1.711
1000	1	1.164	1.217	1.338
2500	2.5	1.066	1.087	1.133
5000	5	1.033	1.043	1.067
10,000	10	1.016	1.022	1.033

For a d_p of 0.4 μm, the Cunningham correction factor should be included. Employing Equation (7.53), one obtains

$$\begin{aligned} A &= 1.257 + 0.40e^{-1.10d_p/2\lambda} \\ &= 1.257 + 0.40 \exp - \left(\frac{1.10(0.4)}{(2)(6.53 \times 10^{-2})} \right) \\ &= 1.2708 \end{aligned}$$

Therefore

$$\begin{aligned} C &= 1 + \frac{2A\lambda}{d_p} \\ &= 1 + \frac{(2)(1.2708)(6.53 \times 10^{-2})}{(0.4)} \\ &= 1.415 \end{aligned}$$

The results clearly demonstrate that the CCF becomes more pronounced for micrometer-sized particles in the 0.01–1.0 range. In addition, an increase in temperature also leads to an increase in this effect. The reader should also realize that a comparable effect does not exist for particles settling in liquids until the diameter becomes less than 0.01 μm.

7.22 Calculated versus Experimental Values for the Drag Coefficient

Compare calculated versus experimental values of the drag coefficient.

Solution: Five equations for the drag coefficient were provided in Section 7.3: Equations (7.8)–(7.10), (7.16), and (7.17). The drag coefficient numbers derived from these equations are given in Table 7.7, over a wide range of Reynolds numbers.

7.23 Appropriate Drag Force Equation I

In order to calculate the terminal settling velocity of a particle in a gravity field, one must decide which of the three approximate drag force equations (Stokes, Intermediate, or Newton) is applicable. Explain why, when all three equations

TABLE 7.7 Calculated vs. Experimental Values for the Drag Coefficient as Function of the Reynolds Number

Range	Re	C_D					
		Equation (7.8)	Equation (7.9)	Equation (7.10)	Equation (7.16)	Equation (7.17)	Experiment
Stokes' law	0.01	2400	—	—	—	2364	2100
	0.02	1200	—	—	—	1195	1050
	0.03	820	—	—	—	803	700
	0.05	490	—	—	—	488	420
	0.07	350	—	—	—	352	300
	0.10	290	—	—	—	249	240
	0.20	120	—	—	—	129	120
	0.30	82	—	—	—	88.3	80
	0.50	49	—	—	—	55.0	49.5
	0.70	35	—	—	—	40.5	36.5
	1.00	24	—	—	22.4	29.5	26.5
Intermediate	2	—	12.0	—	—	16.2	14.4
	3	—	9.5	—	10.5	11.6	10.4
	4	—	7.8	—	—	9.2	8.2
	5	—	6.9	—	—	7.7	6.9
	6	—	—	—	6.23	—	5.9
	7	—	5.4	—	—	5.97	5.4
	10	—	4.5	—	4.26	4.61	4.1
	20	—	3.0	—	—	2.90	2.55
	30	—	2.4	—	—	2.26	2.0
	40	—	2.0	—	1.71	1.93	1.6
	50	—	1.8	—	—	1.71	1.5
	70	—	1.5	—	—	1.45	1.27
	90	—	—	—	1.11	—	1.14

	100	—	1.2	—	—	1.23	1.07
	200	—	0.80	—	0.785	0.93	0.77
	300	—	0.68	—	—	0.82	0.65
	400	—	0.60	—	—	0.76	0.57
Newton's law	500	—	0.56	—	0.568	0.71	0.55
	700	—	—	0.44	—	0.66	0.50
	900	—	—	0.44	0.482	—	0.46
	1000	—	—	0.44	—	0.61	0.45
	2000	—	—	0.44	—	0.54	0.42
	3000	—	—	0.44	0.393	0.51	0.40
	4000	—	—	0.44	—	0.49	0.39
	5000	—	—	0.44	—	0.48	0.385
	7000	—	—	0.44	0.386	0.47	0.39
	9000	—	—	0.44	0.398	—	0.40
	10,000	—	—	0.44	—	0.46	0.405
	20,000	—	—	0.44	0.451	0.44	0.45
	30,000	—	—	0.44	—	0.43	0.47
	40,000	—	—	0.44	—	0.42	0.48
	50,000	—	—	0.44	—	0.42	0.49
	60,000	—	—	0.44	0.520	—	0.49
	70,000	—	—	0.44	—	0.42	0.50
	100,000	—	—	0.44	0.466	0.42	0.48
	200,000	—	—	0.44	—	0.41	0.42
	300,000	—	—	—	—	—	0.20
	400,000	—	—	—	—	—	0.084
	600,000	—	—	—	—	—	0.10
	1,000,000	—	—	—	—	—	0.13
	3,000,000	—	—	—	—	—	0.20

are used to calculate values of the terminal velocities for a given Reynolds number, the correct value is always the smallest of the three.

Solution: Refer to Figure 7.6, which is a slight modification of Figure 7.3. Irrespective of whether one is in region I, II, or III, the calculated drag coefficient from any of these describing equations produces the highest drag for the correct (and applicable) drag force equation. The higher drag provides greater resistance to flow, which in turn corresponds to a smaller (or lower) velocity.

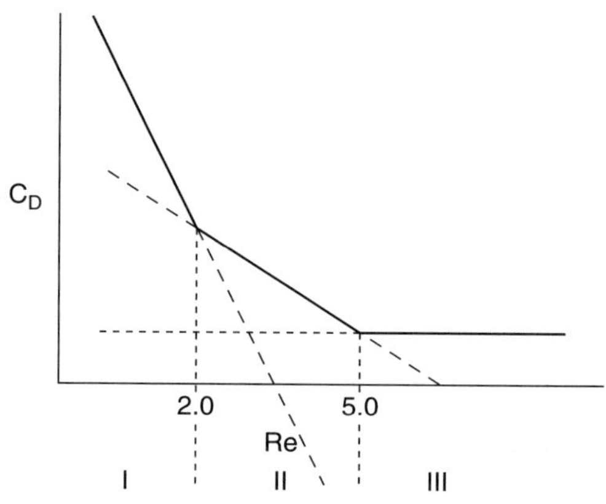

Figure 7.6 Drag Force–Reynolds number regimes.

7.24 Appropriate Drag Force Equation II

Illustrate the answer to Problem 7.23 by calculating the settling velocities of a series of charcoal particles of various diameters. Use a charcoal density of 56.2 $\mathrm{lb/ft^3}$ and an air temperature of 20°C.

Solution: The pertinent calculations and results are provided below after substituting the above data.

Stokes (Equation 7.49):

$$v_t = 8.88 \times 10^{-5}\, d_p^2; \quad d_p = \mu\mathrm{m}$$

$$= 8.25 \times 10^{6}\, d_p^2; \quad d_p = \mathrm{ft}$$

Intermediate (Equation 7.50):

$$v_t = 4.84 \times 10^{-3}\, d_p^{1.14}; \quad d_p = \mu\mathrm{m}$$

$$= 8.67 \times 10^{-3}\, d_p^{1.14}; \quad d_p = \mathrm{ft}$$

Newton (Equation 7.51):

$$v_t = 4.89 \times 10^{-1}\, d_p^{0.5}; \quad d_p = \mu\mathrm{m}$$

$$= 270 d_p^{0.5}; \qquad d_p = \mathrm{ft}$$

Calculations are provided in Table 7.8 with the correct velocities shown in the boxes. Note that the velocities are in ft/s.

TABLE 7.8 Velocity Calculations for Different Regimes (all values in ft/s)

d_p, μm	Stokes, v	IRL, v	Newton, v
1	8.88×10^{-5}	4.84×10^{-3}	4.88×10^{-1}
10	8.88×10^{-3}	6.68×10^{-2}	1.546
100	8.88×10^{-1}	9.22×10^{-1}	4.88
200	3.55	2.03	6.91
400	14.2	4.48	9.775
600	31.97	7.109	11.97
800	56.84	9.868	13.82
1000	88.8	12.73	15.46
1400	174.1	18.68	18.29
2000	355.3	28.05	21.86

IRL = Intermediate range law.

7.25 Particle Drag Calculations

A spherical particle having a diameter of 0.0093 inch and a specific gravity of 1.85 is placed on a horizontal screen. Air is blown through the screen vertically at a temperature of 20°C and a pressure of 1.0 atm. Calculate the following:

(**a**) The velocity required to just lift the particle
(**b**) The particle Reynolds number at this condition
(**c**) The drag force in both engineering and cgs units
(**d**) The drag coefficient C_D

Solution: At 20°C or 68°F

$$\rho = \frac{P(\text{MW})}{RT} = \frac{(1)(29)}{(0.73)(68+460)} = 0.0752\ \text{lb/ft}^3$$

$$\mu = 1.23 \times 10^{-5}\ \text{lb/(ft} \cdot \text{s)}$$

(**a**) K is given by

$$K = d_p\left(\frac{g(\rho_p - \rho)\rho}{\mu^2}\right)^{1/3} \qquad (7.33)$$

$$= \frac{9.3 \times 10^{-3}}{12}\left((32.2)\frac{[(1.85)(62.4) - 0.0752](0.0752)}{(1.23 \times 10^{-5})^2}\right)^{1/3} = 9.51$$

The intermediate range equation (Equation 7.50) for v should be used:

$$v = 0.153\frac{g^{0.71}d_{\mathrm{p}}^{1.14}\rho_{\mathrm{p}}^{0.71}}{\mu^{0.43}\rho^{0.29}}$$

$$= 0.153\frac{(32.2)^{0.71}[(9.3\times10^{-3})/(12)]^{1.14}[(1.85)(62.4)]^{0.71}}{(1.23\times10^{-5})^{0.43}(0.0752)^{0.29}} \qquad (7.50)$$

$$= 4.08\ \mathrm{ft/s}$$

(b) The Reynolds number is

$$\mathrm{Re} = \frac{d_{\mathrm{p}}v\rho}{\mu}$$

$$= \frac{(9.3\times10^{-3}/12)(4.08)(0.0752)}{1.23\times10^{-5}}$$

$$= 19.33$$

(c) The drag force is calculated from Equation (7.14).

$$F_{\mathrm{D}} = \frac{2.31\pi(vd_{\mathrm{p}})^{1.4}\mu^{0.6}\rho^{0.4}}{g_{\mathrm{c}}}$$

$$= \frac{2.31\pi[(4.08)(9.3\times10^{-3})/12]^{1.4}(1.23\times10^{-5})^{0.6}(0.0752)^{0.4}}{32.2}$$

$$= 2.866\times10^{-8}\ \mathrm{lb_f}$$

Using the conversion factor, 4.448×10^{5} dyn/lb$_{\mathrm{f}}$, yields

$$F_{\mathrm{D}} = 0.0127\ \mathrm{dyn}$$

(d) The value of C_{D} may be calculated from Equation (7.3):

$$C_{\mathrm{D}} = \frac{F_{\mathrm{D}}/A_{\mathrm{p}}}{\rho v^2/2g_{\mathrm{c}}}$$

where A_{p}, the projected area, is given by

$$A_{\mathrm{p}} = \frac{\pi d_{\mathrm{p}}^2}{4} = \frac{\pi(9.3\times10^{-3}/12)^2}{4} = 4.72\times10^{-7}\ \mathrm{ft}^2$$

Therefore

$$C_D = \frac{(2.866 \times 10^{-8})/(4.72 \times 10^{-7})}{(0.0752)(4.08)^2/[(2)(32.2)]}$$
$$= 3.12$$

7.26 Particle Settling Velocity

Three different-sized fly ash particles settle through air. You are asked to calculate the particle terminal velocity and determine how far each will fall in 30 s. Assume that the particles are spherical. Data are provided below:

Fly ash particle diameters = 0.4, 40, 400 μm
Air temperature and pressure = 238°F, 1 atm
Specific gravity of fly ash = 2.31

Solution: For the problem at hand, the particle density is calculated using the specific gravity given:

$$\rho_p = (2.31)(62.4)$$
$$= 144.14\,\text{lb/ft}^3$$

The density of air is

$$\rho = P(\text{MW})/RT$$
$$= (1)(29)/(0.7302)(238 + 460)$$
$$= 0.0569\,\text{lb/ft}^3$$

The viscosity of air is

$$\mu = 0.021\,\text{cP}$$
$$= 1.41 \times 10^{-5}\,\text{lb/(ft} \cdot \text{s)}$$

The value of K for each fly ash particle size setting in air may be calculated. Note that $\rho_p - \rho \approx \rho_p$. For a d_p of 0.4 μm:

$$K = \frac{0.4}{(25{,}400)(12)}\left(\frac{(32.2)(144.1)(0.0569)}{(1.41 \times 10^{-5})^2}\right)^{1/3} = 0.0144$$

For a d_p of 40 μm:

$$K = \frac{40}{(25{,}400)(12)}\left(\frac{(32.2)(144.1)(0.0569)}{(1.41 \times 10^{-5})^2}\right)^{1/3} = 1.44$$

For a d_p of 400 μm:

$$K = \frac{400}{(25{,}400)(12)}\left(\frac{(32.2)(144.1)(0.0569)}{(1.41\times10^{-5})^2}\right)^{1/3} = 14.4$$

Therefore

For $d_p = 0.4$ μm; Stokes' law range
For $d_p = 40$ μm; Stokes' law range
For $d_p = 400$ μm; intermediate law range

For a d_p of 0.4 μm (without the Cunningham correction factor):

$$v = \frac{g d_p^2 \rho_p}{18\mu} = \frac{(32.2)[(0.4)/(25{,}400)(12)]^2(144)}{(18)(1.41\times10^{-5})}$$
$$= 3.15\times10^{-5}\ \text{ft/s}$$

For a d_p of 40 μm:

$$v = \frac{g d_p^2 \rho_p}{18\mu} = \frac{(32.2)[(40)/(25{,}400)(12)]^2(144)}{(18)(1.41\times10^{-5})}$$
$$= 0.315\ \text{ft/s}$$

For a d_p of 400 μm:

$$v = 0.153\frac{g^{0.71} d_p^{1.14} \rho_p^{0.71}}{\mu^{0.43}\rho^{0.29}}$$
$$= 0.153\frac{(32.2)^{0.71}[(400)/(25{,}400)(12)]^{1.14}(144.1)^{0.71}}{(1.14\times10^{-5})^{0.43}(0.0569)^{0.29}}$$
$$= 8.90\ \text{ft/s}$$

The distance that the fly ash particles will fall in 30 s may also be calculated using Equation (7.1).

For a d_p of 40 μm:

$$\text{Distance} = (30)(0.315)$$
$$= 9.45\ \text{ft}$$

For a d_p of 400 μm:

$$\text{Distance} = (30)(8.90)$$
$$= 267\ \text{ft}$$

For a d_p of 0.4 μm, the Cunningham correction factor should be included.

$$A = 1.257 + 0.40e^{-1.10d_p/2\lambda}$$
$$= 1.257 + 0.40\exp - \left(\frac{1.10(0.4)}{(2)(6.53 \times 10^{-2})}\right)$$
$$= 1.2708$$

$$C = 1 + \frac{2A\lambda}{d_p} \quad (7.53)$$
$$= 1 + \frac{(2)(1.2708)(6.53 \times 10^{-2})}{0.4}$$
$$= 1.415$$

The revised velocity and distance are

$$\text{Corrected } v = (3.15 \times 10^{-5})(1.415)$$
$$= 4.45 \times 10^{-5}\ \text{ft/s}$$
$$\text{Distance} = (30)(4.45 \times 10^{-5})$$
$$= 1.335 \times 10^{-3}\ \text{ft}$$

7.27 Particle Size Determination from Particle Settling Velocity

Refer to Problem 7.26. Calculate the size of a fly ash particle that will settle with a velocity of 1.384 ft/s.

Solution: For a particle traveling with a velocity of 1.84 ft/s, first calculate the dimension-less number, *W*. Employ Equation (7.36).

$$W = \frac{v^3\rho^2}{g\mu\rho_p} \quad (7.36)$$
$$= \frac{(1.384)^3(0.0569)^2}{(32.2)(144.1)(1.41 \times 10^{-5})}$$
$$= 0.1312$$

Since $W < 0.222$, Stokes' law applies, and

$$d_p = \left(\frac{18\mu v}{g\rho_p}\right)^{0.5}$$
$$= \left(\frac{(18)(1.41 \times 10^{-5})(1.384)}{(32.2)(144.1)}\right)^{0.5}$$
$$= 2.751 \times 10^{-4}\ \text{ft}$$
$$= 83.9\ \mu\text{m}$$

7.28 Particle Diameter Classification

When collecting particle-size data using an in-stack inertial impactor, the diameter of the particle collected is given as the

(**a**) Aerodynamic diameter of the particle
(**b**) Geometric diameter of the particle
(**c**) Martin diameter
(**d**) Extended diameter of the particle

Solution: Since the particle is captured and/or collected under aerodynamic conditions, the particle size is classified as the aerodynamic diameter. The correct consider is therefore (a).

7.29 Rectangle Equivalent Diameter

Provide an equation describing the equivalent diameter of a rectangle.

Solution: For a rectangular duct cross section, the equivalent diameter (ED) is determined from:

$$\begin{aligned} \text{ED} &= \frac{4(\text{area})}{\text{perimeter}} \qquad (7.45) \\ &= \frac{4(H)(W)}{(2H + 2W)}; \quad H = \text{height},\ W = \text{width} \\ &= \frac{2HW}{L + W} \end{aligned}$$

The selection of a sampling site and the number of sampling points in a stack are based on attempts to obtain representative samples. To accomplish this, the sampling site should be at least eight stack or duct diameters downstream and two diameters upstream from any bend, expansion, construction, valve, fitting, or visible flame. Once the sampling location is chosen, the duct cross section is laid out in a number of equal areas, the center of each being the point where the measurement is to be taken. For rectangular stacks, the cross section is divided into equal areas of the same shape, and the traverse points are located at the center of each equal.

7.30 Tyler Screen Scales

Calculate the average arithmetic size of an 8 × 14 mesh particle size.

Solution: An 8 × 14 mesh particle indicates that the particle will pass through an 8-mesh screen but not pass (be captured) through the 14-mesh screen. Since *one* size cannot be specified for the particle in question, the particle is in the size range 1168–2362 μm (see Table 7.2). The average arithmetic size is therefore 1765 μm.

7.31 Aerodynamic Diameter Calculation

Calculate the aerodynamic diameter (μm) for the following three particles:

(**a**) Solid sphere, equivalent diameter = 1.4 μm, specific gravity = 2.0
(**b**) Hollow sphere, equivalent diameter = 2.8 μm, specific gravity = 0.5
(**c**) Irregular sphere, equivalent diameter = 1.3 μm, specific gravity = 2.35

Solution: As defined earlier, the aerodynamic diameter is defined as the diameter of a sphere of unit density (specific gravity = 1.0) having the same falling speed in air as the particle. The aerodynamic diameter is a function of the physical size, shape, and density of the particle, and is defined by

$$d_{\rm p,a} = d_{\rm p}\sqrt{\rho_{\rm p} C} \tag{7.37}$$

where $d_{\rm p,a}$ = aerodynamic diameter, consistent units;
$d_{\rm p}$ = actual (equivalent) diameter, consistent unit;
$\rho_{\rm p}$ = particle specific gravity, dimensionless; and
C = Cunningham correction factor, dimensionless.

For this purpose and analysis assume $C = 1.0$.

(**a**) For the solid sphere:

$$\begin{aligned} d_{\rm p,a} &= 1.4(2.0)^{0.5} \\ &= 1.98\ \mu\text{m} \end{aligned}$$

(**b**) For the hollow sphere:

$$\begin{aligned} d_{\rm p,a} &= 2.80(0.51)^{0.5} \\ &= 2.0\ \mu\text{m} \end{aligned}$$

(**c**) For the irregular shape:

$$\begin{aligned} d_{\rm p,a} &= 1.3(2.35)^{0.5} \\ &= 1.99\ \mu\text{m} \end{aligned}$$

Conversely, particles with different specific gravity but the same equivalent size have different aerodynamic diameters. For example, if $d_{\rm p} = 2\ \mu$m, the reader is left the exercise of showing that the aerodynamic diameter for particles with specific gravity 1.00, 2.00, 4.00, and 8.00 is 2.00, 2.83, 4.00, and 5.66 μm, respectively. Thus, and as noted earlier, particles of different size and shape can have the same aerodynamic diameter while particles of the same size can have different aerodynamic diameters.

7.32 Duct Flow Equation Derivation

Derive the volumetric flow equation

$$q = \pi R^2 \int_0^{1.0} v d\left(\frac{r}{R}\right)^2 \tag{7.58}$$

for flow in a pipe of radius R and where v represents the local velocity at point r.

Solution: Consider the differential element dA in Figure 7.7. The differential volumetric flow rate (dq) passing differential area dA is given by

$$dq = v\,dA$$

Since

$$dA = r\,dr\,d\phi \text{ (in cylindrical coordinates)}$$

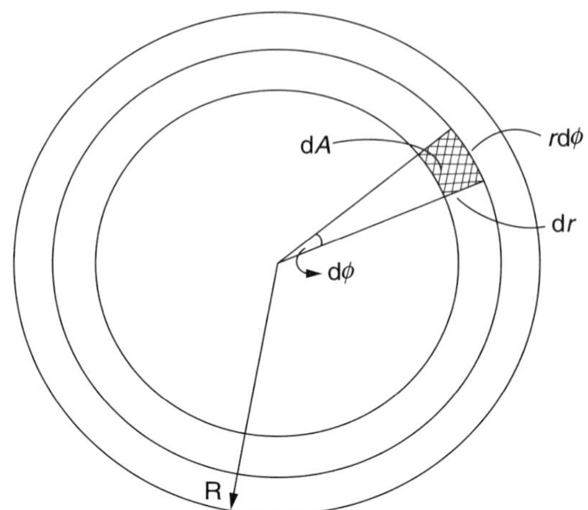

Figure 7.7 Differential element in cylindrical coordinates.

then

$$dq = vr\, dr\, d\phi$$

This equation may be integrated across the entire area to obtain the total volumetric rate (q):

$$q = \int\int dq = \int_0^{2\pi} \int_0^R vr\, dr\, d\phi$$

Since

$$\frac{dr^2}{2} = \frac{2r\, dr}{2} = r\, dr$$

the preceding equation may be rewritten as

$$q = \int_0^{2\pi} \left[\int_0^R v\, d\left(\frac{r^2}{2}\right) \right] d\phi$$

If the outside integral is evaluated, then

$$q = 2\pi \int_0^R v\, d\left(\frac{r^2}{2}\right)$$

Noting that R (or R^2) is a constant, one obtains

$$q = (\pi R^2) \int_0^1 v\, d\left(\frac{r^2}{R}\right)^2; \quad r = R, \left(\frac{r^2}{R}\right)^2 = 1; \quad r = 0, \left(\frac{r^2}{R}\right)^2 = 0$$

The term in parentheses is the area available for flow. The average velocity is the integral; thus, the area under a plot of v versus $(r/R)^2$ represents the average velocity. A smooth curve can be drawn from v versus r data to obtain volumetric flow rate (see Figure 7.8).

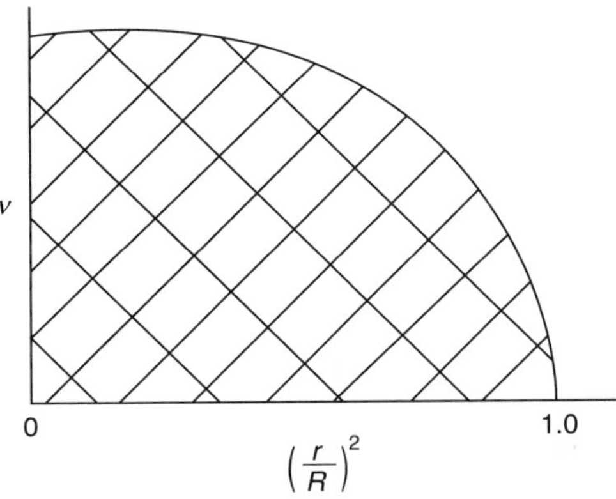

Figure 7.8 Volumetric flow rate calculation.

7.33 Industrial Particle Size Distributions

Which of the following particle size distributions most closely represents discharges from industrial processes:

(**a**) Weibull
(**b**) Normal
(**c**) Lognormal
(**d**) Exponential

Solution: Although all four distributions can be used to describe particle size distributions, the lognormal representation is the best match. The correct answer is therefore (c).

7.34 Lognormal Distribution Plot

A lognormal distribution plot is a straight line on

(**a**) Arithmetic graph paper
(**b**) Semilog paper
(**c**) Log-probability paper
(**d**) Log–log paper

Solution: As discussed in Section 7.5, a lognormal distribution plots out as a straight line on log-probability graph paper. Refer to Section 7.5 for additional details. The correct answer is therefore (c).

7.35 Geometric Mean Particle Diameter

The geometric mean particle diameter occurs at the:

(**a**) 15.87% fraction
(**b**) 50% fraction
(**c**) 84.13% fraction
(**d**) 97.72% fraction

Solution: As one would suppose, the locations of the mean particle size occurs at the 50% weight fraction point. The correct answer is therefore (b).

7.36 Geometric Standard Deviation

The geometric standard deviation for a lognormal distribution is calculated by dividing

(a) 84.13% size/50% size
(b) 50% size/84.13% size
(c) 15.87% size/2.28% size
(d) d_p max/Δlog d_p max

Solution: This is a tricky question, the answer depends on whether the percentages listed above refer to percent less than stated size (%LTSS) or percent greater then stated size (%GTSS); answer (a) would apply to the former, and answer (b) would apply to the latter. The correct answer is therefore either (a) or (b).

7.37 Median and Mean Particle Size

The following particles sizes (in μm) were recorded:

22, 10, 8, 15, 13, 18

Find the median, the arithmetic mean, and the geometric mean of these particle sizes.

Solution: One basic way of smmarizing data is by the computation of a central value. The most commonly used central value statistic is the arithmetic average, or the mean. This statistic is particularly useful when applied to a set of data having a fairly symmetric distribution. The mean is an important statistic in that it summarizes all the data in the set and because each piece of data is taken into account in its computation. The formula for computing the mean is

$$\bar{X} = \frac{X_1 + X_2 + X_3 + \cdots + X_n}{n} = \frac{\sum_{i-1}^{n} X_i}{n} \tag{7.38}$$

where $\bar{X}$ = arithmetic mean
X_i = any individual measurement
n = total number of observations
$X_1, X_2, X_3 \ldots$ = measurements 1, 2, and 3, respectively.

The *arithmetic* mean is not a perfect measure of the true central value of a given data set. Arithmetic means can overemphasize the importance of one or two extreme data points. Many measurements of a normally distributed data set will have an arithmetic mean that closely approximates the true central value. When a distribution of data is asymmetric, it is sometimes desirable to compute a different measure of central value. This second measure, known as the *median*, is simply the middle value of a distribution, or the quantity above which half the data lie and below which the other half lie. If n data points are listed in their order of magnitude, the median is the $[(n+1)/2]$th value. If the number of data is even, then the numerical value of the median is the value midway between the two data nearest the middle. The median, being a positional value, is less influenced by extreme values in a distribution than the mean. However, the median alone is usually not a good measure of central tendency. To obtain the median, the particle size data provided (in μm) should first be arranged in order of magnitude:

8, 10, 13, 15, 18, 22

Thus, the median is 14 μm, or the value halfway between 13 μm and 15 μm since this data set has an even number of measurements. Another measure of central tendency used in specialized applications is the *geometric mean* $\bar{X}_G$. The geometric mean can be calculated using the following equation:

$$\bar{X}_G = {}^n\sqrt{(X_1)(X_2)\cdots(X_n)} \tag{7.59}$$

For the given above particle sizes (substituting d_G (for X), one obtains

$$\bar{d}_G = [(8)(10)(13)(15)(18)(22)]^{1/6} = 13.54\ \mu\text{m}$$

while the arithmetic mean $\bar{d}$ is

$$\begin{aligned}\bar{d} &= (8 + 10 + 13 + 15 + 18 + 22)/6 \\ &= 14.33\ \mu\text{m}\end{aligned}$$

7.38 Standard Deviation

Refer to Problem 7.37. Calculate the standard deviation of the six particle sizes.

Solution: The most commonly used measure of dispersion, or variability, of sets of data is the *standard deviation* σ. Its defining formula is given by expression

$$\sigma = \sqrt{\frac{\sum (X_i - \bar{X})^2}{n-1}} \tag{7.39}$$

where σ = standard deviation (always positive)
X_i = value of ith data point
$\bar{X}$ = mean of data sample
n = number of observations

The expression $(X_i - \bar{X})$ shows that the deviation of each piece of data from the mean is taken into account by the standard deviation. Although the defining formula for the standard deviation gives insight into its meaning, the following algebraically equivalent formula makes computation much easier (now applied to the particle diameter d):

$$\sigma = \sqrt{\frac{\sum (d_i - \bar{d})^2}{n-1}} = \sqrt{\frac{n\sum d_i - \left(\sum d_i\right)^2}{n(n-1)}} \tag{7.39}$$

The standard deviation may be calculated for the data at hand:

$$\sum d_i^2 = (8)^2 + (10)^2 + (13)^2 + (15)^2 + (18)^2 + (22)^2 = 1366$$

$$\left(\sum d_i\right)^2 = (8 + 10 + 13 + 15 + 18 + 22)^2 = 7396$$

Thus

$$\sigma = \sqrt{\frac{6(1366) - 7396}{(6)(5)}} = 5.16\ \mu\text{m}$$

The subject of the standard deviation as well as the mean (see Problem 7.37) is revisited in a later problem.

7.39 The Normal Distribution

Describe the normal distribution and briefly discuss its importance.

Solution: One reason why the normal distribution is so important is that a number of natural phenomena involving particulates are normally distributed or closely approximate it. In fact, many experiments when repeated a large number of times, will produce values that approach the normal distribution curve. In its pure form, the normal curve is a continuous, symmetrical, smooth curve shaped like the one shown in Figure 7.9. Naturally, a finite distribution of discrete data can only approximate this curve.

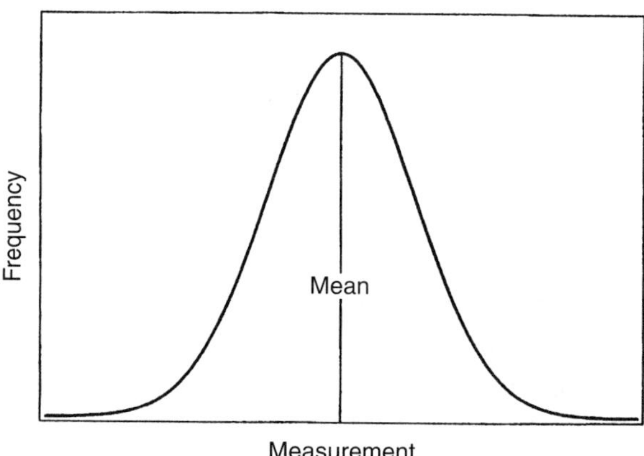

Figure 7.9 Gaussian distribution curve ("normal curve").

The normal curve has the following definite relations to the descriptive measures of a distribution. The normal distribution curve is symmetric; therefore, both the mean and the median are always to be found in the middle of the curve. Recall that, in general, the mean and median of an asymmetric distribution do not coincide. The normal curve ranges along the x axis from minus infinity to plus infinity. Therefore, the range of a normal distribution is infinite. The standard deviation σ becomes a most meaningful measure when related to the normal curve. A total of 68.2% of the area lying under a normal curve is included in the part ranging from one standard deviation below to one standard deviation above the mean. A total of 95.4% lies between -2 to $+2$ standard deviations from the mean (see Figure 7.10). By using tables found in standard statistics texts and handbooks, one can determine the area lying under any part of the normal curve.

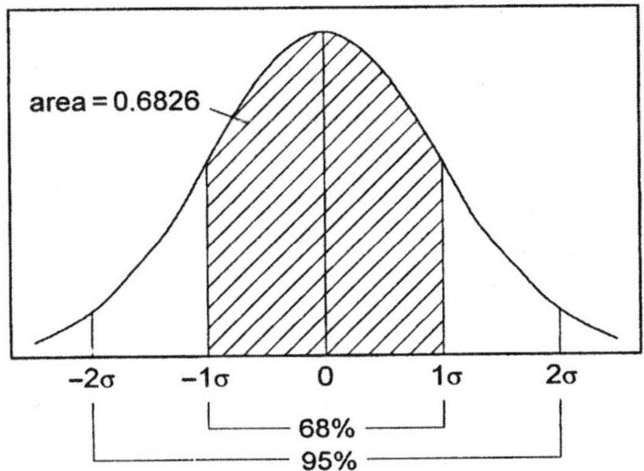

Figure 7.10 Characteristics of the Gaussian distribution.

These areas under the normal distribution curve can be given probability interpretations. For example, if an experiment yields a nearly normal distribution with a mean equal to 30 and a standard deviation of 10, one can expect about 68% of a large number of experimental results to range from 20 to 40, so that the fractional probability of any particular experimental result having a value between 20 and 40 is about 0.68.

Applying the properties of the normal curve to the testing of data and/or readings, one can determine whether a change in the conditions being measured is shown or whether only chance fluctuations in the readings are represented. For a well-established set of data, a frequently used set of control limits is ± 3 standard deviations. Thus, these limits can be used to determine whether the conditions under which the original data were taken have changed. Since the limits of three standard deviations on either side of the mean include 99.7% of the area under the normal curve, it is very unlikely that a reading outside these limits is due to the conditions producing the criterion set of data. The purpose of this technique is to separate the purely chance fluctuations from other causes of variation. For example, if a long series of observations of a measurement yield a mean of 50 and a standard deviation of 10, then control limits can be set up as 50 ± 30, i.e., ± 3 standard deviations, or from 20 to 80. A value above 80 would therefore suggest that the underlying conditions have changed and that a large number of similar observations at this time would yield a distribution of results with a mean different (larger) than 50.

Another important property of normal distributions is as follows. If T is normally distributed with mean μ and standard deviation σ, then the random variable $(T-\mu)/\sigma$ is normally distributed with mean 0 and standard deviation 1. The term $(T-\mu)/\sigma$ is called a *standard normal curve* that is represented by Z. Table 7.9 is a tabulation of areas under a standard normal curve to the right of z_0 for nonnegative values of z_0. Probabilities about a standard normal variable Z can be determined from this table. For example,

$$p(z > 1.54) = 0.062$$

is obtained directly form Table 7.9 as the area to the right of 1.54. As pictured in Figure 7.11, the symmetry of the standard normal curve about zero implies that the area to the right of zero is 0.5, and the area to the left of zero is 0.5. Plots demonstrating the effect of μ and σ on the bell-shaped curve are provided in Figure 7.12 and Figure 7.13. Consequently, one can deduce from Table 7.9 and Figure 7.11 that

$$P(0 < Z < 1.54) = 0.5 - 0.062 = 0.438$$

Also, because of symmetry

$$P(-1.54 < Z < 0) = 0.438$$

and

$$P(Z < -1.54) = 0.062$$

The following probabilities can also be deduced by noting that the area to the right of 1.54 in Figure 7.11 is 0.062. Thus,

$$\begin{aligned} P(-1.54 < Z < 1.54) &= 0.876 \\ P(Z < 1.54) &= 0.938 \\ P(Z > -1.54) &= 0.938 \end{aligned}$$

7.40 Electrostatic Precipitator Failure

A temperature excursion has occurred at an electrostatic precipitator and, on the basis of earlier experiences, the unit will fail in approximately 3 months. Calculate the probability that the unit will fail between 98 and 104 days.

If T, the time to failure in days is normally distributed with mean $\mu = 100$ and standard deviation $\sigma = 2$, then $(T - 100)/2$ is a standard normal variable and one may write

$$P(T_1 < T < T_2) = P\left(\frac{T_1 - \mu}{\sigma} < \frac{T - \mu}{\sigma} < \frac{T_2 - \mu}{\sigma}\right)$$

where

$$\frac{T_1 - \mu}{\sigma} = Z = \text{standard normal variable}$$

TABLE 7.9 Standard Normal, Cumulative Probability[a]

	Next Decimal Place of z_0									
z_0	0	1	2	3	4	5	6	7	8	9
0.0	0.500	0.496	0.492	0.488	0.484	0.480	0.476	0.472	0.468	0.464
0.1	0.460	0.456	0.452	0.448	0.444	0.440	0.436	0.433	0.429	0.425
0.2	0.421	0.417	0.413	0.409	0.405	0.401	0.397	0.394	0.390	0.386
0.3	0.382	0.378	0.374	0.371	0.367	0.363	0.359	0.356	0.352	0.348
0.4	0.345	0.341	0.337	0.334	0.330	0.326	0.323	0.319	0.316	0.312
0.5	0.309	0.305	0.302	0.298	0.295	0.291	0.288	0.284	0.281	0.278
0.6	0.274	0.271	0.268	0.264	0.261	0.258	0.255	0.251	0.248	0.245
0.7	0.242	0.239	0.236	0.233	0.230	0.227	0.224	0.221	0.218	0.215
0.8	0.212	0.209	0.206	0.203	0.200	0.198	0.195	0.192	0.189	0.187
0.9	0.184	0.181	0.179	0.176	0.174	0.171	0.189	0.166	0.164	0.161
1.0	0.159	0.156	0.154	0.152	0.149	0.147	0.145	0.142	0.140	0.138
1.1	0.136	0.133	0.131	0.129	0.127	0.125	0.123	0.121	0.119	0.117
1.2	0.115	0.113	0.111	0.109	0.107	0.106	0.104	0.102	0.100	0.099
1.3	0.097	0.095	0.093	0.092	0.090	0.089	0.087	0.085	0.084	0.082
1.4	0.081	0.079	0.078	0.076	0.075	0.074	0.072	0.071	0.069	0.068
1.5	0.067	0.066	0.064	0.063	0.062	0.061	0.059	0.058	0.057	0.056
1.6	0.055	0.054	0.053	0.052	0.051	0.049	0.048	0.047	0.046	0.046
1.7	0.045	0.044	0.043	0.042	0.041	0.040	0.039	0.038	0.038	0.037
1.8	0.036	0.035	0.034	0.034	0.033	0.032	0.031	0.031	0.030	0.029
1.9	0.029	0.028	0.027	0.027	0.026	0.026	0.025	0.024	0.024	0.023
2.0	0.023	0.022	0.022	0.021	0.021	0.020	0.020	0.019	0.019	0.018
2.1	0.018	0.017	0.017	0.017	0.016	0.016	0.015	0.015	0.015	0.014
2.2	0.014	0.014	0.013	0.013	0.013	0.012	0.012	0.012	0.011	0.011
2.3	0.011	0.010	0.010	0.010	0.010	0.009	0.009	0.009	0.009	0.008
2.4	0.008	0.008	0.008	0.008	0.007	0.007	0.007	0.007	0.007	0.006
2.5	0.006	0.006	0.006	0.006	0.006	0.005	0.005	0.005	0.005	0.005
2.6	0.005	0.005	0.004	0.004	0.004	0.004	0.004	0.004	0.004	0.004
2.7	0.003	0.003	0.003	0.003	0.003	0.003	0.003	0.003	0.003	0.003
2.8	0.003	0.002	0.002	0.002	0.002	0.002	0.002	0.002	0.002	0.002
2.9	0.002	0.002	0.002	0.002	0.002	0.002	0.002	0.001	0.001	0.001
	Details of Tail ($._2135$, (e.g.) = 0.00135)									
2.	$0._1228$	$0._1179$	$0._1139$	$0._1107$	$0._2820$	$0._2621$	$0._2466$	$0._2347$	$0._2256$	$0._2187$
3.	$0._2135$	$0._3968$	$0._3687$	$0._3483$	$0._3337$	$0._3233$	$0._3159$	$0._3108$	$0._4723$	$0._4481$
4.	$0._4317$	$0._4207$	$0._4133$	$0._5854$	$0._5541$	$0._5340$	$0._5211$	$0._5130$	$0._6793$	$0._6479$
5.	$0._6287$	$0._670$	$0._7996$	$0._7579$	$0._7333$	$0._7190$	$0._7107$	$0._8599$	$0._8332$	$0._8182$
0	1	2	3	4	5	6	7	8	9	

[a]Standard normal, cumulative probability is shown in right-hand tail in diagram above the table; for negative values of z, areas are found by symmetry.

[b]*Source*: R. J. Woonacott and T. H. Woonacott, *Introductory Statistics*, 4th ed., John Wiley & Sons, New York, 1985.

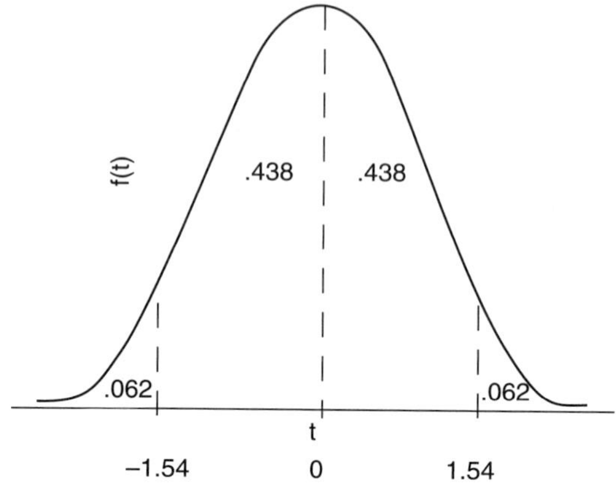

Figure 7.11 Areas under a standard normal curve.

Solution: Set $T_1 = 98$ and $T_2 = 104$ for this application. The describing equation given above becomes

$$
\begin{aligned}
P(98 < T < 104) &= \left(\frac{98-\mu}{\sigma} < \frac{T-\mu}{\sigma} < \frac{104-\mu}{\sigma}\right) \\
&= P\left(\frac{98-100}{2} < \frac{T-\mu}{2} < \frac{104-100}{2}\right) \\
&= P\left(-1 < \frac{T-100}{2} < 2\right) \\
&= P(-1 < Z < 2)
\end{aligned}
$$

From Table 7.9,

$$
\begin{aligned}
P(98 < T < 104) &= 0.341 + 0.477 \\
&= 0.818 \\
&= 81.8\%
\end{aligned}
$$

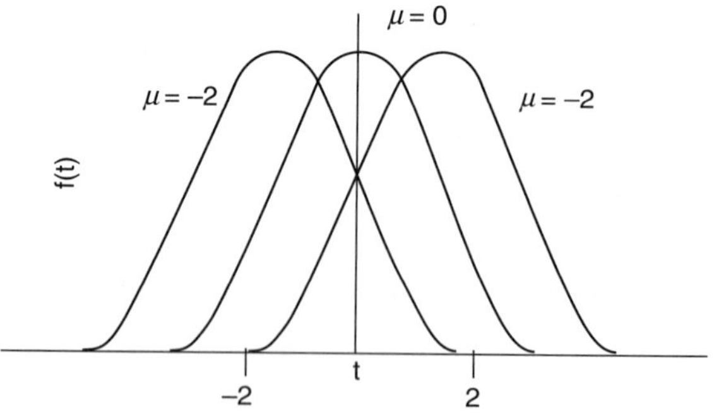

Figure 7.12 Normal probability distribution function (pdf)—varying μ.

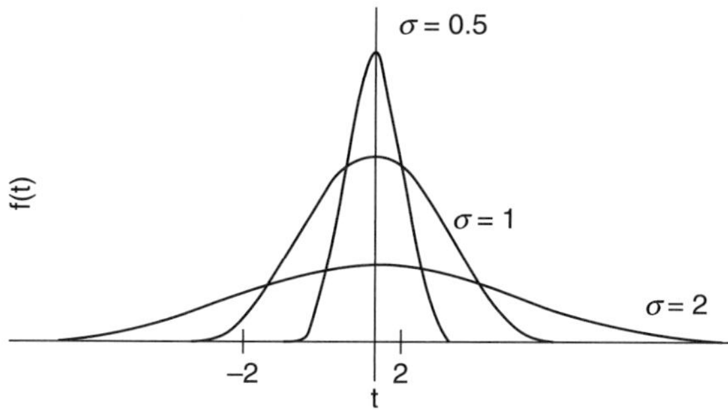

Figure 7.13 Normal pdf—varying

7.41 Lognormal Distribution

You have been requested to determine whether the particle size distribution given in Table 7.10 is lognormal.

TABLE 7.10 Particle Size Distribution Data

Particle Size Range d_p, μm	Distribution, μg/m^3
<0.62	25.5
0.62–1.0	33.15
1.0–1.2	17.85
1.2–3.0	102.0
3.0–8.0	63.75
8.0–10.0	5.1
>10.0	7.65
Total	255.0

Solution: Cumulative distribution plots are generated by plotting particle diameter versus cumulative percent. For lognormal distributions, plots of particle diameter versus either percent less than stated size (%LTSS) or percent greater than stated size (%GTSS) produce straight lines on log-probability coordinates.

Cumulative distribution information can be obtained from the calculated results provided in Table 7.11. The cumulative distribution in Table 7.11 can

TABLE 7.11 Information for Particle Size Distribution Plot

d_p, μm	% Total	Cumulative %GTSS
<0.62	10	90
0.62–1.0	13	77
1.0–1.2	7	70
1.2–3.0	40	30
3.0–8.0	25	5
8.0–10.0	2	3
>10.0	3	0

be plotted on log-probability paper. The cumulative distribution curve is shown in Figure 7.14. Since a straight line is obtained on log-normal coordinates, the particle size distribution is log normal.

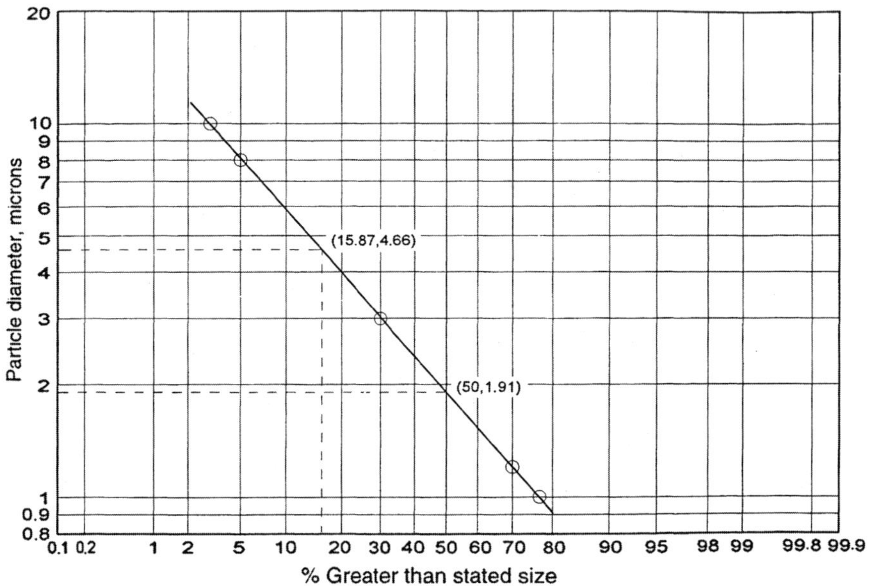

Figure 7.14 Cumulative distribution curve.

7.42 Mean and Standard Deviation of Particle Size Distribution

With reference to Problem 7.41, estimate the mean and standard deviation from the size distribution information available.

Solution: By definition, the size corresponding to the 50% point on the probability scale is the *geometric mean diameter*. The geometric standard deviation is given (for %LTSS) by

$$\sigma = 84.13\% \text{ size}/50\% \text{ size} \tag{7.40}$$

or

$$\sigma = 50\% \text{ size}/15.87\% \text{ size}$$

For %GTSS,

$$\sigma = 50\% \text{ size}/84.13\% \text{ size} \tag{7.40}$$

or

$$\sigma = 15.87\% \text{ size}/50\% \text{ size}$$

The mean, as read from the 50% GTSS point on the graph in Figure 7.14 is approximately 1.9 μm. A value of 1.91 μm us obtained from an expanded plot. From the diagram, the particle size corresponding to the 15.87% point is

$$d_p(15.87\%) = 4.66\ \mu\text{m}$$

The standard deviation may now be calculated.

$$\begin{aligned}\sigma &= d_p(15.87\%)/d_p(50\%)\\ &= 4.66/1.91\\ &= 2.44\end{aligned}$$

7.43 Varying Particle Size Distributions

Qualitatively describe various lognormal particle size distributions on log-probability coordinates.

Solution: The representation of particle size analyses on probability coordinates has the advantage of simple extrapolation or interpolation from a minimum of data. The slope of the line is a measure of the breadth of the distribution of sizes in the sample. For example, in Figure 7.15, curve A represents a batch of uniform-size particles; curve B has the same mean particle size with perhaps a 10-fold variation in diameter between the largest and smallest particles, whereas curve C may have a 20-fold variation. Curve D has the same 10-fold range of sizes as curve B, but in each weight percent category the particles are larger, so that curve D is a coarser grind than curve B but equal in range of sizes.

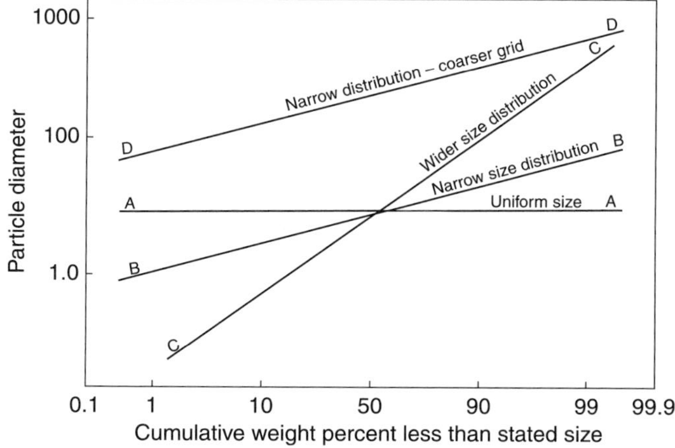

Figure 7.15 Effect of size distribution on relative probability plots.

7.44 Anderson 2000 Sampler

Given Anderson 2000 sampler data from a particulate sidestream emission from a nanoprocess, you have been requested to plot a cumulative distribution curve on log-probability paper and determine the mean diameter and geometric standard deviation of the particulate emission. Pertinent data are provided in Table 7.12.

TABLE 7.12 Anderson 2000 Sampler Data

Plate Number	Tare Weight, g	Final Weight, g
0	20.48484	20.48628
1	21.38338	21.38394
2	21.92025	21.92066
3	21.55775	21.55817
4	11.40815	11.40854
5	11.61862	11.61961
6	11.76540	11.76664
7	20.99617	20.99737
Backup filter	0.20810	0.21156

$$\text{Sampler volumetric flow rate } q = 0.5 \text{ cfm}$$

See also Figure 7.16 for aerodynamic diameter vs. flow rate data for an Anderson sampler.

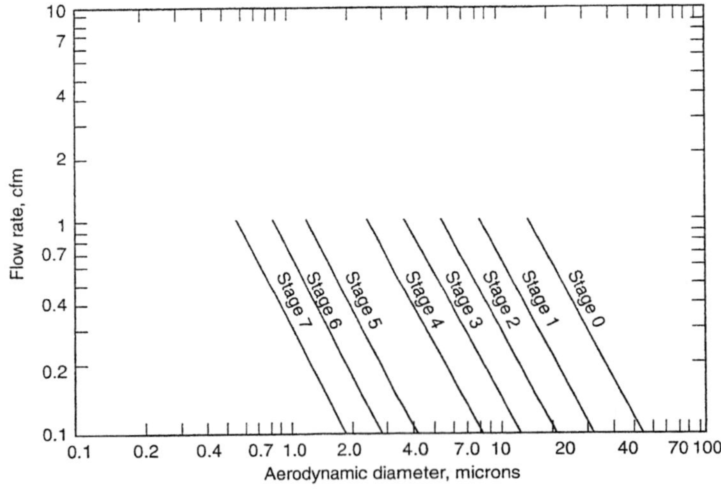

Figure 7.16 Aerodynamic diameter vs. flow rate through Anderson sampler for an impaction efficiency of 95%.

Solution: The Anderson sampler consists of a series of stacked stages and collection surfaces. Depending on the calibration requirements, each stage contains 150–400 precisely drilled jet orifices, identical in diameter in each stage but decreasing in diameter on each succeeding stage. A constant flow of air is drawn through the sampler so that as the air passes from stage to stage through the progressively smaller holes, the velocity increases as the air stream makes a turn at each stage; thus, the particle gains enough inertia to lose its aerodynamic drag. It is "hurled" from the air stream and impacted on the collection surface. In effect, the particle is considered aerodynamically sized the moment it leaves the turning air stream. Adhesive, electrostatic, and van der Waal forces hold the particles to each other and to the collection surface.

Table 7.13 provides the net weight (in mg), percent of total weight, and cumulative percent for each plate. A sample calculation (for plate 0) follows:

$$\begin{aligned}\text{Net weight} &= (\text{final weight}) - (\text{tare weight})\\ &= 20.48628 - 20.48484\\ &= 1.44 \times 10^{-3}\ \text{g}\\ &= 1.44\ \text{mg}\\ \text{Percent of total wt} &= (\text{net wt}/\text{total net wt})(100)\\ &= (1.44/10.11)(100\%)\\ &= 14.2\%\end{aligned}$$

Calculate the cumulative percent for each plate. Again for plate 0,

$$\begin{aligned}\text{Cumulative }\% &= 100 - 14.2\\ &= 85.8\%\end{aligned}$$

For plate 1, one obtains

$$\begin{aligned}\text{Cumulative }\% &= 100 - (14.2 + 5.5)\\ &= 80.3\end{aligned}$$

TABLE 7.13 Anderson 2000 Sampler Data

Plate	Number Net Weight, mg	Percent of Total Weight
0	1.44	14.2
1	0.56	5.5
2	0.41	4.1
3	0.42	4.2
4	0.39	3.9
5	0.99	9.8
6	1.24	12.3
7	1.20	11.9
Backup filter	3.46	34.2
Total	10.11	100.0

Table 7.14 shows the cumulative percent for each plate.

TABLE 7.14 Cumulative Percent for Each Plate

Plate No.	Cumulative Percent
0	85.8
1	80.3
2	76.2
3	72.0
4	68.1
5	58.3
6	46.0
7	34.1
Backup filter	—

Using the Anderson graph in Figure 7.16 determine the 95% aerodynamic diameter at $q = 0.5$ cfm for each plate or stage (see Table 7.15). The cumulative distribution curve is provided on log-probability coordinates in Figure 7.17. The mean particle diameter is the particle diameter corresponding to a cumulative percent of 50%.

$$\text{Mean particle diameter} = d_{50} = 1.6\,\mu\text{m}$$

TABLE 7.15 95% Aerodynamic Diameters

Plate No.	95% Aerodynamic Diameter, μm
0	20.0
1	13.0
2	8.5
3	5.7
4	3.7
5	1.8
6	1.2
7	0.78

The distribution approaches lognormal behavior. The particle diameter at a cumulative percent of 84.13 is

$$d_{84.13} = 15\,\mu\text{m}$$

Therefore, the geometric standard deviation is

$$\begin{aligned}\sigma_G &= d_{84.13}/d_{50} \\ &= 15/1.6 \\ &= 9.4\end{aligned}$$

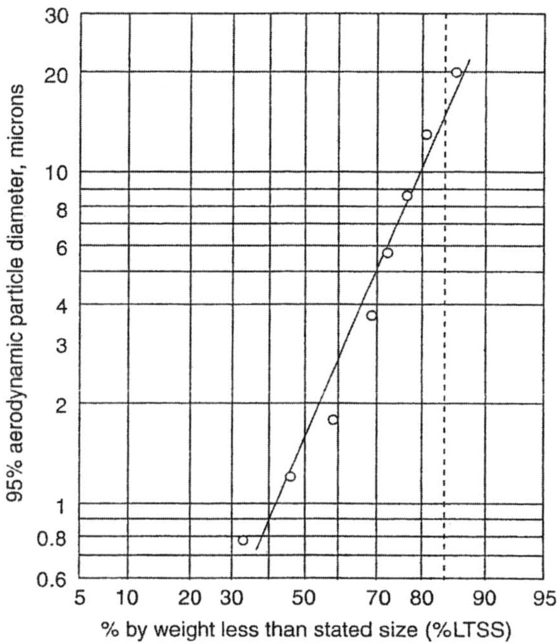

Figure 7.17 Cumulative distribution curve.

7.45 Effectiveness of Control Equipment

The effectiveness of control equipment for different particle sizes is shown by

(**a**) Size–efficiency curves
(**b**) Overall efficiency
(**c**) Log-probability plots
(**d**) Cumulative distribution curves

Solution: Answers (c) and (d) can be eliminated immediately. Although the overall efficiency is indirectly dependent on particle size, it does specifically relate to the relationship between effectiveness and size. The correct answer is therefore (a).

7.46 Collection Efficiency of 50%

If a particulate control device has a collection efficiency of 50%, this generally means

(**a**) The device will remove 50% of the particles.
(**b**) The device will remove 50% of the total mass of particulate matter.
(**c**) Either (a) or (b), since they mean the same thing.
(**d**) None of the above.

Solution: Collection efficiency, unless otherwise stated, is almost always based on mass (or weight). The correct answer is therefore (b).

7.47 Collection Efficiency: Mass Rate

The hazardous particle waste flow rate into a treatment device is 100 lb/hr. Calculate the waste rate leaving the unit to achieve a collection efficiency of

(**a**) 95%
(**b**) 99%
(**c**) 99.9%
(**d**) 99.99%
(**e**) 99.9999%

Solution: The definition of collection efficiency, E, in terms of mass flow rate in $\dot{m}_{\text{in}}$ and mass flow rate out $\dot{m}_{\text{out,}}$ is

$$E = \left(\frac{\dot{m}_{\text{in}} - \dot{m}_{\text{out}}}{\dot{m}_{\text{in}}}\right)(100) \tag{7.42}$$

This equation may be rewritten for $\dot{m}_{\text{out}}$:

$$\dot{m}_{\text{out}} = \dot{m}_{\text{in}}(100 - E)/100$$

(**a**) The mass flow rate out $\dot{m}_{\text{out}}$ for an E of 95% is

$$\begin{aligned}\dot{m}_{\text{out}} &= 100(100 - 95)/100\\ &= 5\ \text{lb/hr}\end{aligned}$$

(**b**) The mass flow rate out $\dot{m}_{\text{out}}$ for an E of 99% is

$$\begin{aligned}\dot{m}_{\text{out}} &= 100(100 - 99)/100\\ &= 1\ \text{lb/hr}\end{aligned}$$

(**c**) The mass flow rate out $\dot{m}_{\text{out}}$ for an E of 99.9% is

$$\begin{aligned}\dot{m}_{\text{out}} &= 100(100 - 99.9)/100\\ &= 0.1\ \text{lb/hr}\end{aligned}$$

(**d**) The mass flow rate out $\dot{m}_{\text{out}}$ for an E of 99.99% is

$$\begin{aligned}\dot{m}_{\text{out}} &= 100(100 - 99.99)/100\\ &= 0.01\ \text{lb/hr}\end{aligned}$$

(**e**) The mass flow rate out $\dot{m}_{\text{out}}$ for an E of 99.9999% is

$$\begin{aligned}\dot{m}_{\text{out}} &= 100(100 - 99.9999)/100\\ &= 0.0001\ \text{lb/hr}\end{aligned}$$

7.48 Penetration

Define penetration.

Solution: An extremely convenient efficiency-related term employed in particulate control calculations is the penetration P. By definition:

$$P = 100 - E; \quad \text{percent basis}$$
$$P = 1 - E; \quad \text{fractional basis} \tag{7.60}$$

Note that there is a 10-fold increase in P as E goes from 99.9% to 99%. For a multiple series of n collectors, the overall penetration is simply given by

$$P = P_1 P_2 \cdots P_{n-1} P_n \tag{7.61}$$

For particulate control, penetrations and/or efficiencies can be related to individual size ranges. The overall efficiency (or penetration) is then given by the contribution from each size range, obtained from the summation of the product of mass fraction and efficiency for each size range. This is examined in more detail later in Problems 7.49–7.52.

7.49 Penetration Calculation: Percent Basis

The efficiency of an air pollution control device is 99.52%. Calculate the penetration in percent, for the unit.

(**a**) 0.48
(**b**) 0.0118
(**c**) 0.68
(**d**) None of the above

Solution: By definition, the penetration is related to the efficiency through Equation (7.60).

$$P = 1.0 - E \tag{7.60}$$

This equation applies on a fractional basis, and can be written as

$$P = 100 - E \tag{7.60}$$

if based on percentages. On the basis of the problem statement, apply Equation (7.60).

$$P = 100 - 99.52$$
$$= 0.48$$

The correct answer is therefore (a).

7.50 Efficiency Calculation Based on Loading Data

The inlet and outlet loading of a particulate pollutant in an air pollution control device have been measured experimentally to be 2.70 and 0.036 gr/ft^3 (grains per cubic foot), respectively. Calculate the efficiency of the unit.

(**a**) 98.56%
(**b**) 99.84%
(**c**) 98.67%
(**d**) None of the above

Solution: Apply Equation (7.42).

$$E = \left(\frac{2.70 - 0.036}{2.70}\right)100 \tag{7.42}$$
$$= (0.9867)100$$
$$= 98.67\%$$

The corresponding penetration is 1.33% or 0.0133 on a fractional basis. The correct answer is therefore (c).

7.51 Efficiency of Two Control Devices in Series

Two particulate air pollution control devices operate in series with collection efficiencies of 90% and 99.5%, respectively. Calculate the overall efficiency of the two units.

(**a**) 99.52%
(**b**) 98.67%
(**c**) 99.84%
(**d**) 99.95%

Solution: This problem is best solved by noting the overall penetration given by the product of the penetration for each device:

$$P = P_1 P_2; \quad \text{fractional basis} \tag{7.61}$$

According to the data given

$$P_1 = 1 - 0.9$$
$$= 0.1$$

and

$$P_2 = 1 - 0.995$$
$$= 0.005$$

Therefore

$$P = (0.1)(0.005)$$
$$= 0.0005; \quad \text{fractional basis}$$
$$= 0.05; \quad \text{percent basis}$$

Finally

$$\begin{aligned} E &= 100 - P \\ &= 100 - 0.05 \\ &= 99.95\% \end{aligned}$$

The correct answer is therefore (d).

7.52 Efficiency of Multiple Collectors

A cyclone is used to collect particulates with an efficiency of 60%. A venturi scrubber is used as a second downstream control device. Given a required overall efficiency of 99.0%, determine the minimum operating efficiency of the venturi scrubber.

Solution: Calculate the mass of particulate leaving the cyclone using a rate basis of 100 lb of particulate entering the unit. Use the efficiency equation:

$$E = (\dot{m}_{\text{in}} - \dot{m}_{\text{out}})/(\dot{m}_{\text{in}}) \tag{7.42}$$

Rearranging and substituting gives

$$\dot{m}_{\text{out}} = (1 - E)(\dot{m}_{\text{in}}) = (1 - 0.6)(100) = 40\,\text{lb}$$

Calculate the mass of particulate leaving the venturi scrubber using an overall efficiency of 99.0% (0.99, fractional basis):

$$\dot{m}_{\text{out}} = (1 - E)(\dot{m}_{\text{in}}) = (1 - 0.99)(100) = 1.0\ \text{lb}$$

Calculate the required efficiency of the venturi scrubber using $\dot{m}_{\text{out}}$ from the cyclone as $\dot{m}_{\text{in}}$ for the venturi scrubber. Use the same efficiency equation above and convert to percent efficiency:

$$E = (\dot{m}_{\text{in}} - \dot{m}_{\text{out}})/(\dot{m}_{\text{in}}) = (40 - 1.0)/(40) = 0.975 = 97.5\%$$

The corresponding penetration is 2.5%.

7.53 Collection Efficiency: Surface Area Basis

The size–collection efficiency information for a control device is provided in Table 7.16. Calculate the overall efficiency for the unit on a

(a) Number basis
(b) Mass basis
(c) Volume basis
(d) Surface area basis

Solutions: The reader is referred to Section 7.6 for more details.

TABLE 7.16 Size–Collection Efficiency Data

d, μm	Collection Efficiency, %	Number of Particles
1.0	0	10,000
100	100	1

(**a**) On a number basis

$$E_{\mathrm{N}} = \frac{(10{,}000)(0) + (1)(1.0)}{10{,}000 + 1}$$

$$= -\frac{1}{10{,}000 + 1} = \frac{1}{10{,}001}$$

$$= 0.0001 = 0.01\%$$

(**b**) On a mass basis

$$E = \frac{(1)(100)^3(1.0) + (10{,}000)(1.0)^3(0)}{(1)(100)^3 + (10{,}000)(1.0)^3}$$

$$= \frac{10^6 + (0)}{10^6 + 10^4}$$

$$= 0.99 = 99\%$$

(**c**) On a volume basis

$$E_{\mathrm{V}} = 0.99 = 99\%$$

since volume is proportional to mass.

(**d**) On a surface area basis, one notes that the surface area of the particle is proportional to the square of the diameter; therefore

$$E_{\mathrm{SA}} = \frac{(1)(100)^2(1.0) + (10{,}000)(1.0)^2(0)}{(1)(100)^2 + (10{,}000)(1.0)^2}$$

$$= \frac{10^4}{10^4 + 10^4}$$

$$= 0.5 = 50\%$$

7.54 Particle Size Distribution/Size–Efficiency Calculation

The following data are provided in Table 7.17. Calculate the overall collection efficiency.

TABLE 7.17 Particle Size Distribution and Efficiency Information

PSR, μm	% in PSR	%LTSS	%GTSS	$\bar{d}_p$, μm	E_i, %
0–5	13	13	87	2.5	0.6
5–10	12	25	75	7.5	7.7
10–15	9	34	66	12.5	19.2
15–20	8	42	58	17.5	38.3
20–30	9	51	49	25.0	56.2
30–50	11	62	38	40.0	79.1
50–100	13	75	25	75.0	98.9
100+	25	—	—	100.0+	100

[a]*Key*:PSR = particle size range; % in PSR = percent in particle size range; %LTSS = percent less than stated size; %GTSS = percent greater than stated size; $\bar{d}_p$ = average particle size in PSR; E_i = mass efficiencies in PSR. Note that columns 1–4 represent particle size distribution data (see Section 7.4). Columns 5 and 6 provide size–efficiency information.

Solution: Refer to Table 7.17. The overall collection efficiency is obtained from the cross-product of columns 2 and 6 for each PSR and summing the results:

$$E = \sum_{i=1}^{n} [(2) \times (6)]/100; \quad \text{percent basis}$$

The calculated results are provided in Table 7.18.

TABLE 7.18 Calculation of Overall Collection Efficiency

PSR, μm	2 × 6
0–5	7.8
5–10	92.4
10–15	172.8
15–20	306.4
20–30	505.8
30–50	870.1
50–100	1285.7
100+	2500
	5741

From Table 7.18, one obtains

$$E = \frac{5741}{100} = 57.41\%$$
$$= 0.5741; \quad \text{fractional basis}$$

7.55 Check for Emission Standards Compliance: Numbers Basis

As a consulting engineer, you have been contracted to modify an existing control device used in a nano feedstock byproduct emission removal. The federal standards for emissions have been changed to a total numbers basis. Determine if the unit will meet an effluent standard of $10^{5.7}$ particles/acf. Data for the unit are given below.

Average particle size, $d_p = 10\ \mu m$; assume constant
Byproduct specific gravity = 2.33
Inlet loading = $3.0\ gr/ft^3$
Efficiency (mass basis), $E = 99\%$

Solution: The outlet loading (OL) is

$$\begin{aligned} OL &= (1.0 - 0.99)3.0 \\ &= 0.030\ gr/ft^3 \end{aligned}$$

Assume a basis of $1.0\ ft^3$ so that the outlet mass is 0.03 gr.

$$\begin{aligned} \text{Particle mass} &= \rho_p V_p = \rho_p \frac{\pi d_p^3}{6} \\ &= \frac{(\pi[(10\ \mu m)(0.328 \times 10^{-5}\ ft/\mu m)]^3(2.33)(62.4\ lb/ft^3))}{6} \\ &\quad \times (7000\ gr/lb) \\ &= 1.880 \times 10^{-8}\ gr/particle \\ \text{Number of particles} &= (0.03\ gr)/(1.880 \times 10^{-8}\ gr/particle) \\ &= 1.596 \times 10^6 \text{ particles in } 1\ ft^3 \end{aligned}$$

$$\begin{aligned} \text{Allowable number of particles}/ft^3 &= 10^{5.7} \\ &= 5.01 \times 10^5 \end{aligned}$$

Therefore, the unit will not meet a numbers standard.

NOTE: Additional problems are available for all readers at www.wiley.com. Follow links for this title.

8

GRAVITY SETTLING CHAMBERS

8.1 INTRODUCTION

The gravity settler was one of the first devices used to control particulate emissions. It is an expansion chamber in which the gas velocity is reduced, thus allowing the particle to settle out under the action of gravity. One primary feature of this device is that the external force causing separation of particles from the gas stream is provided free by nature. This chamber's use in industry, however, is generally limited to the removal of larger-sized particles, e.g., 40–60 μm in diameter. Settling chambers have also been used to study the flow of particles in a gas stream. The data generated from these studies can be useful in the design of other particulate emission control devices.

Today's demand for cleaner air and stricter emission standards has relegated the settling chamber to use in research for testing or as a precleaner/postcleaner for other particulate control devices (cyclones, electrostatic precipitators, scrubbers, and fabric filters).

Inertial collectors, on the other hand, depend on another effect, in addition to gravity, to lead to a successful separation process. This other mechanism is an inertial or momentum effect. It arises by changing the direction of the velocity of the gas and imparting a downward motion to the particle. From a calculational point of view, this induced particle motion is superimposed on the motion arising as a result of gravity.

Air Pollution Control Equipment Calculations. By Louis Theodore

These control devices are also reviewed in this chapter since this method of particle collection can be applied to other control devices, some of which are reviewed in later chapters.

There are basically two types of dry gravity settlers: the simple expansion chamber and the multiple-tray settling chamber. A typical horizontal flow (simple expansion) gravity settling chamber is presented in Figure 8.1. The unit is constructed in the form of a long horizontal box with an inlet, an outlet, and dust collection hoppers. These units primarily depend on gravity for collection of the particles. The particle-laden gas stream enters the unit at the gas inlet. The gas stream then enters the expansion section of the duct. Expansion of the gas stream causes the gas velocity to be reduced. All particles in the gas stream are subject to the force of gravity. However, at reduced gas velocities (in the range of 1.0–10.0 ft/s) the larger particles are acted on preferentially by gravity and fall into the dust hopper(s). Theoretically, a settling chamber of infinite length could collect even the very small particles ($<10\,\mu m$).

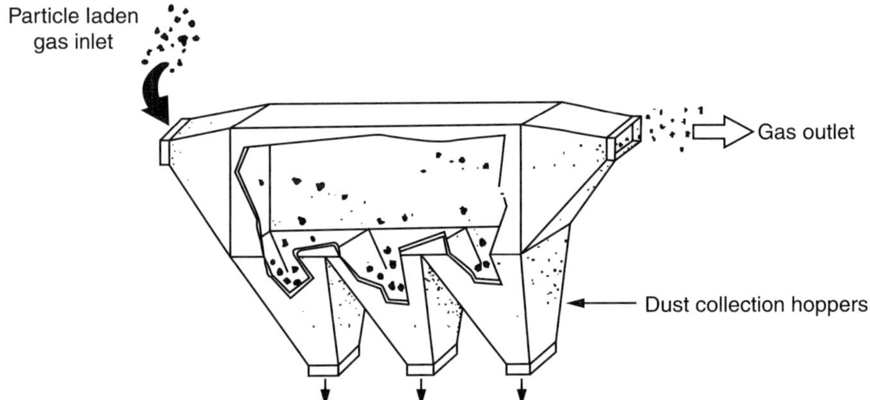

Figure 8.1 Horizontal flow settling chamber.

Since the effective settling rate of the dust decreases with increasing gas turbulence, the velocity of the gas stream in the settling chamber is normally kept as low as possible. For practical purposes, the velocity must not be so high that settled particles are reentrained or so low that the chamber volume becomes excessive. In actual practice, the gravitational settling velocities used in the design must be based on experience or tests conducted under actual conditions because the terminal settling velocity may be influenced by factors such as agglomeration and electrostatic charge. Pressure drops experienced in settling chambers are quite low, generally less than 0.2 in H_2O.

The physical description of a gravity settler includes (1) length, (2) width, (3) height, (4) number of shelves (if applicable), and (5) ancillary equipment—inlet and outlet ducts, cleaning mechanisms, hoppers, etc. Regarding the latter ancillary equipment, the collection hoppers (located at the bottom of the settler) are usually designed with positive-seal valves and must be emptied as dust buildup occurs. Dust buildup will vary depending on the concentration levels of particulate matter in the gas streams, especially in the case of heavy concentrations of particles greater than $60\,\mu m$ in diameter.

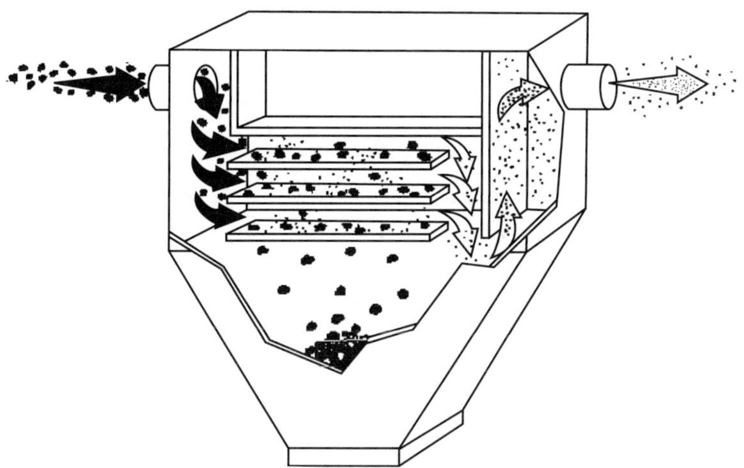

Figure 8.2 Howard settling chamber (multiple tray).

The multiple-tray settling chamber also called the *Howard settling chamber*, is shown in Figure 8.2. Several horizontal collection plates N are introduced to shorten the settling path of the particle and to improve the collection efficiency of small particles (as small as 15 μm in diameter). Each shelf or tray in the unit can collect dust that settles out by gravitational force. Since the vertical distance that a particle must fall to be captured is less than the distance in a standard horizontal settling unit, the overall collection efficiency of the Howard settling chamber can be greater than that of the horizontal chamber. The gas must be uniformly distributed as it passes over each tray throughout the chamber. Uniform distribution is usually achieved by the use of gradual transitions, guide vanes, distributor screens, and perforated plates. The particles settle on the individual trays, which must be cleaned periodically. The vertical distance between trays may be as little as 1 inch, making cleaning much more difficult than with the horizontal settling chamber. Other disadvantages include the tendency of trays to warp during high-temperature operations, and the inability of the unit to handle dust concentrations exceeding approximately 1 gr/ft^3 (2.29 g/m^3). For these reasons, the Howard settling chamber is rarely used for particulate emission control.

A variation of the gravity settling chamber is a baffle chamber, sometimes referred to as an *inertial separator*. These units have baffles within the chamber to enhance particle separation and collection. This arises by changing the direction of the gas velocity and imparting a downward motion to the particle. This induced motion is superimposed on the motion due to gravity. Thus, particle collection is accomplished by gravity and an inertial or momentum effect. Particles as small as 20–40 μm can be collected. An example of this device is shown in Figure 8.3. These units are more compact and require less space than gravity settling chambers. The pressure drops are slightly higher, ranging from 0.1 to 1.0 in H_2O (0.25–2.5 cm H_2O).

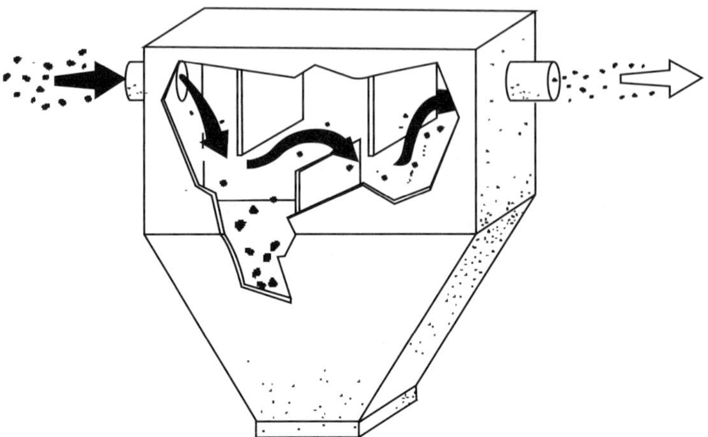

Figure 8.3 Baffle chamber.

An elutriator is a slight modification of the gravity settler. The unit consists of one or more vertical tubes or towers through which the dust-laden gas passes upward at a given velocity. The larger particles that settle at a velocity higher than that of the rising air are collected at the bottom of the tube, while the smaller particles are carried out the top. In order to vary the air velocity, several columns of different diameters are used in series to bring about more refined separation.

Another modification of the basic gravity settling chamber is the gravity spray tower, a wet unit that employs water. As discussed in Section 7.2, when a moving dust-laden gas stream approaches a body, such as spherical droplet of water, the gas will be deflected around the droplet. The dust particles, by virtue of their greater inertia, may impact and be collected on the surface of the droplet that is falling under the influence of gravity. The design and operation of the gravity spray tower utilizes this collection phenomenon. This unit consists of a vertical tower or column. The particulate-laden gas is fed to the bottom of the column. A liquid stream (usually water) is introduced at the top of the column after passing through spray nozzles. The water droplets produced fall vertically downward through the column under the influence of gravity. During their fall the droplets contact and capture the particles in the gas stream. The system is capable of treating larger volumetric gas flow rates and has few mechanical problems. It operates at low pressure drops (usually less than 1 in H_2O) with air velocities in the neighborhood 2–5 ft/s and residence times of approximately 15–30 s. It is capable of handling relatively high dust loadings (greater than 5 gr/ft^3). In addition to the collection of particles (down to the 10–20 μm range), the gravity spray tower can also be utilized to absorb noxious pollutant gases (depending on the liquid reagent used and solubility characteristics) and is often employed as a precooler. The main disadvantage centers on the liquid effluent treatment and/or waste disposal as the air pollution problem is replaced by a water pollution problem. Water usage typically ranges between 5 and 20 gal/1000 ft^3 gas treated (with a typical value around 18 gal/1000 ft^3). Another disadvantage is the possibility of liquid entrainment in the gas stream leaving the tower. However, a demister, mist eliminator, or entrainment separator installed above the spray nozzles (especially when gas velocities are greater than 6 ft/s), will often eliminate this liquid entrainment problem. Similar to the gravity settlers, gravity spray towers have rather large space requirements.

8.2 DESIGN AND PERFORMANCE EQUATIONS

The fundamentals governing particle collection in a gravity settler are now presented. Although the settler is initially assumed to contain no trays, the analysis can easily be extended to a multiple-tray unit since the capture height between the trays can be treated as an individual collector. In addition to operating conditions and physical characteristics, this section will discuss the effect of Reynolds number on particle drag and particle size distribution. Reentrainment effects are also briefly considered.

The analysis begins by examining the behavior of a single spherical particle in a settler in which the bulk flow air velocity profile is plug, i.e., the gas flow over the whole chamber is uniform (see Figure 8.4). The particle is assumed to be located at the top of the unit—the most difficult inlet (or initial) position for particle capture. At the inlet condition, it is assumed that the vertical velocity component of the particle is at its terminal settling velocity, i.e., the particle requires a negligible period of time (and consequently travels a negligible distance) to achieve a value approaching this constant terminal velocity.

For capture to occur, the particle must reach collection surface $a'b'c'd'$ during its residence time t_r in the unit. For plug flow, t_r is given by

$$t_r = L/u \tag{8.1}$$

with

$$u = q/BH \tag{8.2}$$

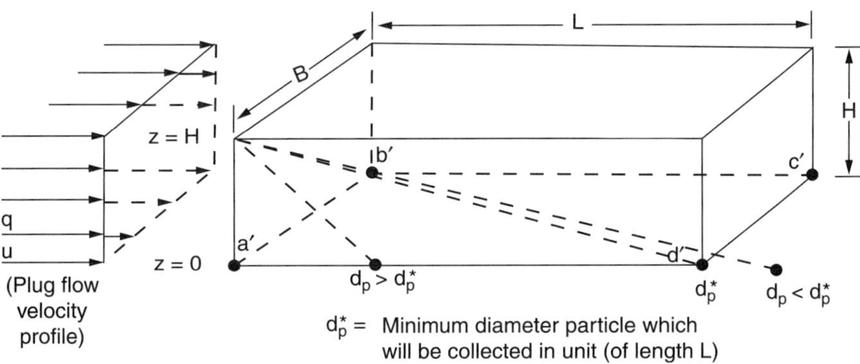

Figure 8.4 Gravity settler nomenclature.

Combining Equations (8.1) and (8.2) gives

$$t_r = LBH/q \tag{8.3}$$

where u is the bulk flow velocity and q is the volumetric flow rate of gas.

The time required for the particle to settle, t_s, a distance, H, while moving at its terminal velocity v_t is

$$t_s = H/v_t \tag{8.4}$$

For capture to occur

$$t_s \leq t_r \tag{8.5}$$

In the limit,

$$t_s = t_r \tag{8.6}$$

Substituting Equations (8.3) and (8.4) into Equation (8.6) gives

$$H/v_t = LBH/q$$

or

$$v_t = q/LB \tag{8.7}$$

The terminal velocity has previously been determined in Chapter 7 for the three Reynolds number ranges. For Stokes' law (without the Cunningham correction factor)

$$v_t = gd_p^2\rho_p/18\,\mu \tag{8.8}$$

Substituting Equation (8.8) into Equation (8.7) and solving for the particle diameter d_p gives

$$d_p = (18\,\mu q/g\rho_p BL)^{0.5} \tag{8.9}$$

For the intermediate range

$$v_t = 0.153(g\rho_p)^{0.71}d_p^{1.14}/\rho^{0.29}\mu^{0.43} \tag{8.10}$$

so that

$$d_p = q^{0.88}\rho^{0.254}\mu^{0.377}/0.193(g\rho_p)^{0.623}(LB)^{0.88} \tag{8.11}$$

Finally, for Newton's law range, one obtains

$$v_t = 1.74\,gd_p\rho_p/\rho \tag{8.12}$$

and

$$d_p = 0.333(\rho/g\rho_p)(q/LB)^2 \tag{8.13}$$

The particle diameters calculated above represent limiting values since particles with diameters equal to or greater than this value will reach the collection surface and particles with diameters less than this value will escape from the unit. This limiting particle diameter may ideally be thought of as the minimum diameter of a particle that will automatically be captured for the above conditions. This diameter is denoted by d_p^* or d_p (min).

The development described above assumes the bulk gas velocity profile to be plug. However, the velocity profile can be parabolic, as in laminar flow, or, in the general case, arbitrarily distributed. The calculation for the minimum particle diameter is not affected, provided that the volumetric flow rate of gas processed remains the same and the particle concentration remains uniform. This can be observed directly from Equation (8.7). The terminal settling velocity for total collection is the same as a hypothetical volume of fluid passing through LB, the collector area $a'b'c'd'$. From this analysis, it can be concluded that the performance of a settler depends only on its collection area and is independent of its height. The minimum height, however, is usually established by requiring that the gas velocity through the chamber be low enough to prevent reentrainment of the collected particles.

Collection efficiencies of 100% were used to derive the equations for d_p^*. The collection efficiency, E, for a monodispersed aerosol (particulates of one size) can be shown to be

$$E = (v_t BL/q); \text{ fractional basis} \tag{8.14}$$

The validity of this equation is observed by again noting that $v_t BL$ represents the hypothetical volume rate of flow of gas passing the collection area, while q is the total volumetric flow rate of gas entering the unit to be treated. An alternate but equivalent form of Equation (8.14) is

$$E = (v_t/u)\,(L/H) \tag{8.15}$$

or

$$E = (H^*/H) \tag{8.16}$$

where H^* is the maximum height above the collection area, at inlet conditions, for which particle capture is assured (see Figure 8.5).

If the gas stream entering the unit consists of a distribution of particles of various sizes, then frequently a fractional or grade efficiency curve is specified for the settler. This is simply a curve describing the collection efficiency for particles of various sizes. If a particle of size d_p will settle a distance H^* in time t, then (H^*/H) represents the fraction of particles of this size that will be collected. If H^* is equal to or greater than H, all the particles of that size or larger will be collected in the settling chamber. A curve of (H^*/H) vs. d_p over the particle size distribution range of the incoming particles entrained in the gas stream is the fractional or size–efficiency curve. Note that H^* represents the distance a particle will settle in residence time t_r.

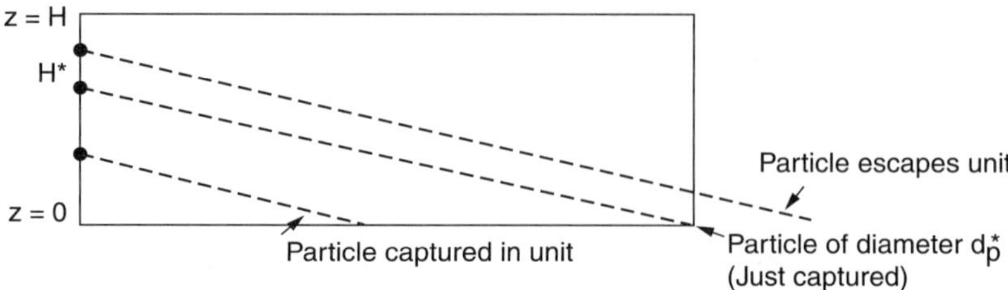

Figure 8.5 Particle collection in a gravity settler.

Combining Equations (8.8) and (8.14) leads to the aforementioned size-efficiency relationship.

$$E = \left[\frac{g\rho_p BLN}{18\,\mu q}\right] d_p^2 \tag{8.17}$$

The term in brackets in Equation (8.17) is often multiplied by a dimensionless empirical factor to correlate theoretical efficiencies with experimental data. If no information is available, it is suggested that 0.5 be used. Thus, Equation (8.17) can be written

$$E = 0.5\left[\frac{g\rho_p BLN}{18\,\mu q}\right] d_p^2 \tag{8.18}$$

The equations above do not apply to large particles. If the Intermediate law (not Stokes' law) applies, Equation (8.18) becomes (where N has been disregarded)

$$E = 0.153\left(g\rho_p\right)^{0.71} d_p^{1.14}(BL)/\rho^{0.29}\mu^{0.43}q \tag{8.19}$$

Finally, for the Newton's law range, the equation becomes

$$E = 1.74\left[d_p g\rho_p/\rho\right]^{0.5}(BL/q) \tag{8.20}$$

All of the preceding equations were developed with the assumption that the gas flow through the settling chamber is *laminar*. The equation for determining efficiency when the flow is turbulent can be shown to be

$$E = 1 - e^{-[Lu/Hv]} \tag{8.21}$$

The process design variables for a settling chamber consist of length (L), width (B), and height (H). These parameters are usually chosen by the chamber manufacturer in order to remove all particles above a specified size. The chamber's design must provide conditions for sufficient particle residence time to capture the desired particle size range. This can be accomplished by keeping the velocity of the exhaust gas through the chamber as low as possible. If the velocity is too high, dust reentrainment will occur. However, the design velocity must not be so low as to cause the design of the chamber volume to be exorbitant. Consequently, the units are designed for gas velocities in the range 1–10 ft/s (0.305–3.05 m/s).

The development in this section also provides the theoretical collection efficiency of a settling chamber for a single-sized particle. Since the gas stream entering a unit consists of a distribution of particles of various sizes, a fractional efficiency curve must be used to determine the overall collection efficiency. This is simply a curve or equation describing the collection efficiency for particles of various sizes (see Chapter 7, Section 7.6). The overall efficiency can then be calculated using

$$E = \sum (E_i)(w_i) \tag{8.22}$$

where E = overall collection efficiency
E_i = fractional efficiency of specific size particle
w_i = mass fraction of specific size particle (in range)

As discussed earlier, the collection of particles from a gas stream involves three distinct phases: deposition on the collecting surface, retention on the surface, and finally, removal of the collected material. Thus far, it has been assumed that once a particle settles to the bottom of the chamber (or onto a tray within the chamber), it is retained there, i.e., reentrainment effects have been neglected. If the bulk flow gas velocity is sufficiently high, a collected particle will not be retained by the surface, but will be picked up and reentrained into the moving air stream. This can substantially reduce the collection efficiency; the consideration of pickup or suspension velocity for horizontal flow is indeed important. Therefore, in a simple gravity settler, the bulk gas velocity should not exceed this pickup or suspension velocity. Typical pickup velocities are given in Table 8.1. Values for this limiting velocity can also be obtained from semitheoretical considerations. By equating the various forces acting on the particle while neglecting interparticle friction, one can derive pickup velocity equations. If no data for determining the pickup velocity are available, it should be assumed to be 10 ft/s. In this case, the velocity of the gas through the settling chamber (throughput velocity) must be less than 10 ft/s.

TABLE 8.1 Pickup Velocities of Various Materials

Material	Density, g/cm^3	Median Particle Size, μm	Pickup Velocity, ft/s
Aluminum chips	2.72	335	14.2
Asbestos	2.20	261	17.0
Nonferrous foundry dust	3.02	117	18.8
Lead oxide	8.26	14.7	25.0
Limestone	2.78	71	21.0
Starch	1.27	64	5.8
Steel shot	6.85	96	15.2
Wood chips	1.18	1370	13.0
Wood sawdust	—	1400	22.3

8.3 OPERATION AND MAINTENANCE; AND IMPROVING PERFORMANCE

Gravity settlers are designed for a particular fluid throughput. Generally, a reasonable overload can be tolerated without causing damage. If operated at excessive flow rates, erosion or vibration can result. Erosion can also occur at normally acceptable flow rates if other conditions vary. Evidence of erosion should be investigated to determine the cause. Vibration can be propagated by problems other than flow overloads, e.g., improper design, fluid maldistribution, or corrosion/erosion of internal flow-directing devices such as baffles.

Recommended maintenance of a gravity settler requires regular inspection to ensure mechanical soundness of the unit and a level of performance consistent with the original design criteria. A brief general inspection should be performed on a regular basis while the unit is operating. Vibratory disturbance, excessive pressure drop, and decreased efficiency are all signs that thorough inspection and maintenance procedures are required.

Complete inspection requires shutdown of the unit for access to the internals. Scheduling can be determined only from experience and general inspections. Internals and exteriors, where accessible, should be visually inspected for fouling, corrosion, or damage. The nature of any metal deterioration should be investigated to properly determine the anticipated life of the equipment or possible corrective action. Possible causes of deterioration include general corrosion, intergranular corrosion, stress cracking, galvanic corrosion, impingement, or erosion attack.

Within the constraints of the existing system, improving operation and performance refers to maintaining operation at the original or a consistent performance level. There are several factors, previously mentioned, which are related to the design and performance of this unit. Any pressure drop in the settler should be minimized. Flow rates and temperature should be checked regularly to ensure that they are in accordance with the original design criteria. Decreased performance due to fouling will generally be exhibited by a gradual decrease in efficiency and should be corrected as soon as possible. Mechanical malfunctions can also be gradual, but will eventually be evidenced by a decrease in or lack of performance.

More often than not, a gravity settler is a secondary control device since it cannot achieve the collection efficiency required by industry using ordinarily available procedures. One advantage of settlers is that they can reduce the load, improve the performance, and extend the operating life of more expensive devices. Gravity settlers frequently return product or an intermediate to be recycled into the process.

Despite the fact that settling chambers are simple in design and can be manufactured from almost any material, they are infrequently used because of their extremely large space requirements and relatively low efficiency. When settling chambers are used, they are normally followed by a more efficient collection device. They have, on rare occasions, been used following a high-efficiency control device (e.g., an electrostatic precipitation) to capture some reentrained particles.

Efforts have been made to improve the efficiency of gravity settling chambers by the use of baffles and various other methods. As described earlier, the Howard settling chamber is an example of these efforts. But this type of settling chamber was never widely used because of the difficulty in removing the settled dust from the horizontal trays.

Some industrial applications include the following:

1. The combination settling chamber and cooling device has been used in the metal refining industry to partially collect large particulates and to reduce the gas temperature before entering the final collection device. One of the more common types is the "hairpin cooler."
2. Arsenic trioxide from smelting arsenical copper ores has often been collected in brick settling chambers known as "kitchens."
3. In the manufacture of various foodstuffs, simple settling is the first step in dust recovery, achieved by spraying the condensed liquids into large chambers. The effluent air is then passed to second-stage cleaners (cyclones) and the exhaust recirculated to the spray chambers (see Chapter 9).
4. Power and heating plants may employ settling chambers upstream of multiple cyclone units. Quite often they are used to collect large unburned carbon particles and reinject them into the boiler.
5. As described above, the gravity settler has occasionally been used as a postcleaner (as opposed to a precleaner) with another particulate control device.

Typical installation costs range from \$0.10 to \$0.40 per acfm. Operating costs are typically less than \$0.01 per acfm per year.

There is little to report on recent developments on gravity settlers. As noted above, the unit has losts its favor in the industry and is rarely, if ever, used today.

PROBLEMS

8.1 Gravity Settler Capture Mechanism

A particle can be collected in a settling chamber when (select one)

(a) the centrifugal force pushes the particle against the wall of the chamber
(b) the gravitational force is less than the drag and buoyant force causing particle collection
(c) the buoyant force causes the particle to be collected in the chamber
(d) the particle is moving at the terminal or setting velocity

Solution: Collection or capture occurs because of the presence of gravity, which in turn leads to a particle settling under the influence of gravity. The correct answer is therefore (d).

8.2 Gravity Settler Throughput Velocity

A typical throughput velocity in a gravity settler is

(**a**) 10 ft/min
(**b**) 60 ft/min
(**c**) 100 ft/min
(**d**) None of the above

Solution: As noted in Sections 8.2 and 8.3, typical velocities are in the 10 ft/s range. The correct answer is therefore (d).

8.3 Particle Size with 50% Efficiency

The term used to describe the size of a particle that is removed with 50% efficiency is the

(**a**) Mean size
(**b**) Critical size
(**c**) Cut size
(**d**) Mode

Solution: Answer (a) refers to the particle size distribution. Answer (b) is the size associated with 100% efficiency. Answer (d) has nothing to do with the question. Finally answer (c) serves as the definition of cut size. The correct answer is therefore (c).

8.4 Effect of Particle Size on Efficiency

The larger the mean particle size of a dust flowing through a gravity settler, the higher (generally) the value of the

(**a**) Pressure drop
(**b**) Inlet velocity
(**c**) Dust concentration
(**d**) Efficiency

Solution: Refer to Equations (8.18)–(8.20). Any increase in diameter results in an increase in efficiency. The correct answer is therefore (d).

8.5 Effect of Throughout Velocity on Settler Efficiency

As the throughput volumetric flow rate of a gas increases, the collection efficiency of a gravity settler will

(**a**) Increase
(**b**) Decrease
(**c**) Remain relatively constant
(**d**) Vary sinusoidally

Solution: On inspecting Equations (8.18)–(8.20), one notes that the efficiency is inversely proportional to the flow rate. The correct answer is therefore (b).

8.6 Effect of Throughput Velocity on Settler Pressure Drop

As the throughput velocity of a gas increases, the pressure drop of a gravity settler will

(**a**) Increase
(**b**) Decrease
(**c**) Remain relatively constant
(**d**) Vary sinusoidally

Solution: As one would intuitively expect, the pressure drop increases approximately as the velocity squared for turbulent flow and velocity to the first power for laminar flow, and the flow of gases is almost always characterized by turbulent flow because of the low viscosity of the gas. The correct answer is therefore (a).

8.7 Effect of Settler Height on Efficiency

If all other operating and design features remain constant, as the height increases, the efficiency of a gravity settler

(**a**) Increases
(**b**) Decreases
(**c**) Remains relatively constant
(**d**) Varies exponentially

Solution: As can be seen in Equations (8.17)–(8.20), the height H does not appear in the efficiency equations. The correct answer is therefore (c).

8.8 Effect of Height on Settler Pressure Drop

If all other operating and design factors remain constant, as the height increases, the pressure drop of a gravity settler

(**a**) Increases slightly
(**b**) Decreases
(**c**) Remains constant
(**d**) Varies exponentially

Solution: The pressure drop across a process unit is a function of the velocity and resistance to flow. With a larger opening for the gas to flow through, the velocity will decrease while the resistance will decrease slightly, resulting in a reduction in pressure drop. The correct answer is therefore (b).

8.9 Effect of Temperature on Settler Efficiency

All other variables being held constant, as the temperature of the throughput gas increases, the overall efficiency of a gravity settler will

(**a**) Increase
(**b**) Decrease
(**c**) Remain relatively constant
(**d**) None the above

Solution: As can be seen in Equations (8.17), (8.19), and (8.20), temperature has essentially no effect on the efficiency. The efficiency is affected by a change in the density term in the denominator of Equations (8.19) and (8.20), but that effect is small since the density is raised to the 0.23 power. Thus, as the temperature increases, the density of the gas decreases; this will result in a slight increase in efficiency. The end result is that the efficiency will be relatively unaffected. The correct answer is therefore (c).

8.10 Gravity Settler Advantages

List some of the advantages of a gravity settler.

Solution: In terms of overall design considerations for gravity settlers, advantages include

1. low cost of construction and maintenance
2. few maintenance problems
3. relatively low operating pressure drops in the range of approximately 0.1 in H_2O
4. temperature and pressure limitations imposed only by the materials of construction used
5. dry disposal of solid particulates

8.11 Gravity Settler Disadvantages

List some of the disadvantages of a gravity settler.

Solution: The disadvantages include

1. large space requirements
2. relatively low overall collection efficiencies (typically ranging within 20–60%)
3. exhibits relatively poor performance

8.12 Minimum Particle Size

A hydrochloric acid mist in air at 25°C is to be collected in a gravity settler. You are requested to calculate the smallest mist droplet (spherical in shape) that will definitely be collected by the settler. Assume the acid concentration to be uniform through the inlet cross section of the unit and Stokes' law applies. Operating data and information on the gravity settler are given below.

Dimensions of gravity settler = 30 ft wide, 20 ft high, 50 ft long
Actual volumetric flowrate of acidic gas = $50\ \text{ft}^3/\text{s}$
Specific gravity of acid = 1.6
Viscosity of air = 0.0185 cP = $1.243 \times 10^{-5}\ \text{lb/ft}\cdot\text{s}$
Density of air = $0.076\ \text{lb/ft}^3$

Solution: For the problem at hand, first determine the density of the acid mist:

$$\rho_p = (62.4)(1.6)$$

$$= 99.84\ \text{lb/ft}^3$$

Calculate the minimum particle diameter in both feet and micrometers using Equation (8.9). This assumes that Stokes' law applies.

$$d_p = \left(\frac{(18)(1.243 \times 10^{-5})(50)}{(32.2)(99.84)(30)(50)}\right)^{1/2}$$
$$= 4.82 \times 10^{-5}\ \text{ft}$$

There are 3.048×10^5 μm in 1 ft. Therefore

$$d_p = 14.7\ \mu\text{m}$$

As noted earlier, the particle diameter calculated above represents a limiting value since particles with diameters equal to or greater than this value will reach the settler collection surface and particles with diameters less than this value *may* escape from the unit. This limiting particle diameter can ideally be thought of as the minimum diameter of a particle that will automatically be captured for the conditions described above. This diameter is denoted by d_p^* or d_p (min). One should verify that Stokes' law applies. As a rule, one can generally assume that when acted on by gravity, particles smaller than 100 μm obey Stokes' Law. Therefore, the Stokes' law assumption is valid.

8.13 Sodium Hydroxide Application

A sodium hydroxide spray in air at 30°C is to be collected in a gravity settler. The unit is 30 ft wide, 15 ft high, and 40 ft long. The volumetric flow rate of the gas is 42 ft^3/s. Calculate the smallest mist droplet (spherical in shape) that will be entirely collected by the settler. The specific gravity of the mist droplets may be assumed to be equal to 1.21.

Solution: The important property data are tabulated below:

$$\mu = 0.0185\ \text{cP} = 1.245 \times 10^{-5}\ \text{lb/ft}\cdot\text{s}$$
$$\rho = 0.0728\ \text{lb/ft}^3$$

Assume that Stokes' law again applies, to be checked later with the calculation of *K*. Then, the describing equation is

$$d_p(\text{min}) = \left(\frac{18\,\mu q}{g\rho_p\,BL}\right)^{0.5} \tag{8.9}$$

Substituting the data, one obtains

$$d_p(\text{min}) = \left[\frac{18(1.245 \times 10^{-5}\ \text{lb/ft}\cdot\text{s})(42\ \text{ft}^3/\text{s})}{(32.2\ \text{ft/s}^2)\ (1.21 \times 62.4\ \text{lb/ft}^3)(30\ \text{ft})(40\ \text{ft})}\right]^{0.5}$$
$$= 5.68 \times 10^{-5}\ \text{ft}$$

Checking the K value, one employs (see Chapter 7)

$$K = d_p(\text{min})\left[\frac{g\,\rho_p\,\rho}{\mu^2}\right]^{1/3}$$

Substituting the data, one finally obtains

$$K = 5.68 \times 10^{-5}\,\text{ft}\left[\frac{(32.2\,\text{ft/s}^2)(1.21 \times 62.4\,\text{lb/ft}^3)(0.0728\,\text{lb/ft}^3)}{(1.245 \times 10^{-5}\,\text{lb/ft}\cdot\text{s})^2}\right]^{1/3}$$

$$= 0.594$$

Therefore, the d_p(min) calculated is correct.

8.14 Effect of Reducing Flow Rate

Calculate the smallest droplet in Problem 8.13 assuming the flow rate is halved.

Solution: If q is half of that in Problem 8.13, then simply divide the answer by $\sqrt{2}$; the basis for this is provided below.

$$d_p(\text{min}_1) = \left(\frac{18\,\mu\,q_1}{g\,\rho_p\,BL}\right)^{0.5}$$

$$d_p(\text{min}_2) = \left(\frac{18\,\mu\,q_2}{g\,\rho_p\,BL}\right)^{0.5}$$

with

$$q_2 = \tfrac{1}{2}q_1$$

Substituting above yields

$$d_p\,(\text{min}_2) = \left(\frac{18\,\mu\,q_1}{2\,g\,\rho_p\,BL}\right)^{0.5} = \frac{1}{\sqrt{2}}\left(\frac{18\,\mu\,q_1}{g\,\rho_p\,BL}\right)^{0.5}$$

Therefore,

$$d_p(\text{min}_2) = \frac{d_p(\text{min}_1)}{\sqrt{2}}$$

$$= \frac{5.68 \times 10^{-5}\,\text{ft}}{\sqrt{2}} = 4.016 \times 10^{-5}\,\text{ft}$$

The K value in this case is the previous K value multiplied by the ratio of the two calculated diameters:

$$K_2 = K_1\left(\frac{d_p(\min_2)}{d_p(\min_1)}\right) = 0.594\left(\frac{4.016\times10^{-5}\text{ ft}}{5.68\times10^{-5}\text{ ft}}\right) = 0.42$$

Therefore, the calculated value for $d_p(\min_2)$ is again correct.

8.15 Effect of Temperature Increase

If the temperature in Problem 8.12 is 25°C, what is the smallest droplet that can be controlled if the gas is at 200°C? Assume that at 200°C, $\mu = 2.0\times10^{-4}$ g/cm · s.

Solution: At 200°C and converting

$$\mu = 1.344\times10^{-5}/\text{lb/ ft}\cdot\text{s}$$

According to Equation (8.9), multiply the answer to Problem 8.12 by the ratio of viscosities to the $\frac{1}{2}$ power.

$$\begin{aligned} d_p(\min_2) &= d_p(\min_1)\left(\frac{\mu_2}{\mu_1}\right)^{1/2} \\ &= 5.68\times10^{-5}\text{ ft}\left(\frac{1.344\times10^{-5}\text{ lb/ ft}\cdot\text{s}}{1.245\times10^{-5}\text{ lb/ ft}\cdot\text{s}}\right)^{1/2} \\ &= 5.9\times10^{-5}\text{ ft} \end{aligned}$$

The K value in this case will be the K value from Problem 8.12 multiplied by

$$\left(\frac{d_p(\min)_2}{d_p(\min)_1}\right)\left[\left(\frac{\mu_1}{\mu_2}\right)^2\right]^{1/3}$$

or

$$\begin{aligned} K_2 &= K_1\left(\frac{d_p(\min_2)}{d_p(\min_1)}\right)\left[\left(\frac{\mu_1}{\mu_2}\right)^2\right]^{1/3} \\ &= 0.594\left(\frac{5.9\times10^{-5}}{5.68\times10^{-5}}\right)\left[\left(\frac{1.245\times10^{-5}}{1.344\times10^{-5}}\right)^2\right]^{1/3} \\ &= 0.586 \end{aligned}$$

Therefore, the $d_p(\min_2)$ calculated is correct.

8.16 Effect of Reducing Length

Calculate the smallest droplet in Problem 8.13, assuming that the length of the unit is halved.

Solution: If the length of the unit is halved, then, according to Equation (8.9), multiply the answer in Problem 8.12 by $\sqrt{2}$:

$$\begin{aligned} d_p(\min_2) &= d_p(\min_1)\sqrt{2} \\ &= 5.68 \times 10^{-5}\,\text{ft}\,(\sqrt{2}) = 8.03 \times 10^{-5}\,\text{ft} \end{aligned}$$

The K value is the value given in Problem 8.12 times $(d_p \min_2/d_p \min_1)$:

$$\begin{aligned} K_2 &= K_1\left(\frac{d_p(\min_2)}{d_p(\min_1)}\right) \\ &= 0.594\left(\frac{8.03 \times 10^{-5}}{5.68 \times 10^{-5}}\right) \\ &= 0.84 \end{aligned}$$

Therefore, the calculation is correct.

8.17 Effect of Plates on Efficiency

A monodispersed aerosol 1.099 μm in diameter passes through a gravity settler 20 cm wide × 50 cm long with 18 plates and channel thickness of 0.124 cm. The gas flow rate is 8.6 L/min, and it is observed that it operates at an efficiency of 64.9%. How many plates would be required to have the unit operate at 80% efficiency?

Solution: The volumetric flow is

$$\begin{aligned} q_1 &= 8.6\ \text{L/min} \times \text{min}/60\ \text{s} \times 1000\,\text{cm}^3/\text{L} \times 1/19\ \text{channels} \\ &= 7.544\ \text{cm}^3/\text{channel} \cdot \text{s}. \end{aligned}$$

The settling velocity can be calculated from Equation (8.14).

$$E_1 = \frac{v_t BL}{q_1}$$

Rearranging, one obtains

$$v_t = \frac{q_1\,E_1}{BL}$$

Substituting the data yields

$$v_t = \frac{(7.544\ \text{cm}^3/\text{s})(0.649)}{(20\ \text{cm})\ (50\ \text{cm})} = 4.896 \times 10^{-3}\ \text{cm/s}$$

At $E_2 = 0.8$, the new volumetric flow is

$$\begin{aligned} q_2 &= \frac{v_t BL}{E_2} \\ &= \frac{(4.896 \times 10^{-3} \text{cm/s})(20\,\text{cm})(50\,\text{cm})}{0.8} \\ &= 6.12\,\text{cm}^3/\text{channel} \cdot \text{s} \end{aligned}$$

Then, the number of channels necessary is

$$\begin{aligned} \text{Number of channels} = \frac{q_1}{q_2} &= (8.6\ \text{L/min}) \times (\text{min}/60\,\text{s}) \\ &\quad \times (10^3\ \text{cm}^3/\text{L})/6.12\,\text{cm}^3/\text{channel} \cdot \text{s} \\ &= 23.42 \approx 24 \end{aligned}$$

The number of plates should be 24.

8.18 Two Particle Size Distributions

An aerosol consisting of particles 0.63 and 0.83 μm in diameter in equal mass amounts passes through a gravity settler at a flow rate of 3.60 L/min. Given the following data, use Stokes' law with the Cunningham correction factor to calculate the efficiency of the settler.

Length = 50 cm
Width = 20 cm
Height of channel = 0.124 cm
Number of channels = 19
$\rho = 1.05\ \text{g/cm}^3$
$\lambda = 0.1\ \mu\text{m}$
$\mu = 0.0182$ cp

Solution: The approach that will be used is to analyze both particle sizes separately. For

$$d_{p_1} = 0.63\ \mu\text{m} = 0.63 \times 10^{-4}\text{cm}$$

the Cunningham correction factor (see Chapter 7) is given by

$$C = 1 + \frac{2A\lambda}{d_p}$$

where

$$A = 1.257 + 0.40e^{(-1.10\,d_p/2\lambda)}$$

Substituting the data, one obtains

$$A_1 = 1.257 + 0.40e^{\left(-\frac{1.10\times 0.63\ \mu\text{m}}{2\times 0.1\ \mu\text{m}}\right)} = 1.2695$$

and

$$C_1 = 1 + \frac{2 \times 1.2695 \times 0.1\ \mu\text{m}}{0.63\ \mu\text{m}} = 1.403$$

The settling velocity is given by

$$v_{t_1} = \frac{d_{p_1}^2\ \rho_p g C_1}{18\mu} \tag{8.9}$$

Substituting the data, one obtains

$$\begin{aligned} v_{t_1} &= \frac{(0.63 \times 10^{-4}\text{cm})^2 (1.05\ \text{g/cm}^3)\ (980\ \text{cm/s}^2)\ (1.403)}{18(1.82 \times 10^{-4}\ \text{g/cm} \cdot \text{s})} \\ &= 1.75 \times 10^{-3}\ \text{cm/s} \end{aligned}$$

If

$$\begin{aligned} q &= (3.60\ \text{L/min}) \times (\text{min}/60\ \text{s}) \times (10^3\ \text{cm}^3/\text{L}) \times (1/19\ \text{channels}) \\ &= 3.16\ \text{cm}^3/\text{channel} \cdot \text{s} \end{aligned}$$

then

$$\begin{aligned} E_1 &= \frac{v_{t_1}\ BL}{q} \qquad (8.14) \\ &= \frac{(1.75 \times 10^{-3}\ \text{cm/s})\ (20\ \text{cm})\ (50\ \text{cm})}{3.16\ \text{cm}^3} \\ &= 0.554 = 55.4\% \end{aligned}$$

For

$$\begin{aligned} d_{p_2} &= 0.83\ \mu\text{m} = 0.83 \times 10^{-4}\ \text{cm} \\ A_2 &= 1.257 + 0.40\ e^{\left(-\frac{1.10\times 0.83\ \mu\text{m}}{2\times 0.1\ \mu\text{m}}\right)} = 1.2611 \\ C_2 &= 1 + \frac{2 \times 1.2611 \times 0.1\ \mu\text{m}}{0.83\ \mu\text{m}} = 1.3038 \end{aligned}$$

Then

$$v_{t_2} = \frac{(0.83 \times 10^{-4}\ \text{cm})^2\ (1.05\ \text{g/cm}^3)\ (980\ \text{cm/s}^2)\ (1.3038)}{18(1.82 \times 10^{-4}\ \text{g/cm} \cdot \text{s})}$$

$$= 2.821 \times 10^{-3}\ \text{cm/s}$$

Therefore

$$E_2 = \frac{v_{t_2}\ BL}{q} \tag{8.14}$$

$$= \frac{(2.821 \times 10^{-3}\ \text{cm/s})\ (20\ \text{cm})\ (50\ \text{cm})}{3.16\ \text{cm}^3/\text{s}}$$

$$= 0.893 = 89.3\%$$

Then

$$E = \Sigma w_i E_i = 0.5(0.554) + 0.5(0.893) = 0.7235$$

or

$$E = 72.35\%$$

8.19 Daily Discharges

Con Edison's electric power plant in Astoria (Queens, NY) requires the installation of a gravity settler to remove dust particles (density 144 lb/ft^3) from an air stream at ambient conditions prior to being treated by an ESP. The following data are available:

Mean particle diameter = 70 μm
Standard deviation = 2.0
Particle density = 144 lb/ft^3
Inlet loading = 20 gr/ft^3
Volumetric flow rate = 5400 acfm
Settler dimensions
  Width = 10 ft
  Length = 50 ft
  Height = 20 ft
Overall efficiency = 92.9%

Calculate outlet dust loading, the daily mass of dust collected, and the daily mass of dust discharged from the unit.

Solution:

$$\begin{aligned}\text{Outlet dust loading (OL)} &= (1 - E)\,(\text{loading in})\\ &= (1 - 0.929)\ 20\\ &= 1.42\ \text{gr/ft}^3\\ \text{Daily mass collected} &= (0.929)(20)(5400)(60)(24)/7000\\ &= 20{,}640\ \text{lb/day}\\ \text{Daily mass discharged} &= (0.071)(20{,}640)/(0.929)\\ &= 1577\ \text{lb/day}\end{aligned}$$

8.20 Gravity Settler Design

As a recently hired engineer for an equipment vending company, you have been requested to design a gravity settler to remove all iron particulates from a dust-laden gas stream. The following information is given:

$d_p = 35\ \mu$m; uniform, i.e., no distribution
gas = air at ambient conditions
$q = 130\ \text{ft}^3/\text{s}$
u, throughput velocity = 10 ft/s
$\rho_p = 7.62\ \text{g/cm}^3$

Solution: First convert d_p and ρ_p to engineering units:

$$d_p = (35\ \mu\text{m})(3.281 \times 10^{-6}\ \text{ft}/\mu\text{m}) = 11.48 \times 10^{-5}\ \text{ft}$$

$$\rho_p = (7.62\ \text{g/cm}^3)(1\ \text{lb}/453.6\ \text{g})(28{,}316\ \text{cm}^3/\text{ft}^3) = 475.7\ \text{lb/ft}^3$$

The K value is

$$\begin{aligned}K &= d_p\left(\frac{g(\rho_p - \rho)\rho}{\mu^2}\right)^{1/3}\\ &= (11.48 \times 10^{-5})\left(\frac{(32.2)(475.7 - 0.0775)(0.0775)}{(1.23 \times 10^{-5})^2}\right)^{1/3}\\ &= 2.28 < 3.3\end{aligned}$$

Stokes' law applies, and the collection area required can be calculated from Equation (8.9):

$$d_p = \left(\frac{18\mu q}{g\rho_p BL}\right)^{0.5}$$

Solving for BL, one obtains

$$BL = \frac{18\mu q}{g\rho_p d_{\rm p}^2}$$

$$= \frac{18(1.23 \times 10^{-5})(130)}{(32.2)(475.5)(11.48 \times 10^{-5})^2}$$

$$= 142.5 \ \text{ft}^2$$

The cross-sectional area for $u = 10$ ft/s is

$$BH = q/u = 130/10$$

$$= 13 \ \text{ft}^2$$

On the basis of the minimum required height for cleaning purposes, H is usually set at 3 ft. Then

$$B = (13)/H = 13/3$$

$$= 4.33 \ \text{ft}$$

and

$$L = (142.5)/B = 142.5/4.33$$

$$= 32.9 \ \text{ft}$$

The total volume of the settler is therefore

$$V = (BL)H = (142.5)(3)$$

$$= 427.5 \ \text{ft}^3$$

This design is revisited in Problems 8.32 and 8.33.

8.21 Gravity Settler Design with Shelves

Design a gravity-settling chamber to remove (with 100% efficiency) particles larger than 60 μm from an air stream of 24,000 acfm at 77°F. The inlet concentration is 15 gr/ft^3, and the outlet concentration is to be 1.0 gr/ft^3. Shelves of 10 ft length and 5 ft width are to be employed. The superficial velocity should be less than 10 ft/s (use this value in the calculations). The properties of air at 77°F are

Viscosity $= 1.23 \times 10^{-5}$ lb/ ft · s
Density $= 0.074$ lb/ ft^3
Specific gravity of particles $= 1.8$

Solution: Pertinent calculations are provided below:

$$K = d_p\left[g\rho_p\ \rho/\mu^2\right]^{1/3}$$

$$= 1.968 \times 10^{-4}\left[(32.2)(1.8)(62.4)(0.074)/(1.23 \times 10^{-5})^2\right]^{1/3}$$

$$= 2.38;\ \text{Stokes' law applies}$$

$$v = g\rho_p\ d_p^2/18\ \mu$$

$$= 0.633\ \text{ft/s}$$

$$E = (15 - 1)/15 = 0.933$$

Capture area: $L = 10$ ft, $B = 5$ ft

$$E = vLB/q_1;\ q_1 = \text{flow rate for one passage} \tag{8.14}$$

$$q_1 = vLB/E$$

$$= (0.633)(10)(5)/0.933$$

$$= 33.92\ \text{acfs}$$

Set n equal to the number of passages. Then

$$q = nq_1$$

Solving for n yields

$$n = 24{,}000/(60)(33.92)$$

$$= 11.8;\ \text{use } 12$$

The unit therefore needs 13 shelves if there are 12 passages or trays. To calculate the tray spacing (h), first note that

$$H = q/uB;\ u = \text{throughput velocity}$$

$$= 24{,}000/(60)(10)(5)$$

$$= 8\ \text{ft}$$

Since there are 12 passages, then

$$h = 8/12$$

$$= 0.667\ \text{ft} = 8\ \text{in}$$

8.22 Overall Collection Efficiency

A settling chamber is installed in a small heating plant that uses a traveling grate stoker. You are requested to determine the overall collection efficiency of the settling chamber given the following operating conditions, chamber dimensions, and particle size distribution data:

Chamber width = 10.8 ft
Chamber height = 2.46 ft

Chamber length = 15.0 ft
Volumetric flow rate of contaminated air stream = 70.6 scfs
Flue gas temperature = 446°F
Flue gas pressure = 1 atm
Particle concentration = 0.23 gr/scf
Particle specific gravity = 2.65
Standard conditions = 32°F, 1 atm

Particle size distribution data of the inlet dust for the traveling grate stoker are given in Table 8.2. Assume that the actual terminal settling velocity is one-half of the Stokes law velocity.

TABLE 8.2 Particle Size Distribution Data For Problem 8.22

Particle Size Range, μm	Average Particle Diameter, μm	c_i, gr/scf	w_i, wt%
0–20	10	0.0062	2.7
20–30	25	0.0159	6.9
30–40	35	0.0216	9.4
40–50	45	0.0242	10.5
50–60	55	0.0242	10.5
60–70	65	0.0218	9.5
70–80	75	0.0161	7.0
80–94	87	0.0218	9.5
+94	+94	0.0782	34.0
		0.2300	100.0

Solution: The collection efficiency E for a monodispersed aerosol (particulates of one size) in laminar flow was shown be

$$E = \frac{v_t BL}{q} (100\%) \tag{8.14}$$

where v_t is the terminal settling velocity of the particle. As noted in Section 8.2, the validity of this equation is observed by noting that $v_t BL$ represents the hypothetical volume rate of flow of gas passing the collection area, while q is the volumetric flow rate of gas entering the unit to be treated. An alternate but equivalent form of Equation (8.14) is

$$E = \left(\frac{v_t}{u}\right)\left(\frac{L}{H}\right)(100\%) \tag{8.15}$$

where H is the height of the settling chamber and u is the gas velocity.

If the gas stream entering the unit consists of a distribution of particles of various sizes, then a fractional or grade efficiency curve is frequently specified for the settler. This is simply a curve describing the collection efficiency for particles of various sizes. The dependency of E on d_p arises because of the v_t term in the above equations. Since the actual terminal settling velocity is assumed to be one-half of the Stokes law velocity (according to the problem statement), then (from Equation 8.8)

$$v = \frac{1}{2}\frac{gd_p^2\rho_p}{18\mu}$$

and therefore

$$E = \frac{gd_p^2\rho_p BL}{36\ \mu q} \tag{8.17}$$

The viscosity of the air in lb/(ft · s) is

$$\mu\ (446°\mathrm{F}) = 1.75 \times 10^{-5}\ \mathrm{lb/(ft \cdot s)}$$

The particle density in lb/ft^3 is

$$\begin{aligned}\rho_p &= (2.65)(62.4)\\ &= 165.4\ \mathrm{lb/ft^3}\end{aligned}$$

To calculate the collection efficiency of the system at the operating conditions, the standard volumetric flow rate of contaminated air of 70.6 scfs is converted to the actual volumetric flow using Charles' Law:

$$\begin{aligned}q = q_s\left(\frac{T_a}{T_s}\right) &= (70.6)\left(\frac{446+460}{32+460}\right)\\ &= 130\ \mathrm{acfs}\end{aligned}$$

The collection efficiency in terms of d_p, with d_p in micrometers, is given below. *Note*: To convert d_p from ft to μm, d_p is divided by (304,800) μm/ft.

$$\begin{aligned}E = \frac{v_t BL}{q} &= \frac{g\rho_p BLd_p^2}{36\ \mu q}\\ &= \frac{(32.2)(165.4)(10.8)(15)\ d_p^2}{(36)(1.75\times10^{-5})(130)(304{,}800)^2}\\ &= 1.14\times10^{-4}\ d_p^2\end{aligned} \tag{8.17}$$

where d_p is in micrometers (μm).

Thus, for a particle diameter of 10 μm, one obtains

$$\begin{aligned} E &= 1.14 \times 10^{-4} d_{\mathrm{p}}^2 \\ &= (1.14 \times 10^{-4})(10)^2 \\ &= 1.1 \times 10^{-2} \\ &= 1.1\% \end{aligned}$$

Table 8.3 provides the collection efficiency for each particle size. The size–efficiency curve for the settling chamber is shown in Figure 8.6. The collection efficiency of each particle size may be read from the size–efficiency curve. The product, w_iE_i, is then calculated for each size. The overall efficiency is equal to Σw_iE_i. These calculations are provided in Table 8.4. The overall collection efficiency for the settling chamber E is 58.7%.

TABLE 8.3 Size–Efficiency Results

d_{p}, μm	d_{p}^2, μm^2	E, %
93.8	8800	100.0
90	8100	92.0
80	6400	73.0
60	3600	41.0
40	1600	18.2
20	400	4.6
10	100	1.1

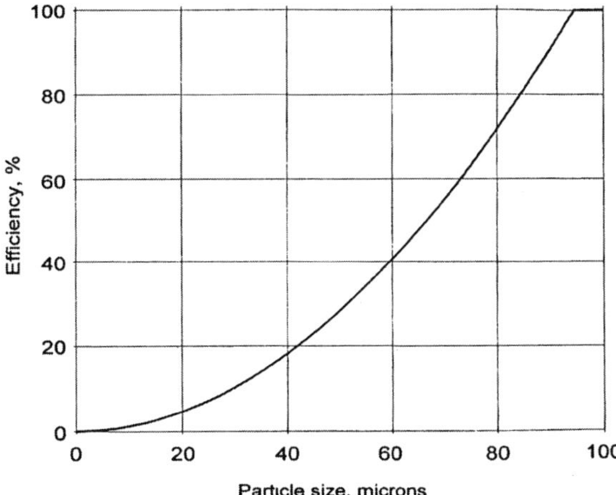

Figure 8.6 Size–efficiency curve for settling chamber.

TABLE 8.4 Overall Collection Efficiency

Average d_p, μm	Weight Fraction w_i	E_i, %	w_iE_i, %
10	0.027	1.1	0.030
25	0.069	7.1	0.490
35	0.094	14.0	1.316
45	0.105	23.0	2.415
55	0.105	34.0	3.570
65	0.095	48.0	4.560
75	0.070	64.0	4.480
87	0.095	83.0	7.885
+94	0.340	100.0	34.000
Total	1.000		58.7

8.23 Parallel versus Series Arrangement

A company has two gravity settlers and they want to use them for removing fly ash particles from an air stream at 60°F and 1 atm. The volumetric flow rate is 50 ft^3/s. The gravity settlers have the same dimensions: $L = B = 10$ ft, $H = 15$ ft. The inlet size distribution of the fly ash is known. The company wants to use the two gravity settlers in a manner that provides the maximum efficiency. They have two alternatives:

1. Put them in parallel (see Figure 8.7)
2. Put them in series (see Figure 8.8)

What is your recommendation in order to obtain the maximum efficiency?

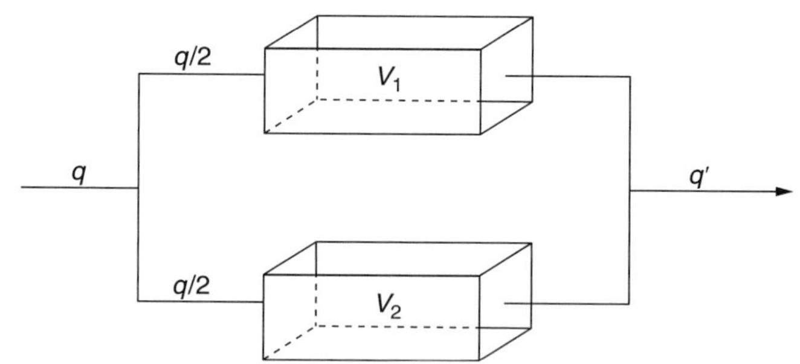

Figure 8.7 Parallel arrangement of two settlers (A).

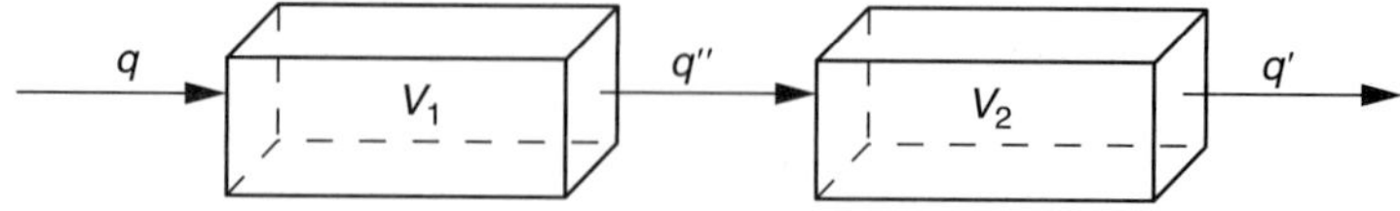

Figure 8.8 Series arrangement of two settlers (B).

Solution: The volume of each settler is

$$V = (10)(10)(15) = 1500\,\text{ft}^3$$
$$V = V_1 = V_2$$

For arrangement A, the average residence time in each settler is

$$t_{rA} = \frac{1500}{25} = 60\,\text{s}$$

since

$$\frac{q}{2} = 25\,\text{ft}^3/\text{s}$$

This residence time dictates the efficiency for each unit.

For arrangement B one notes that this effectively operates with q flowing through a settler with volume $V_1 + V_2$. Therefore, the residence time for the series arrangement here is based on $V_1 + V_2$, so that

$$t_{rB} = \frac{3000}{50}$$
$$= 60\,\text{s}$$

Therefore, and as expected, both arrangement will operate with the same efficiency.

8.24 Quantitative Analysis of Parallel versus Series Arrangement

Refer to Problem 8.22. The particle size distribution entering the settler is given in the first two columns of Table 8.2. Verify the conclusion drawn in the previous problem by calculating and comparing the efficiencies of both arrangements.

TABLE 8.5 Efficiency Calculation for Parallel Arrangement

Size Range, μm	Mass w_i,%	$<d_p>_i$, μm	K_i	v_{t_i}, ft/s	H_i^*, ft	E_i	w_iE_i, %
0–10	2	5	0.2221	0.0058	0.3487	2.32	0.05
10–20	4	15	0.6664	0.0523	3.3185	20.92	0.84
20–30	3	25	1.1107	0.1453	8.7180	58.12	1.74
30–40	4	35	1.5550	0.2848	17.0874	100.00	4.00
40–50	9	45	1.9992	0.4708	28.2465	100.00	9.00
50–60	7	55	2.4435	0.7033	42.1954	100.00	7.00
60–70	11	65	2.8878	0.9822	58.9340	100.00	11.00
70–80	18	75	3.3321	1.3025	78.1518	100.00	18.00
80–90	16	85	3.7763	1.5023	90.1377	100.00	16.00
90–100	12	95	4.2206	1.7054	102.3232	100.00	12.00
100–200	14	150	6.6641	2.8705	172.2317	100.00	14.00
							93.63

Solution: The efficiency calculation for the parallel arrangement is provided in Table 8.5. The overall efficiency is 93.63%. The calculation of the efficiency for the series system is left as an exercise for the reader, but an efficiency in the 93–94% range should result.

8.25 Compliance Calculation for a Gravity Settler

A gravity settler is 15 ft wide × 15 ft high × 40 ft long. In order to meet required ambient air quality standards, this unit must remove 90% of the fly ash particles entering the unit. Planned expansion will increase the flowrate to 4000 acfm with a dust loading of 30 gr/ft^3. The specific gravity of fly ash is 2.31 and the process gas stream is air at 20°C and 1.0 atm. The inlet size distribution of the fly ash is given in Table 8.6.

TABLE 8.6 Particle Size Data for Problem 8.25

Size Range, μm	Mass Percent
0.0–10.0	1.0
10–20	1.0
20–30	3.0
30–40	15.0
40–60	20.0
60–80	25.0
80–100	20.0
100–150	15.0

Will the unit meet the specification?

The following data have been provided:

$B = 15$ ft
$H = 15$ ft
$L = 40$ ft
$q = 4000$ acfm
Loading = 30 gr/ft^3
Specific gravity = 2.31; $\rho_p = 144$ lb/ft^3
$\mu = 1.23 \times 10^{-5}$ lb/(ft · s)

Solution: The throughout velocity u is

$$u = q/BH = 4000/[(15)(15)] = 17.778 \text{ ft/min} = 0.296 \text{ ft/s}$$

The fractional efficiency equation may again be written as

$$E = \left(\frac{g d_p^2 \rho_p}{18\mu}\right)\left(\frac{LB}{q}\right) \tag{8.18}$$

Substituting

$$E = \left(\frac{(32.2)(144)}{(18)(1.23 \times 10^{-5})}\right)\left(\frac{(40)(15)}{(4000/60)}\right) d_p^2$$

$$= 1.88 \times 10^8\, d_p^2; \qquad d_p,\ \text{ft}$$

$$= 0.00202\, d_p^2; \qquad d_p,\ \mu\text{m} \tag{8.18}$$

The tabulation shown in Table 8.7 may now be generated. The overall efficiency is 98.51%. As expected, the efficiency is high because of the coarse dust and the size of the unit. The reader should also check to ensure that Stokes' law does in fact apply to those size ranges where the efficiency is less than 100%.

TABLE 8.7 Overall Efficiency Calculation

Size Range, μm	Average Particle Size, μm	Mass Fraction w_i	E_i, %	$w_i\ E_i$, %
0.0–10.0	5	0.01	5.1	0.051
10–20	15	0.01	45.45	0.455
20–30	25	0.03	100	3.0
30–40	35	0.15	100	15
40–60	50	0.20	100	20
60–80	70	0.25	100	25
80–100	90	0.20	100	20
100–150	125	0.15	100	15
				$\sum = 98.51$

8.26 Check on the Efficiency of a Gravity Settler

A salesperson from Bogus, Inc. suggests a gravity settler for a charcoal dust-contaminated air stream that you must preclean. Your supervisor has provided the particle size distribution shown in Table 8.8. The inlet loading is 20.00 gr/ft^3, and the required outlet loading is 5.00 gr/ft^3. Will the settler that the salesperson has suggested do the job?

TABLE 8.8 Particle Size Data For Problem 8.26

Size Range, μm	Weight Percent
0–10	5
10–20	11
20–40	10
40–60	9
60–90	22
90–125	23
125–150	10
150 +	10

Use the specified critical diameter ($d_p^* = 80$ μm) to calculate the size–efficiency data from the equation

$$E = kd_p^2 \tag{8.23}$$

Solution: First calculate k:

$$k = \frac{E}{d_p^2} = \frac{100}{(80)^2}$$
$$= 0.01563$$

Thus, for $d_p = 5$ μm (average diameter for the first size range)

$$E = 0.01563(5)^2$$
$$= 0.39\%$$

The tabulation shown in Table 8.9 may be generated using the approach outlined above. Therefore, $E = 67.7\%$. The required efficiency E_{reg} is

$$E_{reg} = (I - O)/1$$
$$= (20 - 5)/20 = 75\%$$

Since $67.7 < 75\%$, the gravity settler will *not* do the job.

TABLE 8.9 Size–efficiency Results

Size Range, μm	Average Particle Size, μm	Weight Percent	E_i, %	w_iE_i, %
0–10	5	5	0.39	—
10–20	15	11	3.5	0.4
20–40	30	10	14	1.4
40–60	50	9	39	3.5
60–90	75	22	88	19.4
90–125	107.5	23	100	23
125–150	137.5	10	100	10
150 +	130 +	10	100	10
				$\sum = 67.7$

8.27 Sundberg's Method I

Consider the particle size distribution and fractional efficiency curves presented in Figure 8.9 for a certain gravity settler. Determine the overall collection efficiency.

Solution: From the particle size distribution data presented in Figure 8.9, one obtains

Mass median diameter $d_{p50} = 68$ μm
Geometric standard deviation $\sigma_g = 108/68 = 1.588$ μm

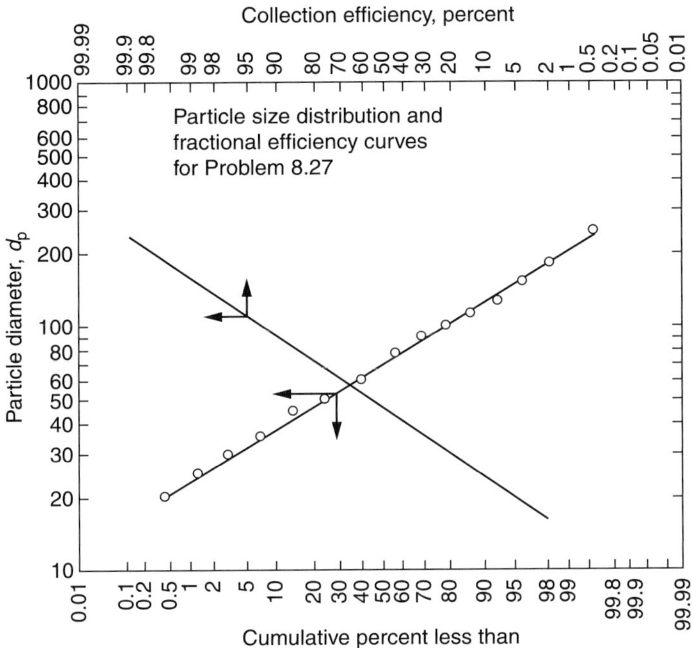

Figure 8.9 Particle size distribution and fractional efficiency curves for the gravity settler in Problem 8.27.

From the size–efficiency graph

$$d'_{p50} = 47\ \mu\text{m}$$

$$\sigma'_g = 78/47 = 1.66\ \mu\text{m}$$

Since both distributions are lognormal, one may apply Sundberg's method [see Equation (7.44)]. Substituting, one obtains

$$E = \text{erf}\left[\frac{\ln(68/47)}{\{(\ln 1.588)^2 + (\ln 1.66)^2\}^{1/2}}\right]$$

$$= \text{erf}(0.538)$$

From Table 7.5, one reads (linearly interpolating)

$$E_T = 70.4\%$$

8.28 Sundberg's Method II

Particle size distribution and size–efficiency data for a gravity settler are provided in Table 8.10. Apply Sundberg's method and calculate the overall efficiency of the unit.

TABLE 8.10 Data For Problem 8.28

PSR, μm	$<d_i>$, μm	w_i	%GTSS	E_i, %	$E_i\ w_i$, %
<20	10	0.032	—	2.0	0.064
20–40	30	0.17	97	10.0	1.70
40–60	50	0.20	80	29.0	5.80
60–80	70	0.16	60	55.0	8.80
80–100	90	0.14	44	92.0	12.88
>100	—	0.30	30	100	30.00
		1.00			59.34

Solution: Plotting of the two graphs on log-probability coordinates is left as an exercise for the reader. Both graphs require "ramming" a straight line through the data. The final results will depend in part on how the line is generated. The author's numbers are provided below:

$$d_{p50} = 72\ \mu\text{m}$$

$$\sigma_{gm} = 72/35 = 2.06;\ \sigma_{gm} = 145/72 = 2.02$$

$$\sigma_{gm} = (2.06 \times 2.01)^{1/2}$$
$$= 2.03$$

$$d'_{p50} = 66\ \mu\text{m}$$

$$\sigma'_{gm} = 85/66 = 1.29,\ \sigma'_{gm} = 66/36 = 1.83$$

$$\sigma'_{gm} = (1.29 \times 1.83)^{1/2}$$
$$= 1.54$$

Applying Equation (7.44), one obtains

$$E = \text{erf}\left[\frac{\ln(72/66)}{\{(\ln 2.03)^2 + (\ln 1.536)^2\}^{0.5}}\right]$$

$$= \text{erf}\left[\frac{0.087}{(0.501 + 0.184)^{0.5}}\right]$$

$$= \text{erf}(0.105)$$

From Table 7.5,

$$E = 54.0\%\ \text{(linear interpolation)}$$

Note that this result is in reasonable agreement with that provided in Table 8.10.

8.29 Atmospheric Discharge Calculation

LT and associates engineers have been requested to determine the minimum distance downstream from a cement source emitting dust that will be free of cement deposit. The source is equipped with a gravity settler. The discharge point from the settler is located 150 ft above ground level. Assume that ambient conditions are at 60°F and 1 atm and neglect meteorological considerations. Additional data are given below.

Particle size range of cement dust = 2.5–50.0 μm
Specific gravity of the cement dust = 1.96
Wind speed = 3.0 mi/hr (mph)

Solution: Note that this can be viewed and/or solved as a fluid particle dynamics problem. A particle diameter of 2.5 μm is used to calculate the minimum distance downstream free of dust since the smallest particle will travel the greatest horizontal distance. To determine the value of K for the appropriate size of the dust, first calculate the particle density using the specific gravity and determine the properties of the gas (assume air).

$$\begin{aligned}\rho_p &= (1.96)(62.4)\\ &= 122.3\ \text{lb/ft}^3\\ \rho &= P(\text{MW})/RT\\ &= (1)(29)/[(0.73)(60+460)]\\ &= 0.0764\ \text{lb/ft}^3\end{aligned}$$

Viscosity of air μ at 60°F = 1.22×10^{-5} lb/(ft · s)
The value of K is

$$\begin{aligned}K &= d_p\left(\frac{g(\rho_p-\rho)\rho}{\mu^2}\right)^{1/3}\\ &= \frac{2.5}{(25{,}400)(12)}\left(\frac{(32.2)(122.3-0.0764)(0.0764)}{(1.22\times10^{-5})^2}\right)^{1/3} = 0.104\end{aligned}$$

The velocity is therefore in the Stokes' law range. Using the appropriate terminal settling velocity equation, the terminal settling velocity is

$$v = \frac{g d_p^2 \rho_p}{18\mu} \tag{8.8}$$

$$\begin{aligned}&= \frac{(32.2)[(2.5)/(25{,}400)(12)]^2(122.3)}{(18)(1.22\times10^{-5})}\\ &= 1.21\times10^{-3}\ \text{ft/s}\end{aligned}$$

The approximate time for descent is

$$\begin{aligned} t &= h/v \\ &= 150/1.21 \times 10^{-3} \\ &= 1.24 \times 10^{5}\ \text{s} \end{aligned}$$

Thus, the distance traveled L is

$$L = tu \tag{8.1}$$

$$\begin{aligned} &= (1.24 \times 10^{5})(3.0/3600) \\ &= 103.3\ \text{min} \end{aligned}$$

8.30 Coal Dust Deposition Volume

An air stream at 100°F, containing 0.13 lb/ft^3 of 75 μm pulverized coal particles, enters a boiler through a 10 ft length of 2 × 2 ft square ductwork, at the rate of 40 acfs. The coal particulate has an actual density of 49.9 lb/ft^3; the bulk density may be assumed equal to one half the actual density. Calculate the weight of coal deposited in the inlet duct after 30 min and comment on the validity of the answer. What fraction of the duct volume will be occupied by the deposited dust after this time period? Assume the throughput velocity to be unaffected by the dust buildup.

Solution: Key calculations are presented below:

$$\begin{aligned} \rho &= 0.071\ \text{lb/ft}^3 \\ \mu &= 1.285 \times 10^{-5}\ \text{lb/ft} \cdot \text{s} \\ d_{\text{p}} &= 75\mu = 2.46 \times 10^{-4}\ \text{ft} \\ K(100^{\circ}\text{F}) &= 2.17 \end{aligned}$$

The calculation for the efficiency is given by [a modified form of Equation (8.15)]

$$\begin{aligned} E &= vL/uH \\ u &= 40/(2)(2) \\ &= 10\ \text{ft/s} \\ v &= 0.42\ \text{ft/s} \\ E &= (0.42)(10)/(10)(2) \\ &= 0.21 \\ &= 21\% \end{aligned}$$

The mass deposited in 30 min m is given by

$$m = qctE$$
$$= (40)(60)(0.013)(30)(0.21)$$
$$= 196.6 \text{ lb}$$

The total (actual) volume of particles V_{act} is

$$V_{\text{act}} = 196.6/49.8$$
$$= 3.94 \text{ ft}^3$$

The volume of the "settler" is

$$V = (10)(2)(2)$$
$$= 40 \text{ ft}^3$$

The volume fraction occupied is

$$VF = 3.94/40$$
$$= 0.1$$
$$= 10\% \text{ of total volume}$$

The total volume occupied by particles V_B is calculated by employing the bulk density (see Section 3.3):

$$V_B = (196.6)/(49.8/2)$$
$$= 7.88 \text{ ft}^3$$

and

$$VF_B = 20\% \text{ of unit}$$

8.31 Dispersion of Soap Particles

A plant manufacturing soap detergent explodes one windy day. It disperses 100 tons of soap particles (specific gravity = 0.8) into the atmosphere (70°F, $\rho = 0.0752 \text{ lb/ft}^3$). If the wind is blowing 20 mph from the west and the particles range in diameter from 2.1 to 1000 μm, calculate the distances from the plant where the soap particles will start to deposit and where they will cease to deposit. Assume that the particles are blown vertically 400 ft in the air before they start to settle. Also, assuming even ground-level distribution through an average 100-ft-wide path of settling, calculate the average height of the soap particles on the ground in the settling area. Assume the bulk density of the particles equal to half of the actual density.

Solution: In a very real sense, this is also a fluid particle dynamics problem. The smallest particle will travel the greatest distance, while the largest will travel the least distance. For the minimum distance, use the largest particle:

$$d_p = 1000\,\mu\text{m} = 3.28 \times 10^{-3}\text{ ft}$$

$$K = d_p\left(\frac{g(\rho_p - \rho)\rho}{\mu^2}\right)^{1/3}$$

$$= (3.28 \times 10^{-3})\left(\frac{(32.2)[(0.8)(62.4) - 0.0752](0.0752)}{(1.18 \times 10^{-5})^2}\right)^{1/3}$$

$$= 31.3$$

Using the intermediate range equation, one obtains

$$v = 0.153\frac{g^{0.71} d_p^{1.14} \rho_p^{0.71}}{\mu^{0.43}\ \rho^{0.29}} \tag{8.10}$$

$$= 0.153\frac{(32.2)^{0.71}(3.28 \times 10^{-3})^{1.14}[(0.8)(62.4)]^{0.71}}{(1.18 \times 10^{-5})^{0.43}(0.0752)^{0.29}}$$

$$= 11.9\text{ ft/s}$$

The descent time t is

$$t = H/v \tag{8.1}$$

$$= 400/11.9 = 33.6\text{ s}$$

The horizontal distance traveled L is

$$L = (33.6)\left(\frac{20}{(60)(60)}\right)(5280)$$

$$= 986\text{ ft}$$

For the maximum distance, use the smallest particle:

$$d_p = 2.1\,\mu\text{m} = 6.89 \times 10^{-6}\text{ ft}$$

$$K = (6.89 \times 10^{-6})\left(\frac{(32.2)[(0.8)(62.4) - 0.0752](0.0752)}{(1.18 \times 10^{-5})^2}\right)^{1/3}$$

$$= 0.066$$

The velocity v is in the Stokes regime and is given by

$$v = \frac{dg_p^2\rho_p}{18\mu} \tag{8.8}$$

$$= \frac{(32.2)(6.89 \times 10^{-6})^2(0.8)(62.4)}{(18)(1.18 \times 10^{-5})}$$

$$= 3.59 \times 10^{-4} \text{ ft/s}$$

The descent time t is

$$t = H/v = 400/3.59 \times 10^{-4} \tag{8.1}$$

$$= 1.11 \times 10^6 \text{s}$$

The horizontal distance traveled L is

$$L = (1.11 \times 10^6)\left(\frac{20}{(60)(60)}\right)(5280)$$

$$= 3.26 \times 10^7 \text{ ft}$$

To calculate the depth D, the volume of particles (actual) V_{act} is first determined:

$$V_{act} = (100)(2000)/[(0.8)(62.4)]$$

$$= 4006 \text{ ft}^3$$

The bulk volume V_B is (see Section 3.3)

$$V_B = 4006/0.5 = 8012 \text{ ft}^3$$

The length of the drop area L_D is

$$L_D = (3.2 \times 10^7) - 994$$

$$= 3.2 \times 10^7 \text{ ft}$$

Since the width is 100 ft, the deposition area A is

$$A = (3.2 \times 10^7)(100) = 3.2 \times 10^9 \text{ ft}^2$$

Note that

$$V_B = AD$$

so that

$$8012 = (3.2 \times 10^9)D$$

solving

$$D = 2.5 \times 10^{-6}\ \text{ft}$$
$$= 0.76\ \mu\text{m}$$

The deposition can be, at best, described as a "sprinkling."

8.32 Optimizing Settler Design

The key process design variable for gravity settlers is the capture area, which is given by

$$A = BL$$

Explain why, once the capture area is calculated, this base area usually takes the form of a square, i.e., $B = L$.

Solutions: Once the capture area is calculated, the cost of the gravity settler internals may be assumed fixed. Because of material costs, however, the larger the outer casing of the physical system, the higher will be the total cost, all other factors being equal. These material costs generally constitute a significant fraction of the total cost of the settler. If the thickness of the outer casing is the same for alternative physical designs and if labor costs are linearly related to the surface area, then the equipment cost will roughly be a linear function of the outer surface area of the structural shell of the unit. This essentially means that the cost is approximately linearly related to the perimeter P, where

$$P = 2L + 2B \tag{8.23}$$

To help minimize the cost (by minimizing the perimeter), one can equate the derivative of the perimeter to zero. Thus, setting

$$B = A/L$$

gives

$$P = 2L + 2A/L$$

and

$$\frac{dP}{dL} = 2 - \frac{2A}{L^2}$$

Setting this derivative equal to zero leads to

$$A = L^2$$

Since

$$A = BL$$

one can conclude

$$L = B$$

Interestingly, most gravity settlers (as well as electrostatic precipitators) are often designed physically in a form approaching a square box.

8.33 Gravity Settler Design Revisited

Refer to Problem 8.20. Redesign the gravity settler using the results of Problem 8.32.

Solution: Employing the results of Problem 8.32, one obtains

$$BL = L^2 = 142.5\ \text{ft}^2$$
$$B = L = 11.94\ \text{ft}$$
$$H = 3\ \text{ft}$$

In this case, the velocity of the gas would be

$$v = \frac{q}{BH} = \frac{(130)}{(11.94)(3)}$$
$$= 3.63\ \text{ft/s}$$

8.34 Effect of Particle Size and Particle Size Distribution on Efficiency

Qualitatively discuss whether the overall efficiency will increase or decrease if the particle size standard deviation is increased. Assume the particle size distribution to be lognormal.

Solution: A typical size–efficiency curve for a gravity settler is presented in Figure 8.10. This is not an easy question to answer because of the discontinuity that the curve exhibits at point A, which represents $d_p(\text{min})$, μm (or d_p^*). From Figure 8.10, this value is 94 μm.

Consider the following scenarios which concern the effect of the particle size standard deviation σ for various average (mean) particle sizes on the efficiency.

1. The mean particle size is d_p^*. For

$$\sigma = 0$$
$$E = 100\%$$

For

$$\sigma > 0$$
$$E < 100\%$$

since particles $<d_p^*$ will be collected with $E < 100\%$, while particles $>d_p^*$ will still be collected with $E = 100\%$. Therefore, $E < 100\%$ (decreases) as $\sigma > 0$.

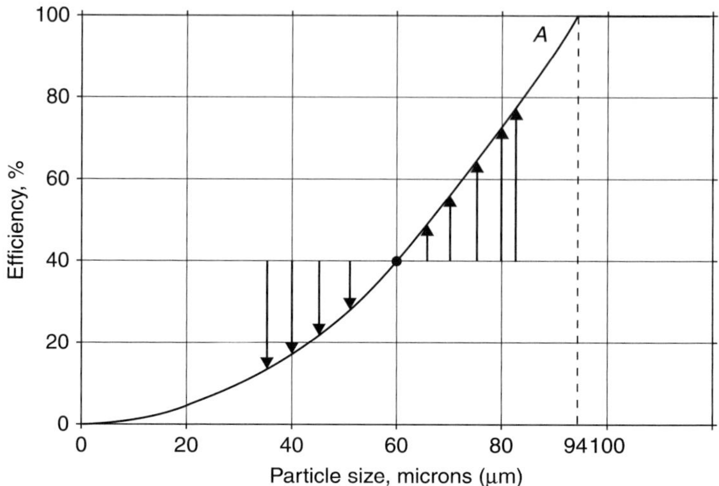

Figure 8.10 Size efficiency curve for a settling chamber.

2. The mean particle size is $>d_p^*$. For

$$\sigma = 0$$
$$E = 100\%$$

If

$$\sigma > 0$$
$$E < 100\%$$

since some particle sizes will be located in a region where $E < 100\%$. Therefore, the effect of increasing σ results in E decreasing; further, the greater σ is, the lower E becomes.

3. The mean particle size is $<d_p^*$. Here is where problems develop. For example, if the mean is 60 μm and

$$\sigma = 0$$
$$E = 40\%$$

If

$$\sigma > 0$$
$$E > 40\%$$

If the particles (by mass) are equally distributed on each side of 60 μm, the larger particles will be collected with a higher net efficiency, while the smaller particles will be collected with a lower net efficiency. Therefore, it appears that the efficiency now *increases* with increasing σ. However, as the mean particle size starts approaching 94 μm, a size will be reached where the efficiency will start *decreasing* since the net efficiency experiences a discontinuity for particles >94 μm, i.e., the efficiency does not continue to increase, but rather remains constant (plateaus) at 100%. The net effect is that for a given mean size <94 μm, the efficiency will experience a maximum for varying σ values. Further, this "maximum" σ will vary with mean particle size.

Since the effect of the lognormal distribution of particle sizes in the three-scenario analysis above is difficult to quantify, the next problem allows one to test this conclusion by performing calculations.

8.35 Analytical Verification of Problem 8.34

Refer to Problem 8.34. Analytically (mathematically) verify the conclusions presented in Problem 8.34.

Solution: The author's mathematical capabilities are limited and cannot provide a determination. This is left as an exercise for those readers who possess strong analytical tools. However, one may perform a trial-and-error verification of the conclusions drawn in Problem 8.34 by varying both the mean particle size and the standard deviation, and applying the calculational procedure given in Problem (8.22).

8.36 Increasing Settler Length

A gravity settler with a width of 7.5 m is designed for 50% removal of particles (SG = 2.5) 100 μm in diameter for an ambient air flow rate of 9.0 L/min. Because of the brittleness of the particle, the actual average diameter was reduced to 40 μm after shearing action in the process. What is the length of settler needed to give the same removal efficiency for the 40 μm particle, and what percent increase in length does this correspond to?

Solution: This is a tricky question. If Stokes' law applies to both particles, one notes from Equation (8.18) that

$$E_1 = K' d_{\mathrm{p}_1}^2 L_1$$

and

$$E_2 = K' d_{\mathrm{p}_2}^2 L_2$$

Since

$$E_1 = E_2 = 0.5_2$$

$$L_2 = (100/40)^2 L_1$$

$$= 6.25\, L_1; \text{ a } 525\% \text{ increase}$$

However, Stokes' law does *not* apply to the 100 μm particle ($K \simeq 4.7$). Since the Intermediate law applies, one applies Equation (8.10) and (8.19).

$$v_2 = 0.153\ (g\rho_p)^{0.71}\ d_p^{1.14}\ (BL)/\rho^{0.29}\ \mu^{0.43} \tag{8.10}$$

and

$$E_2 = 0.153(g\rho_p)^{0.71}\ d_p^{1.14}\ (BL)\ /\rho^{0.29}\mu^{0.43}q \tag{8.19}$$

At "ambient" conditions, assume that

$$\rho = 0.075\ \text{lb/ ft}^3$$

$$\mu = 1.23 \times 10^{-5}\ \text{lb/ft·s}$$

Since (once again)

$$E_1 = E_2$$

and substituting into the equations shown above ultimately leads to

$$L_2 = 2.144\ L_1;\ \text{a 114.4\% increase}$$

This result is reasonable since the 100 μm particle travels with a higher velocity in the Stokes' regime relative to the Intermediate regime.

8.37 Breakeven Pressure Drop/Efficiency

A plant emits 50,000 acfm of gas containing a dust at a loading of 2.0 gr/ft³. A cyclone is employed for particle capture, and the dust captured from the cyclone is worth \$0.01/lb of dust. For the sake of simplified calculation, assume that the efficiency of collection E is related to the system pressure drop ΔP by the formula

$$E = \frac{\Delta P}{\Delta P + 7.5}$$

(where ΔP is in units of lb_f/ft^2). If the fan is 55% efficient (overall) and electric power costs \$0.18/kW · hr, at what collection efficiency is the cost of power equal to the value of the recovered material?

Solution: Take 1.0 minute as a basis.

\$ of recovered particulate, E_R:

$$\begin{aligned} E_R &= 50{,}000\ \text{ft}^3/\text{min} \times 2.0\ \text{gr/ft}^3 \times 1\ \text{lb}/7000\ \text{gr} \times \$0.01/\text{lb} \times E \\ &= 0.143E,\ \$/\text{min} \end{aligned}$$

\$ of power cost to recover, E_P:

$$\begin{aligned} E_P &= 50{,}000\,\text{ft}^3/\,\text{min} \times \Delta P\,\text{lb}_f/\,\text{ft}^2 \times 1\,\text{min}\cdot\text{kW}/44{,}200\,\text{ft}\cdot\text{lb}_f \times 1/0.55 \\ &\quad \times \$0.18/\text{kW}\cdot\text{hr} \times 1\ \text{hr}/60\ \text{min} \\ &= 0.006\ \Delta P,\ \$/\text{min} \end{aligned}$$

However,

$$E = \Delta P/(\Delta P + 7.5)$$

Equating E_R with E_P and replacing E with the equation given above yields

$$(0.143)[\Delta P/(\Delta P + 7.5)] = 0.006\ \Delta P$$

Solving by trial and error (or quadratically) yields

$$\begin{aligned} \Delta P &= 16.3\ \text{lb}_f/\,\text{ft}^2 \\ &= 3.1\ \text{in}\ H_2O \\ E &= 16.3/(16.3 + 7.5) = 0.685 \\ &= 68.5\% \end{aligned}$$

8.38 Optimum Pressure Drop/Efficiency

Refer to Problem 8.37. What pressure drop or efficiency will lead to maximum profit?

Solution: The profit PR is given by

$$PR = 0.143E - 0.006\Delta P$$

with

$$E = \frac{\Delta P}{\Delta P + 7.5}$$

Thus

$$PR = (0.143)\left[\frac{\Delta P}{\Delta P + 7.5}\right] - 0.006\Delta P$$

To obtain either the maximum or minimum, one must set

$$\frac{d(PR)}{d(\Delta P)} = 0$$

Differentiating leads to

$$\frac{d(PR)}{d(\Delta P)} = (0.143)\left[\frac{1}{\Delta P + 7.5} - \frac{\Delta P}{(\Delta P + 7.5)^2}\right] - 0.006 = 0$$

Solving, one obtains

$$\Delta P = 5.9 \text{ lb/ ft}^2$$
$$E = 0.44 = 44\%$$

8.39 Options to Increase the Efficiency of Existing Settler

The collection efficiency of an installed and operating gravity settler has suddenly decreased from a regulatory required 65% to 40%, and leveled off at 40%. Assuming that there are other gravity settlers available at the plant, provide a series of options—with advantages and disadvantages—that may be implemented to bring the discharge back into compliance.

Solution: This is an open-ended problem, and there are several options available for bringing the operation back into compliance:

1. Reduce the process flow rate
2. Extend the length of the settler (if possible)
3. Increase the height of the settler (if possible)
4. Increase the width of the settler (if possible)
5. Insert another settler downstream from the present settler
6. Insert another particle control device downstream from the present unit
7. Reduce the inlet particulate concentration
8. Insert shelves in the unit
9. Reduce the process gas flow temperature (if possible)

NOTE: Additional problems are available for all readers at www.wiley.com. Follow links for this title.

9

CYCLONES

9.1 INTRODUCTION

Cyclones provide a relatively low-cost method of removing particulate matter from exhaust gas streams. Cyclones are somewhat more complicated in design than simple gravity settling systems, and their removal efficiency is accordingly much better than that of settling chambers. However, cyclones are not as efficient as electrostatic precipitators, baghouses, and venturi scrubbers but are often installed as precleaners before these more effective devices.

Cyclones come in many sizes and shapes and have no moving parts. From the small 1- and 2-cm diameter source sampling cyclones used for particle size analysis to the large 5-m diameter cyclone separators used after wet scrubbers, the basic separation principle remains the same. Particles enter the device with the flowing gas (Figure 9.1); the gas stream is forced to turn, but the larger particles have more momentum and cannot turn with the gas. These larger particles impact and fall down the cyclone wall and are collected in a hopper. The gas stream actually turns a number of times in a helical pattern, much like the funnel of a tornado. The repeated turnings provide many opportunities for particles to pass through the streamlines, thus hitting the cyclone wall.

The range of particle sizes collected in a cyclone is dependent upon the overall diameter and relative dimensions of the device. Various refinements such as the use

Air Pollution Control Equipment Calculations. By Louis Theodore

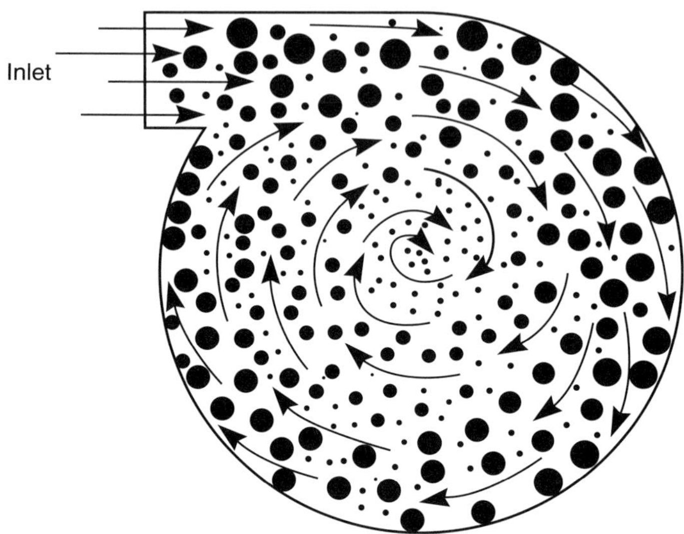

Figure 9.1 Particle collection process.

of skimmers, turning vanes, and water sprays can in some cases improve efficiency. Stacking cyclones in series or in parallel can provide further alternatives for improving overall collection efficiency.

Three types of cyclones are shown in Figure 9.2. The first diagram, Figure 9.2a shows a typical tangential entry cyclone arrangement. These cyclones have a distinctive and easily recognized form and can be found in almost any industrial area of a town or a city—at lumber companies, feed mills, cement plants, power plants, smelters, and at many other process industrial sites. Since top inlet-type cyclones are so widely used, most of this chapter will be devoted to their operational characteristics.

In axial entry cyclones, Figure 9.2b, the gas inlet is parallel to the axis of the cyclone body. Here, the exhaust process gases enter from the top and are directed into a vortex pattern by the vanes attached to the central tube. Axial entry cyclones are commonly used in multicyclone configurations; these units generally provide higher efficiencies.

The larger cyclonic-type separator shown in Figure 9.2c is often used after wet scrubbers to collect particulate matter entrained in water droplets. The gas enters tangentially at the bottom of the unit, forming a vortex. The large water droplets are forced against the walls and are removed from the gas stream.

There are other variations in the design of cyclones. They are usually characterized by where the gas enters and exits the cyclone body (tangentially, axially, or peripherally).

There are four major parts to a cyclone. These are also shown in Figure 9.2. The inlet, the cyclone body, the dust discharge system, and the outlet all affect the overall efficiency of the cyclone. Gas is directed into the cyclone by the inlet, which is instrumental in the formation of the vortex. In the cyclone body the particulate matter is forced to the wall. The gas continues down the cyclone body to the cone, which gives the gas enough rotational

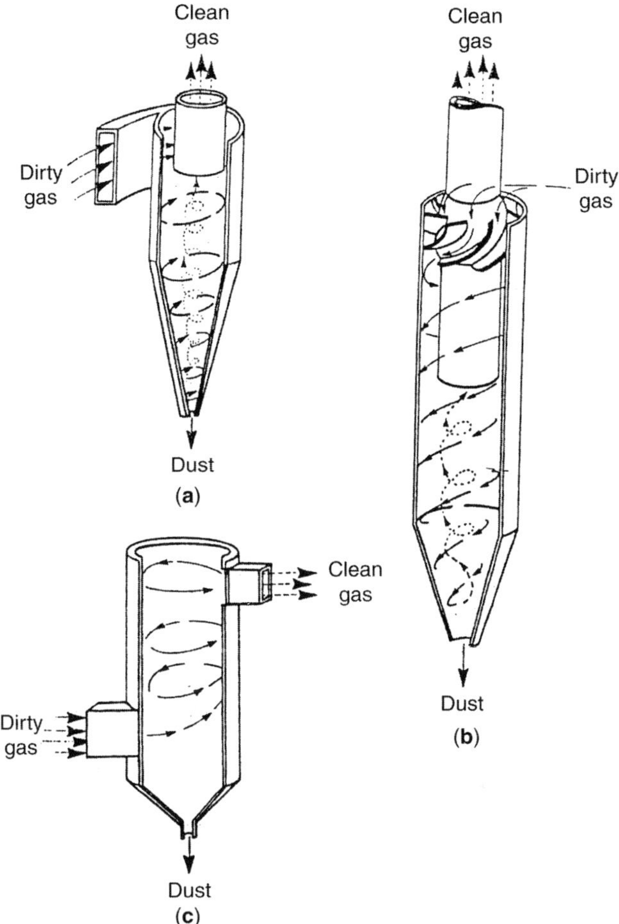

Figure 9.2 Types of cyclones: (**a**) top inlet; (**b**) axial inlet; (**c**) bottom inlet.

velocity to keep the particulates against the wall. At the bottom of the cone, the gas changes direction from downward to upward. The ascending vortex enters a tube extension that is sometimes called a *vortex finder* and exits the cyclone. Meanwhile, the collected particulate matter drops into a hopper, where it is periodically or continuously removed. Each of these parts of a cyclone are discussed in more detail below.

The removal efficiency of a cyclone for a given size particle is very dependent on the cyclone dimensions. The efficiency at a given volumetric flow rate is most affected by the diameter. The overall length determines the number of turns of the vortex. The greater the number of turns, the greater the efficiency. The length and width of the inlet are also important, since the smaller the inlet, the greater the inlet velocity becomes. A greater inlet velocity gives greater efficiency but also increases the pressure drop.

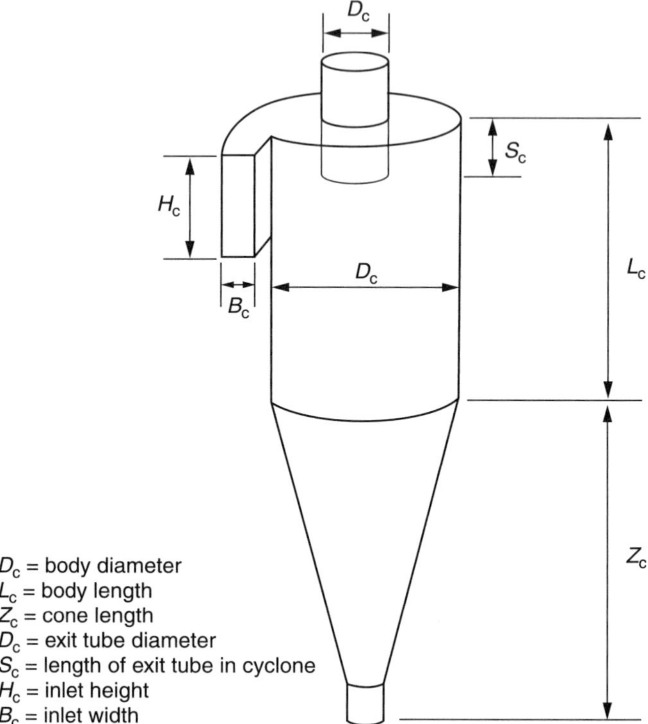

Figure 9.3 Nomenclature for a tangential entry cyclone.

Consider the dimensions shown in Figure 9.3. Many different types of cyclones have been designed by merely varying the dimensions highlighted in Figure 9.3. Table 9.1 gives dimensional characteristics of a number of designs reported in the literature. Dimensions are given relative to the body diameter D_c.

High-efficiency cyclones generally have smaller inlet and exit areas with a smaller body diameter and possibly longer overall length. A conventional cyclone will be

TABLE 9.1 Dimensionless Design Ratios for Tangential Entry Cyclones

		Efficiency		
Symbol	Nomenclature	High	Medium	Conventional
D_c	Body diameter	1.0	1.0	1.0
H_c	Inlet height	0.5	0.75	0.5
B_c	Inlet width	0.2	0.375	0.25
S_c	Outlet length	0.5	0.875	0.625
D_c	Outlet diameter	0.5	0.75	0.5
L_c	Cylinder length	1.5	1.5	2.0
Z_c	Cone length	2.5	2.5	2.0

4–12 ft (1.2–3.6 m) in diameter, with a pressure drop of 2–5 inches (5–13 cm) of water. A high-efficiency cyclone will be less than 3 ft (0.9 m) in diameter with a pressure drop of 2–6 inches (5–15 cm) of water.

It should be apparent from the discussion above that small cyclones are more efficient than large cyclones. Small cyclones, however, have a higher pressure drop and are limited with respect to volumetric flow rates. Smaller cyclones can be arranged either in series or in parallel to substantially increase efficiency at lower pressure drops. These gains are somewhat offset, however, by increased cost and maintenance problems. Multicyclone arrangements also tend to plug more easily. When common hoppers are used in such arrangements, different flows through cyclones can lead to reentrainment problems.

A typical series arrangement is shown in Figure 9.4. Larger particles can be collected in the first cyclone and a smaller, more efficient cyclone can collect smaller particles. Such an arrangement can reduce dust loading in the second cyclone and avoid problems of abrasion and plugging. Also, if the first cyclone should plug, there still will be some collection occurring in the second cyclone. The additional pressure drop produced by the second cyclone adds to the overall pressure drop of the system. The higher pressure drop can be a disadvantage in such a series system design.

Many types of parallel arrangements have been designed for cyclones. An example of a parallel arrangement using tangential entry cyclones is shown in Figure 9.5.

With batteries of cyclones using a common inlet plenum, higher volumes of gas can be treated at reasonable pressure drops. In configurations where a common hopper is

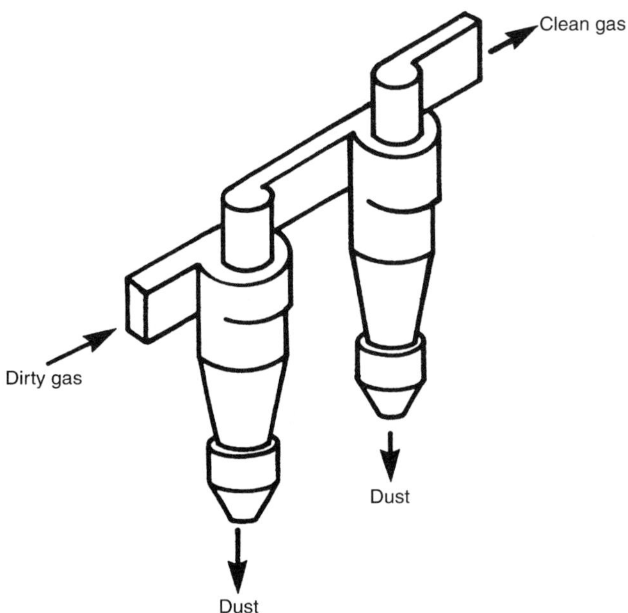

Figure 9.4 Cyclones in series.

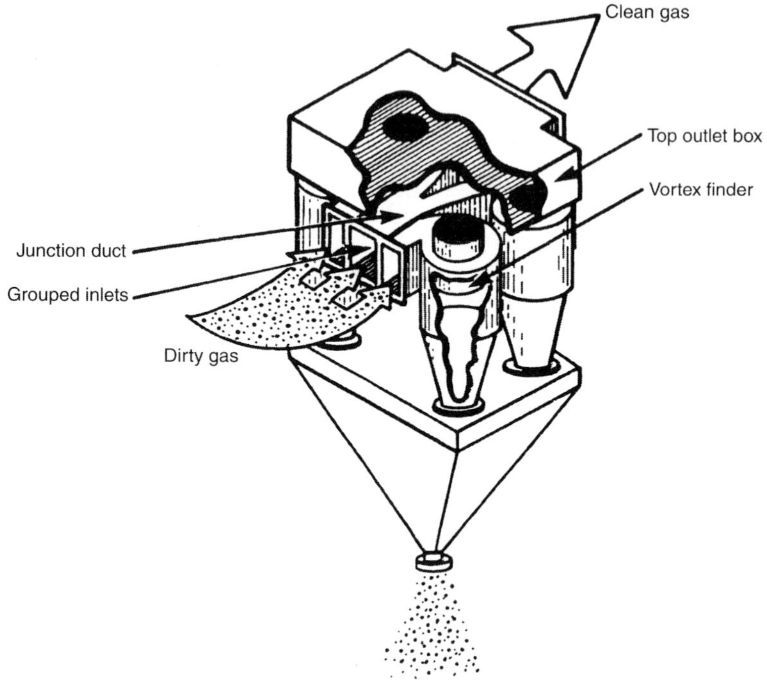

Figure 9.5 Battery of four involute cyclones in parallel.

used, each cyclone should have the same pressure drop or the gas will preferentially channel through one cyclone or several cyclones.

Another type of parallel arrangement uses the axial entry cyclone. Arrangements of high-efficiency, small-diameter axial cyclones can provide increases in collection efficiency with corresponding reductions in pressure drop, space, and cost. Such a *multiclone* arrangement is shown in Figure 9.6. Pressure drops commonly range from 4 to 6 inches (from 10 to 15 cm) of water.

The axial entry minimizes the eddy formation that is common in tangential entry cyclones. Here, the inlet guide vanes create the vortex. Care must be taken in designing the inlet plenum for the multiclone since the inlet exhaust gas should have an even distribution to each individual cyclone. Sticky materials should not be collected using multiclones since the vanes and smaller outlet tubes are prone to plugging.

Reentrainment can be minimized by making the cyclone hopper large in volume and deep enough so that the collected dust level will lie below the point where the vortex ends. The addition of a mechanical valve that can periodically or continually remove the dust from the cyclone can effectively reduce inflow from the hopper. A number of designs are shown in Figures 9.7–9.10. These designs are often employed with electrostatic precipitators and baghouses.

A valve between the cyclone and the bin can be a simple manual device as shown in Figure 9.7, or can provide a continuous discharge as with the rotary valve and screw

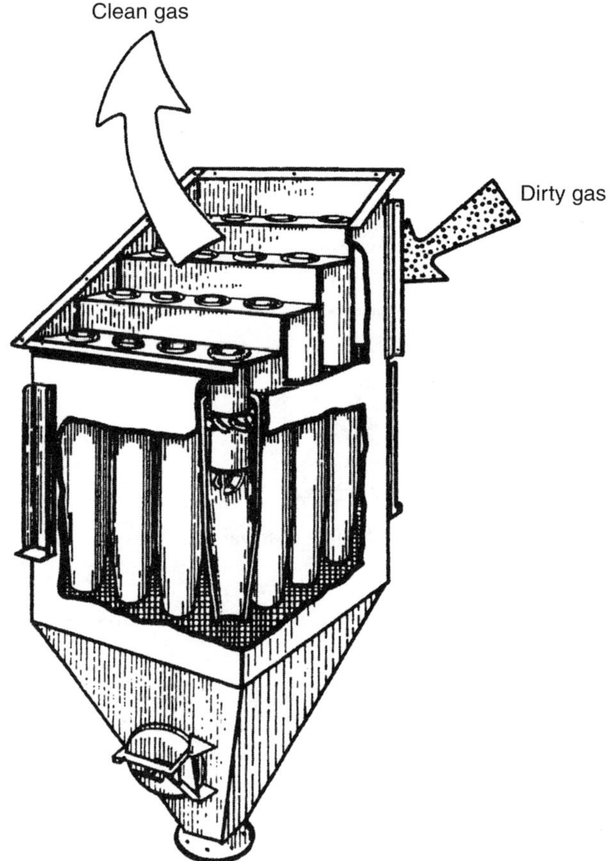

Figure 9.6 Battery of vane axial cyclones.

feeder shown in Figures 9.8 and 9.9. Automatic flap valves shown in Figure 9.10 can periodically swing to discharge accumulated dust in a double-valved arrangement.

Cyclone efficiency can also be improved if a portion of the flue gas is drawn through the hopper. An additional vane or lower pressure duct can provide this flow. However, it may then become necessary to recirculate or otherwise treat this purge exhaust to remove uncollected particulate matter.

9.2 DESIGN AND PERFORMANCE EQUATIONS

Objects moving in circular paths tend to move away from the center of their motion. The object moves outward as if a force is pushing it out. This force is known as *centrifugal force*. The whirling motion of the gas in a cyclone causes particulate

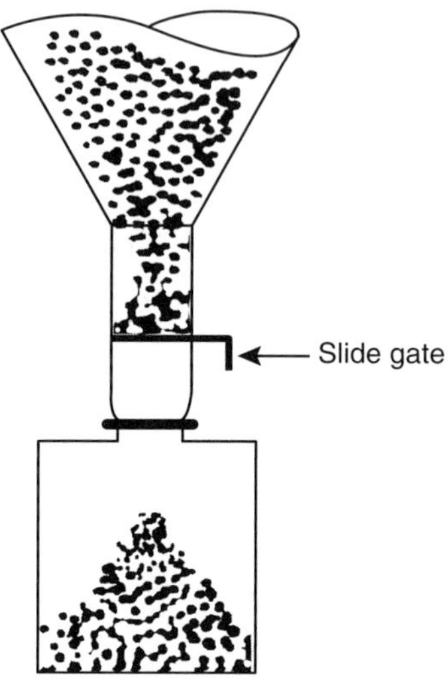

Figure 9.7 Simple manual slide gate.

matter in the gas to sense this force and move out to the walls. An expression for this force is as follows:

$$F = \frac{\rho_p d_p^3 v_p^2}{r} \tag{9.1}$$

where ρ_p = particle density, lb/ft^3 (kg/m^3)
d_p = particle diameter, inches (μm)

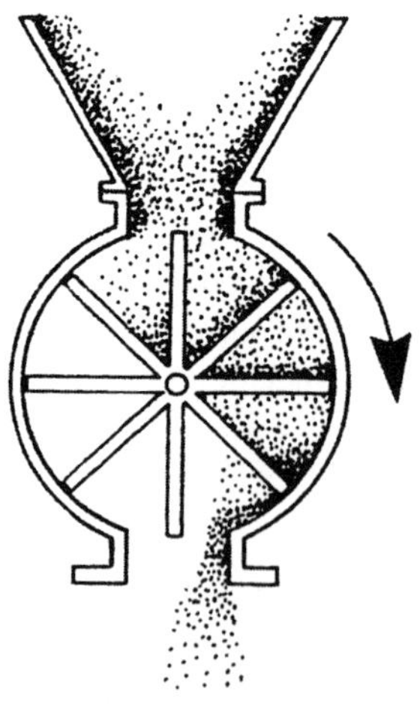

Figure 9.8 Rotary valve.

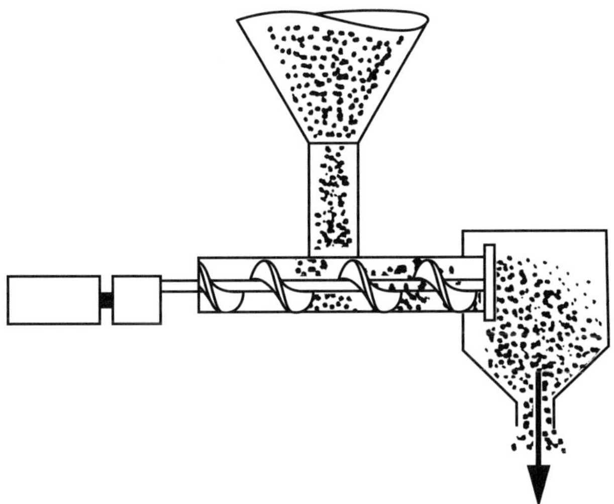

Figure 9.9 Discharge screw feeder.

v_p = particle tangential velocity, ft/s (m/s)
r = radius of the circular path, ft (m)

The term F is the force that the particulate matter experiences in a cyclone. Equation (9.1) explains several of the cyclone characteristics discussed in the previous section. For example $\rho_p d_p^3$ is merely proportional to the mass of the particle. The larger the mass, the greater the force. The tendency to move toward the walls is consequently increased and larger particles are more easily collected. The reason why all of the particles do not move to the wall is because of the drag resistance of the air. The buffeting molecules in the gas resist the outward motion and act like an opposing force. Particles move to the wall when the centrifugal force is greater than the opposing drag force.

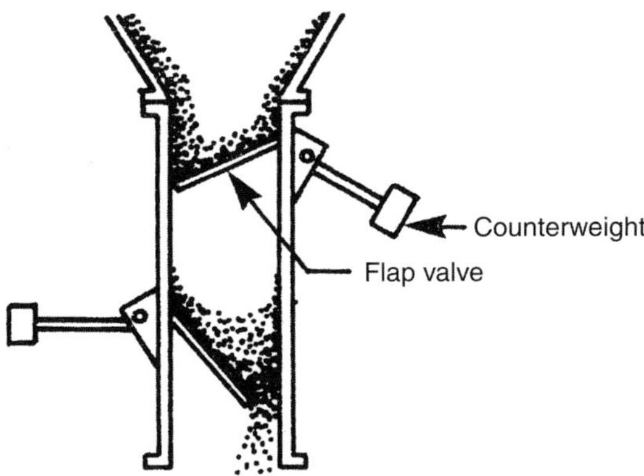

Figure 9.10 Automatic flap valve.

Note also from Equation (9.1) that as r (the radius of the circular path) decreases, the force increases. This is why smaller cyclones are more efficient for the collection of smaller-sized particles than are large cyclones.

These types of considerations, in conjunction with considerations of cyclone geometry and vortex formation, have led to the development of numerous performance equations. These equations attempt to characterize the behavior of cyclones. Some work well, and some do not. None adequately describe performance under all operating conditions such as at high pressure and high temperatures.

Three important parameters can be used to characterize cyclone performance:

$$\begin{aligned} d_{pc} &= \text{cut diameter} \\ \Delta P &= \text{pressure drop} \\ E &= \text{overall collection efficiency} \end{aligned}$$

Equations involving each of these parameters are provided in this section. The equations should be used with caution, however, since there are strict limitations on their applicability.

Cut Diameter

The *cut diameter* is defined as the size (diameter) of particles collected with 50% efficiency. It is a convenient way of defining efficiency for a control device since it provides information on the effectiveness for a particle size range. A frequently used expression for cut diameter is

$$d_{pc} = \left(\frac{9\mu B_c}{2\pi N v_i(\rho_p - \rho)} \right)^{0.5} \tag{9.2}$$

where μ = viscosity, lb/ft · s (Pa · s)
N = effective number of turns (5–10 for the common cyclone)
v_i = inlet gas velocity, ft/s (m/s)
ρ_p = particle density, lb/ft^3 (kg/m^3)
ρ = gas density, lb/ft^3 (kg/m^3)
B_c = inlet width, ft (m)

The cut diameter, d_{pc} or $[d_p]_{cut}$, is a characteristic of the *control device* and should not be confused with the geometric mean particle diameter, d_{gm}, of the size distribution.

Figure 9.11 shows a size efficiency curve and points out the cut diameter and the *critical diameter*, $[d_p]_{crit}$, the particle size collected at 100% efficiency. Values of $[d_p]_{crit}$ are difficult to obtain from such curves so the cut size is often determined instead.

A number of formulas exist for the calculation of the cut diameter and critical diameter. A value of N, the number of turns, must be known in order to solve Equation (9.2)

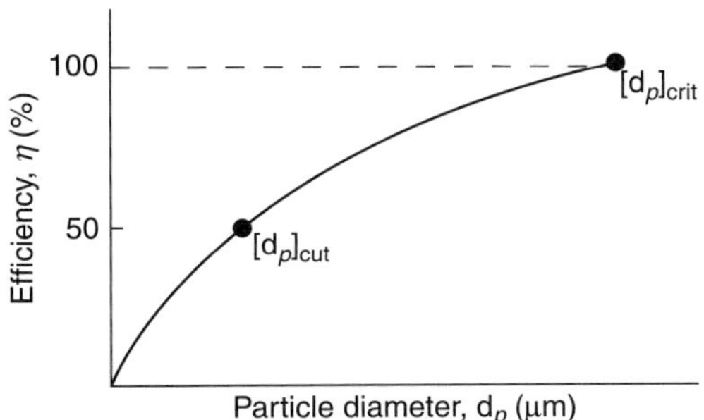

Figure 9.11 Typical size efficiency curve.

for $[d_p]_{cut}$. Given the volumetric flow rate, inlet velocity, and dimensions of the cyclone, N can be easily calculated. Values of N can vary from 1 to 10, with typical values in the 4–5 range.

The expression for the cut diameter [Equation (9.2)] has been found to agree with some experimental data. However, other experimental work has shown limitations to its application. A high-efficiency cyclone can have a cut diameter of typically 5–10 μm. Equation (9.2) is typical of most of those devised for determining the cut or critical diameter. Note that an increase in the number of turns, inlet velocity, or the particle density will decrease the cut size as one would expect. A decrease in viscosity will decrease the drag force opposing the centrifugal force and therefore also reduce the cut size (i.e., smaller particles will be collected).

Collection Efficiency

A number of equations have been developed for determining the fractional cyclone efficiency E_i for a given size particle. As noted earlier, *fractional efficiency* is defined as the fraction of particles of a given size collected in the cyclone, compared to those of that size going into the cyclone.

No efficiency theory or calculation method provides a description for all cyclones. The modification of inlets and outlets, addition of fines educators, etc., introduce variables that are difficult to treat theoretically. Although theoretical efficiencies can give estimations of cyclone performance, it should be kept in mind that designers of equipment commonly rely on comparative evaluations between similar designs and on experience.

This section will describe two methods of calculating cyclone efficiency. The Leith and Licht theory for calculating fractional efficiency will be discussed first. A convenient graphical method for estimating efficiency, developed by Lapple [C. Lapple, *Chem. Eng.*, **58**: 144 (1951)] will also be discussed.

The fractional efficiency equation of Leith and Licht [D. Leith and D. Mehta, "Cyclone Performance and Design," *Atmosph. Environ.*, **7**: 527–549 (1973)] has a form similar to many of those developed for particulate control devices. Fractional efficiency is expressed as

$$E_i = 1 - e^{\left[-2(c\psi)^{1/(2n+2)}\right]} \tag{9.3}$$

where $c =$ cyclone dimension factor
$\psi =$ impaction parameter
$n =$ vortex exponent

The exponential form of this equation is quite common since it is employed with other control devices. In this expression, c is a factor that is a function only of the cyclone's dimensions The symbol ψ expresses characteristics of the particles and gas as

$$\psi = \frac{\rho_p d_p^2 v_i}{18\mu D_c}(n+1) \tag{9.4}$$

and is known as the *inertia* or *impaction parameter*. Note that ρ_p times v_i essentially expresses the particle's initial momentum. The value of n is dependent on the cyclone diameter and temperature of the gas stream. The calculations involved in this method are straightforward, although tedious. The original reference should be consulted if it is to be applied.

An older, and more popular, method of calculating cyclone fractional efficiency and overall efficiency was developed by Lapple (USEPA AP 40, *Air Pollution Engineering Manual*, 1951, pp. 94–99). Lapple first computed the ratios $d_p/[d_p]_{cut}$, the particle diameter versus the cut diameter ratio as determined from Equation (9.2) or Figure 9.12. He found that cyclone efficiency correlates in a general way with this ratio. For a typical cyclone, efficiency will increase as the ratio increases as shown in Figure 9.12.

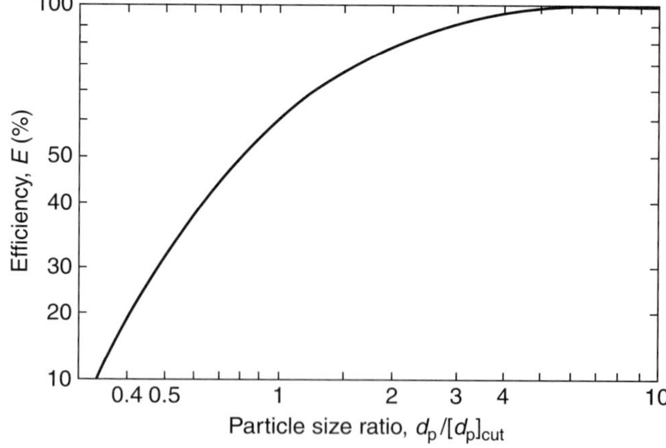

Figure 9.12 Cyclone collection efficiency versus particle size ratio.

Lapple's supposed development of this method receives treatment in the "Problems" section of this chapter.

As a universal curve for common cyclones, the preceding correlation has been found to agree reasonably well with experimental data. To calculate fractional efficiencies, the procedure presented below should be completed.

Lapple Calculation Procedure				
d_p range	wt fraction in range	$d_p/[d_p]_{cut}$	E_i for each d_p from experiment or Lapple's method, %	wt fraction $\times E_i$

The sum of these products in the rightmost section of the box will give the overall efficiency. The following reference provides more detailed information: L. Theodore, "Engineering Calculations: Cyclonic Collection Systems," *Chem. Eng. Progress*, pp. 18–20 (Sept. 2005).

Pressure Drop

The pressure drop across a cyclone is an important parameter to the purchaser of such equipment. Increased pressure drop means greater costs for power to move an exhaust gas through the control device. With cyclones, an increase in pressure drop usually means that there will be an improvement in collection efficiency (one exception to this is the use of pressure recovery devices attached to the exit tube; these reduce the pressure drop but do not adversely affect collection efficiency). For these reasons, there have been many attempts to predict pressure drops from design variables. The idea is that having such an equation, one could work back and optimize the design of new cyclones.

An expression occasionally used is

$$\Delta P = \frac{0.0027\, q^2}{k_c D_c^2 B_c H_c (L_c/D_c)^{1/3} (Z_c/D_c)^{1/3}}; \quad \text{consistent units} \tag{9.5}$$

where $q =$ volumetric flow rate.
(In this case k_c is a dimensionless factor descriptive of cyclone inlet vanes. It is equal to 0.5 for cyclones without vanes, 1.0 for vanes that do not expand the entering gas or touch the outlet wall, and 2.0 for vanes that expand and touch the outlet wall.) This equation, when compared to experimental data, was found to have a poor correlation coefficient.

The most popular of the empirical pressure drop equations has the form

$$\Delta P = K_c \rho v_i^2; \quad \text{consistent units} \tag{9.6}$$

where $K_c =$ a proportionality factor.

If ΔP is measured in inches of water, K_c can vary from 0.013 to 0.024, with 0.024 the norm. Velocities for cyclones range from 20 to 70 ft/s (6–21 m/s), although common velocities range from 50 to 60 ft/s (15–18 m/s). At velocities greater than 80 ft/s (24 m/s), turbulence increases in the cyclone and efficiency will actually decrease. Also, at high loads of particulate matter and high velocities, scouring of the cyclones by the particles will rapidly increase. To minimize erosion in such cases, a cyclone should be designed for lower inlet velocities.

Pressure drops for single cyclones vary depending on both size and design. Common ranges are

Low-efficiency cyclones	2–4 in H_2O (5–10 cm H_2O)
Medium-efficiency cyclones	4–6 in H_2O (10–15 cm H_2O)
High-efficiency cyclones	8–10 in H_2O (20–25 cm H_2O)

9.3 OPERATION AND MAINTENANCE, AND IMPROVING PERFORMANCE

There are many operating variables in cyclone performance. These include characteristics of both the gas and the particles. Gas operating variables include temperature, pressure, and composition. Dust characteristics include size, size distribution, shape, density, and concentration.

As the temperature of a gas increases, its density decreases while its viscosity increases. Since the gas density is negligible compared to the particle density, this has no direct effect on efficiency. A higher temperature will increase the inlet velocity, increasing the particle velocity toward the wall. However, the increasing viscosity decreases the particle velocity toward the wall. The net effect of this is that within the normal operating range of 40–700°F the collection efficiency is essentially constant. As the temperature rises above 1000°F, the viscosity effect dominates, causing a decrease in efficiency. The gas composition can also affect gas viscosity and density.

The influence of gravity on the dust separation in a cyclone is slight, so that the efficiency is almost independent of the position of the cyclone. Separations remain satisfactory whether the cyclone is vertical, horizontal, or even upside-down with the dust carried off upward from the cyclone. A good cyclone separates the dust satisfactorily in any position. However, coarse particles of dust may keep rotating in the conical part of the cyclone, never reaching the outlet. This difficulty can arise no matter how the cyclone is positioned, but less so if the cyclone is in its normal vertical position.

To operate efficiently, cyclone dust collectors should be airtight. Gaskets must be used to close the gaps between flanges. For axial entry collectors, header sheets must be sealed without any breaks in the welds. All doors, ports, and poke holes should also be sealed to prevent reentrainment. Since there are always some fine dust particles suspended in the air, even slight leaks can cause reentrainment and escape of dust particles from the cyclone.

Detection of leaks in cyclones is not very complicated. Bright floodlights can be used to check for leaks. Erosion holes, gasket leakage, and weld breaks can be detected by shining a light up from the bottom of the cyclone. For axial-entry cyclones, breaks in header sheets, decomposed gaskets, and cracks in material can be detected when directing the light up from the hopper.

The prevention of leaks is very important. Because a cyclone operates on an inertial principle and leakage disrupts the flow pattern, fine particles can be reentrained through the outlet vortex. Collection efficiency will decrease due to dust reentrainment, so it is very important to seal the cyclones.

Collection efficiency in a cyclone is primarily determined by the pressure drop and/or the inlet velocity. The pressure drop can be increased or decreased by varying the diameter of the cyclone body or by varying the volumetric flow rate per tube. These features must be designed into the system. If the cyclone is operated at a lower volumetric flow rate, dampers should be used so that the gas velocity will be increased.

Since cyclones have no moving parts, fine tuning them is very difficult. Spiral vanes are used in axial entry cyclones, allowing some control of volumetric flow rate by moving a vane in and out of a constricted opening in the collector element. When fully in, maximum rotation is induced, resulting in greater centrifugal action. When fully out, much of the gas bypasses the vanes.

Dampers at the inlet are used to control turndown of gas flow. This can be accomplished if the collector is sectionalized and allows for the designed pressure drop to be maintained. The excess tubes can be capped off if the turndown is permanent.

Efficiency can be improved by arranging cyclones in series or parallel; however, these gains can be offset by increased maintenance problems. Multicyclone arrangements plug more easily and, when common hoppers are used, uneven flow distribution can lead to reentrainment problems.

TABLE 9.2 Changes in Performance Characteristics

Cyclone and Process Design Changes	Pressure Drop	Efficiency	Cost
Increase cyclone size (D_c)	Decreases	Decreases	Increases
Lengthen cylinder (L_c)	Decreases slightly	Decreases	Increases
Lengthen cone (Z_c)	Decreases slightly	Increases	Increases
Increase exit tube diameter (D_e)	Decreases	Decreases	Increases
Increase inlet area maintaining velocity	Increases	Decreases	Decreases
Increase velocity	Increases	Increases	Operating costs higher
Increase temperature (maintaining velocity)	Decreases	Decreases	No change
Increased dust concentration	Decreases for large increases	Increases	No change
Increasing particle size and/or density	No change	Increases	No change

A summary of changes in performance characteristics produced by changes in cyclone design and exhaust gas properties is given in Table 9.2; quantitative equations are provided in Table 9.3.

TABLE 9.3 Effect of Operation Variables on Cyclone Operation and Performance

Variable	Effect on Efficiency	Effect on Pressure Drop
Flow rate	$\frac{100-E_1}{100-E_2}=\left(\frac{q_2}{q_1}\right)^{0.5}$	$\frac{\Delta P_1}{\Delta P_2}=\frac{q_1^2\rho_1}{T_1}\left[\frac{T_2}{q_2^2\rho_2}\right]$
Gas density	$\frac{100-E_1}{100-E_2}=\left[\frac{\rho_p-\rho_2}{\rho_p-\rho_1}\right]^{0.5}$	Same as above
Particle density	Same as above	Negligible
Gas viscosity	$\frac{100-E_1}{100-E_2}=\left[\frac{\mu_1}{\mu_2}\right]^{0.5}$	Negligible
Dust loading	$\frac{100-E_1}{100-E_2}=\left[\frac{c_{i,2}}{c_{i,1}}\right]^{0.182}$	$\frac{\Delta P_d}{\Delta P_c}=\frac{1}{0.013\,c_i^{0.5}+1}$

Key: 1 = condition 1; 2 = condition 2; q = gas volumetric flow rate, ft^3/s; E = efficiency; ΔP = pressure drop, in H_2O (subscript d indicates the gas stream under loading conditions and subscript c, no dust loading or clean conditions); ρ = gas density, lb/ft^3; T = absolute temperature, °R; μ = gas viscosity, lb/ft · s; c_i = inlet dust concentration, $grains/ft^3$.

There is little to report on recent developments with cyclones. Some research reports indicate that cyclones have been successfully employed to capture fine particulate in the nanometer (nm) range. Actual results have yet to be accepted by industry at the time of preparation of this book.

PROBLEMS

9.1 Typical Cyclone Velocity

A typical throughput velocity in a cyclone is (select one)

(**a**) 10 ft/min
(**b**) 60 ft/min
(**c**) 100 ft/min
(**d**) None of the above

Solution: As indicated in the body of the text for this chapter, typical velocities are 60 ft/s. Answers (a)–(c) do not approach this value. The correct answer is therefore (d).

9.2 Effect of Throughput Velocity on Efficiency

As the throughput velocity of a gas increases, the collection efficiency of a cyclone will

(**a**) Increase
(**b**) Decrease

(**c**) Remain relatively constant
(**d**) Vary sinusoidally

Solution: On inspecting either Equation (9.2) or Tables 9.2 and 9.3, one finds that an increase in velocity results in an increase in efficiency. Note that a decrease in the cut diameter increases efficiency. The correct answer is therefore (a).

9.3 Effect of Throughput Velocity on Pressure Drop

As the throughput velocity of a gas increases, the pressure drop of a cyclone will

(**a**) Increase
(**b**) Decrease
(**c**) Remain relatively constant
(**d**) Vary sinusoidally

Solution: On inspecting either Equations (9.5) and (9.6) or Tables 9.2 and 9.3, one finds that an increase in velocity produces an increase in pressure drop. The correct answer is therefore (a).

9.4 Effect of Temperature on Efficiency

As the temperature of the throughput gas increases, the overall efficiency of a cyclone will

(**a**) Increase slightly
(**b**) Decrease slightly
(**c**) Remain relatively constant
(**d**) Increase exponentially

Solution: On inspecting Equation (9.2), one notes that temperature can affect only the gas viscosity. Since the viscosity of a gas increases with temperature rise (see Section 3.3), the cut diameter would increase slightly, producing a slight decrease in efficiency. The correct answer is therefore (b).

9.5 Particle Size Collected with 100% Efficiency

The smallest particle size that is collected at 100% efficiency by a cyclone is referred to as the

(**a**) Cut size
(**b**) Geometric mean
(**c**) Critical size
(**d**) Design efficiency size

Solution: As noted in this and the previous chapter, this particle is defined as the critical size. The correct answer is therefore (c).

9.6 Cut Size Definition

In a cyclone, the cut size of a particle is the size of the particle

(**a**) Collected with 100% efficiency
(**b**) Less than 20 μm
(**c**) Collected with 50% efficiency
(**d**) That will not be collected

Solution: As defined in this and the previous chapter, the cut size is that particle size that is collected with 50% efficiency. The correct answer is therefore (c).

9.7 Particle Collection Method

The inlet gas velocity in a cyclone is transformed into a vortex that is confined within the unit. The particles are collected when

(**a**) They are thrown against the wall by centrifugal force and fall into the dust hopper
(**b**) The spiral of the vortex changes direction
(**c**) The drag force is greater than the centrifugal force
(**d**) The vortex finder connects with the vortex arrestor

Solution: As described in Section 9.1, it is a combination of centrifugal and gravitational forces that lead to collection. The correct answer is therefore (a).

9.8 Operating Pressure Drop

The operating pressure drop across a cyclone is approximately

(**a**) 4 in H_2O
(**b**) 4 ft H_2O
(**c**) 4 mm H_2O
(**d**) None of the above

Solution: As indicated in Section 9.3, typical pressure drops range from 3 to 4 in H_2O. The correct answer is therefore (a).

9.9 Typical Collection Efficiency

A multicyclone is used in many applications for collecting dust. The efficiency of a multicyclone for collection of particles greater than 10 μm in diameter can be as high as

(**a**) 50%
(**b**) 65%
(**c**) 80%
(**d**) 90%

Solution: Efficiencies in the 80–90% range are not uncommon, but most operate at levels approaching 90%. The "best" answer is therefore (d).

9.10 Doubling Inlet Velocity

The cut diameter for a specific type of dust collected in a cyclone was found to be 25 μm. If the inlet velocity were doubled, what would the cut diameter be?

(**a**) 21.6 μm
(**b**) 14.5 μm
(**c**) 17.7 μm
(**d**) 10.2 μm

Solution: Equation (9.2) indicates that the cut diameter is inversely proportional to the square root of the velocity. Therefore

$$\begin{aligned} d_{p,cut2} &= d_{p,cut1}/\sqrt{2} \\ &= 25/1.414 \\ &= 17.7\ \mu m \end{aligned}$$

The correct answer is therefore (c).

9.11 Gravity Settler Differences

Although both units are described as mechanical collectors, list three major differences between a gravity settler (GS) and a cyclone (CYC).

Solution: Differences include

1. Look different
2. Operate differently—many factors here
3. Costs are different

9.12 Gravity Settler/Cyclone Efficiency Comparison

Explain in layman terms why the collection efficiency of a GS is generally lower than that for a CYC.

Solution: The main reason is that centrifugal forces (CYC) are generally orders of magnitude higher than gravitational forces (GS), producing higher efficiencies for cyclones.

9.13 Effect of Gas Density on Efficiency

A cyclone operates with an efficiency of 78.5%. Estimate a revised efficiency, assuming that the gas density is decreased from 0.081 to 0.075 lb/ft^3. Apply the appropriate equation in Table 9.3.

Solution: One approach to answering this question is to apply the equation

$$\frac{100 - E_1}{100 - E_2} = \left[\frac{\rho_p - \rho_2}{\rho_p - \rho_1}\right]^{0.5}$$

On the basis of this equation, it is reasonable to assume, from an engineering perspective, that the change in gas density has a negligible effect on the efficiency (or penetration). Thus, the revised estimate of the efficiency is 78.5% (i.e., the original estimate has not changed).

9.14 Effect of Changing Flow Rate on Efficiency

A cyclone currently operates with an efficiency of 84%. Assuming that the flow rate is increased by 33%, estimate a revised efficiency. Apply the appropriate equation in Table 9.3.

Solution: Refer to the following equation which appears in Table 9.3.

$$\frac{100 - E_1}{100 - E_2} = \left(\frac{q_2}{q_1}\right)^{0.5}$$

Substituting, one obtains

$$\frac{100 - 84}{100 - E_2} = \left(\frac{1.33}{1}\right)^{0.5}$$

$$16 = (100 - E_2)1.15$$

$$1.15E_2 = 115 - 16 = 99$$

$$E_2 = 86.1\%$$

As one would expect, the efficiency has increased.

9.15 Effect of Inlet Loading on Efficiency

A cyclone operates at an efficiency of 90.1%. Estimate a revised efficiency, assuming that the inlet loading is doubled. Apply the appropriate equation in Table 9.3.

Solution: Refer to the following equation in Table 9.3:

$$\frac{100 - E_1}{100 - E_2} = \left(\frac{c_2}{c_1}\right)^{0.182}$$

Substituting, one obtains

$$\frac{100 - 90.1}{100 - E_2} = \left(\frac{2}{1}\right)^{0.182}$$

$$9.9 = (100 - E_2)(1.134)$$

$$1.134E_2 = 113.4 - 9.9 = 103.5$$

$$E_2 = 91.3\%$$

As expected, the efficiency increases.

9.16 Pressure Drop Across a Cyclone

Estimate the pressure drop across a cyclone treating a gas at ambient conditions with a velocity of 50 ft/s.

Solution: The pressure drop is often described in units of inlet velocity heads. This inlet velocity head, in inches of water, may be expressed as follows:

$$\text{One velocity head} = 0.003\,\rho v_i^2, \quad \text{in } H_2O \tag{9.7}$$

where ρ is the gas density at operating conditions (lb/ft^3) and v_i is the inlet velocity (ft/s). The friction loss through cyclones encountered in practice may range from 1 to 20 inlet velocity heads, depending on the geometric proportions. For most cyclones, the friction loss is approximately 8 inlet velocity heads. This is equivalent to Equation (9.6). For an inlet velocity of 50 ft/s (typical) and a gas density of 0.075 lb/ft^3, the pressure drop across the cyclone is

$$\Delta P = 8(0.003)(0.075)(50)^2$$

$$= 4.5 \text{ in } H_2O$$

9.17 Pressure Drop across Two Units

Assuming that the pressure drops across a GS and CYC are 0.5 in H_2O and 3.5 in H_2O, respectively, calculate the overall pressure drop across the unit. Should this pressure drop be used in sizing the fan? Explain your answer.

Solution: The pressure drops are additive. Thus

$$\begin{aligned}\Delta P_{\text{total}} &= 0.5 + 3.5 \\ &= 4.0 \text{ in } H_2O\end{aligned}$$

This pressure drop can be used if it represents a flange-to-flange value. The fan efficiency is also required in sizing the fan.

9.18 Cyclone Inlet Velocity

The exhaust gas flow rate from an industrial source is 3000 scfm. All of the gas is vented through a cyclone that has an inlet area of 1.2 ft^2. The exhaust gas temperature is 360°F. What is the velocity of the gas through the cyclone inlet in feet per second? Assume standard conditions to be 60°F and 1 atm pressure. Neglect the pressure drop across the cyclone.

Solution: First apply Charles' Law (see Section 3.3) to obtain the actual flow rate:

$$\frac{q_{\text{act}}}{q_{\text{s}}} = \left(\frac{T_{\text{act}}}{T_{\text{s}}}\right)\left(\frac{p_{\text{s}}}{P}\right)$$

$$\begin{aligned}q_{\text{act}} &= 3000\left(\frac{360 + 460}{60 + 460}\right)\left(\frac{1}{1}\right) \\ &= 3000(1.578) \\ &= 4731 \text{ acfm}\end{aligned}$$

$$\begin{aligned}\bar{v} &= \frac{q_{\text{act}}}{A} \\ &= \frac{4731}{1.2} \\ &= 3942 \text{ ft/min} \\ &= 65.7 \text{ ft/s; a reasonable value}\end{aligned}$$

9.19 Cyclone Outlet Diameter

The inlet gas to a wet cyclone is at 400°F and is piped through 3.0-ft ID ductwork at 25 ft/s to the cyclone. The water contact cools the gas to 100°F. In order to maintain a velocity of 50 ft/s, what size ductwork would be needed at outlet conditions? Neglect the pressure drop across the unit and any moisture considerations.

Solution: The inlet volume flow rate is

$$\begin{aligned} q_{\rm i} &= v_{\rm i}A_{\rm i} \\ &= (25)(7.07) \\ &= 176.7\ {\rm ft}^3/{\rm s} \end{aligned}$$

Since there is a temperature change, apply Charles' Law to determine the outlet volumetric flow rate:

$$\begin{aligned} q_0 &= q_{\rm i}\left(\frac{100+460}{400+460}\right)\left(\frac{1}{1}\right) \\ &= 176.7\left(\frac{560}{860}\right) \\ &= 115\ {\rm acfs} \end{aligned}$$

The outlet area should then be

$$\begin{aligned} A_o &= \frac{115}{50} \\ &= 2.30\ {\rm ft}^2 \end{aligned}$$

Applying the equation for the area of circle, one obtains

$$\begin{aligned} 2.30 &= \pi D^2/4 \\ D &= 1.71\ {\rm ft} \\ &= 20.5\ {\rm in} \end{aligned}$$

9.20 Estimating Cyclone Pressure Drop

A large conventional cyclone (no vanes) handles 13,000 acfm (60°F, 1 atm) of a particulate-laden gas. The cyclone dimensions are as follows:

Cyclone diameter = 8 ft
Diameter of gas outlet = 4 ft
Inlet width = 2 ft
Inlet height = 4 ft
Length of cylinder = 16 ft
Height of cone = 16 ft

What is the estimated pressure drop across the cyclone?

(a) 2.5 in H_2O
(b) 0.62 in H_2O
(c) 1.35 in H_2O
(d) 0.312 in H_2O

Solution: An estimate of the pressure drop is based on the inlet velocity to the cyclone:

$$\begin{aligned} v &= q/A \\ &= 13{,}000/(2)(4) \\ &= 1625 \text{ ft/min} \\ &= 27.1 \text{ ft/s} \end{aligned}$$

Apply Equation (9.6):

$$\begin{aligned} \Delta P &= (0.024)(\rho)(v)^2; \ \rho = 0.075 \text{ lb/ft}^3 \\ &= (0.024)(0.075)(27.1)^2 \\ &= 1.32 \text{ in } H_2O \end{aligned} \tag{9.6}$$

The correct answer is therefore (c).

9.21 Overall Efficiency of Two Units in Series

A GS is followed in series by a CYC. If the collection efficiency of the GS is 55% and the penetration of the CYC is 15%, what is the overall efficiency of the unit? Also indicate the overall penetration.

Solution: Apply a penetration equation to the process.

$$P = P_1 P_2; \text{ fractional basis} \tag{9.8}$$

with

$$\begin{aligned} P_1 &= 100 - 55 \\ &= 45\% \\ &= 0.45 \end{aligned}$$

and

$$\begin{aligned} P_2 &= 15\% \\ &= 0.15 \end{aligned}$$

Substituting into Equation (9.8) gives

$$\begin{aligned} P &= (0.45)(0.15) \\ &= 0.0675 \\ &= 6.75\% \end{aligned}$$

Furthermore,

$$\begin{aligned} E &= 1 - P \\ &= 1 - 0.0675 \\ &= 0.9325 \\ &= 93.25\% \end{aligned}$$

9.22 Three Cyclones in Series

As a graduate student you have been assigned the task of studying certain process factors in an operation in São Paulo, Brazil that employs three cyclones in series to treat catalyst-laden gas at 25°C and 1 atm. The inlet loading to the cyclone series is 8.24 gr/ft^3, and the volumetric flow rate is 1,000,000 acfm. The efficiency of the cyclones are 93%, 84% and 73%, respectively. Calculate the following:

(a) Daily mass of catalyst collected (lb/day).
(b) Daily mass of catalyst discharged to the atmosphere.
(c) Whether it would be economical to add an additional cyclone (efficiency = 52%) costing an additional $300,000 per year. (The cost of the catalyst is $0.75 per pound.)
(d) Outlet loading from the proposed fourth cyclone if the operation is based on 300 days per year.

Solution: The mass rate entering, $\dot{m}_i$, is

$$\begin{aligned} \dot{m}_i &= \left(10^6 \text{ ft}^3/\text{min}\right)(60 \text{ min/hr})(24 \text{ hr/day})\left(8.24 \text{ gr/ft}^3\right)(1 \text{ lb}/7000 \text{ gr}) \\ &= 1{,}695{,}086 \text{ lb/day} \end{aligned}$$

The mass rate collected $\dot{m}_c$ is

$$\begin{aligned} \dot{m}_c &= 1{,}695{,}086(0.93) + 0.84[1{,}695{,}086(1 - 0.93)] \\ &\qquad + 0.73[1{,}695{,}086(1 - 0.93)(1 - 0.84)] \\ &= 1{,}689{,}960 \text{ lb/day} \end{aligned}$$

Thus the mass discharge rate $\dot{m}_d$ is

$$\dot{m}_d = 1{,}695{,}086 - 1{,}689{,}960 = 5126 \text{ lb/day}$$

With a fourth cyclone, the additional mass collected $\dot{m}_4$, is

$$\dot{m}_4 = 5126(0.52) = 2666 \text{ lb/day}$$

and $5126 - 2666 = 2460$ lb/day is discharged. The savings S is

$$\begin{aligned} S &= (2666 \text{ lb/day})(\$0.75\text{ /lb})(300 \text{ day/year}) \\ &= \$600{,}000/\text{year} \end{aligned}$$

Since the cyclone costs \$300,000 annually, purchase it. The outlet loading (OL) is

$$\begin{aligned} \text{OL} &= (2460\,\text{lb/day})(1\,\text{day}/24\,\text{h})(1\,\text{h}/60\,\text{min})\left(1\ \text{min}/10^6\,\text{ft}^3\right)(7000\,\text{gr/lb}) \\ &= 0.012\,\text{gr}/\,\text{ft}^3 \end{aligned}$$

9.23 Four Cyclones in Series

Four cyclones are to be employed in series to treat catalyst-laden gas at 30°C and 1 atm. The inlet loading to the cyclone series is 9.16 gr/ft^3, and the volumetric flow rate is 950,000 acfm. The efficiencies of the cyclones are 92%, 83%, 72%, and 61%, respectively. Calculate the mass of catalyst discharged from each cyclone, including what is discharged from each unit in lb/day, the daily mass of catalyst collected, and the overall collection efficiency.

Solution: In a very real sense, this is an extension of Problem 9.22. Key calculations are provided below:

Outlet Loading and/or Concentration

Unit 1: $(9.16)(1-0.92) = 0.733\ \text{gr/ft}^3$
Unit 2: $(9.16)(1-0.92)(1-0.83) = 0.125\ \text{gr/ft}^3$
Unit 3: $(9.16)(1-0.92)(1-0.83)(1-0.72) = 0.0349\ \text{gr/ft}^3$
Unit 4: $(9.16)(1-0.92)(1-0.83)(1-0.72)(1-0.61) = 0.0136\ \text{gr/ft}^3$

Outlet Discharge

Unit 1: $(0.733)(950{,}000)(60)(24)/(7000) = 1.43\times10^5\ \text{lb/day}$
Unit 2: $(0.125)(950{,}000)(60)(24)/(7000) = 2.45\times10^4\ \text{lb/day}$
Unit 3: $(0.0349)(950{,}000)(60)(24)/(7000) = 6.82\times10^3\ \text{lb/day}$
Unit 4: $(0.0136)(950{,}000)(60)(24)/(7000) = 2.66\times10^3\ \text{lb/day}$

Daily Mass Collected

$(9.16)(950{,}000)(0.9985)(60)(24)/(7000)=1.79\times10^6\ \text{lb/day}$

Overall Efficiency

$$\begin{aligned} P &= (0.08)(0.17)(0.28)(0.39) \\ &= 0.00149 \\ &\simeq 0.0015 \\ &= 0.15\% \\ E &= 0.9985 \\ &= 99.85\% \end{aligned}$$

9.24 Maximum Allowable Inlet Loading

Refer to Problem 9.23. If the system is to be set up so that no more than 1500 lb/day will be discharged, what is the maximum allowable inlet loading in gr/ft^3 when the overall efficiency is assumed to be 99.85%?

Solution: The maximum daily flow rate (MDFR) is

$$\begin{aligned}\text{MDFR} &= 1500/0.0015 \\ &= 10^6 \text{ lb/day} \\ &= 7.0 \times 10^9 \text{ gr/day}\end{aligned}$$

To determine the inlet loading (or concentration), divide the results displayed above by the inlet flow rate on a daily basis. First calculate the daily volumetric flowrate.

$$\begin{aligned}q &= 950{,}000 \text{ acfm} \\ &= \left(9.5 \times 10^5\right)(60) \\ &= 5.7 \times 10^7 \text{ ft}^3/\text{hr} \\ &= (5.7 \times 10^7)(24) \\ &= 1.368 \times 10^9 \text{ ft}^3/\text{day}\end{aligned}$$

The concentration is then

$$\begin{aligned}c &= \text{MDFR}/q \\ &= \frac{7.0 \times 10^9}{1.368 \times 10^9} \\ &= 5.12 \text{ gr/ ft}^3\end{aligned}$$

9.25 Cut Diameter and Overall Collection Efficiency

An engineer was requested to determine the cut size diameter and overall collection efficiency of a cyclone given the particle size distribution of a dust from a cement kiln. Particle size distribution and other pertinent data are provided in Table 9.4.

TABLE 9.4 Particle Size Distribution Data for Problem 9.25

Average Particle Size in Range d_p, μm	Weight Percent
1	3
5	20
10	15
20	20
30	16
40	10
50	6
60	3
>60	7

Additional data:

Gas viscosity = 0.02 cP
Specific gravity of the particle = 2.9
Inlet gas velocity to cyclone = 50 ft/s
Effective number of turns within cyclone = 5
Cyclone diameter = 10 ft
Cyclone inlet width = 2.5 ft

Solution: As noted earlier, the performance of a cyclone is often specified in terms of a cut size d_{pc}, which is the size of the particle collected with 50% efficiency. It may be calculated directly from Equation (9.2).

$$d_{pc} = \left(\frac{9\mu B_c}{2\pi N v_i(\rho_p - \rho)}\right)^{1/2}$$

where d_{pc} = cut size particle diameter (particle collected at 50% efficiency), ft
μ = gas viscosity, lb/(ft · s)
B_c = width of gas inlet, ft
N = effective number of turns the gas stream makes in the cyclone, dimensionless
v_i = inlet velocity, ft/s
ρ_p = particle density, lb/ft^3
ρ = gas density, lb/ft^3

Lapple's method provides the collection efficiency as a function of the ratio of particle diameter to cut diameter, as presented in Figure 9.12. One may also use the equation

$$E = \frac{1.0}{1.0 + \left(d_{pc}/d_p\right)^2} \tag{9.9}$$

in place of Figure 9.12. [For additional details on Equation (9.9), the reader is referred to the article by L. Theodore and V. DePaola, "Predicting Cyclone Efficiency," *J. Air Pollut. Control-Assoc.* **30**: 1132–33, (1980)].

For the problem at hand, determine the value of $(\rho_p - \rho)$:

$$\begin{aligned}\rho_p - \rho &= \rho_p \\ &= (2.9)(62.4) \\ &= 181\ \text{lb/ft}^3\end{aligned}$$

Calculate the cut diameter:

$$d_{pc} = \left(\frac{9\mu B_c}{2\pi N v_i \left(\rho_p - \rho \right)} \right)^{1/2} \tag{9.2}$$

$$= \left(\frac{(9)(0.02)(6.72 \times 10^{-4})(2.5)}{(2\pi)(5)(50)(181)} \right)^{1/2}$$

$$= 3.26 \times 10^{-5}$$

$$= 9.94\ \mu\text{m}$$

Table 9.5 is generated using Lapple's method. Slightly more accurate results can be obtained by employing the Theodore–DePaola equation. The overall collection efficiency is therefore

$$E = \sum w_i E_i = 0 + 4 + 7.5 + 16 + 14.4 + 9.3 + 5.7 + 2.94 + 7$$

$$= 66.84\%$$

$$= 0.6684$$

TABLE 9.5 Results for Problem 9.25

d_p, μm	w_i	d_p/d_{pc}	E_i, %	w_iE_i, %
1	0.03	0.10	0	0.0
5	0.20	0.5	20	4.0
10	0.15	1.0	50	7.5
20	0.20	2.0	80	16.0
30	0.16	3.0	90	14.4
40	0.10	4.0	93	9.3
50	0.06	5.0	95	5.7
60	0.03	6.0	98	2.94
>60	0.07	—	100	7.0

9.26 Cyclone Selection

A recently hired engineer has been assigned the job of selecting and specifying a cyclone unit to be used to reduce an inlet fly ash loading (with the particle size distribution given in Table 9.6) from 3.1 gr/ft^3 to an outlet value of 0.06 gr/ft^3. The flow rate from the coal-fired boiler is 100,000 acfm. Fractional efficiency data provided by a vendor are presented in Figure 9.13 for three different types of cyclones (multiclones).

Which types of cyclones are required to meet the above specifications given above? The optimum operating pressure drop is 3.0 in H_2O at this condition, and the average inlet velocity may be assumed to be 60 ft/s.

As described earlier, multiple-cyclone collectors (multiclones) are high-efficiency devices that consist of a number of small diameter cyclones operating

TABLE 9.6 Particle Size Distribution Data for Problem 9.26

Particle Diameter Range, μm	Weight Fraction, w_i
5–35	0.05
35–50	0.05
50–70	0.10
70–110	0.20
110–150	0.20
150–200	0.20
200–400	0.10
400–700	0.10

in parallel with a common gas inlet and outlet. The flow pattern differs from a conventional cyclone in that instead of bringing the gas in at the side to initiate the swirling action, the gas is brought in at the top of the collecting tube, and swirling action is then imparted by a stationary vane positioned in the path of the incoming gas. The diameters of the collecting tubes usually range from 6 to 24 inches with pressure drops in the 2–6 inch range. Properly designed units can be constructed and operated with a collection efficiency as high as 90% for particulates in the 5–10 μm range. The most serious problems encountered with these systems involve plugging and flow equalization.

Solution: Calculate the required collection efficiency E_R:

$$E_R = [(3.1 - 0.06)/3.1](100)$$
$$= 98\%$$
$$= 0.98$$

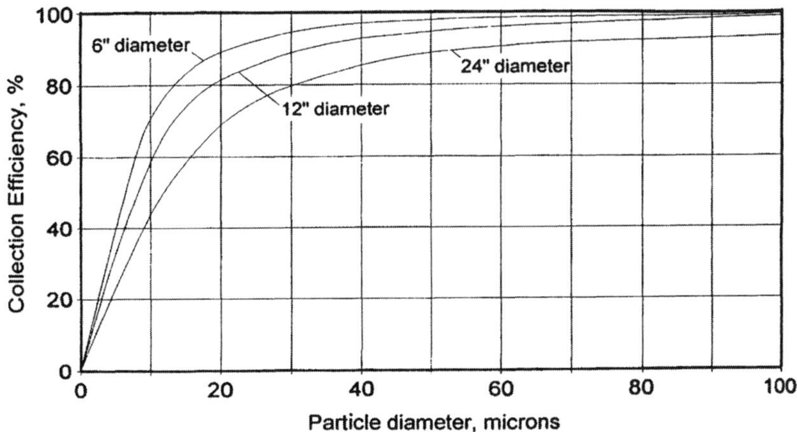

Figure 9.13 Fractional efficiency data.

Calculate the average particle size associated with each size range (see Table 9.7). Table 9.8 provides the overall efficiency E_6 for the 6 inch tubes. Since $E_6 > E_R$, the 6 inch tubes will do the job. Table 9.9 is generated for the 12 inch tubes. Since the overall efficiency $E_{12} < E_R$, the 12 inch tubes will not do the job. Thus, it will be necessary to use the 6 inch tubes for a conservative design.

TABLE 9.7 Average Particle Size

Particle Diameter Range, μm	Average Particle Diameter, μm
5–35	20
35–50	42.5
50–70	60
70–110	90
110–150	130
150–200	175
200–400	300
400–700	550

TABLE 9.8 6 inch Tube Efficiency

Average Particle Diameter, μm	Weight Fraction, w_i	Efficiency for 6 in Tubes, %	E_iw_i for 6 in Tubes, %
20	0.05	89	4.45
42.5	0.05	97	4.85
60	0.10	98.5	9.85
90	0.20	99	19.8
130	0.20	100	20
175	0.20	100	20
300	0.10	100	10
550	0.10	100	10
			$E_6 = 98.95$

TABLE 9.9 12 inch Tube Efficiency

Average Particle Diameter, μm	Weight Fraction w_i	Efficiency for 12 in Tubes, %	E_iw_i for 12 in Tubes, %
20	0.05	82	4.1
42.5	0.05	93.5	4.67
60	0.10	96	9.6
90	0.20	98	19.6
130	0.20	100	20
175	0.20	100	20
300	0.10	100	10
550	0.10	100	10
			$E_{12} = 97.97$

9.27 Cyclone Number Specification

Refer to Problem 9.26. How many cyclones are required to meet the specifications stated in the problem?

Solution: Since the gas flow to a multiclone is axial (usually from the top), the cross-sectional area available for inlet flow is given by the annular area between the outlet tube and cyclone body. The outlet tube diameter is usually one-half the body diameter. Since the outlet tube diameter is one-half the body diameter, the inlet cross-sectional area (for axial flow) for each 6 inch (0.5 ft) tube is

$$\begin{aligned} A &= 0.785\left(0.5^2 - 0.25^2\right) \\ &= 0.147 \text{ ft}^2 \end{aligned}$$

The velocity in each tube is 60 ft/s. The number of tubes n is therefore given by

$$(60)(60)(0.147)(n) = 100{,}000$$

Solving for n, one obtains

$$n = 190 \text{ tubes}$$

required in this multiple-cyclone unit. A 15 × 15, 14 × 14, or 12 × 16 design is recommended.

9.28 Cost of Power Equal to the Value of Recovered Material

A plant emits 50,000 acfm of gas containing a dust at a loading of 2.0 gr/ft^3. A cyclone is employed for particle capture, and the dust captured from the cyclone is worth \$0.01/lb of dust. For the sake of simplified calculation, assume that the efficiency of collection E is related to the system pressure drop ΔP, by the formula

$$E = \frac{\Delta P}{\Delta P + 7.5} \tag{9.10}$$

where ΔP is in units of lb$_f$/ft^2. If the fan is 55% efficient (overall) and electric power costs \$0.18/kW · hr, at what collection efficiency is the cost of power equal to the value of the recovered material? What is the pressure drop in inches of H_2O at this condition?

Solution: Take as a basis 1.0 minute.

Cost of recovered particulate E_R, is

$$\begin{aligned} E_R &= \left(50{,}000 \text{ ft}^3/\text{min}\right)\left(2.0 \text{ gr/ ft}^3\right)(1 \text{ lb}/7000 \text{ gr})(\$0.01/\text{lb})E \\ &= 0.143\, E, \ \$/\text{min} \end{aligned}$$

Cost of power to recover E_p is given by

$$E_p = (50{,}000 \text{ ft}^3/\text{min})(\Delta P \,\text{lb}_f/\text{ft}^2)(1 \text{ min} \cdot \text{kW}/44{,}200 \text{ ft} \cdot \text{lb}_f)(1/0.55)$$
$$\times (\$0.18/\text{W} \cdot \text{hr})(1 \text{ hr}/60 \text{ min})$$
$$= 0.006\Delta P, \$/\text{min}$$

Since

$$E = \Delta P/(\Delta P + 7.5)$$

equate E_R with E_P and replacing E with the equation above. This yields

$$(0.143)[\Delta P/(\Delta P + 7.5)] = 0.006\Delta P$$

Solving by trial and error (or quadratically), one obtains

$$\Delta P = 16.3 \text{ lb}_f/\text{ft}^2$$
$$= 3.1 \text{ in } H_2O$$

The breakeven efficiency is therefore

$$E = 16.3/(16.3 + 7.5)$$
$$= 0.685$$
$$= 68.5\%$$

Does this problem look familiar? The reader might consider returning to Problem 8.37.

9.29 Particle Size Deposition from a Malfunctioning Cyclone

A cyclone on a cement plant suddenly malfunctions. By the time the plant shuts down, some dust has accumulated on parked cars and other buildings in the plant complex. The nearest affected area is 700 ft from the cyclone location, and the furthest affected area measurable on plant grounds is 2500 ft from the cyclone. What is the particle size range of the dust that has landed on plant grounds? On this day, the cyclone was discharging into a 6.0 mph wind. The specific gravity of the cement is 1.96. The cyclone is located 175 ft above the ground. Neglect effects of turbulence.

Solution: A diagram representing the system is provided in Figure 9.14 (S = smaller particle, L = larger particle). For air at ambient conditions, one obtains

$$\rho = 0.0741 \text{ lb}/\text{ft}^3$$

$$\mu = 1.23 \times 10^{-5} \text{ lb}/(\text{ft} \cdot \text{s})$$

$$\text{Wind speed} = (6.0)(5280) = 31{,}700 \text{ ft/hr}$$

Particle traveling times are given by

$$s = ut$$

where s = horizontal distance traveled
u = horizontal velocity
t = travel time

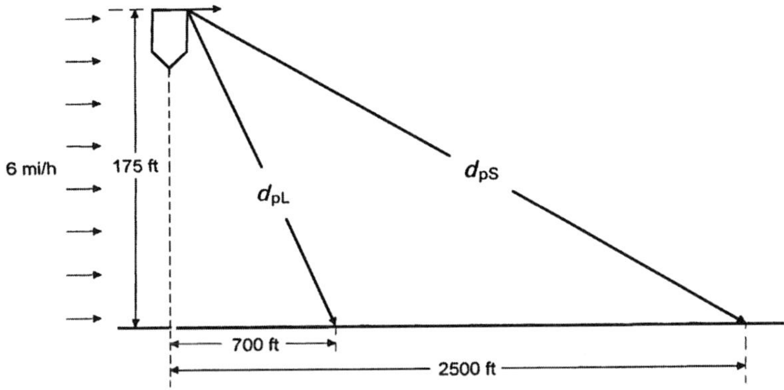

Figure 9.14 Diagram for Problem 9.29.

For the smaller particle, one obtains

$$t_S = 2500/31{,}700 = 0.0789\,\text{hr}$$

and for the larger particle

$$t_L = 700/31{,}700 = 0.0221\,\text{hr}$$

Settling velocities may now be calculated from

$$v = H/t$$

where v = vertical velocity
H = vertical distance traveled

$$v_S = 175/[(0.0789)(3600)] = 0.616\ \text{ft/s}$$
$$v_L = 175/[(0.0221)(3600)] = 2.20\ \text{ft/s}$$

To calculate d_p, assume Stokes' law to apply. Refer to Chapter 7 for more details.

$$\begin{aligned} d_{pS} &= \left(\frac{18\mu v}{g\rho_p}\right)^{0.5} \\ &= \left(\frac{(18)(1.23\times10^{-5})(0.6161)}{(32.2)(1.96)(62.4)}\right)^{0.5} \\ &= 1.86\times10^{-4}\ \text{ft} \end{aligned}$$

Checking the value of K, one obtains

$$K = d_p\left(\frac{g\left(\rho_p - \rho\right)\rho}{\mu^2}\right)^{1/3}$$

$$= \left(1.86 \times 10^{-4}\right)\left(\frac{(32.2)[(1.96)(62.4) - 0.0741](0.0741)}{(1.23 \times 10^{-5})^2}\right)^{1/3}$$

$$= 2.32 < 3.3$$

Therefore, for the smaller particle size, Stokes' law is valid and $d_{pS} = 1.86 \times 10^{-4}$ ft. For the larger particle size

$$d_{pL} = \left(\frac{(18)\left(1.23 \times 10^{-5}\right)(2.200)}{(32.2)(1.96)(62.4)}\right)^{0.5}$$

$$= 3.52 \times 10^{-4} \text{ ft}$$

and

$$K = \left(3.52 \times 10^{-4}\right)\left(\frac{(32.2)[(1.96)(62.4) - 0.0741](0.0741)}{(1.23 \times 10^{-5})^2}\right)^{1/3}$$

$$= 4.39$$

Since $3.3 < 4.39$, Stokes' law is invalid for the larger particles. Assuming that the Intermediate range applies (see Chapter 7), one obtains

$$d_{pL}^{1.14} = \frac{v\mu^{0.43}\rho^{0.29}}{(0.153)g^{0.71}\rho_p^{0.71}} = \frac{(2.200)\left(1.23 \times 10^{-5}\right)^{0.43}(0.0741)^{0.29}}{(0.153)(32.2)^{0.71}[(1.96)(62.4)]^{0.71}}$$

$$= 1.465 \times 10^{-4}$$

$$d_{pL} = 4.33 \times 10^{-4} \text{ ft}$$

Checking on K, one obtains

$$K = \left(4.33 \times 10^{-4}\right)\left(\frac{(32.2)[(1.96)(62.4) - 0.0741](0.0741)}{(1.23 \times 10^{-5})^2}\right)^{1/3}$$

$$= 5.39$$

Since $3.3 < 5.39 < 43.6$, the Intermediate law is valid for the larger particle size.

Therefore, the particle size range is

$$1.86 \times 10^{-4}\ \text{ft} \leq d_\text{p} \leq 4.33 \times 10^{-4}\ \text{ft}$$

or

$$56.7\,\text{m} \leq d_\text{p} \leq 132\,\mu\text{m}$$

9.30 Effect of Particle Size Distribution on Efficiency

Comment on the effect that the particle size distribution has on the overall collection efficiency of a cyclone.

Solution: Unlike the size–efficiency curve for a gravity settler (see Problems 8.34 and 8.35), that for a cyclone is concave downward with no discontinuity, as can be seen in Figure 9.15. This simplifies the discussion that follows. Consider the 24 inch diameter cyclone depicted in Figure 9.15. If all the particles were 30 μm in size (monodispersed), the efficiency of the cyclone would be 80%. This is indicated as point *A* in the figure and represents the efficiency for a standard deviation of zero ($\sigma = 0$). If the standard deviation is greater than zero, and if half the particles by mass are equally distributed above 30 μm, the efficiency would decrease because of the concave downward nature of the curve. This is shown by the shaded areas on both sides of point *A*, and one can note that the efficiencies of the smaller particles have decreased to a greater extent than the efficiencies of the larger particles have increased. One can therefore conclude that it appears that, irrespective of the average size of the distribution, an increase in the standard deviation will result in a decrease in efficiency. Since the effect of the lognormal distribution of the particle sizes in the analysis presented above is difficult to quantify, the next problem outlines how to test this conclusion by performing calculations.

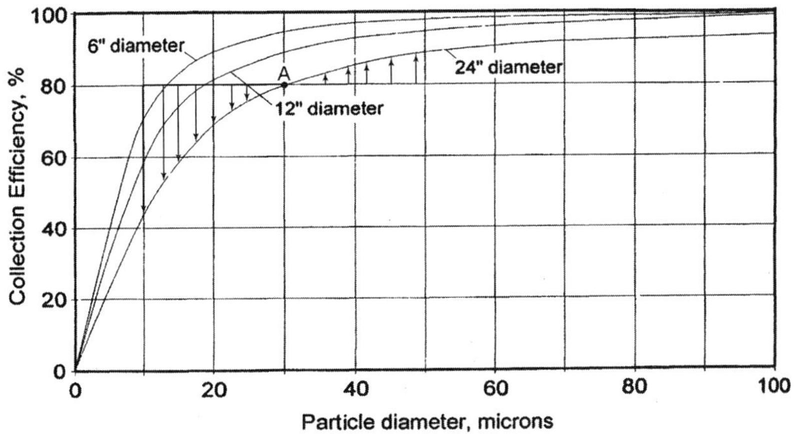

Figure 9.15 Size-efficiency data.

9.31 Verification of Effect of Particle Size Distribution on Efficiency

Outline how to verify the conclusions of Problem 9.30.

Solution: As with Problem 8.35, an analytical solution is beyond the capabilities of the author. However, a trial-and-error procedure can be employed by calculating the efficiency for various particle mean sizes and standard deviations. This is left as an exercise for the reader.

9.32 Open Ended Problem I

Company PAL (Pedro, Alex, and Louie) designs and predicts the performance of a special product line of cyclones using Lapple's "cut" diameter method. Company officials have decided to computerize the preceding calculations. This will require converting Lapple's figure (see Figure 9.12) to equation form.

Dr. Theodore, one of the foremost authorities in the world on air pollution control equipment, has been hired to generate an equation describing fractional collection efficiency E as a function of either d_p or (d_{pc}/d_p). Tied down by numerous other consulting activities, Dr. Theodore has decided to subcontract this job to you. He reasons that since you have taken his air pollution control equipment course, you now qualify as an authority in the field. Present a cyclone efficiency equation to PAL at your earliest convenience. Justify the form of your final equation.

Solution: This is obviously an open-ended problem—a problem for which there are several (if not an infinite number of) answers. One possible answer, which was developed by DePaola and Theodore, was based on the following analysis.

One of the crudest approaches to estimate the overall efficiency of a unit in terms of d_p and d_{pc} ($d_{p,cut}$) is (personal notes: L. Theodore, 1979)

$$E = \frac{d_p}{d_p + d_{pc}} \tag{9.11}$$

If the numerator and denominator are divided by d_p in this equation one obtains

$$E = \frac{1}{1 + \left(d_{pc}/d_p\right)}$$

Setting

$$Z = \frac{d_{pc}}{d_p}$$

Equation (9.11) becomes

$$E = \frac{1}{1 + Z}$$

Assume a model of the form

$$E = \frac{1}{1 + aZ^b}$$

so that

$$E + EaZ^b = 1$$

and

$$\frac{1 - E}{E} = aZ^b$$

Taking logarithms

$$\log\left(\frac{1 - E}{E}\right) = \log a + b \log Z$$

This is an equation of the form

$$Y = B + MX$$

which is a linear equation!

Regressing "data" from Figure 9.12 gives $a = 1.0$, $b = 2.0$. For $a = 1.0$ and $b = 2.0$, a regression coefficient of 0.99946 was obtained. Therefore, one may exactly represent Lapple's supposed experimental field data by

$$E = \frac{1}{1 + \left(d_{pc}/d_p\right)^2} \tag{9.9}$$

This was the procedure employed by DePaola and Theodore to solve the open-ended problem presented in this problem. The reader is left the exercise of obtaining or generating another solution.

9.33 Cyclone Inlet Velocity

A cyclone is designed to run at 60% efficiency. A gas stream consists of mono-dispersed particles with a diameter of 10 μm. The viscosity of the gas, the specific gravity of the particle, the effective number of turns, and width of cyclone inlet are 1.17×10^{-5} lb/ft · s, 2.0, 5.0 and 3 ft, respectively. What is the inlet gas velocity, assuming that the cyclone must operate at its design efficiency? (*Note*: Adopted from M. Barta, homework assignment, 2007.)

Solution: The Theodore–DePaola the equation for the collection efficiency is as follows:

$$E = \frac{1.0}{1.0 + (d_{pc}/d_p)^2} \tag{9.9}$$

The equation for d_{pc} is given as:

$$d_{pc} = \left[9\mu B_c/2\pi N v_i(\rho_p - \rho)\right]^{1/2} \quad (9.2)$$

Because these two equations are complex and interrelated, the value of d_{pc} was determined by trial and error. The calculation produced the following results:

$$v_i = 112\ \text{ft/s (a bit high)}$$

The numerical value of the cut diameter is reasonable.

$$d_{pc} = 2.68 \times 10^{-5}\ \text{ft}$$

NOTE: Additional problems are available for all readers at www.wiley.com. Follow links for this title.

10

ELECTROSTATIC PRECIPITATORS

10.1 INTRODUCTION

Electrostatic precipitators (ESPs) are unique among gas cleaning equipment in that the forces separating the particulates from the gas stream are applied directly to the particulates themselves, and hence the energy required to effect the separation is considerably less than for other types of gas cleaning apparatus. Gas pressure drops through the precipitator may be of the order of 1 inch of water or less as compared with pressures of up to 10–100 inches of water for scrubbers and baghouses. This fundamental advantage of electrostatic precipitation has resulted in its widespread use in applications where large gas volumes are to be handled and high efficiencies are required for collection of small particles.

The removal of fly ash from the discharge gases of electric power boilers is the largest single application of precipitators, both in number of installations and in the volume of gas treated. Modern power boilers are capable of generating 1000 MW or more of electric power, with gas volumes of several million cubic feet per minute.

The widespread use of precipitators in the electric power industry has uncovered problems resulting from variations in the chemical composition of the coal and hence in the properties of the fly ash. Most of the early control installations were made on boilers that burned relatively high-sulfur coal. As new installations were made, boilers

Air Pollution Control Equipment Calculations. By Louis Theodore

burning low-sulfur coal were encountered. This coal produces a fly ash that has a high resistivity and is difficult to collect. The problem, similar to that for the smelter dusts, is insufficient sulfur trioxide in the gas, and it is often corrected by conditioning the gas with this chemical. This resistivity problem is discussed below and later in this chapter.

The electrostatic precipitation process consists of three fundamental steps:

1. Particle charging
2. Particle collection
3. Removal of the collected dust

Particle charging in precipitators is accomplished by means of a corona, which produces ions that become attached to the particles. Generation of a corona requires the development of a highly nonuniform electric field—a condition that occurs near the wire when a high voltage is applied between the wires and collection electrodes. The electric field near the wire accelerates electrons present in the gas to velocities sufficient to cause ionization of the gas in the region near the wire. The ions produced as a result of the corona migrate toward the collection electrode, and in the process collide with and become attached to particles suspended in the gas stream. The attachment of ions results in the buildup of an electric charge, the magnitude of which is determined by the number of ions attached.

The charge on the particles in the presence of an electric field results in a new force in the direction of the collection electrode. The magnitude of the force is dependent on the charge and the field. This force causes particles to be deposited on the collection electrode where they are held by a combination of mechanical, electrical, and molecular forces.

Once collected, particles can be removed by coalescing and draining in the case of liquid aerosols, or by periodic impact or rapping in the case of solid material. In the latter case, a sufficiently thick layer of dust must be collected so that it falls into the hopper or bin in coherent masses (effectively like a sheet) to prevent excessive reentrainment of the material into the gas stream.

Physical arrangement of precipitators differ depending on the type of application. Wire and cylinder electrodes are used in some applications; however, for reasons of space economy, most commercial precipitators use plates as collection electrodes.

The majority of precipitators are constructed so that the charging and collection steps take place within the same region. Precipitators of this type are termed *single-stage*. For some applications charging takes place in one section which is followed by a section consisting of alternately charged plates. The collecting electric field is established independently of the corona field and such precipitators are termed *two-stage*.

Electrostatic precipitation occurs in the space between the discharge electrode and the collection surface. A high-voltage, pulsating direct current is applied to an electrode system consisting of a small-diameter discharge electrode that is usually negatively charged, and a collecting plate electrode that is grounded. This produces a unidirectional, nonuniform electric field whose magnitude is highest near the discharge electrode.

Resistivity is related to the ability of a particle to take on a charge. In most industrial applications, the resistivity of the particle is such that the charge on the particle is only partially discharged on contact with the grounded collection electrode. A portion of the charge is returned and contributes to the intermolecular cohesive and adhesive forces that hold the particles to the collection surfaces. The dust layer builds up on the collection plate to a thickness of 0.5 in (1.27 cm or greater). If the dust layer becomes too thick, it is possible for the accumulated layer to act as an insulator, reducing the flow of the electric field lines.

The negative gas ion movement has two main charging effects on dust particles in the interelectrode region. These effects are called field charging and diffusion charging. Each type of charging is used to some extent in particle charging, but one dominates depending on the particle size. Field charging dominates for particles with a diameter $>1.0\ \mu m$, while diffusion charging dominates for particles with a diameter $<0.3\ \mu m$. A combination of these mechanisms occurs for particles in the range between 0.3 and 1.0 μm in diameter. It is also possible to charge particles by electron charging. In this case, free electrons that do not combine with gas ions are moving at an extremely fast rate. These electrons hit the particle and impart a charge. However, this effect is responsible for very little particle charging.

In field charging, as particles enter the electric field, they cause local dislocation of the field. Negative gas ions traveling along the electric field lines collide with the suspended particles and impart a charge on them. The ions continue to bombard the particle until the charge on the particle is sufficient to divert the electric lines away from the charged particle. This prevents new ions from colliding with the dust particle. When a particle no longer receives an ion charge, it is said to be saturated. Saturated charged particles then migrate to the collection electrode and are collected.

Diffusion charging is associated with the random Brownian motion of the negative gas ions. The random motion is related to the thermal velocity of the gas ions; the higher the temperature, the more movement occurs. Negative ions collide with the particles because of the random thermal motion of the ions and impart a charge on the particle. The charged particles migrate to the collection electrode. The mechanism is important for charging particles in the submicron range. In the intermediate size range from 0.3 to 1.0 μm in diameter, both field and diffusion charging are important.

In practice, the applied voltage is increased until it produces a discharge (corona), which can be seen as a luminous blue glow around the discharge electrode. The corona is a discharge phenomenon in which gaseous molecules are ionized by electron collisions in the region of a high electric field. The intense electric field close to the discharge electrode accelerates the free electrons that are present in the gas. These electrons acquire sufficient velocity to ionize gas molecules upon collision, producing a positive ion and an additional free electron (see Figure 10.1).

The additional free electrons create more positive ions and free electrons as they collide with additional gas molecules. This process is called *avalanche multiplication*, and occurs in the corona glow region (see Figure 10.2). Avalanche multiplication will continue until the local electric field strength decreases to the point where there is insufficient energy to perpetuate ionization. The sluggish positive ions migrate back to the negative discharge electrode and form new free electrons on impaction with the discharge wire or gas space around the wire. The electrons produced during the avalanche multiplication process follow the electric field toward the grounded collection electrode.

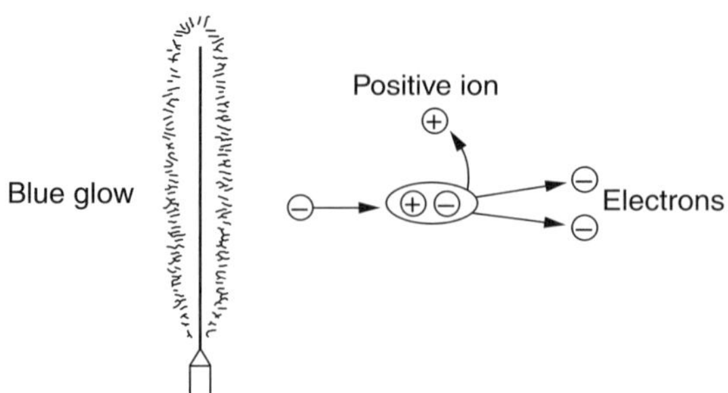

Figure 10.1 Generation of corona.

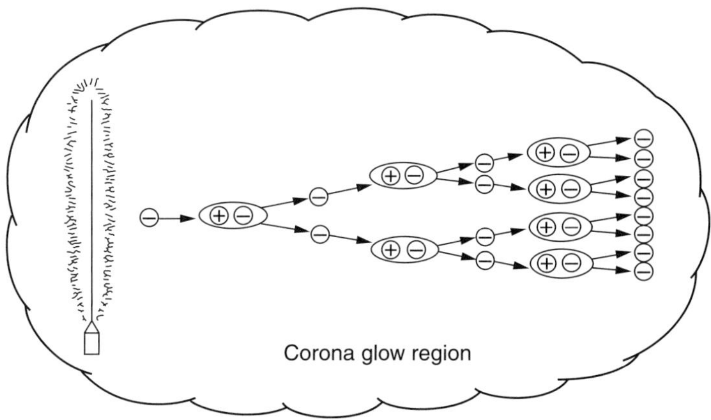

Figure 10.2 Avalanche multiplication.

The electrons leave the corona region and enter the interelectrode region. The magnitude of the electric field is diminished and the free electrons' velocity decrease. When electrons impact on gas molecules in the interelectrode region, they are captured, and negative gas ions are created. These negative ions serve as the principal mechanism for charging the dust.

Negative gas ions migrate toward the grounded collection electrode. A space charge, which is a stable concentration of negative gas ions, forms in the interelectrode region. Increases in the applied voltage will increase the field strength and ion formation until sparkover occurs. *Sparkover* refers to internal sparking between the discharge and collection electrodes. It is a sudden rush of localized electric current through the gas layer between the two electrodes. Sparking causes an immediate short-term collapse of the electric field. In general, it is very desirable to operate at voltages high enough to cause some sparking but not at a frequency such that the electric field constantly collapses. The average sparkover rate for optimum precipitator operation is between 50 and 100 sparks per minute. At this spark rate, the gains in efficiency associated with increased voltage compensate for decreased gas ionization due to the collapse of the electric field. For optimum efficiency the electric field strength should be as high as possible.

This is accomplished by applying a high voltage to the discharge electrode with the consequent high corona current flow from the discharge electrode to the collection electrode.

The heart of the electrostatic precipitation process is the discharge electrode system. It must produce a strong, uniform corona (the *corona* is a discharge phenomenon in which gaseous molecules are ionized by electron collisions in a region of a high electric field), while maintaining the correct distance and alignment with respect to the collecting electrodes to prevent imbalances in the electric field and to avoid unnecessary arcing discharges. Discharge electrode sizes and shapes vary mainly by manufacturer, but variations among different applications or in different sections of the same precipitator are feasible. Round, straight wires about 0.1 inch in diameter are the most common discharge electrodes in use. They may be hung individually and free with a suspension weight at the bottom end, or they may be held in a structural framework that is rigidly attached to the precipitator structure.

High-voltage rectifiers providing pulsating DC waveforms are in use almost exclusively because of the higher voltage and current attainable under sparking conditions when compared to pure direct current. With few exceptions, the discharge electrode system is subdivided into discrete sections, each being energized by a separate transformer–rectifier (TR) set. This "sectionalization" is important in matching corona currents and voltages to the TR set, and to promote reliability and stability under arcing conditions. A TR set consists of a high-voltage transformer and bridge rectifier, with typical secondary winding root mean square (rms) ratings between 53–66 kV and 250–2000 mA. Most TR sets can be connected to the precipitator discharge electrode system in either full-wave or double half-wave.

An important adjunct to the discharge electrode system is the automatic regulation of the high-voltage input to the precipitator, because only in rare cases does nature permit ideal operation. The earliest precipitators had no means of voltage regulation, but state of the art advancements such as the silicon-controlled rectifier, solid-state construction, and digital control circuits have enabled precipitators to perform efficiently under the most adverse conditions.

In dry, parallel-plate precipitators, the collecting electrodes (plates) are suspended from the top of the precipitator, parallel to and in proper alignment with the discharge electrodes. These plates and their attachment must be strong enough to dislodge particulate when rapped, and durable enough to withstand millions of rapping blows without fatigue failure. For these reasons, collecting plates are typically made of light-gauge metal and rigidly fastened to the precipitator structure only at their top ends (see Figure 10.3).

Most designs incorporate baffles to provide quiescent zones where the probability of particle collection is enhanced. These baffles are integrated into vertical stiffeners, which are required because collecting plate heights of up to 50 ft are occasionally in use.

Hoppers are best thought of as temporary holding bins to store collected particulate until permanent disposal can be scheduled. They take a variety of shapes, and are also sectionalized to facilitate handling large quantities of dust. It is a mistake to think the precipitation process stops at the hoppers. Removing collected material from the hopper is just as important as getting the material to fall into the hopper in the first place.

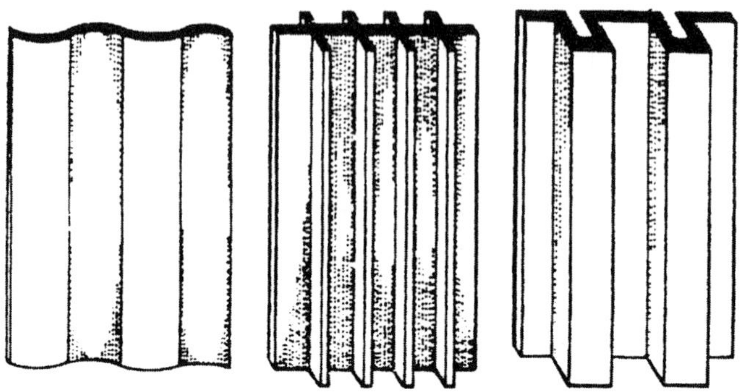

Figure 10.3 Typical collection plates.

The discharge electrodes, collecting plates, and hoppers are all contained and supported by the casing or shell. This structure must provide a gas-tight envelope in which the process takes place and must also hold the two electrode systems in proper alignment, sometimes under cyclical load conditions. In nearly all economical designs, essentially all the important auxiliary equipment is attached somewhere directly to the casing. Insulators that support the discharge electrodes system are made almost exclusively from porcelain of fused alumina and are contained in individual or grouped insulator compartments, or all the high-tension support insulators may be housed in a top housing or penthouse. In many cases, the manufacturer requires these insulators to be heated and ventilated at certain times during the operation of the precipitator.

The widest variation in design among manufacturers comes from rapping. In the simplest sense, rappers are either impulse (single blow) or vibrating (multiple blow) types. One impulse type consists of swing hammers striking the lower portions of the collecting electrodes or an intermediate portion of a rigid frame discharge electrode system. The swing hammers are actuated by a camshaft driven by an electric motor, and swing from the peak of the cam action by gravity to strike the electrode systems in a horizontal direction. Another type of impulse rapper is the drop hammer, which may be actuated by an electromagnetic solenoid located on top of the precipitators, or by a motor-driven mechanical linkage. The rapping blow may be gravity actuated or may be spring-assisted, striking the tops of the system in a vertical direction, or nearly any combination among these practices. Vibrating types are used almost exclusively on the discharge electrode systems, especially on industrial processes where weighted wire geometries are most common. Vibrators may be electric or pneumatic; the latter are extremely powerful and potentially destructive.

Other key components in a precipitator include the high-voltage bus and guard that deliver the TR output to the discharge electrodes; gas distribution differs at the inlet and outlet face; access systems, and a host of other systems, subsystems, and components necessary to support the operation by providing electrical power distribution, control, installation, etc.

As noted above, there are basically two types of electrostatic precipitators (ESPs): high-voltage single-stage and low-voltage double-stage. The high-voltage single-stage precipitator is the more popular type and has been used successfully to collect both solid and liquid particulate matter in industrial facilities such as smelters, steel furnaces, cement kilns, municipal incinerators, and utility boilers. Low-voltage double-stage precipitators are limited almost exclusively to the collection of liquid aerosols discharged from sources such as meat smokehouses, pipe-coating machines, asphalt paper saturators, and high-speed grinding machines.

Low-voltage double-stage precipitators were originally designed for air purification in conjunction with air conditioning systems (they are also referred to as *electronic air filters*). Double-stage ESPs have been used primarily for the control of finely divided liquid particles. Controlling solid or sticky materials is usually difficult, and the collector becomes ineffective for dust loadings greater than 0.4 grains per standard cubic foot (7.35×10^{-4} gr/m^3). Therefore, double-stage precipitators have limited use for particulate emission control. The low-voltage two-stage precipitator differs from the high-voltage single-stage precipitator in terms of both design and amount of applied voltage. As noted earlier, the double-stage ESP has separate particle charging and collection stages. The ionizing stage consists of a series of small (0.007 inch diameter) positively charged wires equally spaced 1 to 2 inches from parallel ground tubes that charge the particles suspended in the airflow through the ionizer. The direct-current potential applied to the wires is approximately 12–13 kV. The second stage consists of parallel metal plates less than 1 inch (2.5 cm) apart. The liquid particles receive a positive charge in the ionizer stage and are collected at the negative plates in the second stage. Collected liquids drain by gravity to a pan located below the plates. The two major types of high-voltage single-stage ESP configurations are *tubular* and *plate*, and particles are both charged and collected in a single stage.

Tubular precipitators consist of cylindrical collection electrodes with discharge electrodes located in the center of the cylinders. Dirty gas flows into the cylinder where precipitation occurs. The negatively charged particulates migrate to and are collected on grounded collecting tubes. The collected dust or liquid is removed by washing the tubes with water sprays located directly above the tubes. These precipitators are generally referred to as *water-walled ESPs*. Tubular precipitators are generally used for collecting mists or fogs. Tube diameters typically vary from 0.5 to 1 ft (0.15–0.31 m), with lengths usually ranging from 6 to 15 ft (1.85–4.6 m).

Summarizing, plate electrostatic precipitators are used more often than tubular ESPs in industrial applications. High voltage is used to subject the particles in the gas stream to an intense electric field. Dirty gas flows into a chamber consisting of a series of discharge electrode (wire) space along the center line of adjacent plates. Charged particles migrate to and are collected at oppositely charged collection plates. Collected particles are usually removed by rapping (dry precipitator) or by a liquid film (wet precipitator). Particles fall by force of gravity into hoppers where they are stored prior to removal and final disposal.

The remainder of the chapter will be devoted primarily to the high-voltage single-stage plate ESP and the descriptive terminology used is defined briefly below.

Active height—vertical length of collecting electrodes (plates)

Active length—horizontal length of energized precipitator, excluding empty spaces between fields

Active surface—total collecting surface area energized

Aspect ratio—active length divided by active height

Bus section—smallest portion of a field that can be deenergized

Chamber—Gastight longitudinal subdivision of a precipitator

Collecting surface area—surface area of energized collecting plates

Distribution plate—a device installed at either the inlet or outlet of a precipitator to achieve optimum gas flow distribution

Effective cross-sectional area—active height × total number of gas passages × gas passage width

Field—transverse subdivision of a precipitator formed by parallel bus sections

Gas passage—volume enclosed by two adjacent collecting plates

Migration velocity—theoretical speed of particulate normal to the direction of flow in the process of collection; defined as the "effective migration" or "drift" velocity when calculated from empirical data, but "collection rate parameter" is less misleading

Power density—ratio of total power input to collecting surface area or gas flow rate, usually expressed in watts (W)/ft^2 or W/1000 ft^3/min

Precipitator—arrangement of electrodes and all other equipment in one independent casing

Specific collecting area—ratio of collecting surface area to gas flow rate, usually expressed as ft^2/1000 ft^3/min

Treatment time—active length divided by treatment velocity

Treatment velocity—Gas flow rate divided by effective cross-sectional area, expressed in ft/s

10.2 DESIGN AND PERFORMANCE EQUATIONS

Once a particle is charged, it migrates toward the grounded collection electrode. An indicator of particle movement toward the collection electrode is denoted by the symbol w and is called the *particle migration velocity* or *drift velocity*. The migration velocity parameter represents the collectability of the particle within the confines of a specific collector. The migration velocity w can be expressed in terms of

$$w = \frac{d_{\text{p}} E_0 E_{\text{p}}}{4(\pi)\mu} \tag{10.1}$$

where d_{p} = diameter of the particle, μm

E_0 = strength of field in which particles are charged, volts per meter (V/m) (represented by peak voltage)

E_{p} = strength of field in which particles are collected, V/m (normally the field close to the collecting plates)

μ = viscosity of gas, Pa · s

The migration velocity is quite sensitive to the voltage since the electric field appears twice in Equation (10.1). Therefore, the precipitator must be designed using the maximum electric field for maximum collection efficiency. The migration velocity is also dependent on particle size; generally, larger particles are collected more easily than smaller ones.

The particle migration velocity can also be determined by the following equation:

$$w = \frac{q_p E_p}{6(\pi)\mu r} \tag{10.2}$$

where q_p = particle charge (charges)
E_p = strength of field in which particles are collected, V/m
μ = gas viscosity, Pa · s
r = radius of the particle, μm

The particle migration velocity can be calculated using either Equation (10.1) or (10.2). However, most ESPs are designed using a particle migration velocity based on field experience rather than theory. Typical particle migration velocity rates such as those listed in Table 10.1 have been published by various ESP vendors. These values can be used to estimate the collection efficiency of the ESP.

TABLE 10.1 Typical Precipitation Rate Parameters for Various Applications

	Particle Migration Velocity	
Application	(ft/s)	(cm/s)
Utility fly ash	0.13–0.67	4.0–20.4
Pulverized coal fly ash	0.33–0.44	10.1–13.4
Pulp and paper mills	0.21–0.31	6.4–9.5
Sulfuric acid mist	0.19–0.25	5.8–7.62
Cement (wet process)	0.33–0.37	10.1–11.3
Cement (dry process)	0.19–0.23	6.4–7.0
Gypsum	0.52–0.64	15.8–19.5
Smelter	0.06	1.8
Open-hearth furnace	0.16–0.19	4.9–5.8
Blast furnace	0.20–0.46	6.1–14.0
Hot phosphorus	0.09	2.7
Flash roaster	0.25	7.6
Multiple hearth roaster	0.26	7.9
Catalyst dust	0.25	7.6
Cupola	0.10–0.12	3.0–3.7

Probably the best way to gain insight into the process of electrostatic precipitation is to study the relationship known as the *Deutsch–Anderson equation*. This equation is used to determine the collection efficiency of the precipitator under ideal conditions.

The simplest form of the equation is

$$E = 1 - e^{-(wA/q)} \tag{10.3}$$

where E = collection efficiency of the precipitator
A = the effective collecting plate area of the precipitator, ft^2 (m^2)
q = gas flow rate through the precipitator, acfs (acms) [actual ft^3/s (actual m^3/s)]
e = base of natural logarithm = 2.718
w = migration velocity, ft/s (m/s)

This equation has been used extensively for many years for theoretical collection efficiency calculations. Unfortunately, while the equation is scientifically valid, there are a number of operating parameters that can cause the results to be in error by a factor of ≥ 2.

The Deutsch–Anderson equation neglects three significant process variables:

1. It completely ignores the fact that dust reentrainment may occur during the rapping process.
2. It assumes that the particle size and, consequently, the migration velocity is uniform for all particles in the gas stream.
3. It assumes that the gas flow rate is uniform everywhere across the precipitator and that particle sneakage through the hopper section does not occur.

Therefore, this equation should be used only for making preliminary estimates of precipitation collection efficiency.

When the desired collection efficiency and gas flow rate are specified, the required collecting area can be determined from the above Deutsch–Anderson equation once an appropriate precipitation rate parameter has been chosen. More recently, better correlations with field data on high-efficiency ESPs have been obtained by raising the exponential term in Equation (10.3) to a power m using existing values of w:

$$E = 1 - e^{-(wA/q)^m} \tag{10.4}$$

This provides a more accurate prediction of performance at high efficiency levels, but can become too pessimistic in certain situations. Typical values of m range between 0.4 and 0.7, with 0.5 as the norm. Equation (10.4) is referred to as the *Matts–Ohnfeldt equation*.

These and many other models have been proposed to predict particulate collection in an electrostatic precipitator, and while special advantages can be argued for each depending on which critical variables they account for, the fact remains that no model exists that accounts for all the variables that describe migration velocity in all situations. This is attributable mainly to the large number of variables that exist and because many of them are interrelated. In the final analysis, the goal is still to determine the correct amount of collecting surface area, and that usually depends on the proper selection of w.

As discussed above, large particles, over 1.0 μm in diameter, are charged mainly by direct collision with ions and free electrons moving toward the collecting plate along electric field lines. This *field charging* mechanism is very efficient for large particles, but the velocity of large charged particles toward the collecting plate is impeded by viscous drag forces. Small particles, less than 0.5 μm, become charged mainly by random thermal motion of ions. This *diffusion charging* mechanism is less efficient, but charged small particles can proceed toward the collecting plate with relative ease since the drag forces opposing their motion are small. The ability of a precipitator to collect the large and small particle sizes is essentially equal, but more difficult for the intermediate-sized particles. A precipitator working on a dust consisting mainly of particles in the intermediate-size range will be less efficient than an equally sized precipitator working on either predominantly large or small particles.

Many parameters must be taken into consideration in the design and specification of electrostatic precipitators. The focus of the remainder of this section will be on typical design parameters such as specific collection area and aspect ratio. Gas flow distribution, resistivity, and electrical sectionalization are treated in the next section.

The specific collection area (SCA) is defined as the ratio of collection surface area to the gas flow rate into the collector. The importance of this term is that it represent the A/q relationship in the Deustsch–Anderson equation:

$$\text{SCA} = \frac{\text{total collecting surface (ft}^2)}{\text{gas flow rate (1000 acfm)}} \tag{10.5}$$

or in metric units:

$$\text{SCA} = \frac{\text{m}^2}{1000\,\text{m}^3/\text{hr}} \tag{10.6}$$

An increase in the SCA of a precipitator design will in most cases increase the collection efficiency of the precipitator. Most conservative designs call for an SCA of 350–400 ft^2 per 1000 acfm (20–25 m^2 per 1000 m^3/hr) to achieve $\geq$99.5% particle removal. The general range of SCA is between 200 and 800 ft^2 per 1000 acfm (11–45 m^2 per 1000 m^3/hr) depending on precipitator design conditions and desired collection efficiency.

The *aspect ratio* (AR), the ratio of the total length to height of the collector surface, can be calculated by

$$\text{AR} = \frac{\text{effective length}}{\text{effective height}} \tag{10.7}$$

Having a precipitator chamber many times larger in length than in height would be ideal. However, space limitations and cost could be prohibitive. The aspect ratio for ESPs can range from 0.5 to 2.0. For $\geq$99.5% collection efficiency, the precipitator design should have an aspect ratio of greater than 1.0.

10.3 OPERATION AND MAINTENANCE, AND IMPROVING PERFORMANCE

In the case of operation, it is probably better to consider the precipitator as a process system rather than as a piece of hardware. Outwardly simple in design, the electrostatic precipitator takes part in a process that is complex. Many things can influence its behavior, but most of these influences can be overcome if the precipitator is well designed and properly operated and maintained. Operation and maintenance, in fact, is equally important to design.

Principal Properties and Characteristics

Three factors involved in the general area of ESP operation, maintenance, and performance—namely, gas flow distribution, resistivity, and electrical sectionalization—are briefly treated below.

Gas Flow Distribution. The best operating condition for a precipitator will occur when the treatment velocity distribution is uniform. When significant maldistribution occurs, the higher velocity in one collecting plate area will decrease efficiency more than a lower velocity at another equal area will increase the efficiency of that area. Small particles tend to follow flow streamlines better than large particles, so a maldistribution in particle size distribution in areas of the precipitator will also occur. Gross flow maldistribution also contributes to reentrainment losses and sneakage of raw gases around collecting plates. These concerns are addressed in a later problem.

Resistivity. Particle resistivity is a condition of the particle in the gas stream that can alter the actual collection efficiency of an ESP design. *Resistivity* is a term that describes the resistance of the collected dust layer to the flow of electrical current. By definition, the resistivity is the electrical resistance of a dust sample 1.0 cm^2 in cross-sectional area, 1.0 cm thick, and with units $\Omega \cdot cm$ (ohms $\times$ centimeters). It can also be described as the resistance to charge transfer by the dust. Dust resistivity values can be classified roughly into three groups:

1. Between 10^4 and 10^7 $\Omega \cdot cm$—low resistivity
2. Between 10^7 and 10^{10} $\Omega \cdot cm$—normal resistivity
3. Above 10^{10} $\Omega \cdot cm$—high resistivity

Particles that have low resistivity are difficult to collect since they are easily charged and lose their charge on arrival at the collection electrode. This happens rapidly, and the particles can take on the charge of the collection electrode. Particles can thus bounce off the plates and become reentrained in the gas stream. Examples of low-resistivity dusts are unburned carbon in fly ash and carbon black. If the conductive particles are coarse, they can be removed upstream of the precipitator with another device such as a cyclone. Baffles are often installed on the collection plates to help eliminate this precipitation-repulsion phenomenon.

Particles that have normal resistivity do not rapidly lose their charge on arrival at the collection electrode. These particles slowly leak their charge to ground and are retained on the collection plates by intermolecular adhesive and cohesive forces. This allows a particulate layer to be built up, which is then dislodged into the hoppers. At this range of dust resistivity (between 10^7 and 10^{10} $\Omega \cdot$ cm) fly ash can be collected most efficiently.

Particles that exhibit high resistivity are difficult to charge. Once they are finally charged, they do not readily give up the acquired negative charge on arrival at the collection electrode. As the dust layer builds up on the collection electrode, the layer and electrode form a high potential electric field. The surface of the dust layer is negatively charged, the interior is neutral, and the collection electrode is grounded. This causes a condition known as *back corona*. Under the influence of the corona discharge, the dust layer breaks down electrically, producing small holes or craters (in the layer) from which back corona discharges occur. Positive ions are generated within the dust layer and are accelerated toward the negative (discharge) electrode. The result of this event would be to counteract the ion generation of the charging electrode with a corresponding reduction in collection efficiency. Disruptions of the normal corona process greatly reduce precipitator collection efficiency which, in severe cases, may fall bellow 50%.

High resistivity can generally be reduced by adjusting the temperature and moisture content of the gas stream. Particle resistivity decreases for both high and low temperatures (see Figure 10.4). The moisture content of the gas stream also affects particle

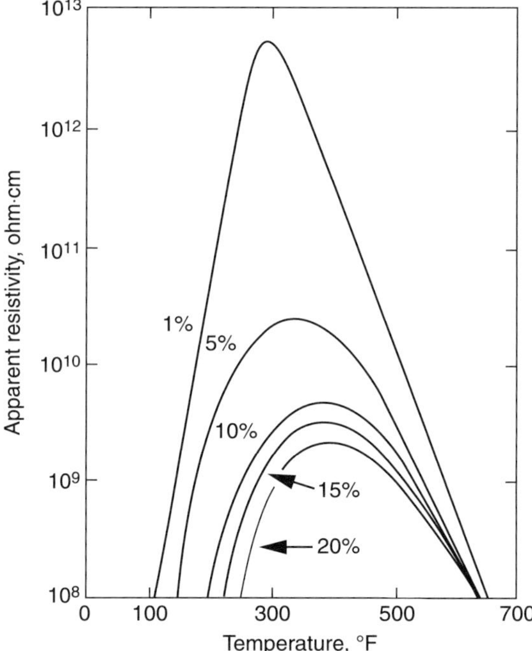

Figure 10.4 Effect of temperature and moisture content on apparent resistivity of precipitated cement dust.

resistivity. Increasing the moisture content of the gas stream lowers the resistivity. This can be accomplished by spraying water or injecting steam into the ductwork preceding the ESP. In both temperature adjustment and moisture conditioning, one must maintain gas conditions above the dew point to prevent corrosion problems.

The presence of SO_3 in the gas stream has been shown to favor the electrostatic precipitation process when problems with high resistivity occur. Most of the sulfur content in the coal burned for combustion sources converts to SO_2. However, approximately 1% of the sulfur converts to SO_3. The amount of SO_3 in the flue gas normally increases with increasing sulfur content of the coal. The resistivity of the particles decreases as the sulfur content of the coal increases.

The use of low-sulfur Western coal for boiler operations has caused fly ash resistivity problems for ESP operations. For coal fly ash dusts, the resistivity can be lowered below the critical level by the injection of as little as 10–20 ppm SO_3 into the gas stream. The SO_3 is injected into the ductwork preceding the precipitator. Other conditioning agents such as sulfuric acid, ammonia, sodium chloride, and soda ash have also been used to reduce particle resistivity.

Two other methods to reduce particle resistivity include: increasing the collection surface area and by inletting the exhaust gas at higher temperatures. Increasing the collection area of the precipitator will increase the overall cost of the ESP. This may not be the most desirable method to reduce resistivity problems. Hot precipitators, which are usually located before the combustion air preheater section of the boiler, are also used to combat resistivity problems. The use of hot precipitators is not discussed in this chapter because of their limited application.

Electrical Sectionalization. Precipitator performance is dependent on the number of individual sections or fields installed (see Figure 10.5). The maximum voltage at which a given field can be maintained depends on the properties of the gas and dust being collected. These parameters may vary from one point to another in the unit. To keep each section of the precipitator working at high efficiency, a high degree of sectionalization is recommended. Multiple fields or stages are used to provide electrical sectionalization (see Figure 10.6). Each field has separate power supplies and controls to adjust for varying gas conditions in the unit.

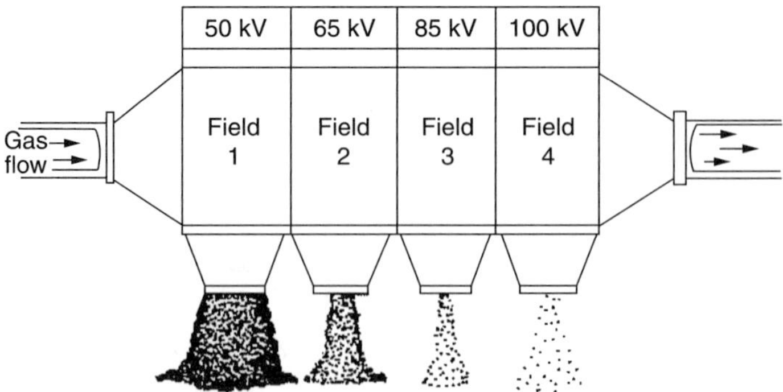

Figure 10.5 Stage or field sectionalization.

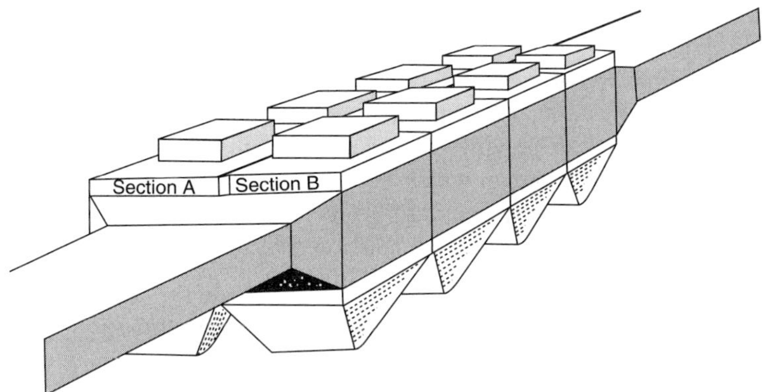

Figure 10.6 Parallel sectionalization.

Modern precipitators have voltage control devices that automatically limit precipitator power input. A well-designed automatic control system keeps the voltage level at approximately the value needed for optimum particle charging by the discharging electrodes.

The voltage control device operates in the following manner. Increases in voltage will cause a greater spark rate between the discharge and collection electrodes. As noted, the occurrence of a spark counteracts high ESP performance since it causes an immediate, short-term collapse of the precipitator field. Consequently, useful power is not applied to capture particles. However, there is an optimal sparking rate where the gain in particle charging is just offset by corona current losses from sparkover. Measurements on commercial precipitators have determined that the optimal sparking rate is between 50 and 150 sparks per minute per electrical section. The objective in power control is to maintain corona power input at this optimal sparking rate. This can be accomplished by momentarily reducing precipitator power whenever excessive sparking occurs.

The need for separate fields arises mainly because power input requirements differ at various locations in a precipitator. The particulate matter concentration is generally higher at the inlet sections of the precipitator. High dust concentrations tend to suppress corona current. Therefore, a great deal of power is needed to generate a corona discharge for optimal particular charging.

In the downstream fields of a precipitator, the dust loading is usually lighter. Consequently, the corona current flows more freely in downstream fields. Particle charging will more likely be limited by excessive sparking in downstream fields than in the inletfields. The power to the outlet sections must still be high in order to collect small particles, particularly if they exhibit high resistivity.

If the precipitator had only one power set, the excessive sparking would limit the power input to the entire precipitator. This would result in a reduction of overall collection efficiency.

The precipitator is divided into a series of independently energized bus sections or fields. Each bus section has individual transformer–rectifier sets, voltage stabilization controls, and high-voltage conductors that energize the discharge electrodes within the section. This allows greater flexibility for individual field energizing for varying conditions in the precipitator. Most ESP vendors recommend that there be at least four or more fields in the precipitator. It might be necessary to design the unit with seven or more fields for $\geq$99.9% collection efficiency.

Parallel sectionalization provides a means of coping with different power input needs due to uneven dust and gas distribution. Uneven gas distributions generally occur across the inlet face of the precipitator. Gains in collection efficiency from parallel sectionalization are likely to be small compared to field or stage sectionalization.

Recent Upgrades and Advances in ESP Technology

In terms of recent advances, there are six areas of contingent technology that have received the greatest attention in upgrading precipitator performance short of retrofitting; two of these are the aforementioned chemical conditioning and pulse energization.

Dust chemical and physical properties can combine to severely limit the performance capacity of a precipitator. Rather than increase the size of a precipitator to treat a difficult ash, chemical conditioning offers a means to modify the ash to suit a given precipitator. Ash conditioning has been used to provide an acceptable solution to difficult problems. Conditioning systems that make use of raw chemical materials should be more cost-effective than methods that use commercially prepared additives, and such systems are available to treat practically every situation that could arise. It is important to know which conditioning agent is best suited for which type of precipitator application. Ammonia appears to have the widest range of use, having been successfully applied to cold-side ESPs with both low- and high-sulfur coals, and on some hot-side units. The conditioning mechanism is not well known, but appears to enhance particle agglomeration and increase space charge. Sulfur trioxide has been most widely used to increase the surface conductivity of low-sulfur coal ashes. Since the mechanism depends on sorption, it cannot be expected to operate well above 190°C. Sodium carbonate addition has seen the most limited use of these three. The charge-carrying capability of the sodium ions appears to predominate in this method, with the carbonate playing no known role.

Pulse energization can enhance the collection efficiency of precipitators where performance limitations are caused by poor energization, arising from either high resistivity or fine-particulate corona quenching. Pulse energization consists of electrical components that may superimpose high-voltage pulses of very short duration on an underlying, relatively constant electric potential, or the pulses may be solely applied and varied automatically with the pulsed parameters, such as voltage, frequency, and width, to tune the precipitator operation and minimize the debilitating effects of resistivity-induced sparking or back corona. The precipitator can then operate with the relatively high charging and collecting fields needed for more effective particulate removal in the precipitator. Pulse energization produces a profuse corona along the discharge electrodes, resulting in uniform corona distribution, which plays an important role in enhancing particulate collection.

As described earlier in this section, for increased performance and reliability, precipitators have been divided into a number of independently energized bus sections. Each bus section has its own transformer-rectifier, voltage stabilization controls, and high-voltage conductors that energize the discharge electrodes within that section. The main reasons for sectionalization are to offset the dampening effects on corona power input of heavy flue dust loadings and to reduce the effect of bus section failures. These heavy flue gas dust loadings occur mainly in the inlet sections of a precipitator. By sectionalization, corona power input and particle charging can be increased in the inlet sections, thereby raising the overall precipitator efficiency. This area continues to provide opportunities for improved performance in the future.

The final three developments with ESPs are

1. Newly designed plates to prevent reentrainment
2. Modification of rapping frequencies and intensity between bus sections
3. Renewed interest in tubular ESPs (reduced reentrainment)

PROBLEMS

10.1 Principal Electrostatic Force

The principal force leading to particle collection in an electrostatic precipitator is (select one)

(**a**) Electrostatic
(**b**) Centrifugal
(**c**) Impaction
(**d**) Gravity

Solution: Obviously, the principal force is electrostatic. The correct answer is therefore (a).

10.2 Effect of Area on Efficiency

As the collection area of an electrostatic precipitator increases, its efficiency generally

(**a**) Decreases
(**b**) Remains the same
(**c**) Increases
(**d**) Varies

Solution: Refer to either Equation (10.3) or (10.4). The correct answer is therefore (c).

10.3 Presence of SO_3 in Gas

The presence of SO_3 in the carrier gas favors the electrostatic precipitation process by

(**a**) Increasing resistivity
(**b**) Aiding surface conduction of electricity while conditioning for high resistivity
(**c**) Improving agglomeration
(**d**) Increasing electrical wind

Solution: The effect of SO_3 on the carrier gas is to decrease the resistivity of the particulates. The correct answer is therefore (b).

10.4 Common Conditioning Agents

Gas conditioning radically affect particle resistivity. The most common conditioning agents are:

(**a**) Steam and low-resistivity particles
(**b**) Steam and as much as 200 ppm H_2SO_4
(**c**) Steam and as much as 20 ppm HNO_3
(**d**) Steam and as little as 20 ppm NH_3 or SO_3

Solution: As discussed earlier, small quantities of either steam, SO_3, or NH_3 can reduce particle resistivity. The correct answer is therefore (d).

10.5 Low-Resistivity Particles

Particles with resistivity below $10^4\ \Omega \cdot \text{cm}$ are difficult to collect because they

(**a**) Rapidly lose their negative charge at the collection electrode but can acquire a strong positive charge and become reentrained with respect to the plate
(**b**) Act as a resistor in series and lower corona current density
(**c**) May experience electrical breakdown and produce back corona
(**d**) Do not readily dissipate negative charges and cling to the collection electrode (eventually affecting the potential difference between electrodes, causing intense sparkover)

Solution: Low-resistivity dusts are highly conductive. As such, they rapidly lose their charge (since they are conductive) to the collecting electrode. The correct answer is therefore (a).

10.6 Dust Hoppers

Briefly describe dust hoppers.

Solution: Dust hoppers collect the precipitated particles from a dry electrostatic precipitator. Typically, the hopper takes the form of an inverted pyramid that converges to a round or square discharge. For best results, particles are removed either intermittently or continuously from the hopper. Either pressure or vacuum systems are used to remove particles from larger hoppers. In smaller installations, a screw conveyor may be used. Wet sluicing is another approach for removal of the solids. In the former cases, the solids can be disposed of in a dry form. Sluiced solids require a pond or other means to remove solids from the liquid stream.

10.7 Effect of Electrical Sectionalization

Electrical sectionalization improves ESP efficiency for which of the following reasons?

(**a**) It assures proper spark rate in all sections of the machine
(**b**) It eliminates problems with strong space charge, lowering the current density in sections near the ESP outlet
(**c**) It maintains optimum voltage and current density in all sections
(**d**) Both (a) and (c)

Solution: Electrical (bus section) sectionalization helps maintain the optimal voltage and current density and helps assume a reasonable spark rate throughout all the sections. In addition, it helps reduce operating problems that can develop if bus section failures arise. The correct answer is therefore (d).

10.8 Electrostatic Precipitator Components

Which of the following is not an integral component in an electrostatic precipitator?

(**a**) A rapper
(**b**) A collection plate
(**c**) A discharge electrode
(**d**) A venturi control rod

Solution: Answers (a)–(c) are components of an ESP. The correct answer is therefore (d).

10.9 Corona Discharge

Which of the following statements do not apply to a description of the corona discharge phenomenon in an electrostatic precipitator?

(**a**) A high DC voltage of negative polarity is applied to the corona discharge wire.
(**b**) The voltage is set for maximum power yet below the level of excessive sparkover.
(**c**) Electrical breakdown of the gas surrounding the discharge wire occurs owing to the action of positive ions striking the discharge wire.
(**d**) The intense electric field near the discharge wire accelerates electrons.

Solution: Answers (a), (b), and (d) do describe, in part, the corona discharge phenomenon. The correct answer is therefore (c).

10.10 Avalanche Multiplication

Avalanche multiplication describes the action of

(**a**) Accelerated positive ions striking the discharge wire and producing free electrons by secondary emissions
(**b**) Corona discharge starting voltage
(**c**) Accelerated electrons ionizing gas molecules by freeing a valence electron
(**d**) Current density gradient between the discharge electrode and the collection electrode

Solution: As described in Section 10.1, accelerated electrons from the discharge wire strike gas molecules and, in the process, split off an election. The correct answer is therefore (c).

10.11 Cause for Migration

Particles subjected to an electric field and ion bombardment in the area near the corona discharge will migrate toward the collection electrode when they reach

(**a**) The proper dielectric constant
(**b**) Saturation charge

(**c**) Field charge
(**d**) Diffusion charge

Solution: The presence of an electric field and charge produces a force that directs the particle to the collection plate. This occurs as the particle approaches the saturation charge. The correct answer is therefore (b).

10.12 Pressure Drop Losses

The ESP has very low draft losses. A designer may ensure proper gas flow into the unit by employing which of the following?

(**a**) Gas turning vanes in the duct elbows
(**b**) Gas turning vanes and an expansion section
(**c**) Turning vanes at duct elbows, an expansion section, and diffusion screens
(**d**) Smaller induced draft fans

Solution: All three options provided in answer (c) can reduce the pressure drop across the ESP. The best answer is therefore (c).

10.13 Design Factors

Provide at least six design factors that require consideration for an ESP specification.

Solution

1. Collection electrodes: type, size (area), mounting, and mechanical and aerodynamic properties
2. Discharge electrodes: type, size, spacing, and method of support
3. Shell: dimensions, insulation requirements, and access
4. Rectifier sets: rating, automatic control system, number, instrumentation, and monitoring provisions
5. Rappers for corona and collecting electrodes: type, size, range of frequency and intensity settings, number, and arrangement
6. Hoppers: geometry, size, insulation requirements, number, and location
7. Hopper dust removal system: type, capacity, protection against air inleakage, and dust blowback
8. Inlet and outlet gas duct arrangements, gas handling, and distribution system
9. Degree of sectionalization
10. Support insulators for high-tension frames: type, number, and reliability

10.14 Electrostatic Precipitation Advantages

List at least six advantages associated with the use of an ESP.

Solution: The electrostatic precipitator has several advantages and disadvantages relative to other particulate collectors. The advantages include:

1. High collection efficiency on removal of submicrometer particulates (as low as 0.01 μm)
2. Low operating costs
3. Low pressure drop (usually below 0.5 in H_2O).

4. Ability to effectively handle relatively large gas flows (to 2,000,000 + acfm)
5. Operation under high pressure (to 150 psi pressure) or vacuum conditions
6. Use under corrosive particulate conditions
7. Removal of precipitator units from operation for cleaning is unnecessary
8. Ability to handle high temperature gases ($\geq$1200°F)

10.15 Electrostatic Precipitator Disadvantages

List at least six disadvantages associated with the use of an ESP.

Solution

1. High capital cost.
2. High sensitivity to fluctuations in gas stream conditions.
3. Certain particulates are difficult to collect because of extremely low or high resistivity characteristics.
4. Relatively large space requirements required for installation.
5. Explosion hazard when treating combustible gases and/or collecting combustible particulates.
6. Special precautions are required to safeguard personnel from high voltages.
7. Ozone is produced by the negatively charged discharge electrode during gas ionization.
8. Relatively sophisticated maintenance personnel is required.

10.16 Matts–Ohnfeldt Equation

Quantitatively discuss the effect of the exponent m in the Matts–Ohnfeldt equation.

Solution: The Matts–Ohnfeldt equation is given as:

$$E = 1 - e^{-(Aq/w)^m} \tag{10.4}$$

Because of the exponent m, the area requirement for a given efficiency increases as m assumes values less than one (unity); i.e., m decreases. This effect is revisited in Problem 10.30.

10.17 Important Gas and Particulate Properties

Briefly discuss some of the more important properties that affect ESP design and performance.

Solution: Important gas stream and particle properties that affect the effectiveness of electrostatic precipitation of particulate matter include particle size distribution, gas flow rate, resistivity, and temperature. The designer should also consider how these characteristics affect the corrosiveness of the particles and the removability of particles from the collection electrodes.

10.18 Dust Buildup

Comment on the impact of the number of rappers on particulate buildup.

Solution: Dust thickness buildup of 0.5 – 1.0 inch prior to dislodging is common. Typical rapper accelerations range from 10 to 100 times that due to gravity (g). With fly ash, for example, accelerations of 10–30 g may be satisfactory, while highly resistive dust (difficult to remove) may require accelerations as high as 200 g. Table 10.2 shows the number of rappers per unit area of collecting electrode and the number per unit length of discharge wire for various applications.

TABLE 10.2 Typical Rapping Practices for Various Applications

Application	Collection Electrodes, Rappers/1000 ft^2	Discharge Electrode, Rappers/1000 ft
Utilities	0.25–0.90	0.09–0.66
Pulp and paper	0.25–0.99	0.21–0.32
Metals	0.11–0.82	0.28–0.50
Cement	0.33–0.52	0.19–0.33

10.19 Particle Maximum Charge

A 5.0 μm radius particle traveling through an ESP is subjected to an electric field of 3 kV/cm. What is the maximum charge in coulombs (C) that can be acquired by the particle? The maximum charge ($q_{p,max}$) can be calculated from the following equation:

$$q_{p,max} = 4\pi \varepsilon_0 p(d_p/2)^2 E_0 \qquad (10.8)$$

where $p = 1.0$ for a particulate without a dielectric constant, dimensionless
$\varepsilon_0 = 8.854 \times 10^{-12}$ F/m (farads per meter)

Solution: Convert micrometers to meters:

$$r_p = d_p/2 = (50)(10^{-6})$$
$$= 5 \times 10^{-6} \text{m}$$

Convert the electric field to V/m:

$$E_0 = (3)(100)(1000)$$
$$= 3 \times 10^5 \text{ V/m}$$

Substitute into Equation (10.8) (above):

$$q_{p,max} = 4\pi(8.854 \times 10^{-12})(5 \times 10^{-6})^2(3 \times 10^5);\ 1\text{F} = 1\text{C/V}$$
$$= 8.34 \times 10^{-16}\text{C}$$

10.20 Electrostatic Force

Refer to Problem 10.19. Assuming that the particle acquires 90% of its maximum charge, what electrostatic force, in newtons (N), is the particle subjected to when the electric field across the ESP is constant at 4 kV/cm?

Solution: By definition

$$F = q_{\mathrm{p}} E_{\mathrm{p}} \tag{10.9}$$
$$= (0.9)(8.34 \times 10^{-16}\,\mathrm{C})(3 \times 10^{5}\,\mathrm{V/m})$$
$$= 2.25 \times 10^{-10}\,\mathrm{C \cdot V/m}$$
$$= 2.25 \times 10^{-10}\,\mathrm{N}$$

10.21 Terminal Velocity Calculation

Refer to Problems 10.19 and 10.20. The particle, which has a density of 150 lb/ft^3, is entrained in an air stream at ambient conditions. What is its terminal drift velocity?

Solution: Assume that Stokes' law applies and solve for the velocity:

$$w = F/3\pi\mu d_{\mathrm{p}} \tag{10.10}$$

At ambient conditions, one obtains

$$\mu = 183\,\mu\mathrm{P}$$
$$= 183 \times 10^{-6}\,\mathrm{P}$$
$$= 183 \times 10^{-6}\,\mathrm{dyn \cdot s/cm^2}$$
$$= 183 \times 10^{-11}\,\mathrm{N \cdot s/cm^2}$$
$$= 183 \times 10^{-7}\,\mathrm{N \cdot s/m^2}$$

Substituting into Equation (10.10), one obtains

$$w = 2.25 \times 10^{-10}/3\pi(183 \times 10^{-7})(10 \times 10^{-6})$$
$$= 0.131\,\mathrm{m/s}$$
$$= 13.1\,\mathrm{cm/s}$$

10.22 Minimum Plate-to-Plate Spacing

Refer again to Problem 10.19. If the ESP length is 8 ft and the air stream velocity is 8 ft/s, what is the minimum plate-to-plate distance that will allow this particle to escape capture?

Solution: The residence time in the ESP is

$$t_{\mathrm{r}} = 8/8$$
$$= 1.0\,\mathrm{s}$$

Employing the w calculated in Problem 10.21, the minimum wire-to-plate distance is then

$$\begin{aligned}\text{WTP} &= (1.0)(13.1)/(2.54)\\ &= 5.14\ \text{inches}\end{aligned}$$

The plate-to-plate spacing is therefore

$$\begin{aligned}\text{PTP} &= (2)(5.14)\\ &= 10.3\ \text{inches}\end{aligned}$$

10.23 Required Applied Voltage

An electrostatic precipitator has six collecting plates 10 ft tall and 10 ft long in the direction of flow. The spacing between the plates is 9 inches (0.229 m). The ESP is to be used to collect particles having a dielectric constant of 4.0 and an effective diameter of 3 μm. The carrier gas (air at 20°C) has a throughput veloclty of 24 ft/s. Calculate the voltage (kV) required for 99.5% collection efficiency. Use the equation

$$w = 1.1 \times 10^{-14} p E_{\mathrm{d}}^2 d_{\mathrm{p}}/\mu \tag{10.11}$$

where w = drift velocity, m/s
E_{d} = field strength at discharge electrode, V/m
d_{p} = particle diameter, μm
μ = gas viscosity in kg/m · hr

and

$$p = 3D/(D+2); \quad D = \text{dielectric constant} \tag{10.12}$$

Solution: Calculate p using Equation (10.12):

$$\begin{aligned}p &= 3D/(D+2) = (3)(4)/(4+2)\\ &= 12/6\\ &= 2.0\end{aligned}$$

The field strength is

$$\begin{aligned}E_{\mathrm{d}} &= (\text{kV})(1000)/(\tfrac{1}{2}\ \text{spacing in meters})\\ &= (\text{kV})(1000)/0.115\\ &= V/0.115\end{aligned}$$

Substituting into Equation (10.11) yields

$$\begin{aligned}w &= 1.1 \times 10^{-14}(2)(E_{\mathrm{d}})^2(3\ \mu\text{m})/(0.0863 \times (0.115)^2); \quad E_{\mathrm{d}} = V\\ &= 5.783 \times 10^{-11}(E_{\mathrm{d}})^2\end{aligned}$$

The area and flow rate are

$$A = 6 \times 10 \text{ ft} \times 10 \text{ ft}$$
$$= 600 \text{ ft}^2$$
$$q = 3(9/12)\,\text{ft}(10 \text{ ft})(24 \text{ ft/s})$$
$$= 540 \text{ ft}^3/\text{s}$$

Substitute into the Deutsch–Anderson equation:

$$E = 1 - e^{-wA/q} \tag{10.3}$$
$$= 1 - \exp[-(600/540)(5.783 \times 10^{-11})(E_\text{d})^2(3.281 \text{ ft/m})^2]$$
$$E = 1 - \exp[-6.917 \times 10^{-10} \times (E_\text{d})^2]$$
$$E_\text{d} = [\ln(1 - E)/(-6.9171 \times 10^{-10})]^{0.5}$$

For $E = 99.5\% = 0.995$, one obtains

$$E_\text{d} = [\ln(1 - 0.995)/(-6.9171 \times 10^{-10})]^{0.5}$$
$$= 87{,}520 \text{ V}$$
$$= 87.52 \text{ kV}$$

10.24 Collection Area

A small coal-fired power plant sends 2400 acfm through its electrostatic precipitation. The particle migration velocity is known to be 0.35 ft/s. What is the collection area if the overall ESP efficiency is 99.78%?

(a) 699.35 ft^2
(b) 669 ft^3
(c) 448 ft^2
(d) 288 ft^2

Solution: Apply the Deutsh-Anderson equation:

$$E = 1 - e^{-Aw/q} \tag{10.3}$$

Substituting, one obtains

$$0.9978 = e^{-(A)(0.35)/(2400/60)}$$

Solving for A yields

$$A = 699 \text{ ft}^2$$

The correct answer is therefore (a).

10.25 Single-Duct ESP

A horizontal parallel-plate ESP consists of a single-duct 24 ft high and 20 ft deep with an 11 inch plate-to-plate spacing. A collection efficiency of 88.2% is obtained with a flow rate of 4200 acfm. The inlet loading is 2.82 gr/ft^3. Calculate the following:

(a) The bulk velocity of the gas (assume a uniform distribution)
(b) The outlet loading
(c) The drift velocity for this system
(d) A revised collection efficiency if the flow rate increased to 5400 acfm
(e) A revised collection efficiency if the plate spacing is decreased to 9 inch

Solution

(a) The bulk velocity is given by

$$v = q/A_\mathrm{d} = 4200/24(11/12) = 191 \text{ ft/min}$$

where A_d = duct cross-sectional area (ft^2).

(b) The outlet loading is $2.82(1 - 0.882) = 0.333$ gr/ft^3.

(c) The term ϕ is first calculated from a modified form of Equation (10.3):

$$E = 1 - e^{-\phi};\ \phi = wA/q \tag{10.13}$$

$$0.882 = 1 - e^{-\phi}$$

Solving for ϕ

$$\phi = 2.14$$

Since

$$\phi = wA/q$$
$$2.14 = (w)(24)(20)(2)/(4200/60)$$
$$w = 0.156 \text{ ft/s}$$

(d) If $q = 5400$ acfm, then a new ϕ can be calculated assuming the same drift velocity:

$$\phi = (0.156)(24)(20)(2)/(5400/60) = 1.67$$

Calculate the revised collection efficiency:

$$E = 1 - e^{-1.67} = 0.812 = 81.2\%$$

(e) Since q, w, and A are all constant, the Deutsch–Anderson equation predicts that the efficiency does not change if the plate spacing is 9 inches.

10.26 Calculation of ϕ

Refer to Equation (10.13) in problem 10.25.

(a) What value of ϕ will yield an efficiency of 99%?
(b) What will the efficiency be if the collection area doubles, while the drift velocity and the volume rate of flow remain the same?
(c) What will the efficiency be if the drift velocity doubles, while the collection area and the volume rate of flow remain the same?

(**d**) What will the efficiency be if the volume rate of flow doubles, while the drift velocity and the collection area remain the same?

Solution

(**a**) Apply Equation (10.13):

$$E = 1 - e^{-\phi}$$

Substituting, one obtains

$$0.99 = 1 - e^{-\phi}$$
$$\phi = 4.62$$

(**b**) Since

$$\phi = Aw/q$$
$$\phi_{\text{new}} = 2\phi$$
$$= (2)(4.62) = 9.24$$

Therefore

$$E = 1 - e^{-9.24}$$
$$= 1.0 - 0.001$$
$$= 0.9999$$
$$= 99.99\%$$

(**c**) Since ϕ_{new} remains the same, the efficiency is still 99.99%.

(**d**) For this case, one obtains

$$\phi_{\text{new}} = \phi/2$$
$$= 2.31$$

Thus,

$$E = 1.0 - e^{-2.31}$$
$$= 1.0 - 0.1$$
$$= 0.90$$
$$= 90\%$$

10.27 Effect of Altering Parameters

Refer to Problem 10.26. Fill in the chart in Table 10.3 for efficiencies, assuming that the original efficiency 99%.

Solution: This is left as an exercise for the reader. Results are presented in Table 10.4.

10.28 Nanometer-Sized Particles

The data for three nanosized particles are provided in Table 10.5. Calculate the efficiency for the three given particle sizes. Estimate the efficiency for a 15-nm particle.

TABLE 10.3 Altered Parameters—Incomplete Chart

	Altered Parameter					
E	A Doubles	w Doubles	q Halves	A Halves	w Halves	q Doubles
Efficiency, %	99.99	99.99	(?)	(?)	(?)	90

TABLE 10.4 Altered Parameters—Completed Chart

	Altered Parameter					
E	A Doubles	w Doubles	q Halves	A Halves	w Halves	q Doubles
Efficiency, %	99.99	99.99	99.99	90	90	90

TABLE 10.5 Nanoparticle Data

$\phi = Aw/q$	Average Dust Particle Size (nanometers, nm)
6.95	5.0
5.94	7.5
5.02	10.0

Solution: Apply the Deutsch–Anderson equation:

$$E = 1 - e^{-\phi} \tag{10.13}$$

For $\phi = 6.95$:

$$\begin{aligned} E &= 1 - e^{-6.95} \\ &= 1 - 0.00096 \\ &= 0.99904 \\ &= 99.904\% \end{aligned}$$

For $\phi = 5.94$:

$$\begin{aligned} E &= 1 - 0.00263 \\ &= 0.9974 \\ &= 99.74\% \end{aligned}$$

For $\phi = 5.02$:

$$E = 1 - 0.00660$$
$$= 0.9934$$
$$= 99.34\%$$

For a 15 nm particle, linearly extrapolate for a rough estimate.

$$E = 1 - 0.0136$$
$$= 98.64\%$$
$$= 98.64\%$$

10.29 Cunningham Correction Factor Effect

A consultant has been hired to determine the effect of including the Cunningham correction factor (CCF) on the collection efficiency for a 10 μm particle. Calculate the percent change in the efficiency if the CCF is included in the calculation. Also calculate the percent change in the area requirement if the efficiency remains constant. The consultant has estimated that the CCF is 1.0164 for a 10 μm particle.

Solution: Referring to Equation (10.10) in Problem 10.21, one notes that the drift velocity is linearly related to the particle size through the drag force. Thus, from the Deutsch–Anderson equation ($w_2 = 1.0164\, w_1$, if the area is constant)

$$\frac{1 - E_2}{1 - E_1} = \frac{e^{-(Aw_2)/q}}{e^{-(Aw_1)/q}}$$

$$\frac{P_2}{P_1} = e^{-0.0164}$$
$$= 0.9837$$
$$P_2 = 0.9837P_1$$

The effect on efficiency is not as simple as that for penetration. Here

$$\frac{1 - E_2}{1 - E_1} = 0.9837$$

Rearranging this equation leads to

$$E_2 = 0.9837E_1 + 0.0163$$

Thus, no percent change can be specified for the efficiency. However, the penetration change (PC) is

$$PC = \frac{P_2 - P_1}{P_1}$$
$$= \frac{0.9837P_1 - P_1}{P_1}$$
$$= 0.0164 = 1.64\%$$

If the efficiency remains constant, one has

$$\frac{1 - E_2}{1 - E_1} = \frac{e^{-(A_2 w_2)/q}}{e^{-(A_1 w_1)/q}}$$

one concludes

$$\begin{aligned} A_2 w_2 &= A_1 w_1 \\ A_2 &= (w_1/1.0164 w_1) A_1 \\ &= 0.9839 \end{aligned}$$

Thus, the area decreases by 1.61%.

10.30 Design Equations

An electrostatic precipitator (ESP) is being used to clean fly ash from a gas. The precipitator contains 30 ducts, with plates 12 ft high and 12 ft long. The spacing between the plates is 8 inches. The gas is evenly distributed through all of the ducts. The following information as provided:

Gas volumetric flow rate = 40,000 acfm
Particle drift velocity = 0.40 ft/s

Use the Deutsch–Anderson (DA) equation to calculate the efficiency of the precipitator. Also, use the modified D–A equation (i.e., the Matts–Ohnfeldt equation) with a range of exponents varying from 0.4 to 0.7 (in increments of 0.05) to calculate the efficiency of the electrostatic precipitator.

Solution: As noted earlier, an attempt to account for the sensitivity of w on process variables, especially for small particle size distributions, appeared in 1957 and was later revised by Allander, Matts, and Ohnfeldt, who derived the expression

$$E = 1 - e^{-(wA/q)^m} \tag{10.4}$$

The second exponent (in this, the so-called *modified Deutsch equation*) provides a more accurate prediction of performance at high efficiency levels but can become too pessimistic in certain situations. Typical values of m range between 0.4 and 0.7, with 0.5 as the norm. This and the DA equation are employed in the solution that follows. The collection surface area per duct A is

$$\begin{aligned} A &= (12)(12)(2) \\ &= 288\ \text{ft}^2 \end{aligned}$$

The volumetric flow rate through each duct in acfs is

$$\begin{aligned} q &= 40{,}000/[(30)(60)] \\ &= 22.22\ \text{acfs} \end{aligned}$$

The collection efficiency using the D–A model can now be calculated:

$$E = 1 - e^{-(wA/q)} = 1 - e^{-((288)(0.4)/22.22)}$$
$$= 0.9944 = 99.44\%$$

Using the modified DA equation to obtain an expression for the efficiency in terms of the exponent m leads to

$$E = 1 - e^{-(wA/q)^m} = 1 - e^{-((288)(0.4)/22.22)^m} = 1 - e^{-5.184^m}$$

Table 10.6 provides E for various values of m.

TABLE 10.6 Matts–Ohnfeldt Equation Calculation

m	E, %
0.40	85.51
0.45	87.72
0.50	89.74
0.55	91.56
0.60	93.17
0.65	94.58
0.70	95.78

While the DA equation predicts an efficiency of 99.44%, it can be seen that the efficiency is probably somewhat lower than that. At a value for the exponent of 0.5, it appears that the ESP operates at 89.7% efficiency. If expressed in terms of penetration, the DA equation gives a value of 0.0056 while the modified DA gives a value of 0.1026. This means that 18.3 times more fly ash is passing through (not collected) the precipitator than that predicted by the DA equation. Because of this, it can be seen that the design of an ESP can be somewhat tricky.

10.31 Three Fields in Series

An electrostatic precipitator is to be used to treat 100,000 acfm of a gas stream containing particulates from a hazardous waste incineration. The proposed precipitator consists of three bus sections (fields) arranged in series, each with the same collection surface. The inlet loading has been measured as 40 gr/ft^3, and a maximum outlet loading of 0.18 gr/ft^3 is allowed by local Environmental Protection Agency (EPA) regulations. The drift velocity for the particulates has been experimentally determined in a similar incinerator installation with the following results:

First section (inlet): 0.37 ft/s
Second section (middle): 0.35 ft/s
Third section (outlet): 0.33 ft/s

(a) Calculate the total collecting surface required based on the *average* drift velocity and the required total efficiency.

(b) Find the total mass flowrate (lb/min) of particulates captured by each section using the above drift velocities.

Solution: (a) Calculate the required total collection efficiency on the basis of the given inlet and outlet loading:

$$\begin{aligned} E &= 1 - \frac{\text{outlet loading}}{\text{inlet loading}} \\ &= 1 - \frac{0.18}{40} \\ &= 0.9955 = 99.55\% \end{aligned}$$

Calculate the average drift velocity w:

$$\begin{aligned} w &= (0.37 + 0.35 - 0.33)/3 \\ &= 0.35 \text{ ft/s} \end{aligned}$$

Calculate the total surface area required using the DA equation:

$$\begin{aligned} A &= -\frac{\ln(1 - E)}{w/q} \qquad (10.3) \\ &= -\frac{\ln(1 - 0.9955)}{0.35/1666.7} \\ &= 25.732 \text{ ft}^2 \end{aligned}$$

(b) Calculate the collection efficiency of each section. Assume that each section has the same surface area ($A/3$) but employs individual section drift velocities:

$$\begin{aligned} E_1 &= 1 - e^{-(Aw_1/3q)} = 1 - e^{-((25{,}732)(0.37)/[(3)(1666.7)])} \\ &= 0.851 \end{aligned}$$

$$\begin{aligned} E_2 &= 1 - e^{-(Aw_2/3q)} = 1 - e^{-((25{,}732)(0.35)/[(3)(1666.7)])} \\ &= 0.835 \end{aligned}$$

$$\begin{aligned} E_3 &= 1 - e^{-(Aw_3/3q)} = 1 - e^{-((25{,}732)(0.33)/[(3)(1666.7)])} \\ &= 0.817 \end{aligned}$$

Calculate the mass flow rate of particulates captured by each section using the collection efficiencies calculated above:

$$\begin{aligned} \dot{m}_1 &= (E_1)(\text{inlet loading})(q) \\ &= 3.404 \times 10^6 \text{ gr/min} \\ &= 486.3 \text{ lb/min} \end{aligned}$$

$$\dot{m}_2 = (1 - E_1)(E_2)(\text{inlet loading})(q)$$
$$= 4.977 \times 10^5 \text{ gr/min}$$
$$= 71.1 \text{ lb/min}$$
$$\dot{m}_3 = (1 - E_1)(1 - E_2)(E_3)(\text{inlet loading})(q)$$
$$= 8.034 \times 10^4 \text{ gr/min}$$
$$= 11.48 \text{ lb/min}$$

Note that nearly fifty times as much particulates are captured in the first section relative to the third section.

10.32 ESP Retrofit

An electrostatic precipitator is to be used to clean 100,000 acfm of a particulate-laden gas stream from a blast furnace. The proposed design of the precipitator calls for three bus sections (fields) arranged in series. The inlet particulate loading has been measured as 3.77 gr/acf, and an outlet loading of 0.05 gr/acf must be achieved in order to comply with state regulations. Each section consists of 12 grounded plates, 8 ft high and 20 ft long, spaced 10 in apart and the gas passes horizontally through the precipitator. Answer the following questions:

(a) What is the minimum collection efficiency that will satisfy the regulation?
(b) If the drift velocity w for each section is 0.41 ft/s, what is the collection efficiency of the proposed unit?
(c) How many identical sections will have to be added (in series) to bring the proposed unit into compliance? Assume that the added sections also have a drift velocity of 0.41.
(d) If the proposed unit is limited to a maximum of three sections, how many plates (additional area) must be added to each section to ensure compliance? Assume that the drift velocity is unaffected by the addition of more plates.

Solution

(a) The required overall penetration is given by

$$P = 0.05/3.77 = 0.0133$$
$$= 1.33\%$$

and the efficiency is

$$E = 1 - (0.0133) = 0.9867$$
$$= 98.67\%$$

(b) There are various ways to solve this part. For 12 plates, one obtains

$$A = (2)(11)(8)(20) = 3520 \text{ ft}^2$$
$$q = 100{,}000/60 = 1{,}667 \text{ acfs}$$

The single section P_1 is therefore

$$P_1 = e^{-[(0.41)(3520)/(1667)]}$$
$$= 0.421 = 42.1\%$$
$$E = 0.579 = 57.9\%$$

The overall penetration, P_0 is

$$P_0 = P_1^3 = (0.421)^3 = 0.0745$$
$$E_0 = 0.925 = 92.5\%$$

(c) Since the maximum allowable penetration is 0.0133, one obtains

$$(0.421)^n = 0.0133$$
$$n = 4.99$$
$$= 6.0$$

the required number of sections

(d) With only three sections, the individual penetration required is

$$P_{\text{req}} = (0.0133)^{1/3}$$
$$= 0.2369 = 0.237$$

The required area per section is then given by the following equation:

$$0.2369 = \exp[(-A)(0.41)/(1667)]$$
$$A = 5855 \text{ ft}$$

The number of surfaces N becomes

$$N = 5855/(8)(20)$$
$$= 36.6 \Rightarrow 38; \text{ even number required}$$

The number of plates N_p is

$$N_p = (38 + 2)/2 = 20$$

10.33 Four Channels in Parallel

A single-stage duct-type electrostatic precipitator contains five plates that are 10 ft high, 20 ft long, and spaced 9 inches apart. Air contaminated with gypsum dust enters the unit with an inlet loading of 53 gr/ft^3 and a velocity through the unit of 5.0 ft/s. The dust bulk density is 47 lb/ft^3.

(a) Estimate the particle drift velocity w given a efficiency of 99%.
(b) What is the outlet loading?
(c) How many cubic feet of dust are collected per hour?

Solution: (a) The collecting area A and gas flow rate q are given by

$$A = (8\ \text{surfaces})\,(10\ \text{ft})\ (20\ \text{ft}) = 1600\ \text{ft}^2$$

$$q = (4\ \text{channels})(5\ \text{ft/s})(9/12\ \text{ft})(10\ \text{ft}) = 150\ \text{ft}^3/\text{s}$$

Using the DA equation yields

$$w = -\frac{\ln(1-E)}{A/q} \qquad (10.3)$$

$$= -\frac{\ln(1-0.99)}{1600/150}$$

$$= 0.4317\ \text{ft/s}$$

(b) The outlet loading (OL) is

$$\text{OL} = 53(1-E) = 53(1-0.99) = 0.53\ \text{gr/ft}^3$$

(c) The volumetric flow rate of particulates captured per hour q_p is

$$q_\text{p} = \frac{(53.0-0.53)\ \text{gr/ft}^3}{(7000\ \text{gr/lb})(47\ \text{lb/ft}^3)}(150\ \text{ft}^3/\text{s})(3600\ \text{s/hr})$$

$$= 86.12\ \text{ft}^3/\text{hr}$$

10.34 Rapping Frequency

Refer to Problem 10.33. Determine the rapping frequency (intervals) in minutes assuming that the maximum allowable dust thickness on the plates is $\frac{1}{8}$ inch. (Assume this layer to be uniform over the entire plate.)

Solution: From Problem 10.33, the volume rate of dust collected is

$$q_\text{p} = 86.12\ \text{ft}^3/\text{hr}$$

The rapping cycle (RC) time is therefore

$$RC = \frac{[(1/8)/12]\ \text{ft}\,(1600\ \text{ft}^2)}{(86.12\ \text{ft}^3/\text{hr})(1\ \text{hr}/60\ \text{min})}$$

$$= 11.61\ \text{min}$$

The general subject of rapping is revisited in the last problem of this chapter.

10.35 Effect of Inlet Distribution

You have been requested to calculate the collection efficiency of an electrostatic precipitator containing three ducts with plates of a given size, assuming a uniform distribution of particles. Also determine the collection efficiency assuming that one duct is fed 50% of the gas and the other passages 25% each. Operating and design data include:

Volumetric flow rate of contaminated gas = 4000 acfm

Operating temperature and pressure = 20°C and 1 atm, respectively
Drift velocity = 0.40 ft/s
Size of the plate = 12 ft long and 12 ft high
Plate-to-plate spacing = 8 inches

Solution: Considering both sides of the plate, one obtains

$$\begin{aligned} A &= (2)(12 \text{ ft})(12 \text{ ft}) \\ &= 288 \text{ ft}^2 \end{aligned}$$

Remembering the volumetric flow rate through a passage is one-third of the total volumetric flow rate:

$$\begin{aligned} q &= \frac{4000}{(3)(60)} \\ &= 22.22 \text{ acfs} \end{aligned}$$

Calculate the collection efficiency using the DA equation:

$$E = 1 - e^{-(wA/q)} \tag{10.3}$$

$$\begin{aligned} &= 1 - e^{-((288)(0.4)/22.22)} \\ &= 0.9944 = 99.44\% \end{aligned}$$

This efficiency calculation assumes the gas is uniformly distributed at the inlet of the precipitor. A revised efficiency can be calculated if the flow is distributed as specified in the problem statement. First, calculate q_1 in acfs through the middle section:

$$\begin{aligned} q_1 &= \frac{4000}{(2)(60)} \\ &= 33.33 \text{ acfs} \end{aligned}$$

Calculate the collection efficiency, remembering that the collection surface area per duct remains the same:

$$\begin{aligned} E_1 &= 1 - e^{-((288)(0.4)/33.33)} \\ &= 0.9684 = 96.84\% \end{aligned}$$

Calculate q_2 in acfs through an outer section:

$$\begin{aligned} q_2 &= \frac{4000}{(4)(60)} \\ &= 16.67 \text{ acfs} \end{aligned}$$

The collection efficiency in the outside section is

$$\begin{aligned} E_2 &= 1 - e^{-((288)(0.4)/16.67)} \\ &= 0.9990 = 99.90\% \end{aligned}$$

Calculate the new overall collection efficiency:

$$E = (0.5)(E_1) + (2)(0.25)(E_2)$$
$$= 98.37\%$$

Note that the penetration $(100-E)$ has increased by a factor of 3. The calculational procedure to follow if the particle size distribution varies with each inlet duct is treated shortly.

10.36 Linear Drift Velocity Variation with Particle Size

The drift velocity for a 154 μm particle from a smelter has been determined to be 0.605 ft/s. Assuming that the drift velocity varies linearly with particle size, obtain an equation for w in terms of d.

Solution: Since the variation is linear, it follows that

$$w = kd_{\text{p}} \tag{10.14}$$

Since $w = 0.605$ for $d_p = 154$, one obtains

$$0.605 = k(154)$$
$$k = 0.00393$$

Therefore,

$$w(\text{ft/s}) = 0.00393\, d_{\text{p}}(\mu\text{m})$$

10.37 Drift Velocity Variation with Particle Size

Drift velocity–particle size data are provided in Table 10.7. Generate an equation describing $w-d_{\text{p}}$ variation.

TABLE 10.7 Drift Velocity–Particle Size Variation

w, ft/s	d_{p}, μm
0.104	0.5
0.104	2.5
0.173	7.5
0.279	12.5
0.356	17.5
0.384	20+

Solution: This is an open-ended problem since one can select any one of several different models. For example, consider the model expressed in the following equation:

$$w = ad_{\text{p}}^{b} \tag{10.15}$$

This equation can be linearized by taking the natural logarithm of both sides of the equation:

$$\ln w = \ln a + b \ln d_{\text{p}}$$

This is an equation of the form

$$y = a + bx$$

Regressing the data (six points), one obtains

$$a = 2.88$$
$$b = 0.079$$

so that

$$w = 2.88\, e^{0.079 d_p}$$

The corresponding regression coefficient r^2 (or R^2) is 0.98. The reader is left the exercise of solving this problem using a different model.

10.38 Cut Diameter Calculation

Fractional efficiency cures describing the performance of a specific model of an electrostatic precipitator have been compiled by a vendor. Although you do not possess these curves, you are told that the cut diameter for a precipitator with a 10 inch plate spacing is 0.9 μm. The vendor claims that this particular model will perform with an efficiency of 98% under your operating conditions. You are asked to verify this claim and to make certain that the effluent loading does not exceed 0.2 gr/ft^3; the inlet loading is 14 gr/ft^3. The aerosol has the particle size distribution given in Table 10.8.

TABLE 10.8 Particle Size Distribution for Problem 10.38

Weight Range, %	Average Particle Size, μm
0–20	3.5
20–40	8.0
40–60	13.0
60–80	19.0
80–100	45.0

Assume a DA equation of the form

$$E = 1 - e^{-Kd_{\rm p}} \qquad (10.16)$$

to apply.

Solution: Equation (10.16) is to be employed.

$$E = 1 - e^{-Kd_{\rm p}}$$

The cut diameter information can be used to calculate K.

$$0.5 = 1.0 - e^{-K(0.9)}$$

$$K = \frac{-\ln(1 - 0.5)}{0.9}$$

$$= 0.77$$

Therefore

$$E = 1 - e^{-0.77d_p}$$

Table 10.9 can now be generated.

TABLE 10.9 Overall Efficiency Calculation

w_i	d_p	E_i
0.2	3.5	0.43250
0.2	8.0	0.99789
0.2	13	0.99996
0.2	19	0.99999
0.2	45	0.99999

Therefore

$$\begin{aligned} E_o &= 0.2[(0.9325) + (0.99789) + (0.99996) + (0.99999) + (0.99999)] \\ &= 0.9861 \\ &= 98.61\% \end{aligned}$$

and the outlet loading is

$$\begin{aligned} \text{OL} &= 14(1 - 0.9861) \\ &= 0.1951\ \text{gr/ft}^3 \\ &= 0.20\ \text{gr/ft}^3 \end{aligned}$$

The standard *is* met.

10.39 Effect of Particle Size Distribution

The following data are available for the proposed design of an ESP to operate at an efficiency of 97.5%:

Air volumetric flow = 100,000 acfm
Uniform flow distribution through 10 ducts
Duct height = 30 ft with 12 inch plate-to-plate spacing
Plate length = 36 ft
Inlet loading = 14 gr/ft^3
Outlet loading = 0.35 gr/ft^3

In addition, the particle size distribution and drift velocity data listed in Table 10.10 have been provided in terms of the average particle size (d_p) of weight fraction (x_i) in a given size range and the corresponding drift velocity (w). Determine whether the proposed design will meet the desired efficiency. Also, prepare a graph of particle size vs. efficiency for the system and comment on the results.

TABLE 10.10 Particle Size Distribution–Drift Velocity Data

d_p, μm	w, ft/s	x_i
0.1	0.27	0.01
0.25	0.15	0.01
0.5	0.12	0.01
1.0	0.11	0.01
1.5	0.15	0.16
2.0	0.20	0.16
2.5	0.26	0.16
5.0	0.50	0.16
10.0	0.60	0.16
25.0	0.70	0.16

Solution: Check if the average through put velocity is in an acceptable range. The average velocity should be 2–8 ft/s.

$$\begin{aligned} v &= (100{,}000)/[(10)(30)(60)] \\ &= 5.56\,\text{ft/s} \end{aligned}$$

This is in the acceptable range.
The volumetric flow rate through each (one) duct is

$$\begin{aligned} q &= (100{,}000)/10 \\ &= 10{,}000\,\text{acfm} \\ &= 167\,\text{acfs} \end{aligned}$$

Calculate the plate area in square feet for each duct. Note once again that both plates contribute to the collection area:

$$\begin{aligned} A &= 2(30)(36) \\ &= 2160\ \text{ft}^2 \end{aligned}$$

If the Deutsch–Anderson (DA) equation applies, then

$$E = 1 - e^{-(wA/q)} \tag{10.3}$$

$$= 1 - e^{-(w(2160)(60)/10{,}000)} = 1 - e^{-12.96w}$$

Calculate the efficiency for the 0.1 μm particle size:

$$\begin{aligned} E &= 1 - e^{-12.96(0.27)} \\ &= 0.9698 = 96.98\% \end{aligned}$$

Calculate the collection efficiencies for all of the other particle size ranges. The results in Table 10.11 show that the overall efficiency is 95.30%. Since the desired efficiency is 97.5%, the proposed design is insufficient.

TABLE 10.11 Overall Efficiency Results for Problem 10.39

d_p μm	x_i	E_i	x_iE_i
0.1	0.01	0.9698	0.009698
0.25	0.01	0.8569	0.008569
0.5	0.01	0.7889	0.007889
1.0	0.01	0.7596	0.007596
1.5	0.16	0.8569	0.137104
2.0	0.16	0.9251	0.148016
2.5	0.16	0.9656	0.154496
5.0	0.16	0.9985	0.159760
10.0	0.16	0.9996	0.159936
25.0	0.16	0.9999	0.159984
		$E = \sum x_iE_i$	$E_i = 0.953048$

As is typical with particulate control, collection of large particles is highly efficient. A decline in efficiency is seen as particle size decreases. As noted earlier, when the particles become very small, diffusion effects occur that actually raise the collection efficiency for these particles. This is graphically illustrated in Figure 10.7. Refer to the following reference for additional details: L. Theodore, "Ask the Experts: Factors Affecting ESP Performance," *Chem. Eng. Prog.* pp. 20–21 (Oct. 2006).

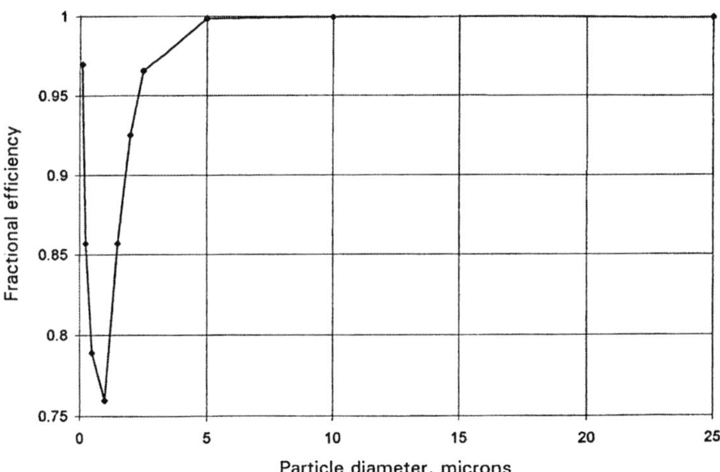

Figure 10.7 Effect of very small particles on collection efficiency.

10.40 Bus Section Failure

A precipitator consists of two sections, each with five plates (four passages) in a field (see Figure 10.8). The corona wires between any two plates are independently controlled so that the remainder of the unit can be operated in the event of a wire failure. The following operating conditions exist:

Gas flow rate: 10,000 acfm
Plate dimensions: 10 ft × 15 ft; four rows per field
Drift velocity: 19.0 ft/min; section 1
16.3 ft/min; section 2

(a) Determine the normal operating efficiency.
(b) During operation, a wire breaks in section 1. As a result, all of the wires in that row are shorted and ineffective, but the others function normally. Calculate the collection efficiency under these conditions. Assume that the gas stream leaving section 1 is uniformly redistributed on entering section 2, i.e., each of the four rows is fed the same volume of gas.
(c) Redo part (b) assuming that the flow in (through) each of the four rows (or passages) acts in a "railroad" manner, i.e., there is no redistribution after section 1.
(d) Calculate a revised efficiency if a second wire fails in a different row. Assume "railroad" flow again.

Solution: (a) Write the equation describing the overall or total efficiency, E_T in terms of the individual section efficiencies, E_1 and E_2:

$$E_T = 1 - (1 - E_1)(1 - E_2)$$

The equation describing the total penetration P_T in terms of the individual section penetrations, P_1 and P_2, is

$$P_T = P_1 P_2$$

Calculate the efficiency of section 1, E_1:

$$E_1 = 1 - e^{-(Aw/q)} \tag{10.3}$$

$$= 1 - e^{-(15)(10)(8)(19.0)/(10,000)}; \qquad w_1 = 19.0 \text{ ft/min}$$

$$= 0.89772 = 89.772\%$$

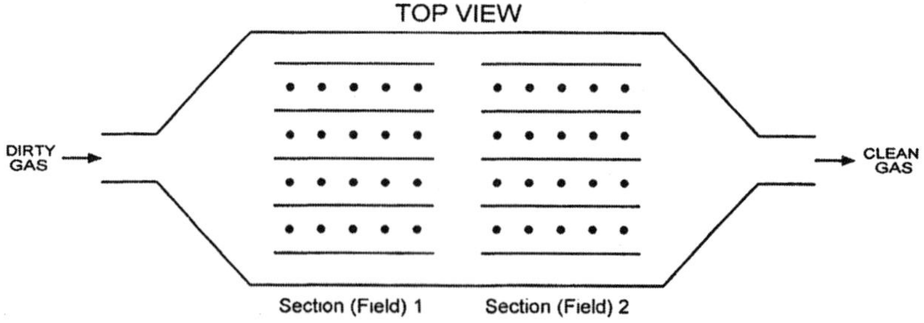

Figure 10.8 ESP with two bus sections and five plates.

Calculate the efficiency of section 2, E_2:

$$E_2 = 0.85858; \quad w_2 = 16.3 \text{ ft/min}$$
$$= 85.858\%$$

The total efficiency E_T is therefore

$$E_T = 1 - (1 - 0.89772)(1 - 0.85858)$$
$$= 0.98554$$
$$= 98.554\%$$

(b) Calculate a revised total efficiency for this part.

$$E_T = 1 - (1 - 3E_1/4)(1 - E_2)$$
$$= 1 - [1 - (0.75)(0.89772)](1 - 0.85858)$$
$$= 0.95380 = 95.380\%$$

(c) Calculate a revised total efficiency for this part.

$$E_T = (1.0)[(0.75)(0.98554) + (0.25)(0.85858)]$$
$$= (1.0)(0.73916 + 0.21464)$$
$$= 0.95380 = 95.380\%$$

(d) The calculation for this part is affected by where the second wire failure occurs. Determine a revised efficiency, assuming that the wire failure is located in section 1:

$$E_T = (1.0)[(0.5)(0.98554) + (0.5)(0.85858)]$$
$$= 0.49277 + 0.42929$$
$$= 0.92206 = 92.206\%$$

Determine a revised efficiency assuming the wire failure in part (d) occurs in a *different* row in section 2:

$$E_T = (1.0)[(0.5)(0.98554) + (0.25)(0.85858) + (0.25)(0.89772)]$$
$$= 0.93185$$
$$= 93.185\%$$

Determine a revised efficiency assuming that the wire failure in part (d) occurs in the *same* row in section 2:

$$E_T = (1.0)[(0.75)(0.98554) + (0.25)(0.0)]$$
$$= 0.73916$$
$$= 73.916\%$$

The reader should note the effect on efficiency of bus section failure and the location of the failure. Parallel sectionalization provides the means for coping with different power input needs due to uneven dust and gas distributions that usually occur across the inlet face of a precipitator. Nevertheless, the gains in collection efficiency from parallel sectionalization are smaller than series sectionalization.

Bus section failure is one of the more important design operating and maintenance variables for ESPs. Detailed calculational procedures for estimating this effect are available in the literature. Two studies addressing this issue are

1. L. Theodore and J. Reynolds, "The Effect of Bus Section Failure on Electrostatic Precipitator Performance," *J. Air Pollut. Control Assoc.* **33**: 1202–1205 (1983).
2. L. Theodore, J. Reynolds, F. Taylor, A. Filippi and S. Errico "Electrostatic Precipitator Bus Section Failure: Operation and Maintenance," *Proc. 5th USEPA Symp. Transfer and Utilization of Particulate Control Technology,* Kansas City, 1984.

10.41 Sectionalized ESPs

A sectionalized electrostatic precipitator (ESP) consists of eight chambers (or parallel channels for airflow) and four independently energized electric fields per chamber. In effect, the unit contains 32 independent "chamber–fields" or "cells." As indicated earlier, ESPs are normally designed to allow for a certain number of cell failures before the unit has to be shut down for maintenance.

(**a**) Assuming that the particulate penetration for all cells is 0.316* and that the incoming gas is perfectly distributed among all eight chambers, calculate the overall collection efficiency of the unit. In order to meet local particulate emission standards, the ESP must operate at a minimum collection efficiency of 95.0%.

(**b**) Assuming that the collection efficiency of a "dead" or failed cell is zero percent (or equivalently, that the cell penetration is 1.00), calculate the overall collection efficiency when four cells in the same chamber fail. Is the unit in compliance?

(**c**) Calculate the overall collection efficiency when four cells, three in the same chamber and one in another chamber, fail. Is the unit in compliance?

(**d**) Calculate the overall collection efficiency when four cells, two in the same chamber and two in two other different chambers, fail. Is the unit in compliance?

(**e**) Calculate the overall collection efficiency when four cells, each in a different chamber, fail. Is the unit in compliance?

(**f**) Comment on the results.

Solution

(**a**) The penetration for one chamber in the product of the cell penetrations. Therefore

$$P_{\text{chamber}} = (0.316)^4$$
$$= 0.00997$$

**Note*: This is not normally the case. Since the larger particles are easier to collect and represent a disproportionate amount of the particulate mass, the upstream cells normally have a higher collection efficiency.

Since there is uniform flow, then

$$P_{\text{overall}} = P_0 = 0.00997 = 0.997\%$$
$$E_{\text{overall}} = E_0 = 0.9900 = 99.00\%$$

(b) When four cells in the same chamber fail, then

$$P_0 = (7/8)(0.316)^4 + (1/8)(1)^4$$
$$= 0.1337$$
$$E_0 = 0.8663 = 86.63\% \text{ (as expected)}$$

The unit is not in compliance.

(c) When three cells fail in one chamber and one cell fails in another chamber, then

$$P_0 = (6/8)(0.316)^4 + (1/8)(0.316)^3(1) + (1/8)(0.316)^1(1)^3$$
$$= 0.0509$$
$$E_0 = 0.9491 = 94.91\%$$

The unit is marginally out of compliance.

(d) For two failures in each of two other chambers

$$P_0 = (5/8)(0.316)^4 + (2/8)(0.316)^3(1) + (1/8)(0.316)^2(1)^2$$
$$= 0.0328$$
$$E_0 = 0.9672 = 96.72\%$$

The unit is in compliance.

(e) With four failures in separate chambers

$$P_0 = (4/8)(0.316)^4 + (4/8)(0.316)^3(1)^1$$
$$= 0.0208$$
$$E_0 = 0.9792 = 97.92\%$$

In compliance!

(f) The effect of failures is minimized when they are distributed and not "railroaded."

10.42 A 6-by-8 ESP

Consider the 6-field $\times$ 8-section precipitator shown in Figure 10.9. Its normal operating efficiency is 99.91125%. Calculate the efficiency of the unit if bus section failures occur at:

Scenario A: (1, 1), (1, 2), (1, 6), (2, 1), (2, 2), (2, 3), (7, 2), (8, 2)
Scenario B: (1, 1), (1, 2), (1, 4), (1, 6), (2, 2), (2, 3), (2, 6), (8, 2)

Assume the efficiency (or penetration) of each field to be equal.

Solution: The failure locations for both scenario A and scenario B have been superimposed on Figure 10.9 and provided in Figure 10.10.

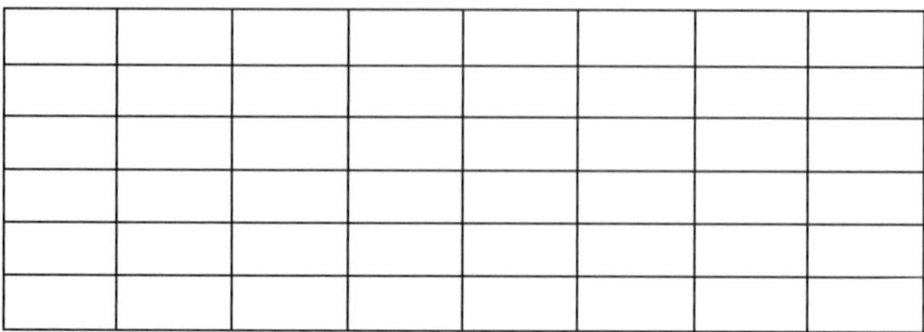

Figure 10.9 A 6 × 8 ESP.

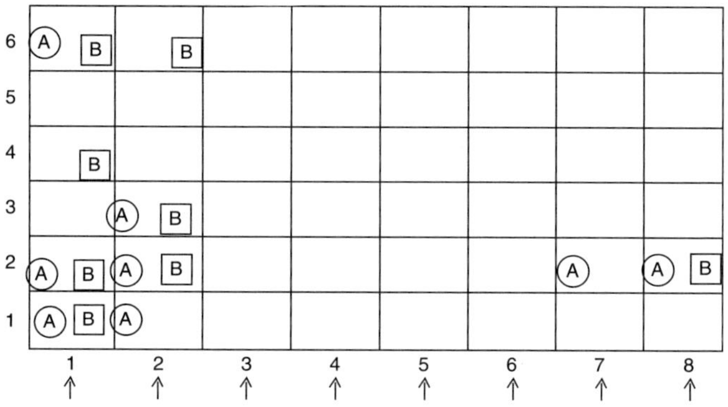

Figure 10.10 Failure locations.

For $E_T = 99.91125\%$:

$$P_T = 0.08875\% = 0.0008875$$

Since $P_1 = P_2 = P_3 = P_4 = P_5 = P_6 = P$, one obtains

$$P = (P_T)^{1/6} = (0.0008875)^{1/6} = 0.31$$

See Figure 10.10 for failure locations for scenario Ⓐ and scenario [B]. For Ⓐ:

$$P_{\textcircled{A}} = \frac{P^3 + P^3 + (4)(P)^6 + 2(P)^5}{8}$$

$$P_{\textcircled{A}} = \frac{(0.31)^3(2) + (4)(0.31)^6 + 2(0.31)^6}{8} = 0.00811$$

$$E_{\textcircled{A}} = 1 - P_{\textcircled{A}} = 0.99189 = 99.189\%$$

For $\boxed{B}$:

$$P_{\boxed{B}} = \frac{P^2 + P^3 + (5)P^6 + P^5}{8}$$

$$P_{\boxed{B}} = \frac{(0.31)^2 + (0.31)^3 + (5)(0.31)^6 + (0.31)^5}{8}$$

$$= 0.01665$$

$$E_{\boxed{B}} = 1 - P_{\boxed{B}}$$

$$= 0.98335 = 98.335\%$$

Comment on why the efficiency in B is lower.

10.43 Bus Section Failures with Varying Efficiencies

(**a**) Consider an ESP designed to operate at an efficiency of 99.55% for a utility boiler. Assume that there are five fields and eight bus sections per field. Calculate a revised efficiency and penetration, given failures occurring at locations (2, 2), (2, 4), (4, 4), and (8, 3). Assume the operating efficiency and/or penetration to be same in each field.

(**b**) Recalculate the efficiency (with the same four bus section failures) assuming that the fractional efficiency varies from field to field.

$$P(J, 1) = 0.10$$
$$P(J, 2) = 0.30$$
$$P(J, 3) = 0.40$$
$$P(J, 4) = 0.50$$
$$P(J, 5) = 0.75$$

Solution: (**a**) For five fields, the overall (or total) efficiency is

$$P_T = \frac{[P^5 + P^3 + P^5 + P^4 + P^5 + P^5 + P^5 + P^4]}{8}$$

Since

$$P = (0.0045)^{1/5}$$
$$= 0.339$$
$$E = 0.661 = 66.1\%$$

The overall efficiency is

$$P_T = \frac{[(5)(0.0045) + (1)(0.339)^3 + (2)(0.339)^4]}{8}$$

$$= \frac{0.022 + 0.0390 + 0.0264}{8}$$

$$= 0.0110$$

$$E = 0.9890 = 98.90\%$$

(**b**) For this case:

$$P(J,1) = 1.10 \quad (90\%\ E)$$
$$P(J,2) = 0.30 \quad (97\%\ E)$$
$$P(J,3) = 0.40 \quad (98.8\%\ E)$$
$$P(J,4) = 0.50 \quad (99.4\%\ E)$$
$$P(J,5) = 0.75 \quad (99.55\%\ E)$$

Thus

$$\begin{aligned} P_{\mathrm{T}} &= [(0.10)(0.40)(0.75) + (0.10)(0.30)(0.40)(0.75). \\ &\quad + (0.10)(0.30)(0.50)(0.75) + (5)(0.0045)]/8 \\ &= \frac{[0.03 + 0.009 + 0.01125 + 0.0225]}{8} = 0.0091 \\ E &= 0.9909 = 99.09\% \end{aligned}$$

10.44 Tubular Precipitation Design

A plant has a 60,000 acfm gas stream containing a hazardous dust with an estimated drift velocity of 0.250 ft/s. The minimum required efficiency is 99.80%. Assume that the Deutsch–Anderson (DA) equation applies. LT Associates has proposed (as a control device) a tubular-type precipitator with tubes 10 inch in diameter and 10 ft height. How many tubes are needed? Approximate the volume occupied by the tubes.

Solution: First apply the DA equation:

$$E = 1 - e^{-wA/q} \tag{10.3}$$

Substitute the data and calculate the required capture area:

$$0.9980 = \exp\left[-\frac{(0.250)A}{60{,}000/60}\right]$$
$$A = 24{,}860\ \mathrm{ft}^2$$

The surface area of one tube is

$$\begin{aligned} A &= \pi DH \\ &= \pi(10/12)10 \\ &= 26.2\ \mathrm{ft}^2 \end{aligned} \tag{10.16}$$

The number of tubes required is

$$\begin{aligned} N &= 24{,}860/26.2 \\ &= 950\ \text{tubes} \end{aligned}$$

The volume occupied by the tubes (alone) is

$$V_t = (\pi/4)(10/12)^2(10)(950)$$
$$= 5180\ \text{ft}^3$$

The reader should note that the actual volume of the ESP is larger because of the void space between the tubes and the outer housing. A procedure is available to minimize the actual volume of the unit (L. Theodore: personal notes, 1990).

10.45 Preliminary ESP Design

As a recently hired engineer at the Elias Cleanup Engineering Company, you have been given a job to prepare a preliminary design of an electrostatic precipitator to treat 175,000 acfm of a gas laden with catalyst dust. The inlet loading of 6.7 gr/ft^3 must be reduced to 0.06 gr/ft^3. Owing to plant space requirements in the catalyst plant, the maximum allowable width of the unit is 30 ft and the maximum length is 36 ft. Present a design that can meet both space and collection efficiency requirements. Assume a plate-to-plate spacing of 10 inches and an effective drift velocity of 0.25 ft/s.

Solution: This is an open-ended design problem that requires specifying certain design variables. Assume the DA equation to apply. First calculate the required efficiency.

$$E = \left(\frac{6.7 - 0.06)}{6.7}\right)100 = 99.10\%$$

Need 3–4 fields: assume 3 fields. Calculate the number of passages:

$$N = 30/(10/12) = 36$$
$$q_{\text{pass}} = 175{,}000/36$$
$$= 4861\ \text{ft}^3/\text{min per passage}$$

Calculate ϕ in the DA equation:

$$0.9910 = 1 - e^{-\phi}$$
$$\phi = 4.71$$

Therefore

$$\phi = \frac{Aw}{q_{\text{pass}}} \tag{10.13}$$

Solving for A, one obtains

$$A_{\text{pass}} = (4.71)(4861)/(0.25)(60)$$
$$= 1526.4\ \text{ft}^2$$

and

$$A_{\text{plate}} = A_{\text{pass}}/2$$
$$= 763.2 \text{ ft}^2$$

Choose three 12-ft-long fields.

$$L = (3)(12) = 36 \text{ ft}$$

For this length

$$A_{\text{plate}} = (H)(36)$$
$$763.2 = (H)(36)$$
$$H = 22 \text{ ft; a reasonable height}$$

Also check the velocity:

$$v = \frac{4861/60}{(22)(10/12)}$$
$$= 4.41 \text{ ft/s}$$

The throughput velocity is also reasonable.

10.46 Price Quotes

The owner of an industrial plant is under considerable pressure from a community that is quickly organizing behind several environmental activists to reduce his emissions. He is considering the option of adding a new pollution control device to replace his antiquated devices. MKT Associates has been hired to assist in his decision. He has decided that an electrostatic precipitator is the way that he wants to go. A cement kiln that operates at his plant at 23,500 acfm and has a 600°F stream temperature. He would like the electrostatic precipitator to operate at 99.5% efficiency. The plant owner has called an ESP vendor and has obtained some price quotes. These quotes are listed in Table 10.12. MKT's first assignment is to convert the information provided in Table 10.12 into an equation of the form

$$P = aA^c \tag{10.17}$$

where P = price, \$
A = plate area, ft^2

TABLE 10.12 Price Quotes

Plate Area, ft^2	Price, \$
10,000	313,400
20,000	483,200
30,000	623,019
40,000	746,500
50,000	850,111

Solution: Convert Equation (10.17) into linear form by taking the natural logarithm of both sides of the equation:

$$\ln P = \ln a + c \ln A$$

This is now linearized since it is of the form

$$y = b + mx$$

Regressing the data gives

$$m = c = 0.622$$
$$\ln a = b = 6.925$$
$$a = 1018$$

The resulting equation is therefore

$$P = 1018(A)^{0.622}; \quad P = \$, \quad A = \text{ft}^2$$

10.48 General Design Procedure

Provide a general design procedure for electrostatic precipitators.

Solution: No critically reviewed design procedure exists for ESPs. However, one suggested "general" design procedure (L. Theodore: personal notes) is provided below.

1. Determine or obtain a complete description of the process, including the volumetric flowrate, inlet loading, particle size distribution, maximum allowable discharge, and process conditions.
2. Calculate or set the overall collection efficiency.
3. Select a migration velocity (based on experience).
4. Calculate the ESP size (capture area).
5. Select the field height (experience).
6. Select the plate spacing (experience).
7. Select a gas throughput velocity (experience).
8. Calculate the number of gas passages in parallel.
9. Select (decide) on bus sections, fields, energizing sets, specific current, capacity of energizing set for each bus section, etc.
10. Design and select hoppers, rappers, etc.
11. Perform a capital cost analysis, including materials, erection, and startup costs.
12. Perform an operating cost analysis, including power, maintenance, inspection, capital and replacement, interest on capital, dust disposal, etc.
13. Conduct a perturbation study to optimize economics.

Refer to the following two references for additional details:

1. L. Theodore, "Ask the Experts: Factors affecting ESP Performance," *Chem. Eng. Prog.* 20–21.
2. J. Reynolds, J. Jeris, and L. Theodore, "Handbook of Chemical and Environmental Engineering Calculations," John Wiley & Sons, Hoboken, NJ, 2004.

10.49 Operation and Maintenance Problem

An electrostatic precipitator's operating collection efficiency has slowly degraded with time. As a plant manager who has recently completed (and passed) an air pollution control equipment (APCE) course given by Dr. Louis Theodore—the supposed foremost authority in the galaxy on APCE—indicate what sound, reasonable engineering steps can be taken to return the precipitator to its original design value.

Solution: Obviously, this is an open-ended problem for which there are many possible solutions. A standard recommendation of the author is to carefully examine the process that is generating the gas treated by the ESP. Any process modification option that can positively impact on the efficiency should be carefully investigated. Options can include changing the final product specification and/or the raw material, changing the operating conditions, including a bypass or recycle line, and reducing throughput requirements.

A second option (which the author has recommended in the past) is to determine the effect of the rappers on the efficiency. Three considerations are as follows:

1. Changing the rapping frequency in one or all fields
2. Changing the rapping intensity in one or all fields
3. Applying considerations 1 and 2 together

Two other rapping considerations—that are probably not attractive financially—are to change either the location of the rappers or the type of rapping employed.

NOTE: Additional problems are available for all readers at www.wiley.com. Follow links for this title.

11

VENTURI SCRUBBERS

11.1 INTRODUCTION

As the name scrubber implies, wet collectors or wet scrubbers are devices that use a liquid for removing particles or polluted gases from an exhaust gas stream. Water sprays can be injected into the gas stream; gas can be forced to pass through sheets or films of liquid; or, the gas can move through beds of plastic spheres covered with liquid. Each of these techniques can effectively remove particulate matter from process exhaust gases. They can also effectively remove gases such as HCl or SO_2, but removal conditions must be right. In addition, gas–liquid contact can bring about gas conditioning, and to a lesser extent, liquid conditioning.

In many cases, the best conditions for removing particulate matter are the poorest for removing pollutant gases. In this chapter, emphasis will be placed on the design and application of wet scrubbers, with particular emphasis on venturi scrubbers for the removal of particulate matter. Optimum operating conditions for particulate matter removal will also be discussed.

Wet collectors provide many options for the control of industrial emissions. As described above, these air pollution control devices can remove both particulate matter and gases from effluent gas streams. They can operate at low removal efficiencies or at high removal efficiencies. They also offer more versatility in design than other air pollution control devices. This versatility, however, does not come without problems. Higher efficiency requires higher operating costs; byproducts are difficult to recover, and an air pollution problem can be transformed into a water pollution problem. In terms of cost, venturi scrubbers are generally more expensive than simple settling chambers and cyclones, but less expensive than high efficiency electrostatic precipitators and baghouses.

A variety of wet collectors are commercially available. The evaluation and selection of a scrubber system is somewhat simplified by the observation that collection efficiency is a function of the amount of energy required to operate the scrubber. This means, simply, that independent of design, the more power put into the system, the greater the collection efficiency. Therefore, for systems of nearly equivalent power inputs, the collection efficiencies should be nearly equivalent. The selection or evaluation procedure between two systems could then concentrate on ease of operation, potential maintenance problems, and comparative costs.

Wet collectors are designed to incorporate small dust particles into larger water droplets. Droplets ranging from 50 to 500 μm in diameter are produced and brought into contact with the particulate matter. These larger droplets containing the captured particles are then collected by simple mechanisms such as gravity, impaction on baffles, or by cyclonic action. Most wet collectors can be represented as having two zones: a contact zone (particle capture) and a separation zone (droplet capture), as described in Figure 11.1.

As indicated above, wet scrubbers have found widespread use in cleaning contaminated gas streams because of their ability to effectively remove both particulate and gaseous pollutant. Specifically, *wet scrubbing* describes the technique of bringing a contaminated gas stream into intimate contact with a liquid. In this chapter, the term *scrubber* will be restricted to those systems that utilize a liquid, usually water, to achieve or assist in the removal of particulate matter from a carrier gas stream.

The particulate collection mechanisms involved in the wet scrubbing operation may include some or all of the following (see Chapter 7 for more details):

1. Inertial impaction
2. Direct interception
3. Diffusion (Brownian motion)

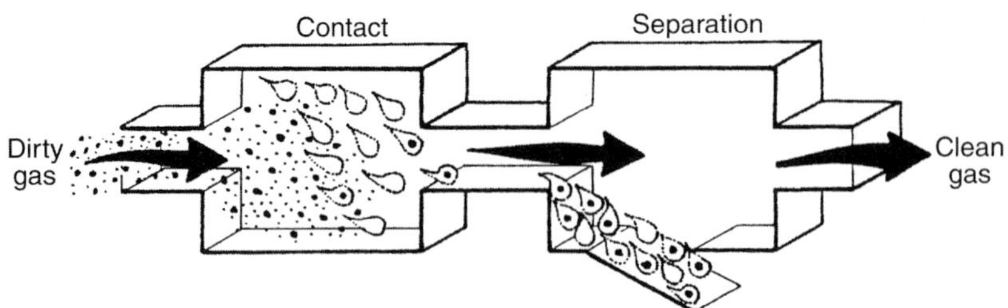

Figure 11.1 Zones of a wet scrubber.

4. Electrostatic forces
5. Condensation
6. Thermal gradients

The two principal collection mechanisms are inertial impaction and diffusion, both of which are again described below:

Impaction. When a gas stream flows around a small liquid droplet, the inertia of the particles causes them to continue to move toward the object, and particles will be collected by the liquid. Inertial impaction customarily describes the effects of small-scale changes in flow direction.

Diffusion. When particles are small enough ($<1.0\,\mu m$), they are "thrashed" about by gas molecules, such that they act similarly to gas molecules. They "diffuse" randomly through the gas. As these small fines are diffusing randomly through the gas, the probability of contact with a liquid droplet increases. This phenomenon explains why the collection efficiency rises for extremely small particle diameters.

The two types of scrubbers for particulate control reviewed are spray towers and venturi scrubbers. More extensive details are presented for the venturi since this control device is most often used for high-efficiency particulate control.

Spray Towers

The most common low-energy scrubbers are *gravity spray towers* in which liquid droplets are made to fall through rising exhaust gases and are drained at the bottom of the chamber (see Figure 11.2). The droplets are usually formed by liquid atomized in spray nozzles. The spray is directed into a chamber shaped to conduct the gas through the finely divided liquid. In a vertical tower, the relative velocity between the droplets and the gas is eventually the terminal settling velocity of the droplets. To avoid spray droplet reentrainment, however, the terminal settling velocity of the droplets must be greater than the velocity of the rising gas stream. In practice, the vertical gas velocity typically ranges from 2 to 5 ft/s. For higher velocities, a mist eliminator must be used at the top of the tower.

Spray towers are suited for both particle collection and mass transfer (gas absorption), as are all wet scrubbers. Operating characteristics include low pressure drop (typically $<1-2$ in H_2O exclusive of mist eliminator and gas distribution plate), ability to handle liquids having a high solids content, and liquid requirements ranging from 3 to $30\,gal/1000\,ft^3$ of gas treated. Spray towers are capable of handling large gas volumes and are often used as quenchers (precoolers) to reduce gas stream temperatures. Gas rates from 800 to $2500\,lb/hr \cdot ft^2$ are typical. Gas retention times within the tower typically range from 2 to 5 s. The chief disadvantage of spray towers is their relatively low scrubbing efficiency for particles in the $0-5\,\mu m$ range.

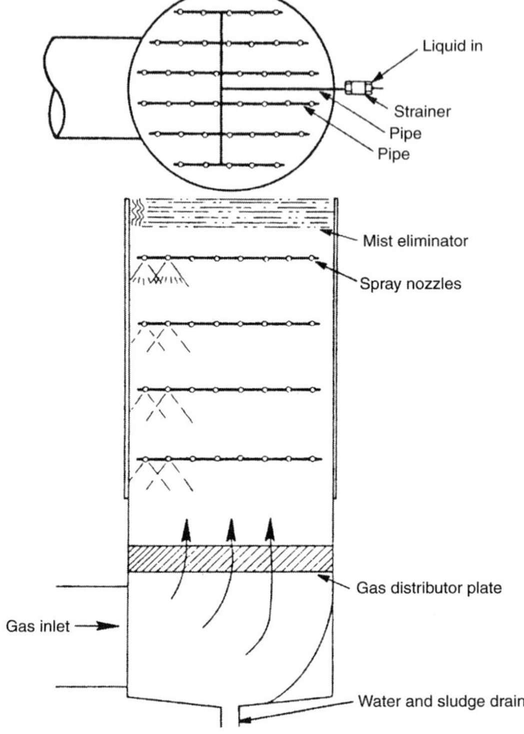

Figure 11.2 Spray scrubber.

Venturi Scrubbers (VSs)

To achieve a high collection efficiency of particulates by *impaction*, a small droplet diameter and high relative velocity between the particle and droplet are required. This is often accomplished in a VS by introducing the scrubbing liquid at right angles to a high-velocity gas flow in the venturi throat (vena contracta). Very small water droplets are formed, and high relative velocities are present.

One of the simplest venturi designs is shown in Figure 11.3. The venturi is designed to fan gas in and out of a constriction. Since the volumetric flow rate of the gas (q) must be the same throughout the system, the velocity of the gas must increase at the throat of the venturi. That is, if

$$\begin{aligned} q_{\text{entrance}} &= q_{\text{throat}} \\ v_{\text{entrance}} A_{\text{entrance}} &= v_{\text{throat}} A_{\text{throat}} \end{aligned} \tag{11.1}$$

where A = area

v = velocity

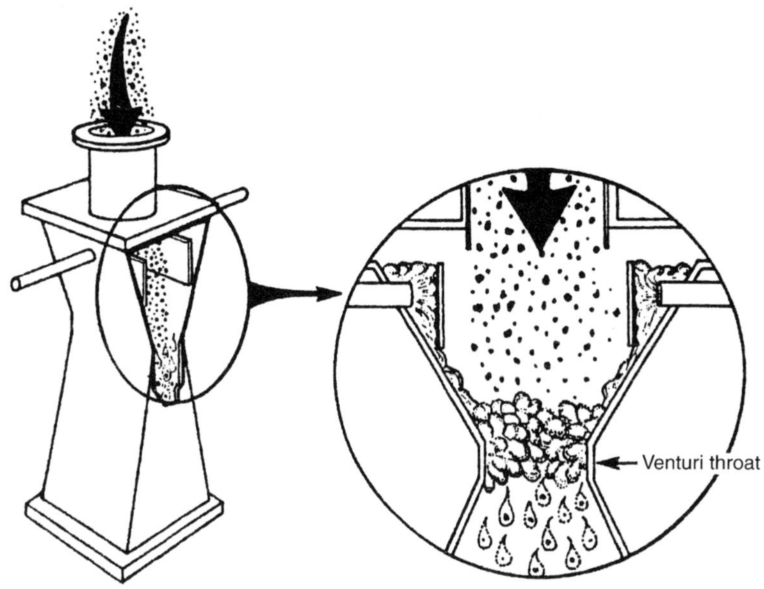

Figure 11.3 Typical venturi scrubber.

The velocity at the throat must increase in order to make up for the decrease in area at the throat. Velocities at such a constriction can range from 200 to 800 ft/s (from 61 to 244 m/s). Now, if water is introduced into the throat, the gas forced to move at high velocity will shear the water droplets. Particles in the gas stream then impact onto the droplets produced. Moving a large volume of gas through a small constriction gives a high-velocity flow, but also a large pressure drop across the system. The collection efficiency for most particles increases with increased velocities (and corresponding increased pressure drops) since the water is sheared into more and smaller droplets than at lower velocities. The large number of small droplets, combined with the turbulence in the throat section, provides numerous impaction targets for particle collection.

11.2 DESIGN AND PERFORMANCE EQUATIONS

A number of performance and design equations have been developed from basic particle movement principles (theory) to explain the action of wet scrubbing systems. Many of these start from firm scientific concepts, but give only qualitative results when predicting collection efficiencies or pressure drops. The interaction of particulate matter having a given particle size distribution with water droplets having another size distribution is not easy to express in quantitative terms. As a result of this complexity, experimentally determined parameters are usually required in order to perform engineering calculations.

One of the more popular and widely used collection efficiency equations is that originally suggested by Johnstone [H. Johnstone et al., *Ind. Chem. Eng.* **46**: 1601 (1954)].

$$E = 1 - e^{-kR(\psi)^{0.5}} \tag{11.2}$$

where E = efficiency (fractional)
ψ = inertial impaction parameter (dimensionless)
R = liquid-to-gas ratio (gal/1000 acf or gpm/1000 acfm)
k = correlation coefficient, the value of which depends on the system geometry and operating conditions (typically 0.1–0.2 acf/gal)

The term ψ is given by

$$\psi = \frac{C d_p^2 \rho_p v_t}{9 \mu_G d_0} \tag{11.3}$$

where d_p = particle diameter (ft)
ρ_p = particle density (lb/ft^3)
v_t = throat velocity (ft/s)
μ_G = gas viscosity (lb/ft · s)
d_0 = mean droplet diameter (ft)
C = Cunningham correction factor

Values for the Cunningham correction factor (see Chapter 7 for more details) are provided in Table 11.1. This correction is usually neglected in scrubber calculations, but the effect becomes more pronounced as the particle size decreases, particularly below 1 μm. The mean droplet diameter d_0 for standard air and water in a venturi scrubber is given by the Nukiyama–Tanasawa relationship:

$$d_0 = \frac{16{,}400}{v_t} + 1.45\, R^{1.5} \tag{11.4}$$

TABLE 11.1 Cunningham Correction Factors for Air at Atmospheric Pressure

	Temperature		
Particle Diameter, μm	70°F	212°F	500°F
0.1	2.88	3.61	5.14
0.25	1.682	1.952	2.525
0.5	1.325	1.446	1.711
1.0	1.160	1.217	1.338
2.5	1.064	1.087	1.133
5.0	1.032	1.043	1.067
10.0	1.016	1.022	1.033

The pressure drop for gas flowing through a venturi scrubber can be estimated from knowledge of liquid acceleration and frictional effects along the wall of the equipment. Frictional losses depend largely on the scrubber geometry and are usually determined experimentally. The effect of liquid acceleration is, however, predictable. Equation (11.5) for estimating pressure drop through venturi scrubbers (given as a function of throat gas velocity and liquid-to-gas ratio) assumes that all the energy is used to accelerate the liquid droplets to the throat velocity.

$$\Delta P = 5 \times 10^{-5} v_t^2 R \tag{11.5}$$

where ΔP = pressure drop (in H_2O).

Another somewhat simpler equation that applies over a fairly wide range of R's is [L. Theodore, *Proc. Environ. Eng. Sci. Conf.* **4**: 365 (1974)].

$$\Delta P^* = 0.8 + 0.12R \tag{11.6}$$

where ΔP^* is a dimensionless pressure drop equal to the pressure drop divided by a velocity head.

Studies by Hesketh [H. Hesketh, *J. Air Pollut. Control Assoc.* **24**: 938–942 (1974)] show that the pressure drop predictions obtained from throat velocity measurements may be subject to error at low velocities if Equation (11.5) is applied for all ranges of velocities. Hesketh's equation for venturi scrubbers that have liquid injected before the throat is given by:

$$\Delta P = \frac{v_t^2 \rho A_t^{0.133} R^{0.78}}{1270} \tag{11.7}$$

where ρ = gas density, lb/ft^3
A_t = throat cross-sectional area, ft^2

Contact power theory [K. Semrau, *J. Air Pollut. Control Assoc.* **10**: 200–202 (1960)] is an empirical approach relating particulate collection efficiency and pressure drop in wet scrubber systems. The concept is an out-growth of the observation that particulate collection efficiency in spray-type scrubbers was determined mainly by pressure drop for the gas plus any power expended in atomizing the liquid. Contact power theory assumes that the particulate collection efficiency in a scrubber is solely a function of the total pressure loss in the unit. The total power loss (p_t) is assumed to be composed of two parts: the power loss of the gas passing through the scrubber (p_G) and the power loss of the spray liquid during atomization (p_L). These two terms are given here in equation form:

$$p_G = 0.157\Delta P \tag{11.8}$$

where p_G = contacting power based on gas stream energy input, HP/1000 acfm
ΔP = pressure drop across the scrubber, in H_2O

and

$$p_{\mathrm{L}} = 5.83 \times 10^{-4} P_{\mathrm{L}} R \tag{11.9}$$

where p_{L} = contacting power based on liquid energy input (HP/1000 acfm)
P_{L} = liquid inlet pressure (psia)
R = liquid-to-gas ratio (gal/1000 acf)

The total power is then

$$p_{\mathrm{t}} = p_{\mathrm{G}} + p_{\mathrm{L}} \tag{11.10}$$

To correlate contacting power with scrubber collection efficiency, the latter is best expressed as the number of transfer units. The number of transfer units is defined by analogy to mass transfer and given by

$$N_{\mathrm{t}} = \ln\left(\frac{1}{1-E}\right) \tag{11.11}$$

where N_{t} = number of transfer units (dimensionless)
E = fractional collection efficiency (dimensionless)

The relationship between the number of transfer units and collection efficiency is by no means unique. The number of transfer units for a given value of contacting power (HP/1000 acfm) or vice versa, varies over nearly an order of magnitude. For example, at 2.5 transfer units ($E = 0.918$), the contacting power ranges from approximately 0.8 to 10.0 hp/1000 acfm, depending on the scrubber and the particulate.

For a given scrubber and particulate properties, there will usually be a very distinct relationship between the number of transfer units and the contacting power. Semrau plotted the number of transfer units (N_{t}) for a series of scrubbers and particulates against total power consumption; a linear relation on a log–log plot was obtained. This relationship is independent of the type of scrubber and can be expressed by (W. Strauss, *Industrial Gas Cleaning*, Pergamon, New York, 1966)

$$N_{\mathrm{t}} = \alpha p_{\mathrm{t}}^{\beta} \tag{11.12}$$

where α, β = parameters for the type particulates being collected (see Table 11.2).

When a combustion gas contains a high particulate loading, as well as one or more of the gaseous pollutants discussed previously, a venturi scrubber is often used in conjunction with a packed-bed or plate tower scrubber. The venturi scrubber removes the particulates from the stream to prevent fouling of the packed-bed or plate tower absorber and may also remove a significant fraction of gases highly soluble in water. However, venturi scrubbers alone are not considered suitable for the removal of low-solubility

TABLE 11.2 Constants for Use with Equation (11.12)

Aerosol	Scrubber Type	α	β
Raw gas (lime dust and soda fume)	Venturi and cyclonic spray	1.47	1.05
Prewashed gas (soda fume)	Venturi, pipeline, and cyclonic spray	0.915	1.05
Talc dust	Venturi	2.97	0.362
Black liquor recovery furnace fume	Orifice and pipeline	2.70	0.362
Cold scrubbing water humid gases	Venturi and cyclonic spray	1.75	0.620
Hot fume solution for scrubbing (humid gases)	Venturi, pipeline, and cyclonic spray	0.740	0.861
Hot black liquor for scrubbing (dry gases)	Venturi evaporator	0.522	0.861
Phosphoric acid mist	Venturi	1.33	0.647
Foundry cupola dust	Venturi	1.35	0.621
Open-hearth steel furnace fume	Venturi	1.26	0.569
Talc dust	Cyclone	1.16	0.655
Copper sulfate	Solivore		
	(A) with mechanical spray generator	0.390	1.14
	(B) with hydraulic nozzles	0.562	1.06
Ferrosilicon furnace fume	Venturi and cyclonic spray	0.870	0.459
Odorous mist	Venturi	0.363	1.41

gases; when water is used as the scrubbing medium, estimated efficiencies are <50–70%. Venturi scrubbers using water are not suitable for highly efficient (more than 99%) removal of either HCl or HF. Thus, although venturi scrubbers are usually designed for particulate collection, they can be used for simultaneous gas absorption as well. There is no satisfactory generalized design correlation for these types of scrubbers, especially when absorption with a chemical reaction is involved. Reliable design should be based on full-scale data or at least laboratory- or pilot-scale data.

Available data indicate that venturis often operate with pressure drops in the 30–100 in H_2O range. Liquid-to-gas ratios for venturi scrubbers are usually in the range of 5–20 gal/1000 ft^3 of gas. At many facilities, liquid-to-gas ratios ranging from 7 to 45 gal/1000 ft^3 of gas have been reported. In many cases, a minimum ratio of 7.5 gal/1000 ft^3 is needed to ensure that adequate liquid is supplied to provide good gas "sweeping". Gas velocities for venturi scrubbers are in the 100–400 ft/s range. The low end of this range, 100–150 ft/s, is typical of power plant applications, while the upper end of the range has been applied to lime kilns and blast furnaces.

11.3 OPERATION AND MAINTENANCE, AND IMPROVING PERFORMANCE

Although wet scrubbers are relatively simple devices, they do require proper care to ensure long service life and trouble-free operation. Properly designed and installed units can do an excellent job in helping plants comply with air pollution control regulations. However, as with most processing equipment, periodic reviews and inspections must be made to ensure that scrubbers are operating most efficiently. Improper performance of a unit can usually be traced to any of the following:

1. Inadequate scrubber system design and/or operation.
2. Process conditions have changed so that the scrubber system cannot operate efficiently.
3. Scrubber system mechanical condition has deteriorated and performance is below design specifications.

Assuming that the proper unit was selected for the given application, most scrubber problems usually involve spray nozzle plugging, liquid circulation restrictions, and/or entrainment of droplets from the unit. Addressing these problems will improve operation and increase performance. Such problems may include the following:

Wet/Dry Zone Buildup. The scrubber design may improperly allow dry, dust-laden gas to contact the juncture of the scrubbing liquid and the vessel, causing dust buildup. Proper design prevents this contact by extending ductwork sections sufficiently into the scrubber and thoroughly wetting all scrubber surfaces through reliable means (usually gravity flush and sometimes sprays).

Nozzle Plugging. Nozzles plug through improper selection, too small orifices through which too dense scrubbing liquid must pass, improper header design, drawing off a sump that also settles (and concentrates) solids, erratic pump operation, chemical scaling, sealing, and mechanical failure.

Flow Imbalance. The headers external to the scrubber must send the required flow to the proper location at the proper rate. Many problems can be solved through the simple adjustment of flow using existing valves or dampers.

Buildup (Scaling). Scaling is the plating out of deposits on a scrubber surface and is most often related to the chemical composition, solubility, temperature, and pH of the scrubbing liquid. Scaling is usually not operationally significant unless the surface in question is a functional one. Proper control begins with the scrubber design and process control.

Localized Corrosion. Corrosion is a major factor in reducing the operating life of a scrubber, whether properly designed or not. Wells or pockets of liquid should be avoided and points of particulate buildup should be adequately flushed.

Instrumentation-Fitting Blockage. One problem that can often cause major trouble is instrumentation blockage or pluggage. Often, a standard fitting is not adequate in a scrubber. Specially designed fittings and connections are often recommended.

Sump Swirling. Sump swirling problems are most often associated with cyclonic devices. The swirling of the scrubbing liquid can cause severe wear and drainage problems unless arrested by antiswirl plates in the scrubber or rapid continuous draining.

Entrainment. Entrainment occurs when the droplet separator is not functioning properly. Nearly all scrubbers produce entrainment; only the properly designed systems reduce it to acceptable levels prior to discharge.

Reentrainment. Reentrainment occurs beyond the droplet removal section through improper draining or erratic flow patterns. It can also occur in stacks with very high velocities or where fittings protrude into a high-velocity area.

Liquid–Gas Maldistribution. The gas and liquid must be properly distributed for the given application. Each affects the other and is aggravated by the influence of baffles (designed or accidental), buildup, mechanical failure, wear, scaling in headers, or improper design.

Thermal Shock. Thermal shock may occur. When hot gases meet cold scrubber liquor, proper design permits gradual cooling rather than abrupt changes. Thermal shock is a relatively simple problem to prevent (i.e., through the use of multiple cooling zones.)

Loss of Seal. The juncture of the scrubber liquid circuit with its surroundings is often a liquid seal. This seal may be at the top of a quencher or from an overflow connection. These lines must have seals able to prevent gas movement to or from the ambient surroundings. Loss of seal can cause entrainment or plugging, and instrumentation malfunction.

Wear. Wear can be tolerated unless it is localized. Unfortunately, a scrubber's functioning parts are also the wear parts. Expect to replace high wear points at frequent intervals. Excessive wear is often a result of excessively high solids concentration.

Vibration. This is most common in pumps and the fans associated with wet scrubber systems. Vibration problems are controlled by monitoring and scheduled preventive maintenance.

As in any air pollution control system, a scrubber system is subjected to a harsh environment. The abrasiveness of some types of particulate causes equipment such as fan wheels and valves to wear away. Stack gases may also contain a variety of components that, when wetted, produce compounds that corrode equipment. This is particularly true of any gas stream from a boiler that fires sulfur-bearing fuel and of gases from many chemical processes. All scrubber components must be designed and maintained with these considerations in mind.

Regarding recent developments, several specialty devices are receiving more attention. Although on the market for many years, their unique collection mechanism has recently attracted more interest. Five such units include the *dynawave unit*, the *multi-microventuri* (MMV), the *electrodynamic venturi* (EDV), the *condensing scrubber*, and the *collision scrubber*. These are briefly described below.

The dynawave unit operates in a manner that creates a "froth" zone through which contaminated gas must pass. There is extreme turbulence in the zone and efficient collection of submicrometer particles occurs. The MMV scrubber consists of a "stacked" bank of staggered tubular elements arranged in a pattern similar to a bank of heat exchanger tubes. The thin spacing causes successive microventuri flows at the "pinch" sections leading to high efficiencies. The EDV concept consists of a sequence of fundamental functions, namely saturation, condensation, ionization, and filtration. In condensing scrubbers, water vapor in the flue gas stream is condensed onto particulates causing the particle to grow to a size that enables easier collection by conventional means. At the same time, the temperature and volume of the contaminated gases is reduced providing a lower capital and operating cost for the downstream collection systems and induced-draft fan. Finally, in the collision scrubber two gas streams are made to collide head-on at the discharge of opposite venturi throats containing water droplets. The collision action shreds the water droplets into finer ones that can more effectively collect submicron particles. The process also produces a larger liquid surface area for gas absorption.

Finally, the author continues to champion the use of extended venturi throats to enhance fine particle capture. The design of such a unit is provided in Problems 11.48 and 11.49.

PROBLEMS

11.1 Collection Mechanism

What is the most common collection mechanism that occurs in wet scrubbers (select one)?

(**a**) Inertial impaction
(**b**) Direct interception
(**c**) Brownian diffusion
(**d**) Gravitation

Solution: The two primary collection mechanisms for wet scrubbers are inertial impaction and Brownian (molecular) diffusion. However, the bulk of the mass collected is by inertial impaction. The correct answer is therefore (a).

11.2 High Collection Efficiency

The collection efficiency of a venturi scrubber can be extremely high because of

(**a**) The low velocity of the gas stream through the throat allowing good gas–liquid contact
(**b**) Low pressure drops through the packing material
(**c**) Increased gas velocity through the throat causing the water to be atomized and providing good gas–liquid contact
(**d**) Use of impingement baffles to increase the likelihood of particle separation by the water droplets

Solution: As discussed in Section 11.1, it is the fast-moving gas stream passing finely atomized droplets that leads to high efficiencies. The correct answer is therefore (c).

11.3 The Nukiyama–Tanasawa Equation

The Nukiyama–Tanasawa (NT) equation is used to estimate

(**a**) Particle size
(**b**) Liquid-to-gas ratio
(**c**) Water droplet size

(**d**) The relative velocity of particulate matter to water droplets

Solution: The NT equation is employed to predict an average droplet diameter. The correct answer is therefore (c).

11.4 Effect of Gas Velocity on Droplet Diameter

As the gas velocity in a venturi scrubber increases, the liquid droplet diameter

(**a**) Increases
(**b**) Decreases
(**c**) Remains the same
(**d**) Increases exponentially

Solution: Refer to Equation (11.4). Since the velocity appears in the denominator on the right-hand side (RHS) of the equation, the droplet diameter decreases with increasing gas velocity—creating a finer mist of liquid collectors. One would also intuitively expect this result. The correct answer is therefore (b).

11.5 Effect of Liquid Flow Rate on Droplet Diameter

As the liquid flow rate in a venturi scrubber increases, the droplet diameter

(**a**) Increases
(**b**) Decreases
(**c**) Remains the same
(**d**) Impossible to tell

Solution: Refer once again to Equation (11.4). The term R on the RHS of the equation represents the liquid-to-gas ratio. Thus, any increase in R corresponds to an increase in the liquid flow rate, which in turn will increase the droplet diameter. The correct answer is therefore (a).

11.6 Transfer Units

Which one of the methods given below uses the number of transfer units N_t to estimate scrubber collection efficiency?

(**a**) The Theodore rule
(**b**) The Johnstone equation
(**c**) The Nukiyama–Tanasawa correlation
(**d**) The contact power theory

Solution: Only a fool would use anything proposed by Theodore. Answer (c) is concerned with droplet diameter, while answer (b) estimates the collection efficiency on the basis of inertial impaction principles. The correct answer is therefore (d).

11.7 Contact Power Theory

Contact power theory states that

(**a**) As pressure drop increases, fractional efficiency increases
(**b**) As pressure drop decreases, fractional efficiency increases
(**c**) Complexity of design increases efficiency
(**d**) Complexity of design decreases efficiency

Solution: Refer to Equation (11.12). Since β is positive—see Table 11.2—N_t increases. Now refer to Equation (11.11). As N_t increases, E also increases. Thus, as p_t increases, E increases. The correct answer is therefore (a).

11.8 Estimating Pressure Drop

Industries contemplating the purchase of a wet scrubber system will most often obtain an estimated value of the pressure drop by

(**a**) Using the Tonry rule
(**b**) Guessing
(**c**) Using data from a pilot plant
(**d**) Using the Johnstone equation

Solution: Who has ever heard of the Tonry rule? The Johnstone equation predicts collection efficiency. Guessing should be a last resort, while pilot plant data are usually reliable. The correct answer is therefore (c).

11.9 Pressure Drop Theories

Which of the following theories expresses the pressure drop across a wet scrubber?

(**a**) Lapple equation
(**b**) Johnstone equation
(**c**) Nukiyama–Tanasawa equation
(**d**) None of the above

Solution: Answer (a) is concerned with cyclones. Answer (b) is concerned with efficiency. Answer (c) is concerned with droplet size. The correct answer is therefore (d).

11.10 Typical Pressure Drops

What would be a characteristic pressure drop for a medium-energy scrubber such as a self-induced spray scrubber (impingement–entrainment scrubber)?

(**a**) 0.1 in H_2O
(**b**) 3 in H_2O
(**c**) 20 in H_2O
(**d**) 100 in H_2O

Solution: Pressure drops across spray scrubbers are typically low and can range between 1 and 5 in H_2O. The correct answer is therefore (b).

11.11 Fine-Particle Capture

To remove particles having 1 μm diameters, venturi scrubbers commonly operate with a pressure drop in the range of

(**a**) 5–10 in H_2O
(**b**) 90–100 ft H_2O
(**c**) 0.5–1.5 in H_2O
(**d**) 60–80 in H_2O

Solution: The capture of particles in the 1 μm range requires high pressure drops; typical values are in the 50–100 in H_2O range. The correct answer is therefore (d).

11.12 Self-Induced Spray Scrubber

In a self-induced spray scrubber

(**a**) Liquid is injected at high pressure
(**b**) The gas atomizes the liquid
(**c**) Particulate matter is removed by cyclonic deposition on the packing
(**d**) Gas flow is counter-current

Solution: Answers (a) and (c) are incorrect, and the flow direction is usually cocurrent. The correct answer is therefore (b).

11.13 Potential Scrubber Applications

List some of the more important conditions that indicate a potential scrubber application.

Solution

1. Introduction of the liquid to the gas is permissible in the process.
2. The liquid can be purged from the process without causing a water pollution problem. Water quality requirements of the receiving water must be considered, and a satisfactory effluent treatment system (or the equivalent) must be provided.
3. The gas may require cooling.
4. Combustible particles or gases must be treated with minimum risk.
5. The particulate matter is rather fine (predominantly under 20 μm in diameter).
6. A high collection efficiency is required.
7. Vapors or gaseous matter must also be removed from the gas.

In general, the factors listed above can be grouped into three categories: economic, environmental, and engineering.

11.14 Venturi Scrubber Disadvantages

List at least five disadvantages associated with venturi scrubbers.

Solution: Although scrubbers have many advantages, they may not be suitable for all applications. Some relative disadvantages follow:

1. *Corrosion problems*—water and absorbed gaseous pollutants can form highly corrosive acid solutions; therefore, choosing proper construction materials for the control system is important.
2. *Meteorological problems*—highly humidified exhaust gases can produce a wet, visible steam plume (a potential esthetic problem), especially during cold weather; fog and precipitation from the plume may cause local meteorological problems.

3. *Difficulty of byproduct recovery*—recovery of dust for reuse is difficult when wet collectors are used; costs associated with dewatering and drying of the scrubber sludge may make other control methods more practical.
4. *Pressure drop and power requirements*—high collection efficiencies for particulate matter are attainable only at high pressure drops; the increased fan power required to move the exhaust gases through the scrubber may result in significant operating costs.
5. *Water pollution*—adequate precautions must be taken before the scrubber liquid waste is disposed; settling ponds and sludge clarifiers are often included in the design of wet collector systems to meet wastewater regulations.

The latter two disadvantages should not be underestimated. Either or both can eliminate the scrubber from consideration for a particular application.

11.15 Efficiency–Penetration Calculation

What is the efficiency of a wet scrubber if the overall fractional penetration was found to be 0.02?

(**a**) 20%
(**b**) 98%
(**c**) 2%
(**d**) 99%

Solution: By definition

$$P = 1 - E; \quad \text{fractional basis}$$

Substituting, one obtains

$$0.02 = 1 - E$$
$$E = 0.98$$
$$= 98\%$$

The correct answer is therefore (b).

11.16 Penetration Calculation

The efficiency of a venturi scrubber is 99.32%. Calculate the penetration, in percent, for the unit.

(**a**) 0.68
(**b**) 0.0068
(**c**) 0.0118
(**d**) None of the above

Solution: For starters, answers (b) and (c) can be eliminated. By definition

$$P = 100 - E$$

Substituting, one obtains

$$P = 100 - 99.32$$
$$= 0.68\% = 0.0068$$

The correct answer is therefore (a).

11.17 Efficiency Calculation

The inlet and outlet loading of a scrubber has been measured experimentally to be 3.2 and 0.036 gr/ft^3, respectively. Calculate the efficiency of the unit.

(**a**) 98.56%
(**b**) 98.88%
(**c**) 99.84%
(**d**) None of the above

Solution: By definition

$$\begin{aligned} E &= (3.2 - 0.036)/3.2 \\ &= 0.98875 \\ &= 98.88\% \end{aligned}$$

The correct answer is therefore (b).

11.18 Total Power Relationship with Transfer Units

The number of transfer units N_t in a venturi scrubber is related to the total power p_t through Equation (11.12):

$$N_t = \alpha p_t^{\beta} \qquad (11.12)$$

Calculate N_t if given $\alpha = 1.47$, $\beta = 1.05$, and $p_t = 2.20$.

(**a**) 1.29
(**b**) 5.23
(**c**) 3.36
(**d**) 2.44

Solution: Substituting into Equation (11.12) gives

$$\begin{aligned} N_t &= (1.47)(2.20)^{1.05} \\ &= 3.36 \end{aligned}$$

The correct answer is therefore (c).

11.19 Transfer Units Calculation

The number of transfer units N_t in a venturi scrubber is related to the fractional efficiency. Calculate E (fractional basis) if $N = 1.47$.

(**a**) 1.20
(**b**) 0.77
(**c**) 0.95
(**d**) 0.86

Solution: Employ Equation (11.11).

$$N_t = \ln\left(\frac{1}{1-E}\right) \tag{11.11}$$

If $N_t = 1.47$, then

$$1.47 = \ln\left(\frac{1}{1-E}\right)$$

$$4.35 = \frac{1}{1-E}$$

$$E = 0.77$$

The correct answer is therefore (b).

11.20 Impaction Collection Efficiency

The efficiency of a venturi scrubber is 95% for 8 μm particles. Everything else being constant, what is the efficiency for 2 μm particles? Assume impaction to be the only collection mechanism.

(**a**) 53%
(**b**) 78%
(**c**) 85%
(**d**) 47.5%

Solution: Refer to Equations (11.2) and (11.3). These equations may be written as

$$E = 1 - e^{-k(\psi)^{0.5}}$$

$$\psi = k' d_p^2$$

Combining these two equations gives

$$E = 1 - e^{-Kd_p}; \quad K = k(k^1)^{0.5}$$

For the initial condition

$$0.95 = 1 - e^{-K(8)}; \quad \text{consistent units}$$

$$K = \tfrac{3}{8}$$

$$= 0.375$$

For the second condition

$$\begin{aligned} E &= 1 - e^{-(0.375)(2)} \\ &= 1 - 0.472 \\ &= 0.528 = 52.8\% \end{aligned}$$

The correct answer is therefore (a).

11.21 Collection Efficiency

A venturi scrubber is employed to reduce the discharge of flyash to the atmosphere. The unit is presently treating 215,000 acfm of gas, with a concentration of 4.25 gr/ft^3, and operating at a pressure drop of 32 in H_2O. Experimental studies have yielded the following particle size collection efficiency data:

Particle Diameter, μm	Weight Fraction, w_i	Collection Efficiency, %
5	0.00	30
10	0.00	42
20	0.02	86
30	0.05	93
50	0.08	97
75	0.10	98.7
100	0.75	99.9+

Estimate the overall collection efficiency of the unit.

Solution: The overall efficiency of the unit can be calculated from

$$E_T = \sum w_i E_i \tag{11.13}$$

The solution is presented in Table 11.3.

TABLE 11.3 Data and Calculations for Problem 11.21

Particle Diameter, μm	Weight Fraction w_i	Collection Efficiency E_i%	w_iE_i%
5	0.00	30	0.00
10	0.00	42	0.00
20	0.02	86	1.72
30	0.05	93	4.65
50	0.08	97	7.76
75	0.10	98.7	9.87
100	0.75	99.9+	75.00
			$E_T = 99.00$

11.22 Discharge from a Venturi Scrubber

With reference to Problem 11.21, calculate the daily mass (in tons) of flyash collected by the scrubbing liquid and discharged to the atmosphere. Also obtain the particle size distribution of the flyash collected and discharged to the atmosphere. Comment on the results.

Solution: The total mass entering is

$$\begin{aligned}
\dot{m}_{\text{in}} &= \left(\frac{4.25\ \text{gr}}{\text{ft}^3}\right)\left(\frac{215{,}000\ \text{ft}^3}{\text{min}}\right)\left(\frac{60\ \text{min}}{\text{hr}}\right)\left(\frac{24\ \text{hr}}{\text{day}}\right) \\
&\quad \times \left(\frac{1\ \text{lb}}{7000\ \text{gr}}\right)\left(\frac{1\ \text{ton}}{2000\ \text{lb}}\right) \\
&= 93.968 \approx 94\ \text{tons/day} \\
\dot{m}_{\text{collected}} &= (0.99)(94) = 93\ \text{tons/day} \\
\dot{m}_{\text{discharged}} &= 94 - 93 = 1.0\ \text{ton/day}
\end{aligned}$$

With regard to the particle size distribution calculations, the results are also presented in Table 11.4.

TABLE 11.4 Mass Collected and Discharged

Particle Diameter, μm	Weight Fraction, w_i	E, %	Mass Entering, tons/day	Mass Collected, tons/day	Mass Discharged, tons/day
5	0.00	30	0.00	0.00	0.00
10	0.00	42	0.00	0.00	0.00
20	0.02	86	1.88	1.62	0.26
30	0.05	93	4.70	4.37	0.33
50	0.08	97	7.52	7.30	0.22
75	0.10	98.7	9.40	9.28	0.12
100	0.75	99.9	70.50	70.43	0.07
$\sum$			94.00	93.00	1.00

The weight fractions, w_i for the collected and discharged streams are given in Table 11.5.

TABLE 11.5 Weight Fractions of Mass Collected and Discharged

Particle Diameter, μm	w_i Collected Mass	w_i Discharged Mass
5	0.000	0.00
10	0.000	0.00
20	0.017	0.26
30	0.047	0.33
50	0.078	0.22
75	0.100	0.12
100	0.757	0.07
$\sum$	0.999	1.00

11.23 Liquid Droplet Size

The empirical relationship of Nukiyama and Tanasawa (NT) is probably the best known and the most widely used to predict the average droplet size in pneumatic (gas-atomized) sprays. In this type of spray, the stream of liquid is broken up or atomized by contact with a high velocity gas stream. The original NT relationship is given by

$$d_0 = \left(\frac{1920}{v_{\rm r}}\right)\left(\frac{\sigma}{\rho'_{\rm L}}\right)^{1/2} + (5.97)\left(\frac{\mu'_{\rm L}}{(\sigma\rho'_{\rm L})^{1/2}}\right)^{0.45}\left(1000\frac{L'}{G'}\right)^{1.5} \qquad (11.14)$$

where d_0 = average surface volume mean droplet diameter, μm
v_r = relative velocity of gas to liquid, ft/s
σ = liquid surface tension, dyn/cm
$\rho'_{\rm L}$ = liquid density, g/cm^3
$\mu'_{\rm L}$ = liquid viscosity, P
L'/G' = ratio of liquid-to-gas volumetric flow rates at the venturi throat

For water scrubbing systems, one may use

$$\sigma = 72\ \text{dyn/cm}$$
$$\rho'_{\rm L} = 1.0\ \text{g/cm}^3$$
$$\mu'_{\rm L} = 0.00982\ \text{Pa}$$

This reduces to Equation (11.4) for "standard" air and water in a venturi scrubber:

$$d_0 = (16{,}400/v) + 1.45\,R^{1.5} \qquad (11.4)$$

where v = gas velocity at venturi throat, ft/s
R = ratio of liquid-to-gas flow rates, gal/1000 actual ft^3

The following data were collected using a bench scale venturi scrubber:

Gas rate = 1.56 ft^3/s
Liquid rate = 0.078 gal/min
Throat area = 1.04 in^2

Estimate the average liquid droplet size in the scrubber. Repeat the calculation using the simplified equation.

Solution: From the given data

$$G' = 1.56\ \text{ft}^3/\text{s}$$

$$L' = (0.078\ \text{gal/min})(1\ \text{ft}^3/7.48\ \text{gal})(1\ \text{min}/60\ \text{s})$$

$$= 1.74 \times 10^{-4}\ \text{ft}^3/\text{s}$$

$$A = 1.04\ \text{in}^2 = 7.22 \times 10^{-3}\ \text{ft}^2$$

The gas velocity is

$$v_{\mathrm{G}} = G'/A = 1.56/7.22 \times 10^{-3}$$
$$= 216\,\mathrm{ft/s}$$

The liquid velocity is

$$v_{\mathrm{L}} = L'/A = 1.74 \times 10^{-4}/7.22 \times 10^{-3}$$
$$= 0.0241\,\mathrm{ft/s}$$

The relative velocity is

$$v_{\mathrm{r}} = v_{\mathrm{G}} - v_{\mathrm{L}} \approx v_{\mathrm{G}} = v = 216\,\mathrm{ft/s}$$

Using the original NT correlation for droplet diameter

$$d_0 = \left(\frac{1920}{v_{\mathrm{r}}}\right)\left(\frac{\sigma}{\rho'_{\mathrm{L}}}\right)^{1/2} + (5.97)\left(\frac{\mu'_{\mathrm{L}}}{(\sigma\rho'_{\mathrm{L}})^{1/2}}\right)^{0.45}\left(1000\frac{L'}{G'}\right)^{1.5}$$
$$= \left(\frac{1920}{216}\right)\left(\frac{72}{1}\right)^{1/2} + (5.97)\left(\frac{0.00982}{[(72)(1)]^{1/2}}\right)^{0.45}\left((1000)\frac{1.74 \times 10^{-4}}{1.56}\right)^{1.5}$$
$$= 75.4\,\mu\mathrm{m}$$

The ratio of liquid-to-gas flow rates is

$$R = \frac{(0.078\,\mathrm{gal/min})(1\,\mathrm{min}/60\,\mathrm{s})(1000)}{(1.56\,\mathrm{ft^3/s})}$$
$$= 0.833\,\mathrm{gal}/1000\,\mathrm{acf}$$

Using the simplified equation gives

$$d_0 = (16{,}400/v) + 1.45\,R^{1.5}$$
$$= (16{,}400/216) + 1.45(0.833)^{1.5}$$
$$= 77.0\,\mu\mathrm{m}$$

There is reasonable agreement between the results of the two equations.

11.24 Pressure Drop

Using the data provided in Problem 11.23, estimate the pressure drop across the bench-scale unit. Use both Equations (11.5) and (11.6).

Solution: Using Equation (11.5) to estimate the pressure drop, one obtains

$$\Delta P = (5 \times 10^{-5})(216)^2(0.833)$$
$$= 1.943\,\mathrm{in}\ \mathrm{H_2O}$$

Using Theodore's equation

$$\Delta P' = 0.8 + (0.12)(0.833) \qquad (11.6)$$
$$= 0.90$$

Thus

$$\Delta P = (0.90(v^2/2g_c)\rho$$
$$= (0.9)\left(\frac{(216)^2}{(2)(32.2)}\right)(0.0775)$$
$$= 50.5\ \text{lb/ft}^2$$

Since 1 $\text{lb/ft}^2 = 0.1922$ in H_2O, it follows that

$$\Delta P = 9.71\ \text{in}\ H_2O$$

The former equation significantly underpredicts the pressure drop at low values of R. Note also that this equation fails when R is zero.

11.25 Power Requirements for a Venturi Scrubber

Calculate the power requirement of a venturi scrubber treating 380,000 acfm of gas and operating at a pressure drop of 60 in H_2O.

Solution: The gas horsepower may be calculated from

$$\text{HP} = \frac{q\Delta P}{6356}; \quad q = \text{acfm}, \quad \Delta P = \text{in } H_2O \qquad (11.15)$$
$$= \frac{(380{,}000)(60)}{6356}$$
$$= 3587\ \text{HP}$$

Alternately, Equation (11.8) may be used:

$$p_G = 0.157\Delta P'$$
$$= 0.157(60)$$
$$= 9.42\ \text{HP}/1000\ \text{acfm}$$

Thus, the power would be

$$(9.42\ \text{HP}/1000\ \text{acfm})(380{,}000\ \text{acfm}) = 3580\ \text{HP}$$

To determine the brake horsepower, the gas horsepower must be divided by the fan efficiency. Assuming a fan efficiency of 60%, the operating brake horsepower would be

$$\text{Brake HP} = 3580/0.60$$
$$= 5970\ \text{HP}$$

11.26 Fan Economics

A 20,000 acfm flyash–laden gas passes through a venturi scrubber with a throat velocity of 250 ft/s. The liquid-to-gas volume ratio is 0.667 L/m^3. Find the annual operating cost, assuming that the fan is 55% efficient and the cost of electricity is \$0.18/kW · hr. Use Equation (11.5) to calculate the pressure drop.

Solution: First, convert the liquid-to-gas ratio to gal/ft^3. Substituting into Equation (11.5), one obtains

$$\begin{aligned}\Delta P &= 5.0 \times 10^{-2}(250)^2(0.50 \times 10^{-2}) \\ &= 15.64 \text{ in } H_2O\end{aligned}$$

Gas horsepower:

$$\begin{aligned}\text{HP} &= 0.0001575(q)(\Delta P) \\ &= 0.0001575(20{,}000 \text{ acfm})(15.64 \text{ in } H_2O) \\ &= 49.27 \text{ HP}\end{aligned}$$

Fan (brake) horsepower:

$$\begin{aligned}\text{BHP} &= 0.0001575(q)(\Delta P)/(\eta); \quad \eta \text{ is fan/motor efficiency} \\ \text{BHP} &= 49.27/0.55 \\ &= 89.57 \text{ HP}\end{aligned}$$

Kilowatts:

$$\text{kW} = 0.746(\text{BHP}) = 0.746(89.57 \text{ HP}) = 66.82 \text{ kW}$$

Kilowatt-hours: (kW · hr = kWh):

$$\begin{aligned}\text{kWh} &= 66.82 \text{ kW}(24 \text{ hr/day})(30 \text{ days/month})(12 \text{ months/year}) \\ &= 577{,}350 \text{ kWh}\end{aligned}$$

Annual cost:

$$\text{AC} = 577{,}350 \text{ kWh}(\$0.18/\text{kWh}) = \$103{,}900/\text{year}$$

11.27 Inertial Impaction Parameter Calculation

Calculate the inertial impaction parameter for particles of 5 μm diameter and a specific gravity of 1.1 in an air stream at ambient conditions (viscosity = 1.21×10^{-5} lb/ft · s). The aerosol is flowing past a 256 μm diameter spherical collector at a relative velocity of 170 ft/s.

Solution: Employ Equation (11.3). Substitution gives

$$\psi = (5.0 \times 3.28 \times 10^{-6})^2(1.1 \times 62.4)(170)/(9)(1.21 \times 10^{-5}) \times (256 \times 3.28 \times 10^{-6})$$

Solving

$$\psi = 34.3$$

11.28 Calculations on a Venturi Scrubber

A venturi scrubber is being designed to remove particulates from a gas stream. The maximum gas flowrate of 30,000 acfm has a loading of 4.8 gr/ft^3. The average particle size is 1.2 μm and the particle density is 200 lb/ft^3. Neglect the Cunningham correction factor. The Johnstone coefficient k for this system is 0.15. The proposed water flow rate is 180 gal/min and the gas velocity is 250 ft/s.

(**a**) What is the efficiency of the proposed system?
(**b**) What would the efficiency be if the gas velocity were increased to 300 ft/s?
(**c**) Determine the pressure drop for both gas velocities. Assume Equation (11.5) to apply.
(**d**) Determine the daily mass of dust collected and discharged for each gas velocity.
(**e**) What is the discharge loading in each case?

Solution

(**a**) The ratio of liquid-to-gas flow rate is given by

$$\begin{aligned} R &= (180)/(30{,}000) \\ &= 6.0\,\text{gal}/1000\,\text{acf} \\ &= 6.0\,\text{gpm}/1000\,\text{acfm} \end{aligned}$$

and

$$\begin{aligned} v &= 250\ \text{ft/s} \\ d_p &= 1.2\,\mu\text{m} = 3.937 \times 10^{-6}\ \text{ft} \\ \rho_p &= 200\,\text{lb/ft}^3 \\ \mu &= 1.23 \times 10^{-5}\,\text{lb/(ft} \cdot \text{s)} \end{aligned}$$

Assume the Nukiyama–Tanasawa (NT) equation to apply:

$$d_0 = (16{,}400/v) + 1.45R^{1.5} \qquad (11.4)$$

$$\begin{aligned} &= (16{,}400/250) + 1.45(6.0)^{1.5} \\ &= 86.91\,\mu\text{m} \\ &= 2.85 \times 10^{-4}\ \text{ft} \end{aligned}$$

Refer to Equation (11.3):

$$\begin{aligned}
\psi &= \frac{d_p^2 \rho_p v}{9\mu d_0} \\
&= \frac{(3.937 \times 10^{-6})^2 (200)(250)}{(9)(1.23 \times 10^{-5})(2.85 \times 10^{-4})} \\
&= 24.56 \\
E &= 1 - e^{-kR\sqrt{\psi}} \\
&= 1 - e^{-(0.15)(6)\sqrt{24.56}} \\
&= 0.9884 = 98.84\%
\end{aligned}$$

(**b**) If v were increased to 300 ft/s, then

$$\begin{aligned}
d_0 &= (16{,}400/300) + 1.45(6.0)^{1.5} \\
&= 75.98\ \mu\text{m} \\
\psi &= \frac{(3.937 \times 10^{-6})^2 (200)(300)}{(9)(1.23 \times 10^{-5})(2.85 \times 10^{-4})} \\
&= 24.93 \\
E &= 1 - e^{-(0.15)(6)\sqrt{24.93}} \\
&= 99.24\%
\end{aligned}$$

(**c**) The pressure drops are given by

$$\Delta P_a = (5 \times 10^{-5})(250)^2(6) = 18.75 \text{ in } H_2O$$

$$\Delta P_b = (5 \times 10^{-5})(300)^2(6) = 27 \text{ in } H_2O$$

(**d**) The total daily loading (TDL) is

$$\begin{aligned}
\text{TDL} &= (4.8\ \text{gr/ft}^3)(30{,}000)(60)(24)/7000 \\
&= 29{,}600\ \text{lb/day} = 14.81\ \text{tons/day}
\end{aligned}$$

For $v = 250$ ft/s:

$$\begin{aligned}
\text{Dust collected} &= (0.9884)(29{,}600) \\
&= 29{,}300\ \text{lb/day} \\
\text{Dust discharged} &= 344\ \text{lb/day}
\end{aligned}$$

For $v = 300$ ft/s:

$$\text{Dust collected} = (0.9924)(29{,}600) = 29{,}400 \text{ lb/day}$$

$$\text{Dust discharged} = 225 \text{ lb/day}$$

(**e**) The discharge loading (DL) for $v = 250$ ft/s is

$$\text{DL} = (4.8)(1 - E) = (4.8)(1 - 0.9884) = 0.056 \text{ gr/ ft}^3$$

and for $v = 300$ ft/s is

$$\text{DL} = (4.8)(1 - E) = (4.8)(1 - 0.9924) = 0.036 \text{ gr/ ft}^3$$

11.29 A flyash–laden gas stream is to be cleaned by a venturi scrubber using a L/G ratio of 8.5 gal/1000 ft^3. The individual efficiency is to be calculated from Equation (11.2):

$$E_i = 1 - e^{-kR(\psi)^{0.5}} \tag{11.2}$$

The flyash has a particle density of 43.7 lb/ft^3 and $k = 200$ ft^3/gal. The throat velocity is 272 ft/s, and the gas viscosity is 1.5×10^{-5} lb/ft · s. The particle size distribution is given in Table 11.6. Make use of the NT relationship and neglect the Cunningham correction factor.

TABLE 11.6 Particle Size Distribution Data for Problem 11.29

d_{pi}, μm	w_i, % by Weight
<0.10	0.01
0.1–0.5	0.21
0.6–1.0	0.78
1.1–5.0	13.0
6.0–10.0	16.0
11.0–15.0	12.0
16.0–20.0	8.0
>20.0	50.0

Solution: Use the NT relationship calculate to the mean droplet diameter (d_0):

$$d_0 = (16{,}400/v_1) + 1.45R^{1.5} \tag{11.4}$$

Substituting, one obtains

$$d_0 = (16{,}400/272) + 1.45(8.5)^{1.5}$$
$$= 96.2\ \mu\text{m}$$

Express the inertial impaction number (ψ) in terms of d_p using Equation (11.3) (assuming $C = 1$)

$$\psi = \frac{d_{pi}{}^2 \rho_p v_t}{9\mu d_1} = d_{pi}^2(43.7)(272)/[(9)(1.5\times 10^{-5})(96.2)(3.048 \times 10^{-5})]$$
$$= 3.003\, d_{pi}^2 (d_{pi}^2,\ \mu\text{m})$$

Obtain the individual collection efficiency in terms of d_{pi} using Equation (11.2):

$$E_i = 1 - e^{-kR(\psi)^{0.5}}$$
$$= 1 - e^{-(2.94)d_{pi}}$$

(*Note*: The numerical value of k in this equation is 0.2.) As noted earlier, the collection efficiency equation given above is referred to in industry as the size efficiency or grade efficiency equation for the system. These individual collection efficiencies are used to calculate an overall efficiency and this is demonstrated in the next step. Calculate the overall collection efficiency. Results are presented in Table 11.7.

TABLE 11.7 Overall Efficiency Calculation[a]

d_p, μm	E_i	w_i, %	w_iE_i, %
0.05	0.1367	0.01	0.001367
0.30	0.586	0.21	0.123
0.80	0.905	0.78	0.706
3.0	0.9998	13.0	12.998
8.0	0.9999	16.0	15.999
13.0	0.9999	12.0	12.000
18.0	0.9999	8.0	8.000
20.0	0.9999	50.0	50.000

[a] $E = \Sigma w_i E_i = 99.928\%$.

11.30 Cunningham Correction Factor Effect

A venturi scrubber is employed for control of the discharge of flyash. Determine the percent difference between the calculation of overall efficiency including the Cunningham correction factor (CCF) and calculations neglecting the CCF. Pertinent system data are provided below along with the particle size distribution

TABLE 11.8 Particle Size Distribution for Problem 11.30

d_p, μm	% by Weight
<0.05	0.03
0.05–0.10	0.30
0.10–0.50	10.5
0.50–1.00	12.0
1.00–5.00	20.0
5.00–10.00	25.0
>10.00	32.17

in Table 11.8. Data:

$$R = 7.0\,\text{gal}/1000\ \text{ft}^3$$

$$k = 350\ \text{ft}^3/\ \text{gal}$$

$$v = 380\ \text{ft/s}$$

$$\rho_p = 50.0\,\text{lb}/\ \text{ft}^3$$

$$\mu = 1.6 \times 10^{-5}\text{lb}/\ \text{ft} \cdot \text{s}$$

The CCF may be calculated from the following equation:

$$C = 1 + (0.172)/(d_p);\ d_p = \mu\text{m} \tag{11.16}$$

Solution: Key calculations are presented below:

$$d_0 = 70.0\,\mu\text{m}$$

Combining Equations (11.3) and (11.16) leads to

$$\psi = (6.18)(C)d_p^2;\ d_p = \mu\text{m}$$

$$= 6.18[1 + (0.172/d_p)]d_p^2$$

Since

$$E = 1 - e^{-kR\sqrt{\psi}}$$

$$= 1 - e^{-(0.35)(7)(6.18)^{0.5}(C)^{0.5}d_p}$$

$$= 1 - e^{-6.09(C)^{0.5}d_p}$$

For $C = 1.0$

$$E = 1 - e^{-6.09d_p}$$

Applying the data in Table 11.8, one ultimately obtains

$$E = 98.73\% = 0.9873$$

For

$$C = 1.0 + (0.172/d_p)$$

one obtains

$$E = 98.01\% = 0.9801$$

The percent difference is

$$\%D = 0.72/98.73$$
$$= 0.0073 = 0.73\%$$

However, the penetration increase is

$$\%P = (1.99 - 1.27)/1.99$$
$$= 0.362 = 36.2\%$$

11.31 Effect of Johnstone Coefficient on Efficiency

A gas stream is to be cleaned by a venturi scrubber using a liquid-to-gas ratio of 9.0 gal/1000 ft^3 and a throat velocity of 300 ft/s. The scrubber is treating 300,000 acfm of gas operating at a pressure drop of 32 in H_2O. The ash has a particle density of 0.9 g/cm^3 and a gas viscosity of 1.5×10^{-5} lb/ft · s. The particle size distribution is provided in Table 11.19.

TABLE 11.9 Particle Size Distribution Data for Problem 11.31

Particle Size Range μm	% by Weight
<0.10	0.01
0.10–0.50	0.21
0.60–1.00	0.78
1.10–5.00	13.0
6.00–10.0	16.0
11.0–15.0	12.0
16.0–20.0	8.0
>20.0	50.0

Assuming that the Nukiyama–Tanasawa (NT) relationship is applicable and the CCF = 1, determine the effect that the correlation coefficient k has on the overall collection efficiency.

Solution: Four different correlation coefficients are employed. The mean droplet diameter is

$$d_0 = \frac{16{,}400}{v} + 1.45(R)^{1.5} \tag{11.4}$$

$$= \frac{16{,}400}{300} + 1.45(9.0)^{1.5}$$

$$= 93.8\ \mu m$$

The inertial impaction parameter ψ is

$$\psi = \frac{d_p^2 \rho_p v}{18\mu\, d_0} \tag{11.3}$$

$$= \frac{d_p^2(0.9)(62.4)(300)}{18(1.5 \times 10^{-5})(93.8)(25{,}400)(12)}$$

$$= 2.18\, d_p^2$$

The individual efficiencies E_i are

$$E_i = 1 - e^{-k_i R \sqrt{\psi}}$$

For $k = 100$:

$$E_{100} = 1 - e^{-\left[(0.1)(100\,\text{ft}^3/\text{gal})(9.0\,\text{gal}/1000\,\text{ft}^3)\sqrt{2.18\, d_p^2}\right]}$$

$$= 1 - e^{-1.33 d_p}$$

For $k = 250$:

$$E_{250} = 1 - e^{-(0.25)(13.3) d_p}$$

$$= 1 - e^{-3.32\, d_p}$$

For $k = 500$:

$$E_{500} = 1 - e^{-(0.5)(13.3) d_p}$$

$$= 1 - e^{-6.64\, d_p}$$

For $k = 1000$:

$$E_{1000} = 1 - e^{-13.3\, d_p}$$

As the results in Table 11.10 display, the variation of the correlation coefficient does not have a major effect on the overall efficiency. The efficiency only deviates 0.62% for k values ranging from 100 to 1000. However, the effect on the penetration for this high-efficiency device is significant.

11.32 Retrofit Application

An organic pigment manufacturer has changed his process for the production of green pigments. However, this resulted in a severe dusting problem. The engineer who had recommended the change and was now faced with correcting this problem remembered having an old venturi scrubber in the basement, which had been used some time ago on another process. The scrubber must be able to capture particles 3.5 μm in diameter with 96% efficiency. Can this scrubber be used with the following operating data?

Throat velocity, $v = 350$ ft^3
Density of dust, $\rho_p = 194$ lb/ft^3
Gas viscosity, $\mu = 1.24 \times 10^{-5}$ lb/ft · s
Liquid-to-gas ratio, $R = 13.8$ gal/1000 ft^3
Johnstone coefficient, $k = 0.2$

TABLE 11.10 Overall Efficiencies For Various Values

d_p, μm	w_i,%	E_{100} ($k = 100$)	$E_{100}\,w_i$	E_{250} ($k = 250$)	$E_{250}w_i$	E_{500} ($k = 500$)	$E_{500}w_i$	E_{1000} ($k = 1000$)	$E_{1000}\,w_i$
0.05	0.01	0.06	0.0006	0.15	0.002	0.28	0.003	0.49	0.005
0.30	0.21	0.33	0.07	0.63	0.13	0.86	0.18	0.98	0.21
0.80	0.78	0.65	0.51	0.93	0.73	0.995	0.78	1.0	0.8
3.0	13.0	0.98	12.8	1.0	13	1.0	13	1.0	13
8.0	16.0	1.0	16.0	1.0	16	1.0	16	1.0	16
13.0	12.0	1.0	12.0	1.0	12	1.0	12	1.0	12
18.0	8.0	1.0	8.0	1.0	8	1.0	8	1.0	8
80.0	50.0	1.0	50.0	1.0	50	1.0	50	1.0	50
			$\Sigma = 99.38$		$\Sigma = 99.86$		$\Sigma = 99.96$		$\Sigma = 100$

Solution: First calculate the droplet diameter:

$$d_0 = (16{,}400/v) + 1.45(R)^{1.5} \quad (11.4)$$

$$= \left(\frac{16{,}400}{350}\right) + 1.45(13.8)^{1.5}$$

$$= 121.2\ \mu\text{m}$$

$$= (121.2\ \mu\text{m})(3.28 \times 10^{-6})$$

$$= 37.64 \times 10^{-4}\ \text{ft}$$

Substituting into the Johnstone equation ultimately gives

$$E = 0.96 = 1 - e^{-3.227\times 10^5 d_p};\ d_p = ft$$

Solving

$$3.227 \times 10^5 d_p = 3.2189$$

$$d_p = 9.97 \times 10^{-6}\text{ft}$$

$$= (9.97 \times 10^{-6}\ \text{ft})/(3.28 \times 10^{-6}\ \mu\text{m/ft})$$

$$= 3.04\ \mu\text{m}$$

Since the unit can capture 3.04 μm particles with 96% efficiency, the scrubber will capture larger particles (including 3.5 μm ones) with higher efficiencies. The unit will therefore work.

11.33 Capture Efficiency for a Given Particle size

A venturi scrubber is designed with the parameters listed below. What is the particle size that can be collected at an efficiency of 97% for these parameters?

Volumetric flow rate of gas stream = 10,000 acfm
Density of particle = 145 lb/ft^3
Liquid-to-gas ratio = 3 $\text{gal}/1000\ \text{ft}^3$
Water droplet size = 50 μm = 1.64×10^{-4} ft
Scrubber coefficient = 0.10
Gas viscosity = 1.23×10^{-5} lb/ft · s
Cunningham correction factor = 1.0
Gas velocity at venturi throat = 18,000 ft/min

Apply the Johnstone equation. (*Note*: Adapted from J. Lehrian, homework problem, 2007.)

Solution: The inertial impaction factor can be expressed in terms of the particle diameter:

$$\psi = c\rho_p v d_p^2/9 d_0 \mu \quad (11.3)$$

$$= (1)(145\ \text{lb/ft}^3)(18{,}000/60\ \text{ft/s})(d_p^2)/9(1.64 \times 10^{-4}\ \text{ft})(1.23\times 10^{-5}\ \text{lb/ft}\cdot\text{s})$$

$$= 2.396 \times 10^{12} d_p^2;\quad d_p = \text{ft}$$

Substitute this into the Johnstone equation and solve for the particle diameter:

$$E = 1 - e^{[-k(q_L/q_G)\psi^{1/2}]} \tag{11.2}$$

$$0.97 = 1 - e^{[-(0.1)(3)(2.396\times10^{12}d_p^2)^{1/2}]}$$

$$0.03 = e^{[-464{,}376\,d_p]}$$

$$d_p = 7.55 \times 10^{-6}\ \text{ft} = 2.30\ \mu\text{m}$$

11.34 Three Venturi Scrubbers in Series

Three identical venturi scrubbers are connected in series. Assuming that each operates at the same efficiency and liquid-to-gas ratio q_L/q_G, calculate the liquid-to-gas ratio assuming the Johnstone equation to apply. Data are provided below:

E_0 (overall) = 99%
Inlet loading = 200 gr/ft^3
Johnstone scrubber coefficient $k = 0.14$
Inertial impaction parameter $\psi = 105$

Solution: First calculate the outlet loading (OL) from the last unit:

$$\begin{aligned} \text{OL} &= \text{IL}(1 - E_0) \\ &= 200(1 - 0.99) \\ &= 2.0\ \text{gr/ft}^3 \end{aligned}$$

Express the individual and overall efficiencies in terms of the penetration P:

$$\begin{aligned} P_0 &= 1 - E_0 = 1 - 0.99 = 0.01 \\ P_1 &= 1 - E_1 \\ P_2 &= 1 - E_2 \\ P_3 &= 1 - E_3 \end{aligned}$$

Calculate the individual efficiency for each venturi scrubber, noting that the efficiencies (or penetrations) are equal:

$$\begin{aligned} P_0 &= P_1P_2P_3 = P^3 \\ P^3 &= 0.01 \\ P &= 0.215 \\ E &= 1 - P \\ &= 1 - 0.215 \\ &= 0.785 = 78.5\% \end{aligned}$$

Using the Johnstone equation, solve for the liquid-to-gas ratio q_L/q_G:

$$\ln(1 - E) = -k\left(\frac{q_L}{q_G}\right)\psi^{0.5}$$

$$\left(\frac{q_L}{q_G}\right) = -\frac{\ln(1 - E)}{k\psi^{0.5}} = -\frac{\ln(1 - 0.785)}{(0.14)(105)^{0.5}}$$

$$= 1.07\ \text{gal}/1000\ \text{acf}$$

$$= 1.07\ \text{gpm}/1000\ \text{acfm}$$

11.35 Overall Efficiency of Multiple Scrubbers in Parallel

Calculate the overall efficiency of N scrubbers in parallel, assuming that the volumetric flow rates in each scrubber are $q_1, q_2, \ldots, q_N$ and the corresponding efficiencies are $E_1, E_2, \ldots, E_N$, respectively. Assume that the gas is sufficiently well mixed that the particle concentration (particles/volume) is the same at the inlet of each scrubber. Express the result in terms of the q values and the corresponding E values (see Figure 11.4).

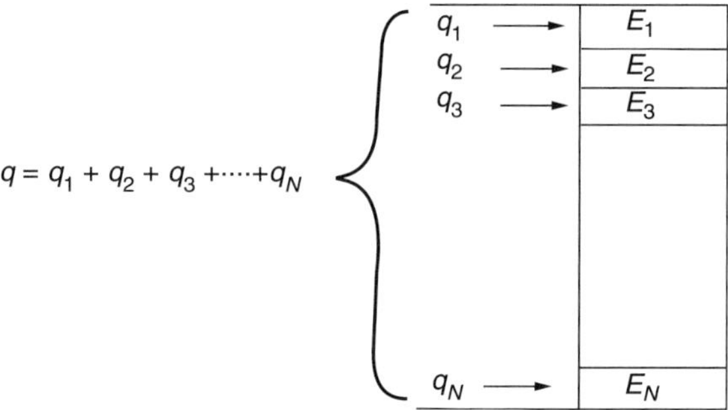

Figure 11.4 N scrubbers in parallel.

Solution: If c_1 is entering and c_{10} is leaving,

$$E_i = 1 - \frac{c_{i0}}{c_i}$$

or

$$c_{i0} = ci(1 - E_i)$$

thus

$$E = 1 - \frac{c_{10}q_1 + c_{20}q_2 + \cdots + c_{N0}q_N}{c_1q_1 + c_2q_2 + \cdots + c_Nq_N}$$

$$= 1 - \frac{c_1(1 - E_1)q_1 + c_2(1 - E_2)q_2 + \cdots + c_N(1 - E_N)q_N}{c_1q_1 + c_2q_2 + \cdots + c_Nq_N}$$

$$= \frac{\Sigma q_ic_iE_i}{\Sigma q_ic_i} \qquad (11.17)$$

If $c_1 = c_2 = \cdots = c_N$ the equation above reduces to

$$E = \frac{\Sigma q_i E_i}{\Sigma q_i} = \frac{\Sigma q_i E_i}{q} \quad (11.18)$$

11.36 Three Venturi Scrubbers in Parallel

Calculate the overall efficiency of three venturi scrubbers operating in *parallel* treating 10,000 acfm of gas. Data are provided in Table 11.11.

TABLE 11.11 Data for Three Scrubbers in Parallel

Scrubber	q, acfm	c, gr/ft^3	E
1	2500	2.0	0.996
2	5000	4.0	0.985
3	2500	2.0	0.996

Solution: Employ a modified form of the equation developed in Problem 11.35:

$$E = 1 - [\Sigma(q_i c_i P_i)/\Sigma(q_i c_i)] \quad (11.19)$$

For the flowrates,

$$\text{Inlet} = (2)(2500)(2.0) + (5000)(4.0) = 30{,}000 \text{ gr/min}$$
$$\text{Outlet} = (2)(2500)(2.0)(0.004) + (5000)(4.0)(0.015) = 340 \text{ gr/min}$$

Thus,

$$E_0 = \frac{30{,}000 - 340}{30{,}000} = 0.9887 = 98.87\%$$

11.37 Varying Particle Size Distribution in Multiple Scrubbers

Refer to Problem 11.36. Suppose that the particle size distributions entering all the scrubbers are not identical. Do the results obtain in problem 11.36 still hold? Explain.

Solution: As demonstrated in the solution to the previous problem, the describing equation is given by

$$E = \frac{\Sigma q_i c_i E_i}{\Sigma q_i c_i} \quad (11.17)$$

Since the particle size distribution does not directly appear in the equation presented above, the earlier result is still applicable. However, in the real world, varying particle size distributions can affect the efficiency E_i, which in turn can affect the final result.

11.38 Throat Area

A consulting firm has been requested to calculate the throat area of a venturi scrubber to operate at a specified collection efficiency. Pertinent data are given below.

Volumetric flow rate of process gas stream = 11,040 acfm (at 68°F)
Density of dust = $187\ \text{lb/ft}^3$
Liquid-to-gas ratio = $2\ \text{gal}/1000\ \text{ft}^3$
Average particle size = $3.2\ \mu\text{m}\ (1.05 \times 10^{-5}\ \text{ft})$
Water droplet size = $48\ \mu\text{m}\ (1.575 \times 10^{-4}\ \text{ft})$
Johnstone scrubber coefficient $k = 0.14$
Required collection efficiency = 98%
Viscosity of gas = $1.23 \times 10^{-5}\ \text{lb/(ft} \cdot \text{s)}$
Cunningham correction factor = 1.0

Solution: Calculate the inertial impaction parameter ψ from Johnstone's equation:

$$E = 1 - e^{-kR\psi^{1/2}} \tag{11.2}$$

$$0.98 = 1 - e^{-(0.14)(2)\psi^{1/2}}$$

Solving for ψ, one obtains

$$\psi = 195.2$$

From the calculated value of ψ above, back calculate the gas velocity v at the venturi throat:

$$\psi = \frac{\rho_p v d_p^2}{9 d_0 \mu} \tag{11.3}$$

$$v = \frac{9\psi d_0 \mu}{\rho_p d_p^2} = \frac{(9)(195.2)(1.575\times10^{-4})(1.23\times10^{-5})}{(187)(1.05\times10^{-5})^2}$$

$$= 165.1\ \text{ft/s}$$

Calculate the throat area S using the gas velocity at the venturi throat v:

$$S = q/v = (11{,}040)/[(60)(165.1)]$$

$$= 1.114\ \text{ft}^2$$

Note that approximately $10\ \text{ft}^3$ of throat area is generally required to treat 10,000–20,000 acfm.

11.39 Recalculation of Throat Area

Recalculation of that area requirement for the system described in problem 11.38, assuming that the inertial impaction parameter is defined as

$$\psi = \frac{C\rho_p v d_p^2}{18 d_0 \mu} \tag{11.20}$$

Solution: The value of ψ remains unchanged. The newly defined ψ leads to a doubling of the velocity:

$$v = 330.2 \text{ ft/s}$$

The area is therefore reduced by a factor of 2 so that

$$S = 0.557 \text{ ft}^2$$

11.40 Throat Diameter for a Given Efficiency

Find the required throat diameter of a venturi scrubber in order to operate at 99% efficiency. The process stream is air and the scrubbing fluid is water. Assume that the Johnstone equation applies. Data are provided below. (*Note*: Adapted from J. DeAngelis, homework problem, 2007.)

Volumetric flow rate of process gas $q = 16{,}384$ acfm
Density of particles $\rho_p = 128 \text{ lb/ft}^3$
Liquid-to-gas ratio $R = 2 \text{ gal/1000 ft}^3$
Average particle diameter $d_p = 2.56 \times 10^{-5}$ ft
Scrubber coefficient $k = 0.16$ (1000 acf/gal)
Gas viscosity $\mu = 1.28 \text{ lb/ft} \cdot \text{s}$
Cunningham correction factor $C = 1$

Solution: Apply the Johnstone equation:

$$E = 1 - e^{-kR(\psi)^{1/2}} \tag{11.21}$$

For

$$E = 0.99,\ R = 2.0,\ \text{and}\ k = 0.16$$
$$\psi = 207$$

Apply the definition of ψ and the NT equation. Two equations and two unknowns result (v and d_0). If ψ is defined as

$$\psi = C\rho_p v d_p^2 / 18 d_0 \mu$$

the following two solutions result:

$$(1):\quad d_0 = 95.8\ \mu\text{m},\ v = 179\ \text{ft/s}$$

and

$$(2):\quad d_0 = 91.7\mu\text{m},\ v = -171\ \text{ft/s}$$

Obviously, only the positive solution is valid. The throat area is therefore

$$A = (16{,}384)/(60)(179)$$
$$= 1.5255 \text{ ft}^2$$

11.41 Particle Size Collected at Specified Efficiency

A venturi scrubber is designed with the parameters listed below. What is the particle size that can be collected at an efficiency of 97% for these parameters?

Volumetric flow rate of gas stream = 10,000 acfm
Density of particle = 145 $\mathrm{lb/ft^3}$
Liquid-to-gas ratio = 3 $\mathrm{gal/1000\ ft^3}$
Water droplet size = 50 μm = 1.64×10^{-4} ft
Scrubber coefficient = 0.10
Gas viscosity = 1.23×10^{-5} lb/ft · s
Cunningham correction factor = 1.0
Gas velocity at venturi throat = 18,000 ft/min
(*Note*: Adapted from J. Lehrian, homework problem, 2007.)

Solution: Express the inertial impaction factor in terms of the particle diameter:

$$\psi = C\rho_p v d_p^2 / 9 d_0 \mu$$

$$\psi = (1)(145\ \mathrm{lb/ft^3})(18{,}000/60\ \mathrm{ft/s})(d_p^2)/9(1.64 \times 10^{-4}\ \mathrm{ft})(1.23 \times 10^{-5}\mathrm{lb/ft \cdot s})$$

$$\psi = 2.396 \times 10^{12} d_p^2$$

Substitute into the Johnstone equation and solve for particle diameter:

$$E = 1 - e^{[-k(q_L/q_G)\psi^{1/2}]}$$

$$0.97 = 1 - e^{[-(0.1)(3)(2.396\times 10^{12} d_p^2)^{1/2}]}$$

$$0.03 = e^{[-464{,}400 d_p]}$$

$$d_p = 7.55 \times 10^{-6}\ \mathrm{ft}$$
$$= 2.30\ \mu\mathrm{m}$$

11.42 Contact Power Theory Calculation

Using contact power theory, estimate the total power loss across a system in inches of H_2O given a gas pressure drop of 5 in H_2O and a liquid-to-gas ratio is 15 gal/1000 $\mathrm{ft_3}$. The liquid inlet pressure is 1000 psi.

(**a**) 8750
(**b**) 1005
(**c**) 8.75
(**d**) 9.53

Solution: Refer to Equation (11.8):

$$p_G = 0.157\Delta P'$$
$$= (0.157)(5)$$
$$= 0.785\ \mathrm{HP/1000\ acfm}$$

Equation (11.9) is employed to calculate p_L:

$$p_L = 5.83 \times 10^{-4} P_L R$$
$$= (5.83 \times 10^{-4})(1000)(15)$$
$$= 8.745\ \text{HP}/1000\ \text{acfm}$$

The total power is obtained from Equation (11.10):

$$p_t = p_G + p_L$$
$$= 0.785 + 8.745$$
$$= 9.53\ \text{HP}/1000\ \text{acfm}$$

The correct answer is therefore (d).

11.43 Compliance Calculations on a Spray Tower

A vendor proposes to use a spray tower on a lime kiln operation to reduce the discharge of solids to the atmosphere to meet state regulations. The vendor's design calls for a certain water pressure drop and gas pressure drop across the tower. You are requested to determine whether this spray tower will meet state regulations that require a maximum outlet loading of 0.05 gr/ft^3. Assume that contact power theory applies. Operating data are provided.

Gas flow rate = 10,000 acfm
Water rate = 50 gal/min
Inlet loading = 5.0 gr/ft^3
Water pressure drop = 80 psi
Gas pressure drop across the tower = 5.0 in H_2O
Maximum gas pressure drop across the unit = 15 in H_2O
Maximum water pressure drop across the unit = 100 psi

The vendor's contact power design data are also available:

$$\alpha = 1.47$$
$$\beta = 1.05$$

Solution: Calculate the contacting power based on the gas stream energy input p_G in HP/1000 acfm:

$$p_G = (0.157)\Delta P \qquad (11.8)$$
$$= (0.157)(5.0)$$
$$= 0.785\ \text{HP}/1000\ \text{acfm}$$

Calculate the contacting power based on the liquid stream energy input, p_L in HP/1000 acfm:

$$p_L = 5.83 \times 10^{-4} P_L (q_L/q_G) \qquad (11.9)$$
$$= (0.583 \times 10^{-3})(80)(50/10{,}000)$$
$$= 0.233\ \text{HP}/1000\ \text{acfm}$$

The total power loss p_T in HP/1000 acfm is then

$$p_T = p_G + p_L \quad (11.10)$$
$$= 0.785 + 0.233$$
$$= 1.018\ \text{HP}/1000\ \text{acfm}$$

The number of transfer units N_t is

$$N_t = \alpha p_T^{\beta} \quad (11.12)$$
$$= (1.47)(1.018)^{1.05}$$
$$= 1.50$$

The collection efficiency can be calculated based on the design data given by the vendor:

$$N_t = \ln\left(\frac{1.0}{1 - E}\right) \quad (11.11)$$

or

$$E = 1 - e^{-N_t} = 1 - e^{-1.50}$$
$$= 77.7\%$$

The collection efficiency required by state regulations E_s is

$$E_s = \frac{(\text{inlet loading}) - (\text{outlet loading})}{\text{inlet loading}}(100)$$
$$= \frac{(5.0) - (0.05)}{5.0}(100)$$
$$= 99.0\%$$

Since $E_s > E$, the spray tower does not meet the regulations.

One may now propose a set of operating conditions that will meet the regulations by reversing the calculational procedure. This is treated in the next problem.

11.44 Adjusted Operating Condition

Refer to Problem 11.43. If the spray tower does not meet state regulations, propose a set of operating conditions that will meet the regulations.

Solution:

$$N_t = \ln\left(\frac{1.0}{1 - E}\right) \quad (11.11)$$
$$= \ln\left(\frac{1.0}{1 - 0.99}\right)$$
$$= 4.605$$

The total power loss p_T in HP/1000 acfm is

$$N_t = \alpha p_T^{\beta} \tag{11.12}$$

$$4.605 = (1.47)(p_T)^{1.05}$$

Solving for p_T, one obtains

$$p_T = 2.96\,\text{HP}/1000\,\text{acfm}$$

Calculate the contacting power based on the stream energy input p_G using a ΔP of 15 in H_2O:

$$\begin{aligned} p_G &= 0.157\Delta P \\ &= (0.157)(15) \\ &= 2.355\,\text{HP}/1000\,\text{acfm} \end{aligned} \tag{11.8}$$

The liquid stream energy input p_L is then

$$\begin{aligned} p_L &= p_T - p_G \\ &= 2.96 - 2.355 \\ &= 0.605\,\text{HP}/1000\,\text{acfm} \end{aligned} \tag{11.10}$$

Calculate q_L/q_G, in gal/acf, using p_L in psi:

$$\begin{aligned} q_L/q_G &= p_L/[(0.583)(P_L)] \\ &= (0.605)/[(0.583)(100)] \\ &= 0.0104 \end{aligned} \tag{11.9}$$

The new water flow rate q'_L in gal/min, is therefore

$$\begin{aligned} q'_L &= (q_L/q_G)(10{,}000\,\text{acfm}) \\ &= (0.0104)(10{,}000\,\text{acfm}) \\ &= 104\,\text{gal/min} \end{aligned}$$

The new water of operating conditions that will meet the regulations are

$$\begin{aligned} \Delta P &= 15\,\text{in}\,H_2O \\ P_L &= 100\,\text{psi} \\ q'_L &= 104\,\text{gal/min} \\ p_T &= 2.96\,\text{HP}/1000\,\text{acfm} \end{aligned}$$

Unlike the Johnstone equation approach, this method requires specifying two coefficients. The validity and accuracy of the coefficients available from the literature for the contact power theory equations have been questioned. Some numerical values of α and β for specific particulates and scrubber devices are provided in Table 11.2.

11.45 Maximum Gas-Side Pressure Drop

A venturi scrubber must operate at 97% efficiency. The following data are given concerning the scrubber:

$$\begin{aligned}&\text{Gas flow rate } q_G = 8000\,\text{acfm}\\&\text{Liquid flow rate } q_L = 60\,\text{gal/min}\\&\text{Liquid inlet pressure } P_L = 50\text{ psi}\\&\alpha = 1.31\\&\beta = 1.12\end{aligned}$$

What must be the maximum gas pressure drop across the scrubber in inches of water to achieve 97% efficiency? Employ contact power theory. (*Note*: Adapted from a homework problem submitted by Michael Barba, 2007.)

Solution: Calculate the contacting power based on the basis of the liquid stream using Equation (11.9):

$$p_L = 5.83 \times 10^{-4} P_L(q_L/q_G) \tag{11.9}$$

$$p_L = 0.219$$

The number of transfer units required to achieve 97% efficiency is calculated using Equation (11.11):

$$N_t = \ln[1.0/(1 - E)]; \quad E = 0.97 \tag{11.11}$$

$$= 3.51$$

The number of transfer units is a function of the total contacting power.

$$N_t = \alpha p_T^{\beta} \tag{11.12}$$

where

$$p_T = p_L + p_G \tag{11.10}$$

Solve for p_T,

$$\begin{aligned}p_T &= (N_t/\alpha)^{1/\beta}\\&= (3.51/1.31)^{1/1.12}\\&= 2.41\end{aligned}$$

Now p_G can be solved for using the above values of p_L and p_T:

$$\begin{aligned}p_G &= p_T - p_L\\&= 2.41 - 0.219\\p_G &= 2.19\end{aligned}$$

The contact power is a function of the gas pressure drop:

$$p_G = 0.157\Delta P \quad (11.8)$$
$$\Delta P = p_G/0.157$$
$$= 2.19/0.157$$
$$= 13.95 \text{ in } H_2O$$

11.46 Open-Hearth Furnace Application

The installation of a venturi scrubber is proposed to reduce the discharge of particulates from an open-hearth steel furnace operation. Preliminary design information suggests water and gas pressure drops across the scrubber of 5.0 psia and 36.0 in of H_2O, respectively. A liquid-to-gas ratio of 6.0 gpm/1000 acfm is usually employed with this industry. Estimate the collection efficiency of the proposed venturi scrubber. Assume contact power theory to apply with α and β given by 1.26 and 0.57, respectively. Recalculate the collection efficiency, neglecting the power requirement on the liquid side.

Solution: Because of the low water pressure drop, it can be assumed that

$$p_G \ggg p_L; \quad p_T \approx p_G$$

with

$$p_G = 0.157(\Delta P) \quad (11.8)$$

Solving for p_G gives

$$p_G = (0.157)(36)$$
$$= 5.65 \text{ hp}/1000 \text{ acfm}$$

The number of transfer units is calculated from

$$N_t = \alpha p_T^{\beta} \quad (11.12)$$
$$= (1.26)(5.65)^{0.57}$$
$$= 3.38$$

The collection efficiency can now be calculated:

$$N_t = 3.38 = \ln\left(\frac{1}{1-E}\right) \quad (11.11)$$

$$E = 0.966 = 96.6\%$$

Since the power requirement on the liquid side is neglected, the efficiency remains the same.

11.47 Applying Sundberg's Method

Your company is utilizing a venturi scrubber to reduce flyash (SG = 0.75) emissions from its coal-fired furnace. It is treating 300,000 acfm of gas at 350°F. The gas throat velocity is 350 ft and the water rate is 8 gal/1000 acfm. Estimate the scrubber efficiency using Sundberg's method. Particle size distribution data and size–efficiency calculations have provided the following:

$$d_{p50} = 22.5\ \mu m$$
$$\sigma = 7.0\ \mu m$$
$$d'_{p50} = 2.6\ \mu m$$
$$\sigma' = 1.6\ \mu m$$

Solution: Apply Equation (7.44):

$$E = \text{erf}\left[\frac{\ln(d_{p50}/d'_{p50})}{[(\ln\ \sigma)^2 + (\ln \sigma')^2]}\right] \tag{7.44}$$

$$= \text{erf}\left[\frac{\ln(22.5/2.6)}{[(\ln 7.0)^2 + (\ln 1.65)^2]}\right]$$

$$= \text{erf}(2.15/2.0)$$
$$= \text{erf}(1.075)$$

From Table 11.5

$$E = 85.88\%$$

11.48 Johnstone Equation Modification

In an attempt to capture very fine talc particles in a contaminated gas (air) stream, it is suggested to lengthen the throat of a venturi scrubber (L. Theodore: personal notes, 1988.) A proposed modification of the famous Johnstone equation is used to calculate the throat length requirement;

$$E = [1 - e^{-kR(\psi)^{0.5}}][e^{-0.0075\,L'}] \tag{11.21}$$

where L' is the venturi throat length in meters. This unit is designed to collect 2.5 μm particles with a 98% efficiency, assuming that water is the scrubbing fluid. How long does the throat length have to be to process 14,500 acfm of a contaminated gas stream? Pertinent data are listed below:

$R = 6.2\ \text{gal}/1000\ \text{ft}^3$
$k = 0.135$
Specific gravity (SG) of particles = 1.7
Viscosity of gas (air) = 1.63×10^{-5} lb/ft · s
Venturi throat = 1.0 ft^2

Solution: The throat velocity is

$$v = 14{,}500/(1.0)(60)$$
$$= 241.7\ \text{ft/s}$$

For an air–water system

$$d_0 = \frac{16{,}400}{v} + 1.45(R)^{1.5} \quad (11.4)$$

$$= \frac{16{,}400}{241.7} + (1.45)(6.2)^{1.5}$$

$$= 90.25\ \mu\text{m}$$

$$= 2.96 \times 10^{-4}\ \text{ft}$$

The inertial impaction parameter is

$$\psi = \frac{C d_p^2 \rho_p v}{9 \mu d_0}; \qquad C = 1.0 \quad (11.3)$$

$$= \frac{(2.5 \times 3.28 \times 10^{-6})^2 (1.7)(62.4)(241.7)}{(9)(1.63 \times 10^{-5})(2.96 \times 10^{-4})}$$

$$= 39.7$$

Equation (11.21) may be rearranged and solved for L':

$$0.0075\, L' = \ln E - \ln\left[1 - e^{-kr(\psi)^{0.5}}\right]$$

Substituting, one obtains

$$(0.0075)L' = 0.0151\ \text{m}$$

$$L' = 2.01\ \text{m}$$

$$= 6.59\ \text{ft} = 79.1\ \text{in}$$

11.49 Fine-Particulate Capture

Provide an explanation as to why fine-particulate capture will increase if the throat of the venturi is increased in length.

Solution: The author has long argued that this statement is true. Here is the basis for his position. Fine particle capture occurs primarily by molecular diffusion. Once the liquid droplets are atomized, the number of the particles striking the droplets is a function of how long a period of time these particles have an opportunity to do so. If the venturi throat is lengthened, the residence time in the throat increases—leading to more capture and a higher efficiency. The statement above has been partially borne out by some field test data that seem to suggest that efficiency decreases at pressure drops above 80–100 inches of H_2O. At these pressure drop levels, the throat velocity is extremely high, providing low residence times.

Please note that Equation (11.21) has never been verified by either laboratory or field testing.

11.50 Total Annual Cost Calculation

Determine the total annual cost and cost per 1000 ft^3 of processed stack gas for a venturi scrubber system with the following specifications and unit costs:

Capacity: 60,000 acfm
Capital cost (including installation and accessories): \$7.0/acfm
Blower specifications: 60,000 acfm at 22 inches of water
Blower efficiency: 0.55
Water pump specification: 420 gal/min at 120 psi
Water pump efficiency: 0.653
Power cost, fan and pump: \$70/HP · yr
Liquid-to-gas ratio: 7.0 gal/1000 ft^3
Water cost: \$0.0168/1000 gal
Maintenance: \$0.105/hr
Operation time: 8000 hr/year
Equipment lifetime (use straight-line depreciation, no salvage value): 5 years
Interest on capital: 7.2% simple

Solution: Costs to be considered:

1. Depreciation
2. Interest
3. Power
4. Water
5. Maintenance

1. Annual depreciation:

$$\text{Capital cost} = 60{,}000 \times 7.0 = \$420{,}000$$

$$\begin{aligned} \text{Annual depreciation (straight line with} \\ \text{no salvage value)} &= \text{Capital cost/years} \\ &= 420{,}000/5 \\ &= \$84{,}000/\text{year} \end{aligned}$$

2. Yearly interest:

$$420{,}000 \times 0.072 = \$30{,}240/\text{year}$$

3. Power cost:

$$\text{Fan HP} = \frac{60{,}000 \times 22 \times 5.2}{60 \times 550 \times 0.55} = 378.45 \text{ HP}$$

$$\begin{aligned} \text{Pump HP} &= \frac{120 \times 144 \times 420}{7.48 \times 60 \times 550 \times 0.653} \\ &= 45.02 \text{ HP} \\ \text{HP} &= 45.02 + 378.45 = 423.47 \text{ HP} \\ \text{Power cost} &= 423.47 \times 70 = \$29{,}643/\text{year} \end{aligned}$$

4. Water cost:

$$\text{Water cost} = 7.0 \times (60{,}000/1000) \times 60 \times 8000 \times (0.0168/1000)$$
$$= \$3387/\text{year}$$

This apparently assumes no recycling and *no* water treatment. For example, it might be sent to a central sewage treatment facility.

5. Maintenance cost (MC):

$$\text{MC} = (0.105)(8000) = \$840/\text{year}$$

The total annual cost (TAC) is therefore

$$\text{TAC} = (1) + (2) + (3) + (4) + (5)$$
$$= 84{,}000 + 30{,}240 + 29{,}643 + 3{,}387 + 840$$
$$= \$148{,}110$$

Note that the power cost is approximately 20% of total cost.

11.51 Optimization Study

You are an engineer in the environmental department for Spillco Chemical Industries, one of the most notorious polluters in the state. The company has just been informed by a state regulatory official that there is a new law on particulate emissions. The company will now be fined at a rate of \$0.03/lb of particulates released. Given the following information, your boss requests you to find the least costly way to handle the problem. Data on a proposed low efficiency scrubber control device are given below:

Capital installed cost: \$8.00/acfm (zero salvage value)
Efficiency vs. ΔP: $E = \dfrac{\Delta P}{(\Delta P + 10)}$; ΔP in H_2O
Useful life of equipment: 10 years at 8000 hr/year (straight line)
Dust stream flowrate: 100,000 acfm
Inlet loading: 5.0 gr/ft^3
Power costs: \$0.06/kW · hr
Overall blower efficiency: 50%

What is the total minimum annual cost (TC), i.e., capital cost (CC), operating (OC) cost (assume only fan electrical cost) and, the cost associated with the fine (FC).

Solution: On the basis of the problem statement

$$\text{TC} = \text{CC} + \text{OC} + \text{FC}$$

Each term on the RHS is now evaluated:

$$\begin{aligned}\text{TC} &= (100{,}000)(8)/10 \\ &= \$80{,}000/\text{year}\end{aligned}$$

For OC;

$$\text{OC} = (\text{HP})(0.746\,\text{kW/HP})(\$0.06/\text{kW}\cdot\text{hr})(8000)\,\text{hr/year})$$

with

$$\begin{aligned}\text{HP} &= q\Delta P/\eta \\ &= (100{,}000)(\Delta P)(5.2)/(33{,}000)(0.5)\end{aligned}$$

and

$$\Delta P = (10E)/(1 - E)$$

Thus, substituting the latter two equations into OC gives

$$\text{OC} = (112{,}804)(E)/(1 - E)$$

For FC;

$$\text{FC} = (100{,}000)(c_0)(60)(8000)(0.03)/7000$$

with

$$c_0 = 5(1 - E)$$

Thus,

$$\text{FC} = 1{,}028{,}570(1 - E)$$

Summing:

$$\text{TC} = 80{,}000 - 112{,}804(E)/(1 - E) + 1{,}028{,}570(1 - E)$$

For minimum cost:

$$\begin{aligned}\frac{d(\text{TC})}{dE} &= 0 \\ &= 112{,}804\left[\frac{E - 1 - E}{(1 - E)^2}\right] + 1{,}028{,}570\end{aligned}$$

Rearranging

$$O = \frac{112{,}804}{(E-1)^2} + 1{,}028{,}570$$

Solving for E, one obtains

$$E^2 - 2E + 0.8903$$
$$E = 1.33 \text{ or } 0.6688; \text{ reject } 1.33$$

Therefore,

$$E = 66.88\%$$

and

$$\begin{aligned} \text{TC(min)} &= 80{,}000 + 227{,}823 + 340{,}620 \\ &= \$648{,}450/\text{year} \end{aligned}$$

Note that the solution to the derivative could also be obtained graphically.

11.52 Optimum Operating Conditions

Refer to Problem 11.52. What is the outlet concentration and pressure drop at the minimum cost point?

Solution: Since

$$E = 0.6688$$

the equation for the pressure drop may be solved

$$\begin{aligned} \Delta P &= (10E)/(1-E) \\ &= (10)(0.6688)/(1-0.6688) \\ &= 20.2\,H_2O \end{aligned}$$

For the outlet concentration:

$$\begin{aligned} c_0 &= (1-E)(5) \\ &= (1-0.6688)(5) \\ &= 1.66\,\text{gr}/\text{ft}^3 \end{aligned}$$

11.53 Improving Performance

A venturi scrubber's operating collection efficiency has slowly degraded with time. As a plant manager who has recently completed (and passed) an air pollution control equipment (APCE) course given by Dr. Lois Theodore—the supposed foremost authority in the galaxy on APCE—indicate what sound, reasonable engineering steps can be taken to return the scrubber to its original design value.

Solution: This is obviously another open-ended question. After the standard maintenance checks, one should check back upstream (in the process) to see what steps can be taken to improve performance. For example, this can include increasing the throughput velocity, assuming that the fan can handle the additional load. Reducing the throat area is another option, but once again, subject to fan considerations. The reader is left to ponder and/or discuss other options.

12

BAGHOUSES

12.1 INTRODUCTION (Adapted with permission from J. McKenna, ETSI, Roanoke, VA)

One of the oldest, simplest, and most efficient methods for removing solid particulate contaminants from gas streams is by filtration through fabric media. The fabric filter is capable of providing high collection efficiencies for particles as small as 0.1 μm and will remove a substantial quantity of those particles as small as 0.01 μm. In its simplest form, the industrial fabric filter consists of a woven or felted fabric through which dust-laden gases are forced. A combination of factors results in the collection of particles on the fabric filters. When woven fabrics are used, a dust cake eventually forms; this, in turn, acts predominantly as a sieving mechanism. When felted fabrics are used, this dust cake is minimal or almost nonexistent and the primary filtering mechanisms are a combination of inertial forces, impingement, and so on. These are essentially the same mechanisms that are applied to particle collection on wet scrubbers, where the collection media is in the form of liquid droplets rather than solid fibers.

As particles are collected, the pressure drop across the fabric filtering media increases. Owing in part to fan limitations, the filter must be cleaned at predetermined intervals. Dust is removed from the fabric by gravity and/or mechanical means. The fabric filters or bags are usually tubular or flat.

Air Pollution Control Equipment Calculations. By Louis Theodore

The structure in which the bags hang is referred to as a *baghouse*; the number of bags in a baghouse may vary from less than a dozen to several thousand. Quite often, when great numbers of bags are involved, the baghouse is compartmentalized so that one compartment may be cleaned while others are still in service.

The basic filtration process may be conducted in many different types of fabric filters in which the physical arrangement of hardware and the method of removing collected material from the filter media will vary. The essential differences may be related, in general, to the following:

1. Type of fabric
2. Cleaning mechanism(s)
3. Equipment geometry
4. Mode of operation

Depending on the above factors, the flow process in the equipment will follow one of three systems as shown in Figure 12.1. Bottom-feed units are characterized by the introduction of dust-laden gas through the baghouse hopper and then to the interior of the filter tube. In top-feed units, dust-laden gas enters the top of the filters to the interior or clean-air side. When the gas flow is from inside the bag to the outside by virtue of the pressure differential, the internal area of the filter element will be open and self-supporting; unsupported filter elements are tubular. When the filtration process is reversed, with the gas flow from outside the bag to inside, it is necessary to support the media against the developed pressures so that the degree of collapse is controlled. Supported filter elements are either of the tubular or envelope shape.

Baghouse collectors are available for either intermittent or continuous operation. Intermittent operation is employed where the operational schedule of the dust generating source permits halting the gas cleaning function at periodic intervals (regularly set by time or by pressure differential) for removal of collected material from the filter

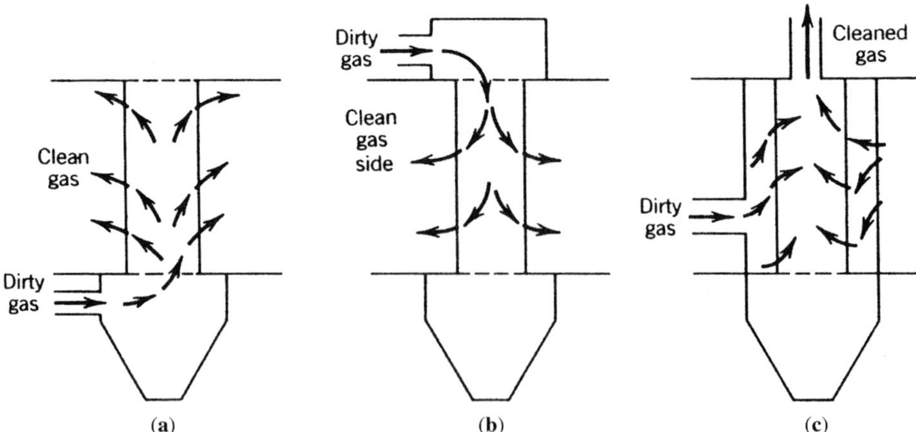

Figure 12.1 Baghouse filtration methods: (a) bottom feed; (b) top feed; (c) exterior filtration.

media (*cleaning*). Collectors of this type are utilized primarily for the control of small-volume operations such as grinding and polishing, and for aerosols of a very coarse nature. For most air pollution control installations and major dust control problems, however, it is desirable to use collectors that allow for continuous operation. This is accomplished by arranging several filter areas in a parallel flow system and cleaning one area at a time according to some preset mode of operation.

Baghouses may also be characterized and identified according to the method used to remove collected material from the bags. Particle removal can be accomplished in a variety of ways, including shaking the bags, blowing a jet of air on the bags from a reciprocating manifold, or rapidly expanding the bags by a pulse of compressed air. In general, the various types of bag-cleaning methods can be divided into those involving flexing and those involving a reverse flow of clean air. Figure 12.2 illustrates four of the fabric-flexing cleaning methods.

There are two basic types of filtration that occur in commercial fabric filters; the first is referred to as *media* or *fiber filtration*, and the second is *layer* or *cake filtration*. In fiber filtration the dust is retained on the fibers themselves by settling, impaction, interception, and diffusion. In cake filtration the fiber acts as a support on which a layer of dust is deposited to form a microporous layer capable of removing additional particles by sieving as well as other basic filtration mechanisms (impaction, interception, diffusion, settling, and electrostatic attraction). In practical industrial cloth filters, both methods occur, but cake filtration is the more important process after the new filter cloth becomes thoroughly impregnated with dust.

Perhaps the outstanding characteristic of fabric filters is their inherent very high collection efficiency with even the finest particles. These units usually have the capability of

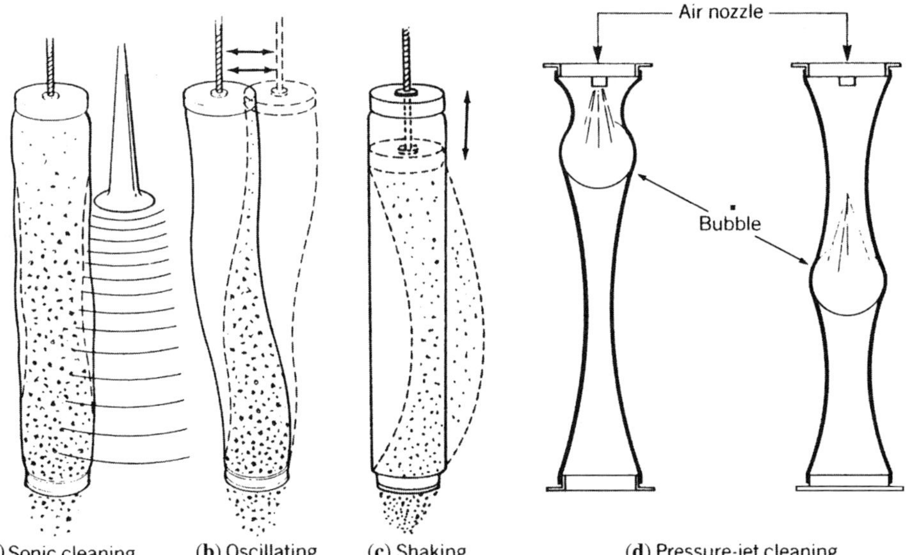

Figure 12.2 Fabric-flexing cleaning methods: (a) sonic cleaning; (b) oscillating; (c) shaking; (d) pressure-jet cleaning.

achieving efficiencies of 99% almost automatically—provided they are properly constructed and maintained in satisfactory operating condition.

The number of variables necessary to design a fabric filter is very large. Since fundamentals cannot treat all of these factors in the design and/or prediction of performance of a filter, this determination is basically left up to the experience and judgment of the design engineer. In addition, there is no one formula that can determine whether a fabric filter application is feasible. A qualitative description of the filtration process is possible, although quantitatively the theories are far less successful. Theory, coupled with some experimental data, can help predict the performance and design of the unit.

The state of the art of engineering process design is the selection of filter medium, superficial velocity, and cleaning method that will yield the best economic compromise. Industry relies on certain simple guidelines and calculations, which are usually considered proprietary information, to achieve this. Despite the progress in developing pure filtration theory, and in view of the complexity of the phenomena, the most common methods of correlation are based on predicting a form of a final equation that can be verified by experiment.

12.2 DESIGN AND PERFORMANCE EQUATIONS

The gas-to-cloth (G/C) ratio is a measure of the amount of gas passed through each square foot of fabric in the baghouse. It is given in terms of the number of cubic feet of gas per minute passing through one square foot of cloth. In other words, the G/C ratio = gas volume rate/cloth area. Also note that this velocity is not the actual velocity through the openings in the fabric, but rather the apparent velocity of the gas approaching the cloth. As the G/C ratio increases, pressure drop (ΔP) also increases. In the United States, pressure drop in baghouse applications is generally measured by inches of water (in H_2O).

Despite several sophisticated formulas that have been developed, there is no satisfactory set of published equations that allows a designer to accurately calculate the efficiency of a prospective baghouse. However, there are three more heuristic formulas worth mentioning that can help baghouse designers:

$$\text{Gross gas/cloth ratio} = \frac{\text{total inlet volume rate}}{\text{total filter cloth area in collector}} \tag{12.1}$$

$$\text{Net gas/cloth ratio} = \frac{\text{total inlet gas volume rate + cleaning volume rate}}{\text{on stream cloth}} \tag{12.2}$$

$$\text{Units gas/cloth ratio} = \frac{\text{gas volume rate}}{\text{cloth area}} = \frac{\text{ft}^3/\text{min}}{\text{ft}^2} = \text{ft/min} \tag{12.3}$$

As mentioned earlier, baghouse design is still very much an artform and nowhere is this more evident than in the selection of G/C ratios. Factors influencing the G/C include the cleaning method, filter media, dust size, dust density, dust loading, and other factors unique to each situation. Because of their variability, it has never been able to satisfactorily quantify all of these factors for every application.

There are two basic design approaches that are routinely employed throughout the industry. One approach is to collect all empirical data available for the source in question. If there are no data for the industry at hand, then one could check with a similar industry which is using a baghouse and determine the G/C range successfully employed in that industry and conservatively apply it. There is a risk in that the subtle but essential differences between the two industries may prohibit the G/C transition. An alternative approach is to run a pilot unit on a slip stream and vary the G/C ratio, and then select and design around the G/C ratio which appears to provide the best results. The pilot plant should run for a few months in order to obtain data representative of a long-term operation.

Once the gas/cloth ratio (G/C) has been selected, the size of the baghouse is nearly defined. Variations also occur in the number of walkways, hopper slope, and other design considerations. These other variations will influence the total size of the baghouse to a limited degree, once the basic needs have been identified by the G/C.

The G/C and cleaning method are tied to each other, as shown in Table 12.1.

Once the G/C ratio and cleaning method have been selected, the next major decision is the selection of the filter medium. At last, selection of the filter medium is more science than art! Temperature limitations and chemical resistance often determine filter medium selection. As the operating temperature rises, options decrease.

The maximum temperature for economical and commercially available filter media is approximately 550°F. If the gas stream is hotter than this level, it is important to cool the gas to within safe parameters before it reaches the filter medium.

Important fiber characteristics are:

1. *Temperature.* The fiber must be able to perform without failure at temperatures higher than those expected during operation. The fiber must also be able to handle temporary heat surges (temperature excursions).
2. *Corrosiveness.* The fiber must be able to resist degradation from exposure to specific acids, alkalies, solvents, or oxidizing agents found in the dust-laden gas stream.
3. *Hydrolysis.* Humidity levels must be accounted for.
4. *Dimensional Stability.* If the fiber is expected to shrink or stretch within the process environment, those effects must be controlled.
5. *Cost.* The least costly selection that will satisfy overall requirements is usually preferred.

Table 12.2 compares the characteristics of the more widely used filtration media. Most bag manufacturers, as well as baghouse vendors, will provide media selection charts similar to Table 12.2. Once the critical parameters of a specific application have been identified, such a chart can be used to make an initial selection of media

TABLE 12.1 Gas-to-Cloth Ratios[a]

	Shaker/Woven; Reverse-Air/Woven	Pulse Jet/Felt; Reverse-Air/Felt
Alumina	2.5	8
Asbestos	3.0	10
Bauxite	2.5	8
Carbon black	1.5	5
Coal	2.5	8
Cocoa, chocolate	2.8	12
Clay	2.5	9
Cement	2.0	8
Cosmetics	1.5	10
Enamel frit	2.5	9
Feeds, grain	3.5	14
Feldspar	2.2	9
Fertilizer	3.0	9
Flour	3.0	12
Fly ash	2.5	5
Graphite	2.0	5
Gypsum	2.0	10
Iron ore	3.0	11
Iron oxide	2.5	7
Iron sulfate	2.0	6
Lead oxide	2.0	6
Leather dust	3.5	12
Lime	2.5	10
Limestone	2.7	8
Mica	2.7	9
Paint pigments	2.5	7
Paper	3.5	10
Plastics	2.5	7
Quartz	2.8	9
Rock dust	3.0	9
Sand	2.5	10
Sawdust (wood)	3.5	12
Silica	2.5	7
Slate	3.5	12
Soap, detergents	2.0	5
Spices	2.7	10
Starch	3.0	8
Sugar	2.0	7
Talc	2.5	10
Tobacco	3.5	13
Zinc oxide	3.0	5

[a]Generally safe design values—application requires consideration of particle size and grain loading; ft/min.

TABLE 12.2 Fabric Selection Chart

Fabric	Maximum Temperature, °F	Acid Resistance	Fluoride Resistance	Alkali Resistance	Flex Abrasion Resistance
Cotton	180	Poor	Poor	Good	Very good
Polypropylene	200	Excellent	Poor	Excellent	Very good
Polyester	275	Good	Poor to fair	Good	Very good
Nomex	400	Poor to fair	Good	Excellent	Excellent
Teflon	450	Excellent	Poor to fair	Excellent	Fair
Fiberglass	500	Fair to good	Poor	Fair to good	Fair

possibilities. A more detailed evaluation of the media possibilities, including cost, availability, and previous application history, should then be made.

Once a preliminary selection of filtering fabric has been made, the media supplier can usually provide additional information that should be considered prior to finalizing the fabric choice. The information below gives basic information that may be used along with the media selection charts.

Cotton—low temperature capability; good flex abrasion resistance; still used in some low temperature applications.

Polypropylene—very sleek fabric; demonstrates good cake release and resistance to blinding. Does not readily absorb moisture and would be a good selection for a low temperature, high moisture condition, e.g., dye production.

Polyester—very sturdy material; good resistance to acids and alkalies, slightly higher temperature capability than polypropylene. Costs about the same as polypropylene. Used in most routine low temperature applications including quarry, woodworking, and sand handling operations.

Nomex—extremely sturdy material with respect to flex abrasion. Superior to glass in resistance to fluorides and abrasion. Good temperature capability. Poor acid resistance; not used in gas streams containing SO_2 and SO_3. Costs about 2.5 times that of polypropylene and polyester. Used in asphalt, steel, carbon black, and cement industries.

Teflon—generally chemically inert and for that reason useful in severe environments. Very expensive, but the cost is usually justified because of superior bag life. Used in the carbon black industry, lead smelting, coal-fired boilers, and various unusual applications. Costs about 10 times that of polyester for the same size and weight bag.

Fiberglass—normally used in high temperature applications. Improvements in finishes and techniques of fabrication, installation, and operation have paid dividends in extended bag life. Fiberglass is the primary fabric used in the boiler market. Cost is between polyester and Nomex.

Once the G/C ratio, the cleaning method, and the filter medium have been selected, the essence of the flange-to-flange design selection process is complete. The only major consideration remaining is the baghouse material of construction. Typically, this is mild steel. In some specialty applications stainless steel units are employed. Thicker house and hopper gauges are employed where acid gases are present, such as in boiler fly ash control systems.

Overall system component selection can be simple, as in the case of a single-point pickup for an ambient application—or it can be complex, as in the case of a high temperature, acid gas, dust control system involving system bypass, auxiliary heat, numerous dampers, temperature and pressure alarms, automatic shutdown, and extra modules for maintenance and off-line cleaning.

However, once a system is designed, it is important to have criteria for performance and reliability. Also, local ordinances as well as state and federal regulations must be adhered to. Flange-to-flange requirements, such as temperature limitations and volume and dew point restrictions, must be considered in advance. Importantly, the successful baghouse must be built as a partner with the production process, not as a restricter of output. Finally, all extreme ranges must be anticipated and accounted for within the design.

An equation that can be used for determining the collection efficiency of a baghouse is (Theodore: personal notes, 1981)

$$E = 1 - e^{-(\psi L + \phi t)} \tag{12.4}$$

where ψ = constant based on fabric ft^{-1}
ϕ = constant based on cake, s^{-1}
t = time of operation to develop the cake thickness, s
L = fabric thickness, ft
E = collection efficiency (dimensionless)

The exit concentration(w_c) for the combined resistance system (the fiber and the cake) is

$$w_e = w_i e^{-(\psi L + \phi t)} \tag{12.5}$$

where w_e = exit concentration (lb/ft^3)
w_i = inlet concentration (lb/ft^3)

A variation on Darcy's formula for the flow of fluid through a porous bed has been developed for the flow of gases through a filter medium. The basic Darcy equation can be used to predict the pressure drop for an operating fabric filter with accumulated dust cake:

$$\Delta P = S_E v + K_2 c_1 v^2 t \tag{12.6}$$

where ΔP = pressure drop, in H_2O
S_E = effective residual drag, in H_2O
v = velocity, fpm
K_2 = specific cake coefficient

The effect of bag failure on baghouse efficiency can be described by the following equation:

$$P_t^* = P_t + P_{tc} \tag{12.7}$$

$$P_{tc} = 0.582(\Delta P)^{1/2}/\phi$$

$$\phi = q/[LD^2(T+460)^{1/2}]$$

where P_t^* = penetration after bag failure
P_t = penetration before bag failure
P_{tc} = penetration correction term; contribution of broken bags to P_t^*
ΔP = pressure drop, in H_2O
ϕ = dimensional parameter
q = volumetric flow rate of contaminated gas, acfm
L = number of broken bags
D = bag diameter, inches
T = temperature, °F

For a detailed development of Equation (12.7), refer to L. Theodore and J. Reynolds, "Effect of Bag Failure on Baghouse Outlet Loading," *J. Air Pollut. Control Assoc.* pp. 870–872 (Aug. 1979) Additional material is available in L. Theodore, "Engineering Calculations: Baghouse Specification and Operation Simplified," *Chem. Eng. Progress* p. 22 (June 2000).

12.3 OPERATION AND MAINTENANCE, AND IMPROVING PERFORMANCE (Adapted with permission from J. McKenna, ETSI, Roanoke, VI)

The award of the year for the least maintained piece of equipment again goes to the fabric filter!!! This unfortunate lack of care not only costs industry millions of dollars each year but also reduces the potential impact of air pollution control on air quality levels.

Normally, a dust collector has been or will be installed at the encouragement of the local, state, or federal regulatory agency, not the production department. Even though it is often tied directly in series with the production process, it is often treated like a stepchild. The first step toward efficient, trouble-free operation of a fabric filter system is to treat it the same as any other piece of production equipment. It is a machine, and machines need "tender loving care." It requires routine inspection, preventive maintenance, and quick response to malfunctions. Routine inspections of key components and operating parameters are the key ingredients to normal maintenance.

A preventive maintenance program is a must for systems requiring continuous duty. The preventive maintenance procedures must be continually updated to reflect actual plant experience, not some typical guidelines. The extent of preventive maintenance during scheduled shutdowns must be evolved, starting with the equipment supplier's recommendations and supplemented by one's own routine inspection and maintenance records.

Detecting and correcting small problems before they snowball into large and expensive crises is the goal of any maintenance program. A major malfuction in the collector system will result in one or more of the following identifying symptoms:

1. Abnormally high pressure drop across the collector
2. Visible emission of dust in the exhaust stack
3. Inadequate face velocity or "puffing" at pickup points and hoods
4. Lower-than-normal dust discharge
5. Loud or unusual noises
6. Severe corrosion of material

If one is unable to identify and correct a problem in a reasonable period of time, the user should contact the system designer of the equipment manufacturer. Often, a phone call to the right person is all that is required. Remember, the manufacturer has the advantage of experience and feedback from many sources to assist in solving problems.

The highest source of maintenance and cost are generally the filter bags. All bag sets have a finite lifetime that will vary by application, installation, operating parameters, fabric type, and so on. Typical causes of bag failure are

1. High G/C abrasion
2. Metal-to-cloth abrasion
3. Chemical attack
4. Bag-to-bag abrasion
5. Inlet velocity abrasion (on inside-out cleaning)
6. Accidents
7. Upset conditions (e.g., temperature)
8. Thread mismatched
9. Cuff mismatched
10. Improper installation

In addition, each bag in a set may have a different life as a result of fabric quality, bag manufacturing tolerances, location in the collector, and variation in the bag-cleaning mechanism. Any one or a combination of these factors can cause bags to fail. This means that a baghouse will experience a series of intermittent bag failures until the failure rate requires total bag replacement. Typically, a few bags will fail initially or after a short period of operation as a result of installation damage or manufacturing defects. The failure rate should then remain very low until the operating life of the bags is reached unless a unique failure mode is present within the system. The failure then increases, normally at a near-exponential rate. Industry often describes this type of failure rate behavior as that similar to a "bathtub" curve. The reader is referred to the following text for additional information: J. Reynolds, J. Jeris, and L. Theodore, *Handbook of Chemical and Environmental Engineering Calculations*, John Wiley & Sons, Hoboken, NJ, 2004.

The importance of when to correct or replace a broken bag will depend on the type of collector and the resultant effect on outlet emissions. In "inside bag collection" types of collectors, it is very important that dust leaks be stopped as quickly as possible to prevent adjacent bags from being abraded by jet streams of dust emitting from the broken bag. This is called the "domino effect" of bag failure. "Outside bag collection" systems do not have this problem, and the speed of repair is determined by whether the outlet opacity has exceeded its limits. Often, it will take several broken bags to create an opacity problem and a convenient maintenance schedule can be employed instead of emergency maintenance.

In either type of collector, the location of the broken bag(s) has to be determined and corrective action taken. In a noncompartmentalized unit, this requires system shutdown and visual inspection. In inside collectors, bags often fail close to the bottoms, near the tube sheet. Accumulation of dust on the tube sheets, the holes themselves, or unusual dust patterns on the outside of the bags often occurs. Other probable bag failure locations in reverse-air bags are near anticollapse rings or below the top cuff. In shaker bags, one should inspect the area below the top attachment. Improper tensioning can also cause early failure.

In outside collectors, which are normally top-access systems, inspection of the bag itself is difficult; however, location of the broken bag(s) can normally be found by looking for dust accumulation on top of the tube sheet, on the underside of the top-access door, or on a blowpipe.

Points to be considered in terms of improving operation and performance are discussed below.

1. Can the system be run at a higher G/C ratio and allow for increased gas volume rate, and therefore increased production or the addition of another source operation?
2. Can the cleaning system be modified to allow for either pressure drop reduction or longer bag life? Bag costs vs. utility costs need to be weighed here. It is at this point that the user recognizes that the original selection of the G/C ratio and cleaning cycle was purely an empirical selection and how, based on the actual operation, this can be modified to arrive at lower-cost operation.
3. Bag types and finishes are evolving rapidly in the current market, and it may be worthwhile to initiate one's own bag-screening program. This can be done by installing one or more test bags and monitoring their performance. Such a program may be simply oriented to monitoring bag life or if conditions allow, provide pressure drop and emission data as well. In cases of very poor bag life (i.e., less than 1 year), excessive pressure drops (e.g., greater than 6–8 in H_2O), or very costly bags, it is recommended that a bag-screening program be instituted.

There have been a few recent developments with baghouses. They include the following improvements:

1. More units are being designed with pulse-jet cleaning
2. Longer bags are being used
3. More bags per module are now available
4. Fabric modification include smaller fibers but different layering (microdeniered)
5. New fabrics, including those with more sophisticated membranes, are available.

PROBLEMS

12.1 Collection Mechanisms

The collection mechanism(s) primarily responsible for the filtering in a fabric filtration system are (select one)

(**a**) Diffusion and centrifugal
(**b**) Impaction and interception
(**c**) Electrostatic attraction
(**d**) Agglomeration and direct interception

Solution: To a certain degree, all four answers are correct. There is capture by diffusion; there is capture by electrostatic attraction; and, there is capture by agglomeration. However, it is impaction that accounts for the bulk of capture from a mass basis. The correct answer is therefore (b).

12.2 Natural Fibers

Natural fibers used for bags in a baghouse such as cotton and wool

(**a**) Can be used for a power plant particle collector
(**b**) Have a low-temperature limitation
(**c**) Are very expensive to purchase
(**d**) Have some resistance to fluorides

Solution: Natural fibers are relatively inexpensive and offer some resistance to fluorides. However, their upper temperature limitation is approximately 200°F, a temperature significantly below the flue gas temperature in a power plant. The correct answer is therefore (b).

12.3 Fiber Resistance to Acidic and Alkaline Attack

A fiber that has very good resistance to acidic and alkaline attack and has a high temperature limitation is

(**a**) Cotton
(**b**) Teflon
(**c**) Fiberglass
(**d**) Wool

Solution: Answers (a) and (d) can be immediately eliminated. Both (b) and (c) offer resistance, but Teflon provides excellent resistance. The correct answer is therefore (b).

12.4 Felt Advantages

An advantage of using felted material for bags in a baghouse is that

(**a**) Larger air-to-cloth ratios are possible
(**b**) It provides lower pressure drops
(**c**) It takes longer for the cake to form
(**d**) It is resistant to acidic gas streams

Solution: Generally, felt bags allow the unit to operate at higher velocities, i.e., air-to-cloth ratios. The correct answer is therefore (a).

12.5 Shake Mechanism

Which control device employs a shake mechanism for cleaning?

(**a**) Venturi scrubber
(**b**) Cyclone
(**c**) Baghouse
(**d**) Electrostatic precipitator

Solution: The obvious answer to this question is the baghouse. The correct answer is therefore (c).

12.6 Shake Motion Description

Bags are cleaned in a baghouse that utilizes a shaking motion by

(**a**) Rapping with a hammer–anvil setup
(**b**) Electrifying the bag cage
(**c**) Sonic horns, oscillating motion, or vertical motion
(**d**) Rinsing the bags with water

Solution: Keep in mind that this is a chapter on baghouses. As discussed in Section 12.1, both vertical and lateral motion as well as horns are employed in shaking. The correct answer is therefore (c).

12.7 Reverse-Air Definition

Reverse air is a type of cleaning mechanism to clean the bag by

(**a**) Reversing the air, causing the bag to collapse
(**b**) Causing the bag to vibrate, releasing the dust
(**c**) Blowing a jet of air, causing the bag to bubble and release the dust
(**d**) Pressurizing the bag

Solution: As its name implies, reverse-air cleaning involves reversing the direction of flow through the bag. The correct answer is therefore (a).

12.8 Pulse-Jet Cleaning Description

In pulse-jet baghouses the cleaning mechanism used for cleaning the bags is

(**a**) Reversing the flow of air through the compartment
(**b**) A blast of air into each bag, knocking the dust away from the bag
(**c**) A blast of air against the outside of the bag
(**d**) Pulsating air, which causes the bags to shake back and forth

Solution: The cleaning mechanism involves supplying a jet of high-pressure air into the inside of the bag. The correct answer is therefore (b).

12.9 Baghouse Cage

In a baghouse, a cage for the bag is used in a pressure jet or pulse-jet unit to

(**a**) Help the bags collapse
(**b**) Help the bag shake
(**c**) Support the bag
(**d**) Keep the squirrels out

Solution: The main purpose of the cage is to provide better physical support for the bag. The correct answer is therefore (c).

12.10 Sieving Action

Sieving action plays an important role in

(**a**) Measurement of pressure drop across a felted filter
(**b**) Eliminating the need for woven fabrics
(**c**) Designing multi-compartment baghouses
(**d**) Collecting large particles to build the cake for subsequent collection of small particles

Solution: Answers (a)–(c) are not applicable, while the latter answer is. The correct answer is therefore (d).

12.11 Air-to-Cloth Ratio Parameter

The air-to-cloth ratio is

(**a**) A measure of the amount of dust deposited on the filter
(**b**) Imperative for good design and prevention of premature bag failure
(**c**) Often referred to as *low filter drag*
(**d**) Is always less than 2 inches of water

Solution: The air-to-cloth ratio is one of the key design parameters for a variety of reasons, including the prevention of bag failure. The correct answer is therefore (b).

12.12 Filter Drag Definition

Filter drag for fabric filters is defined as the filter drag across the fabric–dust layer and is a function of the

(**a**) Quantity of dust accumulated on the filter
(**b**) Resistance to flowing air from the filter
(**c**) Zone of cake repair
(**d**) Force opposing filtration.

Solution: The definition of filter drag can be drawn directly from the two terms *filter* and *drag*. *Drag* implies resistance, and in this case it applies to the filter. The correct answer is therefore (b).

12.13 Baghouse Advantages

List some of the advantages of a baghouse.

Solution

1. Highest efficiency of all particulate control devices
2. Moderate capital cost

3. Moderate operating cost
4. Not affected by dust resistivity
5. Not affected by inlet loading temperature limitations
6. Not significantly affected by flowrate
7. Dry capture

12.14 Baghouse Disadvantages

List some of the disadvantages of a baghouse

Solution:

1. Problems at high temperatures
2. Bag failure
3. Affected by moisture
4. Explosion potential
5. Space requirements

12.15 Gas velocity

The gas velocity in a baghouse has been measured to be 2.63 ft/min during steady-state operation. Covert this velocity to cm/s.

(a) 1.34
(b) 1.10
(c) 1.23
(d) 0.96

Solution: Apply a dimensional analysis to the velocity:

$$\begin{aligned} v &= \left(2.63 \frac{\text{ft}}{\text{min}}\right)\left(\frac{\text{min}}{60\ \text{s}}\right)\left(\frac{30.48\ \text{cm}}{\text{ft}}\right) \\ &= 1.336\ \text{cm/s} \end{aligned}$$

The correct answer is therefore (a).

12.16 Air-to-Cloth Ratio Units

Typical units describing the air-to-cloth ratio are

(a) cfm/ft min
(b) cfm/ft^2
(c) ft^3/ft^2
(d) cfm/ft

Solution: The air-to-cloth ratio is a velocity term and has the units of velocity. The term

$$\left(\frac{\text{ft}^3}{\text{min}}\right) \Big/ (\text{ft}^2)$$

has the units of ft/min. The correct answer is therefore (b).

12.17 Air-to-Cloth Ratio Calculations

A plant has an inlet loading into a baghouse of 10 gr/ft^3. The average filtration velocity is 10 ft/min, and the gas flow rate is 25,000 acfm. What is the air-to-cloth ratio of the system?

(a) 250 ft/min
(b) 10 cfm/ft^2
(c) 2500 ft^2/min
(d) 5 ft/min

Solution: As demonstrated in Problem 12.16, 10 cfm/ft^2 is equal to 10 ft/min. The correct answer is therefore (b).

12.18 Baghouse Area Requirement

Calculate the fabric area, in square feet, required in a baghouse treating 230,000 acfm of particulate-laden gas at an efficiency of 99.87%; the unit operates at an air-to-cloth ratio of 2.3 ft/min.

(a) 143,478
(b) 121,053
(c) 150,000
(d) 100,000

Solution: This area is given by

$$\begin{aligned} A &= 230{,}000/2.3 \\ &= 100{,}000\ \text{ft}^2 \end{aligned}$$

The correct answer is therefore (d).

12.19 Bag Area Requirement

How many cylindrical bags, 6 inches in diameter and 25 ft long, would be needed to filter a particulate-laden gas stream if the total filtering surface area is 4045 ft^2?

(a) 300
(b) 162
(c) 15
(d) 103

Solution: The area of the bag (neglecting the top or bottom area) is given by

$$A = \pi DH$$

Substituting gives

$$\begin{aligned} A &= (\pi)\left(\frac{6}{12}\right)(25) \\ &= 39.25\ \text{ft}^2 \end{aligned}$$

The number of bags is then

$$N = 4045/39.25$$
$$= 103 \text{ bags}$$

The correct answer is therefore (d).

12.20 Bag Requirement

Tests showed that filtration of a dusty air stream containing 2 grains of particulate matter per cubic foot of air gave a maximum pressure drop of 10 inches of water at a flow rate of 3 ft^3/min per square foot of filtering surface. If the exhaust volume rate is 10,000 acfm, what is the number of 1 ft (diameter) by 20 ft (length) filtering bags required?

(**a**) 45 bags
(**b**) 53 bags
(**c**) 100 bags
(**d**) 65 bags

Solution: This is a modification of Problem 12.19. The bag area is

$$A = \pi DH$$
$$= (\pi)(1)(20)$$
$$= 62.8 \text{ ft}^2$$

The area required is

$$A_{\text{req}} = 10{,}000/3$$
$$= 3333 \text{ ft}^2$$

The number of bags is

$$N = A_{\text{reg}}/A$$
$$= 3333/62.8$$
$$= 53$$

The correct answer is therefore (b).

12.21 Pressure Drop Equation

The pressure drop across the baghouse can be calculated by

(**a**) $\Delta P = \Delta P_{\text{filter}} + \Delta P_{\text{cake}}$
(**b**) $\Delta P = QA$
(**c**) $\Delta P = (K_3 - 1)K_3$
(**d**) $\Delta P = S/v_t$

Solution: The pressure drop across a baghouse is given by the sum of the pressure drops across the filter and cake. The correct answer is therefore (a).

12.22 Effect of Velocity on Pressure Drop

The pressure drop through a filter is 2.5 in H_2O with a filter velocity of 3 ft/min. If the velocity dropped to 2.7 ft/min, what is the pressure drop in inches of H_2O assuming that the filter drag remains constant?

(a) 1.67
(b) 1.33
(c) 3.75
(d) 2.25

Solution: For laminar flow (Darcy's law), the pressure drop is proportional to the product of the resistance and the velocity. If the resistance remains constant, and the velocity is decreased by 10% (from 3.0 to 2.7), the pressure drop would correspondingly decrease by 10%. Thus

$$\begin{aligned}\Delta P_{\text{new}} &= \Delta P - 0.1\Delta P \\ &= 0.9\,\Delta P \\ &= (0.9)(2.5) \\ &= 2.25 \text{ in } H_2O\end{aligned}$$

The correct answer is therefore (d).

12.23 Efficiency–Penetration Relationship

What is the efficiency of a baghouse if the fractional overall penetration is 0.005?

(a) 20%
(b) 98%
(c) 2.0%
(d) 99.5%

Solution: By definition

$$P = 1 - E$$

Rearranging and substituting, one obtains

$$\begin{aligned}E &= 1 - 0.005 \\ &= 0.995 \\ &= 99.5\%\end{aligned}$$

The correct answer is therefore (d).

12.24 Particle Mass Collection Calculation

The dimensions of a bag in a filter unit are 8 inches in diameter and 15 feet long. Calculate the filtering area of the bag. The filtering unit consists of 40 such bags and is to treat 480,000 ft^3/hr of gas from an open-hearth furnace. Calculate the "effective" filtration velocity in feet per minute and acfm per square foot of filter area. Also calculate the mass of particles collected daily assuming that the inlet loading is 3.1 gr/ft^3 and the unit operates at 99.99+% collection efficiency.

Solution: Assume once again the bag to be cylindrical in shape with diameter D and height H. The total area of the bag is (including the flat top)

$$\begin{aligned} A &= A_{\text{curved surface}} + A_{\text{flat top}} \\ &= \pi DH + \pi D^2/4 \\ &= \pi\left(\tfrac{8}{12}\right)(15) + \pi\left(\tfrac{8}{12}\right)^2\Big/4 \\ &= 31.43 + 0.34 \\ &= 31.77\ \text{ft}^2 \end{aligned}$$

Obviously, the flat-top area can be safely neglected. The total area for 40 bags is

$$A = (40)(31.77) = 1271\ \text{ft}^2$$

The filter velocity is then

$$\begin{aligned} v &= \frac{q}{A} = \frac{(480{,}000/60)}{1271} \\ &= 6.30\ \text{ft/min} \end{aligned}$$

The calculation for acfm per square foot of filter area is, of course, the same. Assuming 100% collection efficiency, the mass collected daily is

$$\begin{aligned} \dot{m} = qc_i &= (480{,}000)(24)(3.1)/7000 \\ &= 5102\ \text{lb/day} \end{aligned}$$

Note once again that 7000 gr = 1 lb.

12.25 Number of Bags, Pressure Drop, and Cleaning Frequency

A calcium hydroxide plant is required to treat the exhaust "fume" generated from the plant. The ash generated from the system is collected at the bottom of the baghouse. The exhaust gas flow of 350,000 acfm enters the baghouse with a loading of 6.0 gr/ft^3. The air-to-cloth ratio is 8.0, and the operating particulate collection efficiency is 99.3%. The maximum allowable pressure drop is 10 in H_2O. The contractor's empirical equation for the pressure drop is given by

$$\Delta P = 0.3v + 4.0cv^2t$$

where ΔP = pressure drop in inches of water
v = filtration velocity in ft/min
c = dust concentration in lb/ft^3 of gas
t = time in minutes since bags were cleaned

(a) How many cylindrical bags, 12 inches in diameter and 30 ft high, will be needed?

(**b**) The system is designed to begin cleaning when the pressure drop reaches 10.0 in H_2O, its maximum allowable value. How frequently should the bags be cleaned?

Solution: (**a**) Calculate the total required surface area A of the bags with an air-to-cloth ratio of 8.0:

$$\begin{aligned} A &= (\text{volumetric gas flow rate})(\text{filtration velocity}) \\ &= 350{,}000/8 \\ &= 43{,}750\ \text{ft}^2 \end{aligned}$$

Calculate the surface area of each bag a and the number of the bags required N:

$$\begin{aligned} a &= \pi DH + \pi D^2/4 \\ &= (\pi)(12/12)(30) + (\pi)(12/12)^2/4 \\ &= 95\ \text{ft}^2 \\ N &= A/a \\ &= 43{,}750/95 \\ &= 461\ \text{bags} \end{aligned}$$

(**b**) Solve the pressure drop equation explicitly for the time:

$$\Delta P = 0.3v + 4cv^2 t$$

$$t = (\Delta P - 0.3v)/4cv^2$$

The concentration c is given by

$$\begin{aligned} c &= 6.0/7000 \\ &= 8.57 \times 10^{-4}\ \text{lb/ft}^3 \end{aligned}$$

Solving for the time yields

$$\begin{aligned} t &= \frac{10 - 0.3(8)}{(4)(8.57 \times 10^{-4})(8)^2} \\ &= 34.6\ \text{min} \end{aligned}$$

12.26 Time to Cleaning

A small fabric filter system with four bags is used to clean a gas stream containing a dust load of 10 gr/ft^3 with a cake bulk density of 30.0 lb/ft^3. If the system must be cleaned when there is 1.125 inches of cake on the bags in order to retain its near 100% efficiency, what is the filtering time allowed between each cleaning?

Given: Bag diameter = 12 in

Bag height = 10 ft

Flow rate = 4000 ft^3/min

Solution: Neglecting the top area of the bag

$$A = (\pi)(12)\left(\tfrac{1}{12}\right)(10\,\text{ft})(4) = 125.7\,\text{ft}^2$$
$$= 18{,}100\ \text{in}^2$$

At 100% efficiency

$$10\,\text{gr/ft}^3 = 0.00143\,\text{lb/ft}^3$$

so that

$$\dot{m} = (0.00143)(4{,}000\,\text{ft}^3/\text{min}) = 5.72\,\text{lb/min}$$

Convert the units of the bulk density,

$$\rho_B = 30.0/1728$$
$$= 0.01736\,\text{lb/in}^3$$

The maximum mass of particulate collected per 4 bags is

$$m_p = (18{,}100)(0.01736)(1.125)$$
$$= 353\,\text{lb}$$

The time is then

$$t = 353/5.72$$
$$= 61.8\ \text{min or 1 hr}$$

12.27 Effect of Bag Cleaning Frequency on Efficiency

Laboratory tests performed on a filter bag 1 ft × 20 ft indicated that the bag operated at an efficiency of 99.7% with an average outlet dust loading of 0.03 gr/ft^3 over 15 min of operation. The dust-laden gas had a particle bulk density of 10 lb/ft^3 and a volumetric flow rate of 500 acfm. Calculate the thickness of the solids deposited on the bag. How thick would the buildup be after 25 min? After 25 min the average outlet loading was 0.02 gr/ft^3. Compare the changes in efficiency, outlet loading, and solids buildup between 15 and 25 min.

Solution: The area of a bag is

$$A = \pi DH$$
$$= \pi(1)(20)$$
$$= 62.83\ \text{ft}^2$$

The inlet concentration is

$$\begin{aligned} c_i &= c_O/(1 - E) \\ &= 0.03/(1 - 0.997) \\ &= 10\,\text{gr/ft}^3 \end{aligned}$$

A simplified equation is employed to calculate the cake thickness on a bag:

$$\begin{aligned} \Delta z &= qc_it/\rho_B A \\ &= (500)(10)(15)(12)/(10)(62.838)(7000) \\ &= 0.205\text{ in for 15 min} \end{aligned} \tag{12.8}$$

For 25 minutes

$$\begin{aligned} \Delta z &= 0.205(25/15) \\ &= 0.342\text{ min} \end{aligned}$$

The result are presented in Table 12.3.

TABLE 12.3 Cake Thickness Results

Parameter	15 min	25 min	% Change
E, %	99.7	99.8	0.1
OL, gr/ft^3	0.03	0.02	33.3
Δz, in	0.205	0.342	66.8

12.28 Bag Failure

A baghouse has been used to clean a particulate gas steam for nearly 30 years. There are 600 8-in diameter bags in the unit and 50,000 acfm of dirty gas at 250°F enters the baghouse with a loading of 5.0 gr/ft^3. The outlet loading is 0.03 gr/ft^3. Local Environmental Protection Agency (EPA) regulations state that the outlet loading should not exceed 0.40 gr/ft^3. If the system operates at a pressure drop of 6.0 in H_2O, how many bags can fail before the unit is out of compliance? The Theodore–Reynolds equation applies and all contaminated gas emitted through the broken bags may be assumed the same as that passing through the tube sheet thimble. As described in Section 12.2, the effect of bag failure on baghouse efficiency can be described by the following equations:

$$P_t^* = P_t + P_{tc} \tag{12.7}$$

$$P_{tc} = \frac{0.528(\Delta P)^{0.5}}{\phi}$$

$$\phi = \frac{q}{LD^2(T + 460)^{0.5}}$$

where $P_t^* =$ penetration after bag failure
$P_t =$ penetration before bag failure
$P_{tc} =$ penetration correction term; contribution of broken bags to P_t^*
$\Delta P =$ pressure drop, in H_2O
$\phi =$ dimensional parameter
$q =$ volumetric flow rate of contaminated gas, acfm
$L =$ number of broken bags
$D =$ bag diameter, in
$T =$ temperature, °F

For a detailed development of the equation above, refer to L. Theodore and J. Reynolds, "Effect of Bag Failure on Baghouse Outlet Loading," *J. Air Pollut. Control Assoc.* pp. 870–872 (Aug. 1979). Additional information can be found in L. Theodore, "Engineering Calculations: Baghouse Specification and Operation Simplified," *Chem. Eng. Progress* p. 22 (July 2005).

Solution: Calculate the efficiency E and penetration P_t before the bag failure(s):

$$\begin{aligned} E &= [(\text{inlet loading}) - (\text{outlet loading})]/(\text{inlet loading}) \\ &= (5.0 - 0.03)/(5.0) \\ &= 0.9940 = 99.40\% \\ P_t &= 1 - 0.9940 \\ &= 0.0060 = 0.60\% \end{aligned}$$

The efficiency and penetration P_t^* based on regulatory conditions are

$$\begin{aligned} E &= (5.0 - 0.4)/5.0 \\ &= 0.920 = 92.0\% \\ P_t^* &= 1 - 0.920 \\ &= 0.0800 = 8.00\% \end{aligned}$$

The penetration term P_{tc} associated with the failed bags is then

$$\begin{aligned} P_{tc} &= 0.0800 - 0.0060 \\ &= 0.0740 \end{aligned}$$

Write the equation(s) for P_{tc} in terms of the failed number of bags L. Since

$$P_{tc} = \frac{0.528(\Delta P)^{0.5}}{\phi} \tag{12.7}$$

and

$$\phi = \frac{q}{LD^2(T + 460)^{0.5}}$$

then

$$L = \frac{qP_{\text{tc}}}{(0.582)\Delta P^{0.5}D^2(T+460)^{0.5}}$$

The number of bag failures that the system can tolerate and still remain in compliance is now calculated:

$$L = \frac{(50{,}000)(0.074)}{(0.582)(6)^{0.5}(8)^2(250+460)^{0.5}}$$

$$= 1.52$$

Thus, if two bags fail, the baghouse is out of compliance.

12.29 Bag Failure Factors

Discuss some of the factors that affect bag failure.

Solution: It is important to note that each bag in a set may have a different life as a result of fabric quality, bag manufacturing tolerances, location in the collector, and variation in the bag cleaning mechanism. Any one or a combination of these factors can cause bags to fail. This means that a baghouse will experience a series of intermittent bag failures until the failure rate requires total bag replacement. Typically, a few bags will fail initially or after a short period of operation as a result of installation damage or manufacturing defects. The failure rate should then remain very low until the operating life of the bags is approached, unless a unique failure mode is present within the system. The failure then increases, normally at a near-exponential rate. Industry often describes this type of failure rate behavior as a "bathtub" curve. Statistically, it is defined as a *Weibull distribution*. The reader is referred to S. Shaefer and L. Theodore, *Probability and Statistics Applications For Environmental Science*, CRC Press, Boca Raton, FL, 2007 for detailed information on system accidents/failures. The proper time to replace a broken bag depends on the type of collector and the resultant effect on outlet emissions.

12.30 Frequency of Bag Failure

As a recently assigned plant engineer, you are asked to troubleshoot the plant's baghouse. The baghouse in question is used to collect the dust created in the manufature of an extremely expensive drug. The dust is collected and recycled into the main process. Over the past 6 months (since the baghouse was installed) the amount of dust collected has dropped off significantly without any change in the inlet loading. Since the baghouse is operated on a round-the-clock basis, i.e., 24 hr per day, 7 days per week, the bags (for this unit) have not been inspected to find the problem. The following data have collected:

Flow rate = 60,000 acfm (60°F)
Dust loading = 6.00 gr/ft^3
Number of bags = 500
Diameter of bags = 5.0 inches

Pressure drop = 9.1 in H_2O
See Table 12.4 for additional data.

TABLE 12.4 Dust Collected; Monthly Basis

Months of Operation	Amount Collected, lb/hr
New	3054.9
1	2900.4
2	2808
3	2653.8
4	2530.2
5	2376
6	1789.8

Having recently attended a class on the effects of bag failures given by the foremost authority in this field (whom you are already aware of), you are asked to determine whether the loss has been caused by broken bags, and, if so, how many have broken every month.

Solution: First calculate the inlet loading (IL):

$$\begin{aligned} \text{IL} &= (60{,}000)(6.0)(60)/(7000) \\ &= 3086\,\text{lb/hr} \end{aligned}$$

The calculations of the initial efficiency E and penetration P follow:

$$\begin{aligned} E &= 3054.9/3086 = 0.99 = 99\% \\ P &- 1 - 0.99 = 0.01 = 1.0\% \end{aligned}$$

After the first month

$$\begin{aligned} E^* &= 2900/3086 = 0.94 = 94\% \\ P^* &= 1 - 0.94 = 0.06 = 6\% \end{aligned}$$

(* indicates that time has passed since the baghouse installation). The amount of increase in the penetration P_c is

$$P_c = 0.06 - 0.01 = 0.05$$

Using Equation (12.7), one obtains

$$\phi = 35.1 = \frac{q}{LD^2(t + 460)^{1/2}}$$

$$L = \frac{q}{\phi D^2(t + 460)^{1/2}} = \frac{60{,}000}{(35.1)(5)^2(520)^{1/2}} = 2.998 \approx 3$$

Similarly, see Table 12.5 for the remaining months.

TABLE 12.5 Bag Failure Results

Month	P^*, %	P_c %	ϕ	L	ΔL
1	6	5	35.1	3	3
2	9	8	21	5	2
3	14	13	15	7	2
4	18	17	10.5	10	3
5	23	22	7.5	14	4
6	42	41	4.2	25	11

The baghouse problem appears to be bag failure. Note that the efficiency drops below 90% after just 2 months. The reader should consider whether the bag failure distribution with respect to time is reasonable. Refer to the discussion of the Weibull distribution in Problem 12.29.

12.31 Process Modification for Bag Failures

Having just completed the world-famous Manhattan College Air Pollution Course, you have been hired to design a baghouse for treating gypsum.

(a) Given the following information, determine the number of bags needed, the time for cleaning, and the outlet loading:

$q = 150{,}000$ acfm
$E = 99\%$
Inlet loading $= 10$ gr/ft^3
$\Delta P = 10$ in H_2O
$T = 60^\circ$F
A/C ratio $= 2.0$
Bag diameter $= 12$ in
Bag length $= 20$ ft

Pressure drop relationship:

$$\Delta P = 0.2v + 5c_i v^2 t$$

with ΔP in inches of H_2O, v in ft/min, c_i lb/ft^3, and t in min.

(b) The unit is installed and running like a charm until suddenly the efficiency drops. Thorough inspection of the unit reveals that 10 bags are ripped to shreads. At that moment Johnathan Smog, famous EPA shutdown man, informs the gypsum company that he is making an inspection of their unit, and if it doesn't meet standards, he will shut the operation down. A quick call to the Acme Bag Company reveals that replacement bags are not immediately available. You are called to determine the efficiency and outlet loading with 10 bags out.

Solution

(**a**) The total bag area requirement is

$$A = 150{,}000/2$$
$$= 75{,}000\ \text{ft}^2$$

The area per bag is

$$a = \pi DL$$
$$= \pi(1)(20)$$
$$= 62.83\ \text{ft}^2$$

The number of bags is therefore

$$n = 75{,}000/62.83$$
$$= 1193\ \text{bags}$$

Rearranging the pressure drop equation allows one to solve for the time:

$$t = [\Delta P - (0.2)(v)]/5c_i v^2$$
$$= 10 - (0.2)(2)/(5)(10/7000)(2)^2$$
$$= 336\ \text{min}$$

Since the outlet loading is

$$\text{OL} = (1 - 0.99)(10)$$
$$= 0.1\ \text{gr/ft}^3$$

the daily outlet discharge becomes

$$\text{DOD} = (0.1)(150{,}000)(1440)/7000$$
$$= 3086\ \text{lb/day}$$

(**b**) The parameter ϕ is first calculated for 10 bags

$$\phi = \frac{q}{LD^2(T + 460)^{0.5}} \quad (12.7)$$

$$= \frac{150{,}000}{(10)(12)^2(60 + 460)^{0.5}} = 4.5$$

$$P_c = 0.582(\Delta P)^{0.5} = 0.582(10)^{0.5}/4.5$$
$$= 0.409$$

Therefore

$$P = 0.01 + 0.409$$
$$= 0.419$$

and

$$E = 0.581 = 58.1\%$$

The revised outlet loading may now be calculated:

$$\text{OL} = (1 - 0.581)(10)$$
$$= 4.19\ \text{gr/ft}^3$$
$$\text{DOL} = 129{,}300\ \text{lb/day}$$

The bags should be replaced immediately to bring the efficiency back to the design level.

12.32 Bag Failure Associated with a Specified Operating Efficiency

(**a**) Determine the outlet loading (lb/min) from a 200°F operation equipped with a fabric filter baghouse system. The design cleaning frequency of this system is 25 min, and the maximum pressure drop across the unit is 8 in H_2O. Additional data provided below.

Inlet concentration of particulates = 4.0 gr/ft^3
Baghouse efficiency = 98% (normal/design)
Number of bags = 100
Bag diameter = 6 in
Bag length = 20 ft
$\Delta P = 0.2v_f + 5cv_{ft}^2$; units consistent with previous problem

(**b**) If 3 bags fail in this system, what is the new outlet loading?

(**c**) Outline how to calculate the number of bag failures associated with a specified operating efficiency below the normal/design value.

Solution

(**a**) Since

$$E = 0.98$$

the outlet loading is

$$\text{OL} = (0.02)(4)$$
$$= 0.08\ \text{gr/ft}^3$$
$$= (0.08\ \text{gr/ft}^3)(30{,}000\ \text{ft}^3/\text{min})(\text{lb}/7000\ \text{gr})$$
$$= 0.343\ \text{lb/min}$$

(**b**) Apply Equation (12.7):

$$P_t = 0.02$$
$$\Delta P = 8.0$$
$$\phi = 10.8 \text{ (for 3 bag failures)}$$

Therefore

$$P_{tc} = (0.582)(8.0)^{0.5}/10.8$$
$$= 0.152$$

and

$$P_t^* = P_{tc} + P_t$$
$$= 0.152 + 0.02$$
$$= 0.172$$
$$E = 1 - 0.172$$
$$= 0.828$$

The outlet loading with three bag failures is

$$OL_3 = (0.343)(0.172/0.02)$$
$$= 2.95 \text{ lb/min}$$

(**c**) Basically, the calculation should be reversed. From the desired efficiency, generate P_{tc} by subtracting 0.02. Employ the same equation as above and calculate the number of failed bags. Repeat the calculation for other E values.

12.33 Bag Failure Location(s)

Consider two similar baghouses connected in series. Discuss the effect of the location of the bag failures (whether it is the first or second baghouse) on the overall efficiency.

Solution: Consider limiting and/or extreme conditions:

1. If zero bags fail: no difference
2. If all bags fail: no difference
3. If one bag fails: no difference since

$$P_{\text{overall}} = P_1 P_2$$

Since the effect of either P_1 or P_2 will be essentially the same, the order does not matter.

12.34 Maximum Allowable Bag Failure

LT Industries owns and operates a baghouse system consisting of one compartment with 100 bags. The bags are 4 inches in diameter, and the pressure drop across the system is 7.0 in H_2O. The operating temperature and pressure are 70°F and 1 atm, respectively. The inlet load to the baghouse is 4.0 gr/ft^3, and the system is 99.5% efficient, assuming that all bags are completely functional. The filtering area is 5500 ft^2 and the filtering velocity is 420 ft/hr.

(**a**) Calculate the efficiency assuming that 3 bags fail.

(**b**) What is the maximum number of bag failures that can be tolerated to ensure a minimum coeffection efficiency of 91.50%?

Solution

(**a**) Once again, employ Equation (12.7):

$$\begin{aligned} P_t &= 1 - E \\ &= 1 - 0.995 \\ &= 0.005(\text{before failure}) \\ q &= (A)(v) = (5500)(7) \\ &= 38{,}500 \text{ ft}^3/\text{min} \\ \phi &= q/LD^2(T + 460)^{0.5} \\ &= 38{,}500/(3)(4)^2(70 + 460)^{0.5} \\ &= 34.84 \\ P_{tc} &= (0.582)(\Delta P)^{0.5}/\phi \\ &= (0.582)(7.0)^{0.5}/34.84 \\ &= 0.0442 \end{aligned}$$

and

$$\begin{aligned} P_t^* &= P_t + P_{tc} \\ &= 0.005 + 0.0442 \\ &= 0.0492 \\ E^* &= 1 - P_t^* \\ &= 0.951 = 95.1\% \end{aligned}$$

(**b**) For

$$E^* = 0.915 = 91.5\%$$

$$P^* = 1 - 0.915 = 0.085$$

The calculation in part (a) is now reversed.

$$P_{tc} = P_t^* - P_t$$
$$= 0.085 - 0.005$$
$$= 0.08$$

$$\phi = (0.582)(\Delta P)^{0.5}/P_{tc} = (0.582)(7)^{0.5}/0.08 = 19.25$$

One can now solve directly for the number of bag failures:

$$19.25 = 38{,}500/(L)(4)^2(70 + 460)^{0.5}$$
$$L = 5.43$$

Therefore, a maximum of five bags can fail before the efficiency drops below 91.50%.

12.35 Data Inconsistency

There is a major inconsistency in the information provided in the problem statement of Problem 12.34. Indicate what you believe to be wrong.

Solution: If H = bag height one may write

$$(100)(H)(\pi)\left(\frac{4}{12}\right) = 5500$$
$$H = 52.5 \text{ ft}$$

This is an unreasonable bag height. Generally, the recommended maximum height is 36 ft (not a reasonable height).

12.36 A Baghouse and a Scrubber in Series

An extremely high collection efficiency is desired for a chemical process that produces an exhaust containing toxic particles. Because the company in question has a large budget to prevent toxic emissions, they want two APCE devices to be placed in series to clean the exhaust gas. One design that has been proposed is the use of a baghouse followed by a scrubber. The following data for the proposed system are provided:

Operating conditions = 68°F, 1 atm
Inlet loading to baghouse = 7.0 gr/ft^3
Density of dust = 200 lb/ft^3
Pressure drop across baghouse = 8 in H_2O
Volumetric flow rate = 20,000 acfm
Liquid-to-gas ratio (for scrubber) = 2 gal/1000 ft^3
Bag diameter = 6 in
Baghouse efficiency = 97%

Average particle size = 3.2 μm
Water droplet size = 100 μm
Scrubber coefficient = 0.15 (assume Johnstone's equation to apply)
Viscosity of gas at operating conditions = 1.23×10^{-5} lb/ft$^3 \cdot$ s
Area at throat of scrubber = 1.0 ft^2

Calculate the collection efficiency and outlet loading of the baghouse and scrubber (combined).

Solution: For the scrubber (see Chapter 11):

$$E_S = 1 - e^{-k(q_L/q_G)\sqrt{\psi}}$$

$$\psi = C\rho_p v d_p^2/9\mu d_0$$

$$v = 20{,}000/(60)(1) = 333.3 \text{ ft/s}$$

$$d_p = 3.2\ \mu\text{m} = 1.05 \times 10^{-5} \text{ ft}$$

$$d_0 = 100\ \mu\text{m} = 3.28 \times 10^{-4} \text{ ft}$$

Substituting yields

$$\psi = \frac{(1)(200)(333.3)(1.05 \times 10^{-5})^2}{(9)(3.28 \times 10^{-4})(1.23 \times 10^{-5})} = 202.4$$

Thus

$$E_S = 1 - e^{-(0.15)(2)(202.4)^{0.5}} = 1 - 0.014 = 0.986 = 98.6\%$$

The combined efficiency of both units is then

$$E_C = 1 - (1 - E_B)(1 - E_S) = 1 - (1 - 0.97)(1 - 0.986) = 0.99958 = 99.958\%$$

The outlet loading is

$$\text{OL} = (7.0)(0.00042) = 0.00294 \text{ gr/ft}^3$$

12.37 Combined System with Two Bag Failures

Refer to Problem 12.36. Calculate the collection efficiency and outlet loading of the combined system after two bag failures.

Solution: For two bag failures

$$\begin{aligned}\phi &= 20{,}000/(2)(6)^2(68 + 460)^{0.5} \\ &= 12.1\end{aligned}$$

Continuing with Equation (12.7), one obtains

$$\begin{aligned}P_{\text{tc}} &= 0.582(8)^{0.5}/12.1 \\ &= 0.136\end{aligned}$$

$$P_{\text{t}}^{*} = 0.03 + 0.136 = 0.166$$

For the baghouse

$$\begin{aligned}E_{\text{B}} &= 1 - 0.166 \\ &= 0.834\end{aligned}$$

The new combined efficiency is then

$$\begin{aligned}E_{\text{C}} &= 1 - (1 - 0.834)(0.014) \\ &= 0.9977 = 99.77\%\end{aligned}$$

The revised outlet loading is

$$\begin{aligned}\text{OL}_{\text{new}} &= (7)(0.0023) \\ &= 0.0161 \text{ gr/ft}^3\end{aligned}$$

12.38 Normally Distributed Bag Failures

The lifetime T of a bag employed in a baghouse is normally distributed with a mean of 2500 days. What is the largest lifetime standard deviation that the installed bags can have if 95% of them need to last at least 365 days?

Solution: For this application (see Chapter 7) and note that

$$P(T > 365) = 0.95$$

Normalizing gives

$$P\left(\left[\frac{T - \mu}{\sigma}\right] > \left[\frac{356 - 2500}{\sigma}\right]\right) = 0.95$$

$$P\left(Z > \frac{-2135}{\sigma}\right) = -1.645$$

The following equation must apply for this condition:

$$-\frac{2135}{\sigma} = -1.645$$

Solving for the standard deviation, one obtains

$$\sigma = 1298\ \text{days}$$

12.39 Overall Collection Efficiency

Calculate the overall efficiency of N compartments in a baghouse operating in parallel, assuming that the volumetric flow rates and inlet concentrations to each compartment are $q_1, q_2, \ldots, q_N$ and $c_1, c_2, \ldots, c_N$, respectively, and the corresponding efficiencies are $E_1, E_2, \ldots, E_N$.

Calculate the overall efficiency of a baghouse consisting of three compartments treating 9000 acfm of gas with an inlet loading of 4.0 gr/ft^3. The first and third compartments operate at a fractional efficiency of 0.995 while the second compartment operates at a fractional efficiency of 0.990. What is the overall efficiency of the baghouse if the flow and inlet concentration are evenly distributed? Also calculate the efficiency if the flow distribution given in Table 12.6 exists. Express the result in terms of the qs, the Es, and the cs.

TABLE 12.6 Baghouse Compartment Data

Compartment	q, acfm	c, gr/ft^3	E
1	2500	3.8	0.995
2	4000	4.25	0.990
3	2500	3.8	0.995

Solution: Write the equation for the outlet concentration c_{1_o} from module 1 in terms of c_1 and E_1

$$E_1 = 1 - (c_{1o}/c_1)$$

$$c_{1o} = c_1(1 - E_1)$$

The equation for the inlet mass flowrate to module $\dot{m}_1$ is

$$\dot{m}_1 = c_1 q_1$$

The equation for the outlet mass flowrate $\dot{m}_{1_o}$ from module 1 is

$$\dot{m}_{1o} = c_1(1 - E_1)q_1$$

For module i

$$E_i = 1 - (c_{io}/c_i)$$

$$c_{io} = c_i(1 - E_i)$$

$$\dot{m}_i = c_i q_i$$

$$\dot{m}_{io} = c_i(1 - E_i)q_i$$

The equation for the overall efficiency E for modules 1, 2, ..., N is then

$$E = 1 - \frac{\sum \dot{m}_{io}}{\sum \dot{m}_i}$$
$$= 1 - \frac{c_1(1 - E_1)q_1 + c_2(1 - E_2)q_2 + \cdots + c_N(1 - E_N)q_N}{c_1q_1 + c_2q_2 + \cdots + c_Nq_N}$$
$$= 1 - \frac{\sum c_i(1 - E_i)q_i}{\sum c_iq_i} \qquad (12.9)$$

The companion equation for the penetration P is

$$P = \frac{c_1P_1q_1 + c_2P_2q_2 + \cdots + c_NP_Nq_N}{c_1q_1 + c_2q_2 + \cdots + c_Nq_N}$$
$$= \frac{\sum c_iP_iq_i}{\sum c_iq_i} \qquad (12.10)$$

If the inlet concentrations c to each module are equal, i.e.,

$$c_1 = c_2 = \cdots = c_N = c$$

the c terms can be factored out from the above equation for the efficiency to yield

$$E = 1 - \frac{(1 - E_1)q_1 + (1 - E_2)q_2 + \cdots + (1 - E_N)q_N}{q_1 + q_2 + \cdots + q_N}$$

Note that the total volumetric flow rate q is given by

$$q = q_1 + q_2 + \cdots + q_N$$

Therefore

$$E = 1 - \frac{(1 - E_1)q_1 + (1 - E_2)q_2 + \cdots + (1 - E_N)q_N}{q}$$
$$= 1 - \frac{\sum (1 - E_i)q_i}{q} \qquad (12.11)$$

Equivalently

$$P = 1 - \frac{\sum q_iP_i}{q} \qquad (12.12)$$

Using the data provided in the problem statement, calculate the efficiency for the situation where the flow is equally distributed with the same inlet loading. Since $q_i = 3000$ acfm for all modules, then

$$E = 1 - \frac{(2)(3000)(1 - 0.995) + (1)(3000)(1 - 0.99)}{9000}$$
$$= 0.9933 = 99.33\%$$

If neither the flow nor the concentration is uniformly distributed, the general equation for the efficiency in a compartmentalized baghouse is used:

$$E = 1 - \frac{\sum c_i(1 - E_i)q_i}{\sum c_i q_i} \tag{12.9}$$

Substituting (see data), one obtains

$$E = 1 - \frac{(2)(3.8)(2500)(1 - 0.995) + (1)(4.25)(4000)(1 - 0.99)}{36{,}000}$$

$$= 0.9926 = 99.26\%$$

Note that the penetration has increased by slightly over 10%.

12.40 Collection Efficiency Model

Consider the situation where 50,000 acfm of gas with a dust loading of 5.0 gr/ft^3 flows through a baghouse with an average filtration velocity of 10 ft/min. The pressure drop is given by

$$\Delta P = 0.20v + 5.0\, c_i v^2 t$$

Where ΔP = pressure drop, in H_2O
v = filtration velocity, ft/min
c_i = dust concentration, lb/ft^3 gas
t = time after bags were cleaned, min

The fan can maintain the volumetric flowrate up to a pressure drop of 5.0 inches of water. Show that the baghouse can be operated for 8.40 min between cleanings.

In an attempt to determine the efficiency of this unit at "terminal" conditions, both the fabric and deposited cake (individually) were subjected to laboratory experimentation. The following data were recorded (see Figure 12.3). Using the Theodore–Reynolds collection efficiency model, determine the overall efficiency at the start and end of a cleaning cycle.

Solution: Solve the equation for the pressure drop explicitly for t:

$$t = \frac{\Delta P - 0.2v}{5c_i v^2} = \frac{5.0 - 0.2(10)}{5(5/7000)(10)^2}$$

$$= 8.4\,\text{min}$$

Refer to Equation (12.4). Evaluate the parameter ψ with units of ft^{-1} and parameter ϕ with units of min^{-1}:

$$\ln(c_o/c_i) = -\psi L$$
$$\ln(0.1245/1.0) = -\psi(0.1/12)$$

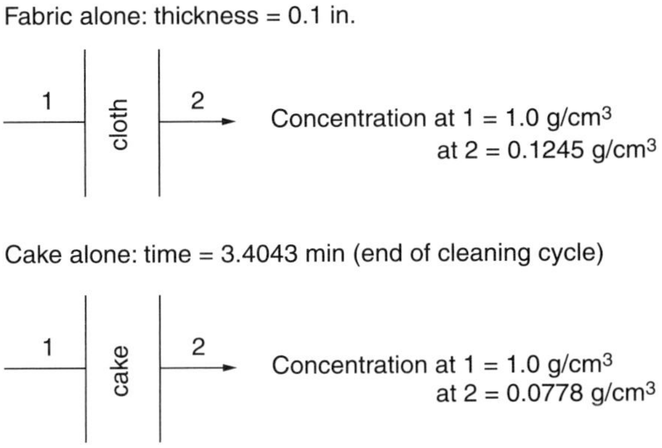

Fabric alone: thickness = 0.1 in.

Cake alone: time = 3.4043 min (end of cleaning cycle)

Figure 12.3 Cloth–cake efficiency data.

$$\begin{aligned}
\psi &= 250\,ft^{-1} \\
\ln(c_o/c_i) &= -\phi t \\
\ln(0.0778/1.0) &= -\phi(8.4) \\
\phi &= 0.304\,\text{min}^{-1}
\end{aligned}$$

Calculate the efficiency at the end of the filtering (cleaning) cycle:

$$\begin{aligned}
E &= 1 - e^{-\psi L}e^{-\phi t} \\
&= 1 - e^{-(250)(0.1/12)}e^{-(0.304)(8.4)} \\
&= 1 - (0.1245)(0.7780) \\
&= 1 - 0.0097 \\
&= 0.9903 \\
&= 99.03\% \text{ (at the end of cleaning cycle)}
\end{aligned}$$

Also calculate the efficiency at the start of the filtering (cleaning) cycle:

$$\begin{aligned}
E &= 1 - (0.1245)(1) \quad t = 0 \\
&= 0.8755 \\
&= 87.55\% \text{ (at start of cycle)}
\end{aligned}$$

12.41 Theodore Model Justification

Explain in your own words whether the Theodore bag collection efficiency model can and/or should be used to estimate the efficiency of a baghouse. Also, explain whether the Theodore *bag failure* model can and/or should be used to estimate the effect on the efficiency of a bag house.

Solution: The reader should note that ϕ and ψ in Equation (12.4) are modified inertial impaction numbers based on the cake and fabric, respectively. Unfortunately, this key equation has been totally ignored by responsible EPA individuals in this field. Taxpayers' dollars continue (for over a dozen years) to be provided to contractors whose research efforts have produced little, if any, usable results in developing quantitative equations to describe collection efficiency. Using the preceding model, it was also shown that the exit concentration (w_e) for the combined resistance system (the fiber and the cake) is

$$w_e = w_i e^{-(\psi L + \phi t)} \tag{12.5}$$

where w_e = exit concentration; units consistent with w_i
w_i = inlet concentration

On the other hand, the bag failure model has been used by industry to determine the effect of bag failure on efficiency. It was recently the subject of further study and some controversy. Details are available in the following two references:

1. W. Quin et al., "Production of Particulate Loading in Exhaust From Fabric Filter Baghouses with one or more Failed Bags," *J. Air Waste Manage. Assoc.* **56**: 1177–1183 (2006).
2. L. Theodore, Letter to the Editor, *J. Air Waste Manage. Assoc.* **56**: 1367 (2006).

12.42 Baghouses in Series

A chemical company has hired you to determine the efficiency of its baghouse under worst-case conditions consisting of three failed bags. The baghouse presently consists of two compartments with 50 bags in each compartment. You are also asked to determine whether the addition of a third compartment of 50 new bags *in series* with the first two compartments will have a marked effect on the collection efficiency under worst-case conditions, i.e., will the required outlet loading the met?

Data:

Required conditions
Inlet loading = 9.8 gr/ft^3
Outlet loading = 0.05 gr/ft^3
Pressure drop: $\Delta P = 3$ in H_2O
Volumetric flow: $q = 10{,}000$ ft^3/min
Temperature: $T = 90°$F
Bag diameter = 8 in
New bag data
Fabric thickness, L = 0.2 in
$\psi = 260\ (\text{ft})^{-1}$

Solution

Before bag failure

$$\begin{aligned} E &= (9.8 - 0.05)/9.8 \\ &= 0.99485 = 99.489\% \\ P_t &= 0.00511 = 0.511\% \end{aligned}$$

Calculate ϕ without the third compartment and three bags broken. Employ Equation (12.7):

$$\begin{aligned} \phi &= q/[LD^2(T + 460)^{1/2}] \\ &= 10{,}000/[(3)(8)^2(90 + 460)^{1/2}] \\ &= 2.22 \\ P_{tc} &= (0.582)(\Delta P)^{0.5}/\phi \\ &= (0.582)(3)/2.22 \\ &= 0.454 \\ P_t^* &= P_t + P_{tc} \\ &= 0.00511 + 0.454 \\ &= 0.4591 \\ E^* &= 0.5409 = 54.05\% \text{ (with 3 bag failures)} \end{aligned}$$

The new outlet loading becomes

$$0.5409 = (9.8 - \text{OL})/9.8$$

$$\text{OL} = 4.49 \text{ gr/ft}^3$$

The required efficiency for the third compartment (or the second unit) is

$$\begin{aligned} E_3 &= (4.49 - 0.05)/4.49 \\ &= 0.9888 = 98.88\% \end{aligned}$$

The minimum efficiency of the third unit is given by

$$\begin{aligned} E &= 1 - \exp(-\psi L) \\ &= 1 - \exp[(-260)(0.2/12)] \\ &= 1 - 0.0131 \\ &= 0.9869 = 98.69\% \end{aligned}$$

This is marginally close to the required efficiency of 98.88%.

12.43 Filter Bag Fabric Selection

It is proposed to install a pulse-jet fabric filter system to clean an air stream containing particulate pollutants. You are asked to select the most appropriate filter bag fabric considering performance and cost. Pertinent design and operating data, as well as fabric information, are given below (see also Table 12.7):

Volumetric flow rate of polluted air stream = 10,000 scfm (60°F, 1 atm)
Operating temperature = 250°F
Concentration of pollutants = 4.00 gr/ft^3
Average, ACR (G/C) = 2.5 cfm/ft^2 cloth
Collection efficiency requirement = 99%

TABLE 12.7 Fabric Information

Filter Bag	A	B	C	D
Tensile strength	Excellent	Above average	Fair	Excellent
Recommended maximum temperature, °F	260	275	260	220
Resistance factor	0.9	1.0	0.5	0.9
Cost per bag, $	26	38	10	20
Standard size	8 in × 16 ft	10 in × 16 ft	1 ft × 16 ft	1 ft × 20 ft

Note: No bag has an advantage from the standpoint of durability under the operating conditions for which the bag was designed.

Solution: Consider the bag types listed in Table 12.7. Bag D is eliminated since its recommended maximum temperature (220°F) is below the operating temperature of 250°F. Bag C is also eliminated since a pulse-jet fabric filter system requires the tensile strength of the bag to be at least above average. Consider the economics for the two remaining choices. The actual gas flow rate and filtration velocity are (applying Charles' Law)

$$q = 10{,}000\left(\frac{250 + 460}{60 + 460}\right)$$

$$= 13{,}654 \text{ acfm}$$

$$v = 2.5 \text{ cfm}/\text{ ft}^2 \text{ cloth}$$

$$= 2.5 \text{ ft min}$$

The filtering (bag) area is then

$$A = q/v$$

$$= 13{,}654/2.5$$

$$= 5{,}462 \text{ ft}^2$$

For bag A, the area and number N of bags are

$$a = \pi DH$$

$$= \pi\left(\frac{8}{12}\right)(16)$$

$$= 33.5 \text{ ft}^2$$

$$N = A/a = 5462/33.5 = 163$$

For bag B:

$$a = \pi\left(\frac{10}{12}\right)(16)$$

$$= 41.9 \text{ ft}^2$$

$$N = 5462/41.9 = 130$$

The cost per bag is \$26.00 for A and \$38.00 for B. The total cost (TC) for each bag is as follows:

For bag A:

$$TC_A = N(\text{cost per bag})$$

$$= (163)(26.00)$$

$$= \$4238$$

For bag B:

$$TC_B = (130)(38.00)$$

$$= \$4940$$

Since the total cost for bag A is less than bag B, select bag A.

12.44 Platinum Catalyst Application

Platinum is a catalyst used in the reaction of alkenes with elemental hydrogen to form (straight-chain) alkanes. However, in your plant, which produces *n*-propane from propene, 40 lb of catalyst are lost each year. Because the catalyst is a precious metal averaging about \$400 per ounce, operating costs in the plant could be reduced dramatically if the catalyst were captured. It is proposed to install a pulse-jet fabric system (baghouse) to recover most of that valuable platinum. The baghouse must handle 30,000 acfm of platinum-laden gas at 290°F with an overall efficiency of 99.50%. Nomax (nylon aromatic) bags were chosen as the best material for this application. However, they cost \$105.00 each. Bags are purchased in standard dimensions of 1.0 ft × 20.0 ft and normally operate with an air-to-cloth ratio of 3.25. The pressure drop is 1.5 in H_2O for the baghouse and 2.5 in H_2O for the remainder of the system. Annualized installed capital costs amount to \$3.35 per acfm. Other economic data are

Overall fan efficiency: 62.5%
Operating time: 7,000 hr/yr
Electrical power: 0.18 \$/kW · hr

Annual maintenance cost: \$3800 per year plus replacement of 20% of the bags

Will the first year's annual savings in terms of platinum recovery outweigh the first year's combined annual operating, capital, and maintenance costs?

Solution: Assume that all the platinum is recovered if the baghouse is installed. Base the calculations on one year.

$$\text{Profit}: P = (40)(400)(16) = \$256{,}000$$

Cost considerations include

1. Initial bag cost, BC
2. Replacement bag cost, BRC
3. Maintenance cost, MC
4. Installed cost, IC
5. Fan (operating) cost, FC

All other costs, including labor, land, etc., are neglected.

1. Bag area:

$$a = \pi DH = (\pi)(1)(20) = 62.8\ \text{ft}^2$$

Bag area required:

$$A = 30{,}000/3.25 = 9230\ \text{ft}^2$$

Number of bags:

$$N = 9230/62.8 = 147$$

Bag cost:

$$\text{BC} = (147)(105) = \$15{,}434$$

2. Bag replacement cost:

$$\text{BRC} = (15{,}434)(0.2) = \$3086$$

3. Maintenance cost:

$$\text{MC} = 3800$$

4. Installed cost:

$$\begin{aligned} \text{IC} &= (30{,}000)(3.35) \\ &= \$100{,}500 \end{aligned}$$

5. Fan cost:

$$\begin{aligned} \text{HP} &= (q)(\Delta P)/(\eta)(6356) \\ &= (30{,}000)(1.5 + 2.5)/(0.625)(6356) \\ &= 30.2 \\ \text{FC} &= (30)(0.746)(0.18)(7000) \\ &= \$28{,}200 \end{aligned}$$

Total cost:

$$\begin{aligned} \text{TC} &= \text{BC} + \text{BRC} + \text{MC} + \text{IC} + \text{FC} \\ &= 15{,}434 + 3086 + 3800 + 100{,}500 + 28{,}200 = \$151{,}000 \end{aligned}$$

Conclusion: A profit will be realized within a year.

12.45 A textile dye and finishing plant has *two* coal-fired stoker boilers for process steam and space heating. Each boiler is rated at 60,000 lb of steam per hour, with an exhaust volume of 35,000 acfm at 350°F. A baghouse was installed for particulate control employing Teflon felt bags, each with an area of 12 ft^2 (\$75.00 per bag) at an air-to-cloth ratio (acfm/ft^2) of 5.81. Installed capital costs amounted to \$2.536 acfm. The total pressure drop across the system is 1.3 inches of water for the bags themselves, plus 2.0 inches of water for the remainder of the system.

Determine the installed capital, operating, and maintenance costs on an *annualized* basis. The following economic factors exist at the time of purchase:

Overall fan efficiency = 60%
Operating time = 6240 hr/year
Electrical power = \$0.03 per kW · hr
Yearly maintenance cost = \$5000 per year plus replacing 25% of the bags each year
Lifetime of baghouse (m) = 15 years
Interest rate (i) = 8% = 0.08
Salvage value = Zero

$$\textit{Annualized}\ \text{installed capital cost} = (\text{installed cost})\left[\frac{(i)(1+i)^m}{(1+i)^m - 1}\right] \qquad (12.13)$$

Solution: The number of bags is calculated as follows:

$$\begin{aligned} N &= \frac{70{,}000}{(5.81)(12)} \\ &= 1004\ \text{bags} \end{aligned}$$

Installed cost:

$$\text{IC} = (2.536)(70{,}000) = \$177{,}520$$

Annual installed cost:

$$\begin{aligned}\text{AIC} &= 177{,}520\left[\frac{(0.08)(1+0.08)^{15}}{(1+0.08)^{0.5}-1}\right]\\ &= \$20{,}762/\text{year}\end{aligned}$$

Operating cost:

$$\begin{aligned}\text{OC} &= \frac{(70{,}000)(3.3)(6240)(5.2)(0.746)(0.03)}{(0.6)(33{,}000)}\\ &= 8678/\text{year}\end{aligned}$$

Maintenance cost:

$$\begin{aligned}\text{MC} &= 5000 + \frac{(75)(1000)}{4}\\ &= \$23{,}750/\text{year}\end{aligned}$$

Total annualized cost:

$$\begin{aligned}\text{TAC} &= 20{,}762 + 23{,}750 + 8678\\ &= \$53{,}190/\text{year}\end{aligned}$$

12.46 Gypsum Application

The Alex Pedro Gypsum Company has a four(4)-compartment baghouse. Each compartment has a 10×10 array of 8-inch diameter $\times$ 15-foot long Dacron bags. The baghouse is processing 55,565 acfm (70°F,1 atm) at 160°F with an inlet dust loading of 9.0 gr/ft^3 and is operating at a pressure drop of 5.0 in H_2O. The plant runs 24 hr/day on a 330-day year and operates at a profit level (based-on design conditions) of \$25 million per year. The EPA limits the outlet loading to 0.036 gr/ft^3, which corresponds to both the normal and design conditions, and will levy fines of \$10.50/lb for any amount over this limit. Given that the recovered gypsum is worth \$0.10/lb determine how many bags can be allowed to fail before the company will start to lose money. Also, find the outlet dust loading this corresponds to. Use the Theodore–Reynolds method to estimate bag failure discharges.

Solution: The actual flow rate is first calculated employing Charles' Law.

$$q = 55{,}565(620/530)$$
$$= 65{,}000 \text{ acfm}$$

Perform the economic analysis on an *hourly* basis:
Profit:

$$P = 25 \times 10^6/(330)(24)$$
$$= \$3156/\text{hr}$$

Inlet dust loading:

$$\text{IDL} = (9.0)(65{,}000)/7000$$
$$= 83.57 \text{ lb/min} = 5014 \text{ lb/hr}$$

Outlet dust loading:

$$\text{ODL} = (0.036 \text{ gr/ft}^3)(65{,}000)(60)/7000$$
$$= 20.05 \text{ lb/hr} = 20 \text{ lb/hr}$$

Note that EPA fines \$10.50/lb for every pound over 20 lb/hr. Using Equation (12.7), one obtains

$$\phi = \frac{q}{(L)(D^2)(T + 460)^{0.5}}$$
$$= \frac{65{,}000}{(L)(8)^2\sqrt{620}} = \frac{40.8}{L}$$
$$E = 0.036/9.0$$
$$= 99.6\% = 0.996$$
$$P = 1.0 - 0.996 = 0.004$$

The results from the Theodore–Reynolds model are provided in Table 12.8. When the second bag fails, the company will start losing money to the tune of

TABLE 12.8 Bag Failure Cost Effect

L	ϕ	P_c	P^*	gr/ft^3 ODL	lb/hr ODL	Gypsum Cost, \$/hr	EPA Fines, \$/hr	Total Cost, \$/hr
1	40.8	0.028	0.032	0.288	160	14.0	1470	1484
2	20.4	0.065	0.069	0.621	346	32.6	3423	3456
3	13.6	0.10	0.104	0.936	521	50.1	5261	5311

$3456 − $3156 = $300/hr! Therefore, replace the failed bags when two (2) of them break!

12.47 Baghouse Degradation with Time

A baghouse's operating collection efficiency has slowly degraded with time. As a plant manager who has recently completed (and passed) an air pollution control equipment (APCE) course given by Dr. Louis Theodore—the foremost authority in the universe on APCE—indicate what sound, reasonable engineering steps can be taken to return the baghouse to its original design value.

Solution: This is the last of the open-ended problems. The first step is to check the fan. Assuming that the fan is operating according to specifications, check for bag failures. If there are no bag failures, check the process to determine whether anything has changed recently that could account for the decrease in efficiency. If the particles discharged are finer (smaller in size), consideration should be given to decreasing the velocity through the bags. Reducing the velocity provides a longer residence time in the cake or fabric and increases collection efficiency by moleculer diffusion. The author has argued over the years that one way of increasing the efficiency for fine particles is to reduce the velocity. As with electrostatic precipitators (ESPs), the cleaning frequency and intensity should be looked into. Decreasing the frequency of cleaning increases the cake thickness; this leads to an increase in gas residence time, which can also increase efficiency. Reducing the "intensity" of the cleaning could also produce similar results.

NOTE: Additional problems are available for all readers at www.wiley.com. Follow links for this title.

Appendix A

HYBRID SYSTEMS

A.1 INTRODUCTION

The basic design of air pollution control equipment has remained relatively unchanged since first used in the early part of the twentieth century. Some modest equipment changes and new types of devices have appeared in the last few decades, but all have essentially employed the same capture mechanisms used in the past. One area that has recently received some attention is hybrid systems—equipment that in some cases operate at higher efficiency more economically than conventional devices. Tighter regulations and a greater concern for environmental control by society has placed increased emphasis on the development and application of these systems.

Hybrid systems are defined as those types of control devices that involve combinations of control mechanisms—for example, fabric filtration combined with electrostatic precipitation. There are four major hybrid systems. These include wet electrostatic precipitators, ionizing wet scrubbers, dry scrubbers, and electrostatically augmented fabric filtration. Each of these hybrid units is briefly described below.

A.2 WET ELECTROSTATIC PRECIPITATORS

The wet electrostatic precipitator (WEP) is a variation of the dry ESP design. The two major added features in a WEP system are

1. A preconditioning step, where inlet sprays in the entry section are provided for cooling, gas absorption, and removal of coarse particles.
2. A wetted collection surface, where liquid is used to continuously flush away collected materials.

Particle collection is achieved by (1) the introduction of evenly distributed liquid droplets to the gas stream through sprays located above the electrostatic field sections and (2) migration of the charged particles and liquid droplets to the collection plates. The collected liquid droplets form a continuous downward-flowing film over the collection plates, and maintain them clean by removing the collected particles.

The WEP overcomes some of the limitations of the dry ESP. Its operation is not influenced by the resistivity of the particles. Furthermore, since the internal components are continuously being washed with liquid, buildup of tacky/sticky particles is controlled and there is some capacity for removal of gaseous pollutants. In general, applications of the WEP fall into two areas: removal of fine particles and removal of condensed organic fumes. Outlet particulate concentrations are typically in the 10^{-3} to 10^{-2} gr/ft^3 range.

Data on the capability of the WEP to remove acid gases are very limited. This device has been installed to control HF emissions. Using a liquid-to-gas (L/G) ratio of 5 gal/1000 acf and a liquid pH between 8 and 9, fluoride removal efficiencies of $>98\%$ have been reported; outlet concentrations of HF were found to be <1 ppm.

Some of the advantages of a WEP include:

1. Simultaneous gas absorption and dust removal
2. Low energy consumption
3. No dust resistivity problems
4. Efficient removal of fine particles

Disadvantages of the WEP are the following:

1. Low gas absorption efficiency
2. Sensitivity to changes in flow rate
3. Dust collection is wet

A.3 IONIZING WET SCRUBBERS

The ionizing wet scrubber (IWS) is a relatively new development in the technology of removal of particulate matter from a gas stream. These devices have been incorporated in commercial incineration facilities. In the IWS, high-voltage ionization in a charging section places static electrical charge on the particles in the gas stream, which then passes through a crossflow packed-bed scrubber. The packing is normally polypropylene in the form of circular-wound spirals and gearlike wheel configurations, providing a large surface area. Particles with sizes of 3 μm or larger are trapped by inertial impaction within the packed bed. Smaller charged particle pass close to the surface of either the packing material or a scrubbing water droplet. An opposite charge in that surface is induced by the charged particle, which is then attracted to an ion attached to the surface. All collected particles are eventually washed out of the scrubber. The scrubbing water also functions to absorb gaseous pollutants. According to Ceilcote (the IWS vendor), the collection efficiency of a two-stage IWS is greater than that of a baghouse or a conventional ESP for particles in the 0.2–0.6 μm range. For 0.8 μm and above, the ESP is as effective as the IWS. Scrubbing water can include caustic soda or soda ash when needed for efficient absorption of acid gases. Corrosion resistance of the IWS is achieved by fabricating its shell and most internal parts from fiberglass-reinforced plastic (FRP) and thermoplastic materials.

The pressure drop through a single-step IWS is around 5 in H_2O (primarily through the wet scrubber section). All internal areas of the ionizer section are periodically deluge-flushed with recycled liquid from the scrubber recycle system. The advantages of an IWS are those of the combination of a WEP and a packed-bed absorber. The disadvantages include the need for separation of particulates from the scrubbing medium, potential scaling and fouling problems (aggravated by recycle neutralizing solutions), possible damage to the scrubber if the scrubber solution pumps fail, and the need for a downstream mist eliminator. Despite some of these limitations, the IWS is a particulate control device that has worked successfully and efficiently.

A.4 DRY SCRUBBERS

The success of fabric filters in removing fine particles from flue gas streams has encouraged the use of combined dry scrubbing/fabric filter systems for the dual purpose of removing both particulates and acid gases simultaneously. Dry scrubbers offer potential advantages over their wet counterparts, especially in the areas of energy savings and capital costs. Furthermore, the dry scrubbing process design is relatively simple, and the product is a dry waste rather than a wet sludge.

There are two major types of so-called dry scrubber systems: spray drying and dry injection. The first process is often referred to as a *wet–dry* system. When compared to a conventional wet scrubber, it uses significantly less scrubbing liquid. The second process has been referred to as a *dry–dry* system because no liquid scrubbing is involved. The spray-drying system is predominantly used in utility and industrial applications.

The method of operation of the spray dryer is relatively simple, requiring only two major equipment items: (1) a spray dryer similar to those used in the chemical food-processing and mineral-preparation industries and (2) a baghouse (or ESP) to collect the fly ash and/or entrained solids.

In a spray dryer at a coal-fired utility boiler, an alkali sorbent solution, or slurry, is atomized into the incoming flue gas stream to increase the liquid–gas interface and to promote the mass transfer of the SO_2 from the gas to the slurry droplets, where it is absorbed. Simultaneously, the thermal energy of the gas evaporates the water in the droplets to produce a dry powdered mixture of sulfite–sulfate and some unreacted alkali. Because the flue gas is not saturated and contains no liquid carryover, potentially troublesome mist eliminators are not required. After leaving the spray dryer, the solids-bearing gas passes through the fabric filter (or ESP), where the dry product is collected and where a percentage of the unreacted alkali reacts with SO_2 for further removal. The cleaned gas is then discharged through the fabric filter plenum to an induced draft (ID) fan and to the stack.

Among the inherent advantages that the spray dryer enjoys over wet scrubbers are

1. Lower capital costs
2. Lower draft losses
3. Reduced auxiliary power
4. Reduced water consumption, with liquid-to-gas (L/G) ratios significantly lower than those of wet scrubbers
5. Continuous, two-stage operation from liquid feed to dry product

The sorbent of choice for most spray-dryer systems is a lime slurry. One system under development uses a sodium carbonate solution. Although the latter will generally achieve a higher level of SO_2 removal than will a lime slurry at similar operating conditions, the significant cost advantage that lime has over sodium carbonate makes it the overwhelming favorite. Also, when sodium alkalis are used, the products are highly water-soluble and may create disposal problems.

Dry injection processes generally involve pneumatic introduction of a dry, powdery alkaline material, usually a sodium-based sorbent, into the flue gas stream with subsequent fabric filter collection. The injection point in such processes can vary from the boiler–furnace area all the way to the flue gas entrance to the baghouse, depending on operating conditions and design criteria.

A.5 ELECTROSTATICALLY AUGMENTED FABRIC FILTRATION

Advanced electrostatic stimulation of fabric filtration is another example of a hybrid system. It combines the efficiency of fabric filters with the efficiency and low pressure drop of an ESP. This combination is accomplished by placing a high-voltage electrode coaxially inside a filter bag to establish an electric field between the electrode and the bag surface. The electric field alters the dust deposition pattern within the bag, yielding a much lower pressure drop than is found in conventional bags. The bags can also operate at much higher air-to-cloth ratios.

Another system combines fabric filtration with a surface electric field. This version is similar to the above system but uses an array of wires separated by insulating spacers mounted on the clean side of the fabric. These electrostatic forces can be utilized by allowing the natural particle charges to accumulate on the fabric and then collecting the particles, and also by applying an electric field at the fabric surface.

Numerous attempts have been made to use electrostatic effects to improve on the performance of fabric filters. These efforts have met with varying degrees of success. Presently, these techniques are only available in the pilot plant stage. The implementation of some of these new techniques is being generally resisted by most industries. As with any new technology, there is room for improvement, but the results may be promising enough to continue research and development.

Appendix B

SI UNITS

B.1 THE METRIC SYSTEM

The need for a single worldwide coordinated measurement system was recognized over 300 years ago. Gabriel Mouton, Vicar of St. Paul in Lyons, proposed in 1670 a comprehensive decimal measurement system based on the length of one minute of arc of a great circle of the earth. In 1671 Jean Picard, a French astronomer, proposed the length of a pendulum beating seconds as the unit of length. (Such a pendulum would have been fairly easily reproducible, thus facilitating the wide-spread distribution of uniform standards.) Other proposals were made, but over a century elapsed before any action was taken.

In 1790, in the midst of the French Revolution, the National Assembly of France requested the French Academy of Sciences to "deduce an invariable standard for all the measures and weights." The Commission, appointed by the Academy, created a system that was, at once, simple and scientific. The unit of length was to be a portion of the earth's circumference. Measures for capacity (volume) and mass (weight) were to be derived from the unit of length, thus relating the basic units of the system to each other and to nature. Furthermore, the larger and smaller versions of each unit were to be created by multiplying or dividing the basic units by 10 and its multiples. This feature provided a great convenience for users of the system by eliminating the

Air Pollution Control Equipment Calculations. By Louis Theodore

need for such calculations as dividing by 16 (to convert ounces to pounds) or by 12 (to convert inches to feet). Similar calculations in the metric system could be performed simply by shifting the decimal point. Thus the metric system is a *base–10* or *decimal* system.

The Commission assigned the term *metre* (which is now spelt *meter*) to the unit of length. This name was derived from the Greek word *metron* meaning "a measure." The physical standard representing the meter was to be constructed so that it would equal one ten-millionth of the distance from the north pole to the equator along the meridian of the earth running near Dunkirk in France and Barcelona in Spain.

The metric unit of mass, called the *gram*, was defined as the mass of one cubic centimeter (a cube that is $\frac{1}{100}$th of a meter on each side) of water at its temperature of maximum density. The cubic decimeter (a cube $\frac{1}{10}$th of a meter on each side) was chosen as the unit of fluid capacity. This measure was given the name *liter*.

Although the metric system was not accepted with enthusiasm at first, adoption by other nations occurred steadily after France made its use compulsory in 1840. The standardized character and decimal features of the metric system made it well suited to scientific and engineering work. Consequently, it is not surprising that the rapid spread of the system coincided with an age of rapid technological development. In the United States, by Act of Congress in 1866, it was made "lawful throughout the United States of America to employ the weights and measures of the metric system in all contracts, dealings, or court proceedings."

By the late 1860s, even better metric standards were needed to keep pace with scientific advances. In 1875, an international treaty, the "Treaty of the Meter," set up well-defined metric standards for length and mass, and established permanent machinery to recommend and adopt further refinements in the metric system. This treaty, known as the *Metric Convention*, was signed by 17 countries, including the United States.

As a result of the Treaty, metric standards were constructed and distributed to each nation that ratified the Convention. Since 1893, the internationally agreed to metric standards have served as the fundamental weights and measures standards of the United States.

By 1900 a total of 35 nations—including the major nations of continental Europe and most of South America—had officially accepted the metric system. Today, with the exception of the United States and a few small countries, the entire world is using predominantly the metric system or is committed to such use. In 1971 the Secretary of Commerce, in transmitting to Congress the results of a 3-year study authorized by the Metric Study Act of 1968, recommended that the United States change to predominant use of the metric system through a coordinated national program.

The International Bureau of Weights and Measures, located in Sevres, France, serves as a permanent secretariat for the Metric Convention, coordinating the exchange of information about the use and refinement of the metric system. As measurement science develops more precise and easily reproducible ways of defining the measurement units, the General Conference of Weights and Measures—the diplomatic organization made up of adherents to the Convention—meets periodically to ratify improvements in the system and the standards.

B.2 THE SI SYSTEM

In 1960, the General Conference adopted an extensive revision and simplification of the system. The name *Le Systeme International d'Unites* (International System of Units), with the international abbreviation *SI*, was adopted for this modernized metric system. Further improvements in and additions to SI were made by the General Conference in 1964, 1968, and 1971.

The basic units in the SI system are the *kilogram* (mass), *meter* (length), *second* (time), *Kelvin* (temperature), *ampere* (electric current), *candela* (the unit of luminous intensity), and *radian* (angular measure). All are commonly used by the engineer. The Celsius scale of temperature (0°C—273.15 K) is commonly used with the absolute Kelvin scale. The important derived units are the *newton* (SI unit of force), the *joule* (SI unit of energy), the *watt* (SI unit of power), the *pascal* (SI unit of pressure), the *hertz* (unit of frequency). There are a number of electrical units: *coulomb* (charge), *farad* (capacitance), *henry* (inductance), *volt* (potential), and *weber* (magnetic flux). One of the major advantages of the metric system is that larger and smaller units are given in powers of 10. A further simplification is introduced in the SI system by recommending only those units with multipliers of 10^3. Thus, for lengths in engineering, the *micrometer* (previously *micron*), *millimeter*, and *kilometer* are recommended, and the *centimeter* is generally avoided. A further simplification is that the decimal point may be substituted by a comma (as in France, Germany, and South Africa), while the other number, before and after the comma, is separated by spaces between groups of three, i.e., one million dollars is $1 000 000,00. More details are provided below.

B.3 SI MULTIPLES AND PREFIXES

Multiples and Submultiples		Prefixes	Symbols
100 000 000 000	10^{12}	tera	T
100 000 000	10^9	giga	G
100 000	10^6	mega	M
1 000	10^3	kilo	k
100	10^2	hectos	h
10	10^1	deka	da
Base unit 1	10^0		
0.1	10^{-1}	deci	d
0.01	10^{-2}	centi	c
0.001	10^{-3}	milli	m
0.000 001	10^{-6}	micro	μ
0.000 000 001	10^{-9}	nano	n
0.000 000 000 001	10^{-12}	pico	p
0.000 000 000 000 001	10^{-15}	femto	f
0.000 000 000 000 000 001	10^{-18}	atto	a

B.4 CONVERSION CONSTANTS (SI)

Length

To convert from	to	multiply by
m	cm	100
m	mm	1000
m	micrometers (μm)	10^6
m	angstroms (Å)	10^{10}
m	in	39.37
m	ft	3.281
m	mi	6.214×10^{-4}
ft	in	12
ft	m	0.3048
ft	cm	30.48
ft	mi	1.894×10^{-4}

Mass

To convert from	to	multiply by
kg	g	1000
kg	lb	2.205
kg	oz	35.24
kg	ton	2.268×10^{-4}
kg	gr (grains)	1.543×10^4
lb	oz	16
lb	ton	5×10^{-4}
lb	g	453.6
lb	kg	0.4536
lb	gr	7000

Time

To convert from	to	multiply by
s	min	0.01667
s	hr	2.78×10^{-4}
s	day	1.157×10^{-7}
s	week	1.653×10^{-6}
s	year	3.171×10^{-8}

Force

To convert from	to	multiply by
N	$kg \cdot m/s^2$	1
N	dyn	10^5
N	$g \cdot cm/s^2$	10^5
N	lb_f	0.2248
N	$lb \cdot ft/s^2$	7.233
lb_f	N	4.448
lb_f	dyn	4.448×10^5
lb_f	$g \cdot cm/s^2$	4.448×10^5
lb_f	$lb \cdot ft/s^2$	32.17

Pressure

To convert from	to	multiply by
atm	N/m^2 (Pa)	1.013×10^5
atm	kPa	101.3
atm	bars	1.013
atm	dyn/cm^2	1.013×10^6
atm	lb_f/in^2 (psi)	14.696
atm	mm Hg at 0°C (torr)	760
atm	in Hg at 0°C	29.92
atm	ft H_2O at 4°C	33.9
atm	in H_2O at 4°C	406.8
psi	atm	6.80×10^{-2}
psi	mm Hg at 0°C (torr)	51.71
psi	in H_2O at 4°C	27.70
in H_2O at 4°C	atm	2.458×10^{-3}
in H_2O at 4°C	psi	0.0361
in H_2O at 4°C	mm Hg at 0°C (torr)	1.868

Volume

To convert from	to	multiply by
m^3	L	1000
m^3	cm^3 (cc, mL)	10^6
m^3	ft^3	35.31
m^3	gal (US)	264.2
m^3	qt	1057
ft^3	in^3	1728
ft^3	gal (US)	7.48
ft^3	m^3	0.02832
ft^3	L	28.32

Energy

To convert from	to	multiply by
J	N · m	1
J	erg	10^7
J	dyn · cm	10^7
J	kW · hr	2.778×10^{-7}
J	cal	0.2390
J	ft · lb_f	0.7376
J	Btu	9.486×10^{-4}
cal	J	4.186
cal	Btu	3.974×10^{-3}
cal	ft · lb_f	3.088
Btu	ft · lb_f	778
Btu	HP · hr	3.929×10^{-4}
Btu	cal	252
Btu	kW · hr	2.93×10^{-4}
ft · lb_f	cal	0.3239
ft · lb_f	J	1.356
ft · lb_f	Btu	1.285×10^{-3}

Power

To convert from	to	multiply by
W	J/s	1
W	cal/s	0.2390
W	ft · lb_f/s	0.7376
W	kW	10^{-3}
kW	Btu/s	0.949
kW	HP	1.341
HP	ft · lb_f/s	550
HP	kW	0.7457
HP	cal/s	178.2
HP	Btu/s	0.707

Concentration

To convert from	to	multiply by
$\mu g/m^3$	lb/ft^3	6.243×10^{-11}
$\mu g/m^3$	lb/gal	8.346×10^{-12}
$\mu g/m^3$	gr/ft^3	4.370×10^{-7}
gr/ft^3	$\mu g/m^3$	2.288×10^6
gr/ft^3	g/m^3	2.288
lb/ft^3	$\mu g/m^3$	1.602×10^{10}
lb/ft^3	$\mu g/L$	1.602×10^7
lb/ft^3	lb/gal	7.48

Viscosity

To convert from	to	multiply by
P (poise)	g/cm · s	1
P	cP (centipoise)	100
P	kg/m · h	360
P	lb/ft · s	6.72×10^{-2}
P	lb/ft · hr	241.9
P	lb/m · s	5.6×10^{-3}
lb/ft · s	P	14.88
lb/ft · s	g/cm · s	14.88
lb/ft · s	kg/m · hr	5.357×10^{3}
lb/ft · s	lb/ft · hr	3600

Heat Capacity

To convert from	to	multiply by
cal/g · °C	Btu/lb · F	1
cal/g · °C	kcal/kg · C	1
cal/g · °C	cal/gmol · C	Molecular weight
cal/gmol · °C	Btu/lbmol · F	1
J/g · °C	Btu/lb · F	0.2389
Btu/lb · °F	cal/g · C	1
Btu/lb · °F	J/g · C	4.186

Appendix C

EQUIPMENT COST MODEL

[Adapted with permission from J.D. McKenna, ETSI Inc, Roanoke, VA, 2008]

A simple method for determining costs associated with air pollution control equipment is presented below. To simplify the presentation, only costs associated with baghouses will be considered. However, the algorithm can easily be extended to include all other plant or environmental equipment.

As with most economic/cost models involving (plant) equipment, there are two classes of cost that need to be considered.

1. Capital investment
2. Operation and Maintenance (O & M) costs

Specific information follows.

Capital investments can be divided into four general categories.

1. Control equipment hardware costs
2. Auxiliary equipment costs

Air Pollution Control Equipment Calculations. By Louis Theodore

3. Field installation costs
4. Engineering studies, land, preparation, initial inventory, strucure modification(s), and start-up

There are 6 basic factors that must be considered with regard to annual O & M costs.

1. Gas volume
2. Pressure drop
3. On-steam time
4. Electricity
5. Mechanical efficiencies of the fan(s)
6. Filter bag replacement

It is not uncommon to split the O & M costs into two categories: Operation and Maintenance. These two costs may be determined from Equation (C.1).

$$G = A + B + C + D \tag{C.1}$$

where G = Annual operating and maintenance cost
A = Electrical cost
B = Liquid consumption cost
C = Fuel cost
D = Maintenance cost

For a baghouse, the above equation reduces to

$$G = A + D \tag{C.2}$$

since B = 0
C = 0

The standard electrical cost component of the above equation is given by

$$A = \left[\frac{0.7457}{6356E}\right](P)(H)(K)(S) \tag{C.3}$$

where S = Design capacity of baghouse fabric filter in acfm
P = Pressure drop in H_2O
E = Fan efficiency
H = Annual operating time in hr
K = Power cost, \$/kW · hr
0.7457 and 6356 are conversion constants

Thus, the annual O & M costs become

$$G = S\left[\frac{(0.7457)(P)(H)(K)}{(6356)(E)} + M\right] \tag{C.4}$$

Finally, the total annualized cost is given by

$$T = G + X + Y \tag{C.5}$$

where T = Total annualized cost (as described in text)
G = Annual costs for operation maintenance
X = Annualized capital costs
Y = Depreciated capital investment

ILLUSTRATIVE EXAMPLE

Estimate the total annualized cost for a baghouse given the following data.
Capital charges:

Initial capital cost, ICC = $153,700
Annualized capital costs, ACC = 18% of ICC (CRF = 0.18)
Depreciated capital investment = 6.67% of ICC

Operating information:

Gas flowrate = 70,000 acfm
Pressure drop = 6.0 in H_2O
Fan efficiency = 60%
Annual on stream time = 6000 hr
Electrical cost = 0.05 $/kW · hr

Maintenance information:

Number of bags = 1080
Bag life = 4 yr (25% replacement per year)
Bag replacement cost = $100/bag
Routine maintenance time = 4 hr/week
Labor cost = $20/hr

Solution: First calculate the annual maintenance cost in $/acfm.

$$M = M_1 + M_2$$

where M_1 = Bag replacement cost
M_2 = Bag house routine maintenance cost

Employing the maintenance data provided,

$$\begin{aligned} M &= [(1080)(0.25)(100) + (4)(52)(20)]/70{,}000 \\ &= [\$27{,}000 + \$4160]/70{,}000 \text{ acfm} \\ &= 0.45 \ \$/\text{acfm} \end{aligned}$$

Equation (C.4) is employed to obtain the annual O & M cost.

$$\begin{aligned} G &= S\left[\frac{(0.7456)(P)(H)(K)}{(6356)(E)} + M\right] \\ &= (70{,}000)\left[\frac{(0.7456)(6)(6000)(0.05)}{(6356)(0.6)} + 0.45\right] \\ &= \$56{,}138 \end{aligned}$$

The total annualized cost can now be calculated from Equation (5)

$$\begin{aligned} T &= G + X + Y \\ &= 56{,}138 + (0.18)(153{,}700) + (0.067)(153{,}700) \\ &= 56{,}138 + 27{,}666 + 10{,}296 \\ &= \$94{,}000 \end{aligned}$$

INDEX